Weather, Climate and Climate Change

PEARSON
Education

Weather, Climate and Climate Change

Human Perspectives

Greg O'Hare, John Sweeney and Rob Wilby

Harlow, England • London • New York • Boston • San Francisco • Toronto • Sydney • Singapore • Hong Kong
Tokyo • Seoul • Taipei • New Delhi • Cape Town • Madrid • Mexico City • Amsterdam • Munich • Paris • Milan

Pearson Education Limited
Edinburgh Gate
Harlow
Essex CM20 2JE
England

and Associated Companies throughout the world

Visit us on the World Wide Web at:
www.pearsoned.co.uk

First published 2005

ISBN 0130 283193

British Library Cataloguing-in-Publication Data
A catalogue record for this book is available from the British Library

Library of Congress Cataloging-in-Publication Data
O'Hare, Greg.
Weather, climate, and climate change : human perspectives / Greg O'Hare, John Sweeney, and Rob Wilby.
p. cm.
Includes bibliographical references and index.
ISBN 0-13-028319-3
1. Weather. 2. Climatic changes. 3. Climatic changes—Environmental aspects. I. Sweeney, John, 1952 May 22- II. Wilby, R. L. (Robert L.) III. Title.
QC861.3.035 2005
551.6—dc22

2004048650

10 9 8 7 6 5 4 3 2 1
08 07 06 05

Typeset in 10/12½ Minion by 35
Printed and bound by Ashford Colour Press, Gosport

The publisher's policy is to use paper manufactured from sustainable forests.

CONTENTS

CONTENTS

PREFACE

The renaissance of the study of climate in recent decades has largely been powered by concerns regarding what many have claimed to be the greatest challenge facing humankind in the present century – climate change. Today, climate change has moved centre stage in what was formerly, to many students of the atmosphere, the rather unexciting discipline of climatology. In so doing, it has acted to energise and enthuse not just the student, but also the public, and increasingly the decision-makers in our society. As far as the growth of interest in climate change is concerned, there can be few areas of academic endeavour which have made such a dramatic and successful transition in recent years into capturing the imagination of society at large.

Despite this, however, many climatology texts follow a rather traditional format of including climate change considerations as a separate section, frequently towards the end of the book, and proceeded by individual sections on weather and climate respectively. In structuring this work, the authors have sought to integrate the three main aspects of weather, climate and climate change throughout. It must be admitted that sometimes this approach is less seamless than might be wished for, because to fully understand the issues concerned, we believe a three-legged stool is necessary.

The first leg consists of the basic understanding of atmospheric functioning which must underpin any work on this topic. A number of chapters are therefore devoted to providing this rudimentary information. We believe this can be done without emphasising the advanced theoretical science which a mainstream meteorologist or atmospheric physicist might consider desirable. The target audience for this text is instead the environmental science and geography students who comprise the bulk of the community studying the topics concerned. It is such individuals who will in turn be required to use their knowledge of the atmosphere mainly in policy formulation and resource management issues.

The second leg of the stool is the necessity for understanding the regional and spatial scales of climate and climate change. Climate change is inherently a spatial/geographical process. The polar margins, the oceans, the continental interiors, the tropics have fundamentally different climates and thus their own particular sets of signals and responses to climate forcing. To appreciate this requires awareness of their current climate controls and experiences. A number of chapters are therefore devoted to exploring climate and climate change within a regional context.

Of course, strategies for projecting future climate are now maturing to the extent that quite confident statements may be made, at least for the next few decades. The authors consider it timely therefore that as a third leg of the book modelling aspects be introduced into the mainstream curriculum of climate, and serve as an integrating component in the overall work.

In the end, though, there is one final point to stress. We believe that humans should be at the centre of climate change and that the three-legged stool should bear the weight of people using it. The work therefore emphasises wherever appropriate the risk and vulnerability of people (not just regions) to past, present and likely future climate change. A large number of case studies in the book are used to facilitate this goal, reflecting the considerable experience of the authors in many areas of the developed and developing world alike where climate change has a growing societal urgency and relevance.

PUBLISHER'S ACKNOWLEDGEMENTS

We are grateful to the following for permission to reproduce copyright material:

Figures 1.1, 1.2, 6.3 from Hadley Centre, UK crown copyright Met Office; Figure 2.1 from *Climate Change: Causes, Effects and Solutions*, (Hardy, J 2003) © John Wiley & Sons Limited. Reproduced with permission; Figure 2.15 from *Understanding our Atmospheric Environment*, 2/E by Morris Neiburger, et al. © 1973, 1982 by W.H. Freeman and Company. Used with permission; Figure 2.16 from A historical record of blizzards/major snow events in the British Isles, publisher Royal Meteorological Society (Wild, R. O'Hare, G. and Wilby, R 1996); Figures 2.17 and 6.4 from *Global Change and the Irish Environment* publisher Royal Irish Academy (Sweeney J. 1997); Figure 2.18 from *An Introduction to Global Environmental Issues*, publisher Taylor & Francis (Pickering K and Owen L 1994); Figure 2.19 from *Global Change Newsletter 37* (Scholes, B 1999) Reproduced with the permission of IGBP Secretariat, Stockholm; Figure 2.21 from Introductory Remote Sensing, Vol 2, publisher Taylor & Francis (Gibson and Power 2000); Table 2.1, Figures 2.11a and 4.4 from *Meteorology Today: Introduction to Weather, Climate and the Environment, with Info Trac* 6th edition by Ahrens. © 2000. Reprinted with permission of Brooks/Cole, a division of Thomson Learning: www.thomsonrights.com. Fax 800 730-02215; Table 2.5, Table 6.3, Figure 8.15, from Contribution of Working Group I to the Third Assessment Report of the Intergovernmental Panel on Climate Change in *Climate Change 2001* publisher World Meterological Organisation. (Houghton, J.T., 2001); Table 2.4 from The value of the world's ecosystem services and natural capital from *Nature* publisher MacMillan Magazines Limited (Constanza, R. 1997); Figure 3.2 from *Global Warming. The Science of Climate Change* (Frances Drake (2000) reproduced by permission of Hodder Arnold; Figures 3.4(b) and 3.11 from *Physical Climatology* (Sellers W.D. (1965) published by The University of Chicago Press; Figure 3.4c and 3.7a and b from Atmosphere Weather and Climate publisher Taylor & Francis (Barry R.G. and Chorley R.J. 2003); Figure 3.13 from Boundary Layer Climates publisher Taylor & Francis (Oke.T.R. 1990); Table 3.2 from *The Role of the Sun in Climate Change* by Douglas V. Hoyt and Kenneth H. Schatten, copyright – 1996 by Oxford University Press, Inc. Used by permission of Oxford University Press, Inc.; Figures 4.1, 4.12, 4.21 and 11.1 from *Contemporary Cluimatology 2/e* (Henderson-Sellers and Robinson) published by Pearson Education Limited; Figure 4.9 from *Introduction to the* Atmosphere (Neiburger M. (1982) published by W H Freeman & Company; Figure 4.13 from *Global Warming: the complete briefing*, 2nd ed. (Houghton J. (1997) publisher Intergovernmental Panel on Climate Change; Figure 4.14 Atmospheric and Oceanic Circulation from *Christopherson, Robert W., Elemental Geosystems 2 Edn©1998.* Electronically reproduced by permission of Pearson Education Inc., Upper Saddle River, New Jersey; Figure 4.20, from IPCC in *Weather Systems*, publisher World Meterological Organisation (Musk L, 1988); Figure 4.22 from Sverdrup in *The Oceans Johnson & Fleming, Oceans, 1st, © N/A*, Electronically reproduced by permission of Pearson Education, Inc., Upper Saddle River, New Jersey. Figure 4.20 from *The Glacial World According to Wally* (Broecker W.S. (1995) published by Springer-Verlag GmbH; Table 4.2 from Characteristics of Air Masses over the British Isles, in *Geophysical Memoirs* 87 crown copyright Met Office; Figure 5.10 from *Ocean-Atmosphere circulation and global climate* (Dawson and O'Hare (2000) published by Geographical

Association; Figures 6.12, 7.1 from IPCC, publisher World Meterological Organisation; Table 6.1 from *A Climate Modelling Primer*, publisher John Wiley & Sons (Henderson-Sellers A and McGuffie, K. 1987); Table 6.2 from The Complete Briefing in Global Warming, World Meterological Organisation (Houghton, J. 1997); Figure 7.3 from *Climate and Global Environmental Change* (Harvey) published by Pearson Education Limited; Figure 7.5 from *Efficient three dimensional global models for climate studies Model I and II*, (Russell J, Rind G, Stone P, Lacis A, Lebedeff S, Ruedy R and Travis (1983) Permission granted by American Meteorological Society; Figure 7.7 from *Climate 7* (Wilson et al. 1987) permission granted by American Meteorological Society; Figures 7.11 and 7.14 source Climate Impacts LINK Project www.cru.uea.ac.uk/link; Figure 7.12 from *Climate Dynamics 17*, evaluation and intercomparison of coupled climate models, Lambert and Boer (2001), published by Springer-Verlag GmbH & Co.; Figure 8.3 from *Nature Vol 379* (1996) published by Nature Publishing Group; Figures 8.5 and 8.10 publisher *Nature* publisher MacMillan Magazines Limited (Petit 1999); Figures 8.6, 8.7, 8.11 and 11.17 from Climate Change and Life in *The Great Ice* Age publisher Taylor & Francis (Wilson et al. 1999); Figure 8.8 from *Nature Vol. 365* (Bond G. et al. (1993) published by MacMillan Publishers; Figure 8.15 from *British Antarctic Survey (BAS)*, (Vaughan D.G. and Doajke, C.S.M.) published by Nature Publishing Group; Figure 8.16 reprinted from *Science, Vol 291* with permission from American Association for the Advancement of Science, copyright 2001 (Blunier T. & Brooke E.J); Tables 8.1 and 8.2 from Dawson, A. publisher University of Coventry; Figures 9.1 and 10.3 from *Atmosphere, Weather & Climate* publisher Taylor & Francis (Barry, Chorley & Chase 1998). Figure 9.3 from A Mesosynoptic analysis of thunderstorms of 28 August 1958 from *Theory and Observations*, crown copyright Met Office (Barry R, and Chorley R, 1962); Figure 9.4a adapted from Temple, 2003 publisher Royal Meteorological Society; Figure 9.6 source O'Hare and Sweeney published by Geographical Association (1986); Figure 9.7 (a) and (b) from *Geography (78)* (O('Hare and Sweeney J. (1998) publishes Blackwell Publishing Limited; Table 9.1 from British Isles Weather Types and a Register of the daily sequence of circulation patterns in *geophysical Memoir No. 116* (Lamb 1972) crown copyright Met Office; Table 9.2 from Climate change and hydrological stability in *Weather 56* publisher Royal Meteorological Society (Marsh 2001); Table 9.3 from Climate Change Scenarios for the UK in *The UKCIP02 Scientific Report*, publisher Tyndall Centre, University of East Anglia (Hulme et al. 2002); Table 9.4 from A Climate Change impact in London evaluation study in *Final Technical Report, Entec UK Ltd* publisher Greater London Authority, (London Climate Change Partnership 2002); Figure 10.2 adapted from Henderson-Sellers and Robinson (1986) and Shepherd et al. (2002); Figure 10.4 from *Climate Research Vol* 24 (Boruff et al. 2003); Figure 10.7 from Kalkstein et al., 1996 publisher Royal Meteorological Society; Figure 10.8 source: Trenberth and Guillemot (1996) permission granted by American Meteorological Society; Figure 10.9, source Karl and Knight (1998) permission granted by American Meteorological Society; Figure 10.11 from Change in mean annual runoff (mm) publisher Pacific Studies Institute (Wolock and McCabe 1999); Tables 10.2 and 10.3 from HadCM2 climate models by 2030 and 2095, publisher US National Assessment; Figures 11.7 and 11.9 from Fein J.S. and Stephens P.L. in *Monsoons* (Webster, P 1987). This material is used by permission of John Wiley & Sons Inc.; Figure 11.11 from *International Journal of Climatology Vol 11*, published by Royal Meteorological Society; Figure 11.12 from *Meterorological Applications* published by Royal Meteorological Society; Figure 11.14 from Short and long range monsoon prediction in India in *Monsoons* (Das, P. 1987) "This material is used by permission of John Wiley & Sons Inc."; Figure 11.15 from *Monsoons No. 2* in Scientific American, Vol 245 (Alan D Iselin) (Webster P.J. 1981); Figure 12.1 from Impacts, Adaptation and Vulnerability, in *Climate Change*, publisher World Meterological Organisation (McCarthy J 2001); Figure 12.2 'Out of the blue' from *The Economist, July 6th 2002*, (Munich Re); Figure 12.3(b) source R. Pielke (National Centre for Atmospheric Research, Boulder Colorado) and C. Leansea, permission granted by American Meteorological Society; Figure 12.4 from *The Geographical Journal* (O'Hare 2002) publisher Blackwell Publishing Limited; Figure 12.5 from Desertification: exploding the myth, (Thomas D.S.G. and Middleton, N.J. (1994) © John Wiley & Sons Limited, Reproduced with permission; Figure 12.10 We would like to thank Dr. Assaf Anyamba for granting permission to use this figureure; Figure 12.12 source: National Water Service, NOAA, US Government www.noaa.gov. Figure 12.13 from *The Geographical Journal* O'Hare, source Geographical Association (2002); Table 12.2 from Vulnerable People and Groups in Society, publisher University of Derby, (Dina Abbott); Table 12.3 from Social security and the family in rural India: coping with seasonality and calamity from *Journal of Peasant Studies* publisher Taylor & Francis (Agarwal, B. 1990); Table 12.5 from People at risk from drought, flood and famine in southern Africa in 2003 publisher United Nations, World Food Programme.

Plate 2.1, 4.1, 4.2, 5.2, 8.1 and 11.2 from NASA website @www.nasa.gov; Plate 5.1 from www.nede.noaa.gov/reports/mitch publisher National Oceanic and Atmospheric Administration, National Environmental Satellite, Data and Information Service; Plate 6.1 from National Oceanic and Atmospheric Administration Paleoclimatology Program/Department of Commerce, publisher The Ohio State University; Plate 6.2 source University of Colorado-Boulder (Anne Jennings); Plate 6.3 source University College Dublin (Dr. Frank O'Mara); Plate 8.1 The disintegration of the Larsen B ice shelf on the Antarctic Peninsular between 1995 and Spring 2002. source NASA; Plate 9.1, 9.2 and 9.4 source NERC Satellite Receiving Station, University of Dundee; Plate 10.1 from *Linking Climate Change to Land Surface Change* (2000) with kind permission of Kluwer Academic Publishers; Plate 10.2 Dai & Wigley 2000, Reproduced/modified by permission of American Geophysical Union; Plate 11.2 NASA's SeaWIFS satellite on March 3 2000. source: SeaWIFS Project, GSFC, NASA; Plate 11.3 (a) and (b) source Crown copyright Met Office.

In some instances we have been unable to trace the owners of copyright material, and we would appreciate any information that would enable us to do so.

CHAPTER 1

Introduction to systems and the climate system

1.1 Early development of climatology

1.1.1 From classical times to the Age of Discovery

As they sailed up the River Nile, the ancient Greeks noticed that it became warmer. Being logical in their search for explanation, they assumed that the earth must slope upwards towards the sun. The word 'climate' traces its roots from this now discredited deduction, being derived from a Greek word meaning 'slope'. But despite writings on climate by authors such as Aristotle and Hippocrates as early as the fourth century BC, progress in the study of the atmosphere was slow for almost the next two millennia. It was only with the Renaissance, when a renewed interest in exploration and overseas trade occurred, that the need to understand spatial variations in climate became obvious once again. In an Age of Discovery based on sailing ships, understanding the vagaries of the winds became vital. Later on, rainfall reliability, and thus water supply, became relevant matters for public health in urban areas. Amassing of large data collections began, aided by improvements in instrumentation. But, data collection alone does not help understanding, and organising and classifying of all kinds of weather data was the obvious next stage. Classification is never the most glamorous part of an emerging discipline and so climatology tended to become the 'ugly sister' of meteorology, its bookkeeping arm, until well into the twentieth century.

1.1.2 Determinism and possibilism

Making causative linkages between simplistic classifications of any phenomena can be dangerous, no less so with climate. It is tempting to infer that benign climates with few limitations in heat or moisture will be associated in general with productive agricultural or natural vegetation systems, and ultimately with the creation of an agricultural surplus which in the past provided the security for expansionism and economic development of all kinds. There may be considerable justification for this, especially in societies dependent on the annual harvest for their survival. It is entirely different, and misplaced, however, to extend this logic unquestioningly to issues related to cultural and economic development. Yet this was done widely in the early decades of the twentieth century. Such a philosophy, known as *environmental determinism*, implied control of human affairs by an external influence, namely the physical environment. Taken to its extreme, it consigned cultural and economic development to spatial and temporal variations in the physical environment.

Progress, it was argued by determinists, was based on climate, heredity and culture. Mid-latitude climates, with their seasonal contrasts implying a constant need to plan ahead, were implicitly deemed superior as influences promoting well-being and intellectual development by comparison with their equatorial counterparts. In the aftermath of the Second World War this smacked of ethnic prejudice and was rightly considered totally unacceptable. Environmental determinism was discredited.

The pendulum swung towards *possibilism* – a denial of the importance of the physical environment. The featureless plains which characterised key economic and geographic models by Von Thunen, Weber and Chrystaller, pointedly played down the importance of the physical environment to the point of virtually ignoring its existence. Climatology, heavily tarred by the deterministic brush, languished. The proportion of articles dealing with weather and climate in key scientific journals fell from about 35 per cent in the second decade of the twentieth century to 4 per cent in the mid 1960s. However, just like determinism, advocates of possibilism also pushed their logic too far.

Awareness that all human endeavours must be accommodated within a physical setting, and within climatic constraints, dawned slowly. A paradigm that believed that technological advances could 'fix' the physical environment when required was dominant for long periods and remains deeply embedded in the psyche of some ideologies. Gradually, a more balanced perspective on the role of the physical environment has emerged. Concepts of *stewardship*, rather than domination, of the environment were learned the hard way. Some spectacular failures in agricultural expansion into marginal areas were instrumental in this. But gradually the relevance of a climatic perspective has become clear in a number of crucial areas. The 'Green Revolution', the management of urban air quality, the emergence of the ozone hole, the growing toll of climate-related natural disasters, irrigation schemes, deforestation, soil erosion – all had important unanticipated climatic dimensions to them. But perhaps the clinching issue, which has led to climatology emerging as a vital discipline for humankind, has been the realisation that climate is itself changing over short and long timescales. Far from being prisoners of our climates as determinists suggested, climate is increasingly seen as a prisoner of human actions. The role of human action in contributing to most of the temperature change observed in the last 50 years (Figure 1.1) has now been widely accepted. At the annual meeting of the World Economic Forum, a global partnership of business, political, intellectual and other leaders held in Davos, Switzerland in January 2000, it was declared that climate change is the greatest global challenge facing humankind in the twenty-first century.

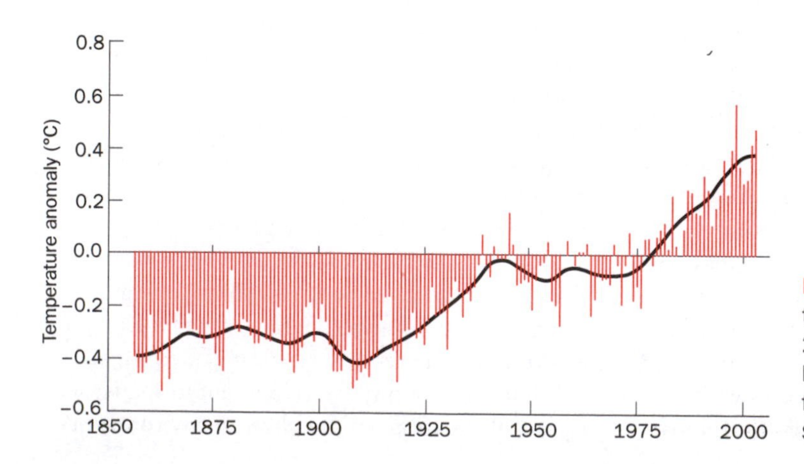

Figure 1.1 Global mean annual temperature trends from 1860 to 2002. The year 2002 was the second highest annual temperature recorded for the period with 1998 the highest. Source: Hadley Centre

Key ideas

1. A reactionary backlash to the philosophy of environmental determinism stifled the growth of modern climatology until recent decades.

2. Awareness of the central importance of atmospheric and climatic considerations in key areas of human endeavour has now grown enormously. This has particularly been demonstrated by the need to address pressing problems of climate change.

1.2 A systems approach

1.2.1 The nature of systems

Complex phenomena can often be better understood using a powerful conceptual tool known as systems theory. As one of the most complex natural phenomena known, the atmosphere has for long been subjected to a systems-based analysis, and conceptualising climate itself as a system provides a means both of understanding its functioning and predicting its potential changes.

A system is an ordered group of interrelated components, linked by flows of energy and material. A clear demarcation between what is inside and outside the system must also exist. Though this sounds complicated, we are only too familiar with everyday examples of the concept, such as a plumbing system or an electrical system, or a central heating system. In each case an organisational entity can be identified with component parts, or even subsystems integrated into the larger entity. Energy and material are moved around between the components and a power source drives the operation.

Systems seldom exist in a totally self-contained manner. Truly isolated systems, with no interchange of energy or materials with their surroundings, do not occur in the natural world. Even closed systems, with energy transfers, but not materials transfer, with their surroundings are scarce. The earth–atmosphere system can be thought of as a closed system, with energy in the form of sunlight (solar radiation) entering and heat (terrestrial and atmospheric thermal radiation) leaving, and negligible interchange of matter between the earth and space involved. Most natural systems are, however, open systems with both energy and matter transferred across the system boundary. The climate system is an excellent example of such an open system (Figure 1.2). Understanding thc pathways and the dynamic response of the climate system to forcing pressures, is central to managing it. Such understanding is as yet elusive, though great advances have been made in recent years.

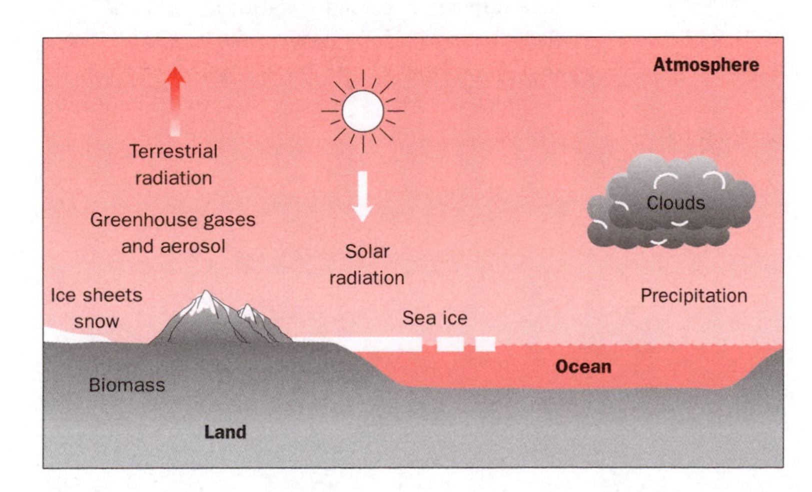

Figure 1.2 The climate system.
Source: Hadley Centre

1.2.2 System regulation

When the average condition of a system is relatively constant over time, an equilibrium exists which reflects the system's ability to cope with forces either external or internal seeking to disturb it. This ability to self-regulate is a characteristic of systems and is frequently controlled by internal readjustments known as feedback mechanisms. Two kinds of feedback mechanisms can be identified. Negative feedback operates to minimise the effect of any disturbance and seeks to return the system to the pre-existing status quo. This 'damping' effect characterises systems or parts of systems resilient to change. For example, it might be hypothesised that global warming would lead to more evaporation and cloud cover. The latter would reflect more incoming solar energy, promoting a tendency for cooling, i.e. negating temperature rises somewhat. Positive feedback, on the other hand, amplifies the effect of a disturbance and may have the effect of destabilising a system. Again it might be hypothesised that global warming might lead to more water vapour in the air. Since water vapour is a greenhouse gas this might be expected to encourage more trapping of outgoing terrestrial radiation, exacerbating the temperature rise. Further examples of both of these types of feedback are discussed in Chapter 6, while the difficulties they pose for climate modelling are discussed in Chapter 7.

Systems-based approaches offer windows on how the atmosphere functions, and how climate might respond to forcing either by natural or human agencies. The conceptual model must never be mistaken for the real thing, however, and the complexity of the atmosphere will ensure that incorporating its processes fully into system models is destined to be an elusive objective for a long time. For the moment we can make informed guesses in answer to the 'what if' questions using a systems approach incorporated into powerful computer models. But uncertainty remains, and just like the imperfect classifications of climate in former times, imperfect atmospheric models will always exist, calling for a cautious approach to managing the atmospheric/climatic system.

Key ideas

1. Systems are conceptual tools composed of elements with linkages between them, along which flows of energy and materials take place. An external power source provides the driving force for systems.

2. Systems are characterised by integration between their components such that a disturbance of any individual component has consequences throughout the system.

3. Systems have a capacity to regulate themselves such that they can adjust to a change imposed either externally or internally. This is accomplished by feedback mechanisms.

Further reading

Christopherson, R. (1995) *Elemental Geosystems*. Prentice-Hall, New Jersey, 540 pp.

Lewthwaite, G. R. (1966) Environmentalism and determinism: a search for clarification, *Annals Assoc. American Geographers*, **56**(4): 1–23.

Phillips, J. D. (1999) *Earth Surface Systems: Complexity, Order and Scale*. Blackwell, Oxford, 320 pp.

CHAPTER 2

Mass components of the climate system

2.1 Atmospheric composition

According to Greek mythology, Icarus escaped imprisonment using wings of feathers and wax made by his father to fly out of the island of Crete. Not taking heed of his father's warnings though, he flew too close to the sun whereby his wings melted and he perished. The myth exemplifies how little was known in ancient times about the atmosphere above the ground – it is one of the most difficult, and dangerous, areas to explore or monitor without sophisticated technology – and it is only in relatively recent times that detailed information about the envelope of gases surrounding the surface has become clearer and the importance of the different components to the climate system and life at the surface apparent.

While both the moon and the planet Mercury have no significant atmospheres, Earth's nearest neighbours Venus and Mars have atmospheres composed almost exclusively of carbon dioxide (CO_2). In the case of Venus, early movement of water vapour into the atmosphere seems to have produced a runaway greenhouse effect (see Chapter 6). Surface temperatures now exceed 500 °C. In the case of Mars, its thin atmosphere also shows a significant greenhouse effect, though not enough to warm the surface to much more than –50 °C.

2.1.1 The early atmosphere

Early in the evolution of the earth it is likely that its atmosphere was dominated by hydrogen and helium, the two most abundant gases in the universe. It would appear that at an early stage these light gases mostly escaped to space and were replaced mainly by water vapour from the interior emitted from volcanic vents. Sometime around 3 billion years ago a large proportion of this water vapour condensed to form the oceans of the world, and more outgassing from the interior of the planet added further components such as CO_2 and sulphur dioxide. Much of this carbon and sulphur became locked up in sedimentary rocks and the chemically inactive gas nitrogen gradually came to dominate the earth's atmospheric composition. Oxygen was a later arrival, only reaching 10 per cent of present concentrations about 1 billion years ago, probably mainly as a result of the sun's energy splitting water vapour molecules into their constituent elements. The hydrogen, being light, could escape more easily to space, leaving oxygen in small concentrations behind. Due to the thinness of the oxygen layer, ultraviolet radiation could penetrate down to the surface, inhibiting the development of life forms, unless they were shielded by a considerable depth of ocean. Only slowly did photosynthesis raise the concentration of oxygen to present levels by about 300 million years ago, by which time terrestrial life forms were established. This sequence of events suggests that the evolution of the atmosphere was the chief determinant of the evolution of life forms.

2.1.2 The Gaia hypothesis

The link between life forms and the atmosphere, particularly the role of vegetation in regulating atmospheric gases such as carbon dioxide and oxygen, has formed the basis of James Lovelock's Gaia hypothesis. The central tenet of this is that the physical and chemical characteristics of the land, atmosphere and oceans have been rendered congenial for life by biological activities. This is in contrast to the conventional idea that life evolved to adapt to changing environments. The Gaia concept suggests a symbiotic, self-regulating mechanism for climate change whereby more CO_2 would provoke more vegetation growth and encourage oxygen formation. This, combined with warmer temperatures, would in turn provoke more fire damage to plants, and thus a return to the status quo. A negative feedback system operating to maintain environmental conditions suitable for life forms was hypothesised. Similarly, a number of other self-regulatory feedback mechanisms can be identified, combining living and non-living components. The Gaia hypothesis has generated a great deal of discussion, but has also been criticised as implying that the earth–atmosphere system was behaving with a sense of purpose, as a kind of life force which was actively controlling the climate and other life-support systems of the planet. Neither did it give a significant role to natural selection based ultimately on adaptation to, and not modification of, environment. While these criticisms have been partially refuted, the main impact of the Gaia hypothesis has been to alert us to the complex interconnections between the atmosphere and life forms. Damage done to the atmospheric system, such as through global warming mechanisms, may be difficult to rectify, and thresholds may be crossed which cannot easily be recovered from.

2.1.3 Vertical structure of the atmosphere

Manned balloon ascents were the earliest means of investigating the higher levels of the atmosphere and by the end of the eighteenth century a fairly uniform composition up to about 3 km was confirmed (Table 2.1). Spatially, there is also little difference apparent in atmospheric composition with (e.g. oxygen, CO_2) concentrations not varying significantly at sea level around the globe. When higher altitudes were investigated using unmanned balloons, rocket flights and satellites, however, more significant variations were discovered. Some of these compositional changes have proven to be important in terms of explaining vertical changes in temperature (Figure 2.1).

Table 2.1 Composition of the earth's atmosphere near the surface

Permanent gases			**Variable gases**			
Gas	**Formula**	**% by volume**	**Gas and particles**	**Formula**	**% by volume**	**ppm**
Nitrogen	N_2	78.08	Water vapour	H_2O	0–4	
Oxygen	O_2	20.95	Carbon dioxide	CO_2	0.036	370
Argon	Ar	0.93	Methane	CH_4	0.00017	1.7
Neon	Ne	0.0018	Nitrous oxide	N_2O	0.00003	0.3
Helium	He	0.0005	Ozone	O_3	0.000004	0.04
Hydrogen	H_2	0.00006	Particles (dust, soot, etc.)		0.000001	0.01–0.15
Xenon	Xe	0.000009	Chlorofluorocarbons		0.00000002	0.0002

Source: Ahrens (2000, p. 3)

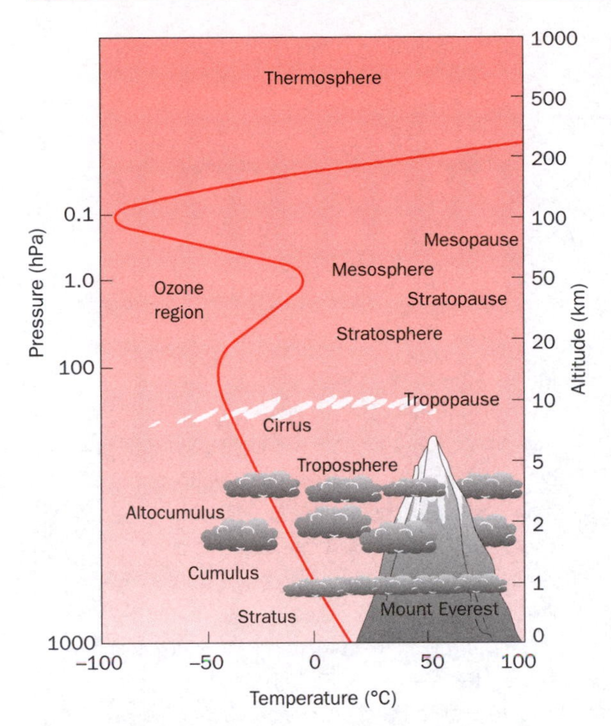

Figure 2.1 Vertical temperature and pressure structure of the atmosphere
Source: Hardy (2003, p. 5)

(a) The troposphere

From the surface up to about 11 km in height, the temperature falls by an average of 6.5 °C/km. On occasion this may not occur, or be reversed, giving rise to isothermal (no change in temperature with height) or inversion (increase in temperature with height) layers respectively. This part of the atmosphere receives the bulk of its heat from the surface of the earth (see Chapter 3), meaning that the warmest layers are closest to the surface and warming air rises and mixes freely. Natural tendencies for mixing and overturning are why this layer is called the troposphere, from the Greek word *tropein* (to turn). About 75 per cent of the mass of the atmosphere and almost all the water vapour and aerosols are contained in this layer. The tropopause marks the upper limit of this layer of mixing, and naturally is located higher over the equator (16 km) than over the poles (8 km). Temperatures at this height are generally below −60 °C.

(b) The stratosphere

Above this point temperature begins to increase with height, slowly at first, then more rapidly, up to a height of about 50 km at which air temperature may even have exceeded 0 °C once again. This upper layer of the atmosphere has its warmest zones overlying its coolest zones and so great stability exists with limited tendencies for overturning. This is the stratosphere, from the Greek word *stratos* (layer). The temperature inversion that the stratosphere exhibits is due primarily to its relatively high ozone content. Although in absolute terms this is small – all of the ozone in the atmosphere would amount to a layer 3 mm thick at sea level – it plays a vital role in shielding the surface from harmful ultraviolet radiation. Stratospheric ozone strongly absorbs UV radiation, leading to a warming of the upper stratosphere, and its recent depletion due to human activities has constituted a major global environmental problem (Box 2.1).

(c) Mesosphere and thermosphere

Ozone concentrations decline above about 50 km in height, a point known as the stratopause, and temperature falls again to around −90 °C at about 80 km. This layer is known as the mesosphere. Although the mix of oxygen and nitrogen is the same as at sea level, the density is extremely low and there are simply not enough oxygen molecules available in each breath to stop the brain becoming oxygen starved. Even at much lower altitudes this condition, called hypoxia, known to afflict high-altitude climbers and pilots, may occur, beginning with feelings of tiredness and ending in unconsciousness and death. In the mesosphere, suffocation would happen in a matter of minutes. Severe burning due to ultraviolet radiation could also be expected as well as loss of blood due to evaporation in the low pressure environment.

BOX 2.1

THINKING FURTHER

Ozone depletion

Natural production and loss

Ozone is a form of oxygen in which each molecule consists of three atoms instead of the more normal two. In the lower atmosphere or troposphere (10 per cent of global total) the ozone gas is a product of the photochemical pollution process and instrumental in producing the smogs that blanket many cities during the summer months. Where it is present in high concentrations it is a bluish-green gas which is highly toxic. Even at lower concentrations it irritates the respiratory tract and can trigger asthma. Ozone also occurs in the stratosphere (90 per cent of global total) at low concentrations of around 10 ppm. Ozone is created here naturally as a result of the bombardment of oxygen molecules by ultraviolet (UV) solar radiation (equation (1)). About 7 per cent of incoming solar radiation possesses wavelengths between 0.2 and 0.4 μm. Were this to reach the earth's surface in its entirety, considerable damage would result to the genetic make-up of life forms. Fortunately, ozone in the stratosphere provides a good absorber of almost all the radiation which arrives in the 0.2–0.3 μm wavelengths and as a result shields the surface from this harmful UV radiation.

The absorption of UV radiation (denoted by *hv* below) by oxygen molecules produces a dissociation of the molecular structure as follows:

$$O_2 + hv \rightarrow 2\,O \qquad (1)$$

The free oxygen atom quickly attaches itself to another oxygen molecule to form tri-atomic oxygen or ozone:

$$O_2 + O \rightarrow O_3 \qquad (2)$$

It can be noted here that ozone is also destroyed by natural means including the absorption of UV radiation, i.e. by the same processes that create it:

$$O_3 + hv \rightarrow O_2 + O \qquad (3)$$

Ozone decline and the ozone hole

Measurements of ozone have been made in Switzerland since 1926. These show a decline of about 4 per cent a decade since the 1970s. More dramatic changes, however, have appeared from monitors established in the Antarctic from the mid 1950s. For the first two decades or so, relatively little change in the ozone column over the Antarctic was noticed. However, from the late 1970s, using an instrument originally developed in the 1920s known as a Dobson spectrometer, scientists at the British Antarctic Survey began to report dramatic declines in ozone during the Antarctic spring in September and October. These findings were initially treated with some scepticism since more sophisticated satellite-based monitoring failed to replicate them. Further analysis of the data quality control programmes used by the satellites revealed that they had been programmed to discard, as unreliable, data which departed from the mean by a substantial amount. Fortunately the raw data had been retained in storage and recalculation confirmed the existence of a major ozone reduction that by 1984 had resulted in October ozone levels 35 per cent lower than the average for the 1960s. The ozone hole has since become a regular springtime occurrence and has increased in size and extent through the 1990s. By 2000 it was over 28 Mkm^2 in extent, covering an area about twice the size of the continent of Antarctica.

The identification of a family of human-made chemicals known as chlorofluorocarbons (CFCs) as the cause of the ozone hole was not immediately achieved. These substances had been first synthesised in the 1920s and developed as ▶

coolants and propellants so that by 1970 industry used about a million tonnes annually. Thought to be inert they were detected in the air by James Lovelock (the originator of the Gaia concept) in 1970. Working in southern Ireland, he found trace amounts of one of the compounds, CFC-11, on air samples in Ireland which had passed over southern England. But he also detected traces in air samples from seemingly uncontaminated sources, leading him to believe that the gases were widely dispersed at a global scale.

Being lighter than air, CFCs rise into the stratosphere where their vulnerability to UV radiation is demonstrated and they are broken apart, liberating chlorine atoms. These chlorine atoms destroy ozone, as shown in equations (4) and (5):

$$Cl + O_3 \rightarrow ClO + O_2 \qquad (4)$$

$$ClO + O \rightarrow Cl + O_2 \qquad (5)$$

The same amount of chlorine is present at the end of the reactions shown in equations (4) and (5), and is available to start the chain again. A single chlorine atom can destroy around 100,000 ozone molecules. This is how a tiny amount of CFCs can ultimately create an ozone hole of continental proportions – a salutary lesson in how a tiny amount of key pollutants could cause potential global catastrophe.

The great vortices of air around the poles are natural collecting zones for atmospheric pollution. In the Antarctic, during the long nights of winter, temperatures can fall to –85 °C and thin, high ice crystal clouds form. These crystals provide a very efficient surface for chemical reactions to take place. During the spring, photochemical activity rates increase as the sunlight arrives and ozone becomes depleted until ozone-rich air is mixed in from further north with the advance of summer.

Globally, stratospheric ozone losses now run at about 10 per cent in winter and spring and 5 per cent in summer and autumn. New areas of depletion have become evident over the high latitudes of the northern hemisphere where the receipt of harmful UV radiation has increased by 5 per cent over the past decade. Malignant skin cancer rates are 10 times higher than in the middle of the twentieth century, and although much of this is not due to the loss of the ozone shield there is a strong connection. For each 1 per cent decline in ozone a 2–3 per cent increase in some skin cancer rates are apparent. The genetic code of all living cells can be damaged by excessive harmful UV radiation.

Montreal Protocol

Concerns over the damaging effects of CFCs led to an international agreement to phase out their production. The Montreal Protocol has been successful in leading to a global ban on the manufacture of the most harmful members of the family. However, the CFCs presently in the system will continue to remove ozone for much of the present century. Little amelioration in the severity of the Antarctic ozone hole is apparent in recent years (Plate 2.1), though the effects of the Montreal Protocol in reducing CFC loading on the atmosphere is now detectable. The agreement represents an important milestone in the global community's willingness to tackle a global pollution problem. However, it must be emphasised that removing CFCs does not entail hardship or lifestyle changes of any significance for most people. This is in contrast to the effort which will be required to tackle the other major global atmospheric problem – global warming. Though in the public psyche these two global problems are often confused, it must in conclusion be emphasised that they are separate issues. CFC-induced ozone depletion below about 25 km height in the atmosphere contributes more to global cooling than global warming.

Some traces of water vapour are occasionally carried up into the high mesosphere where they may condense around dust particles, possibly meteoric in origin, to produce the ice-crystal luminous night clouds known as noctilucent clouds sometimes observed over the high latitudes in summer. Above about 85 km, the mesopause, lies the thermosphere. This is so called because once again temperatures rise with height. The cause of this is the absorption of very short wavelength UV radiation by both molecular and atomic forms of oxygen. However, temperature as such is largely a theoretical concept here as the air is so thin. An air molecule may move a kilometre before encountering another air molecule. Only the slight frictional drag on orbiting satellites allows temperatures to be inferred at the higher levels. It is that drag which causes orbits to decay and satellites to fall back to earth. Above 500 km molecules may escape the gravity of the earth altogether and the upper limits of the atmosphere are encountered, a zone sometimes referred to as the exosphere.

2.1.4 Measuring atmospheric pressure

Since the atmosphere is composed of various gases, each of which has a certain mass, gravity provides the atmosphere with weight. The weight of air in an average living room is about the same as a person. Such is the weight of the overlying air at sea level that a force of about 100,000 newtons (N) is exerted on each square metre of the surface. This is known as barometric pressure and more precisely is quantified as 101,325 pascals (Pa) or 1,013.25 hectopascals (hPa). (A pascal is 1 newton/m^2.) This is taken as standard sea-level pressure and is occasionally still expressed in the older units of millibars (1 mb = 1 hPa). Surface air pressures vary from time to time and from place to place depending on the nature of vertical air movement in the vicinity of the observation. Falling or subsiding air increases the surface pressure, while rising or ascending air diminishes it somewhat. The highest air pressure observed at the surface was 1,084 hPa in the centre of a subsiding air mass over Siberia, while the lowest was 870 hPa in the centre of a rising revolving mass of tropical air in the centre of a typhoon in the Pacific Ocean.

At 3 km high in the atmosphere, the air column above is 3 km shorter than at the surface. Air pressure is thus reduced considerably to about 700 hPa at that height. At 10 km elevation, it is less than 300 hPa. The reduction in pressure with height comes about because the atmosphere obeys the Gas Law which specifies that the pressure of a gas (P) is equal to its temperature (T) times its density (ρ) times a constant (C) as follows:

$$P = T \times \rho \times C$$

A change in either temperature or density can thus produce a change in pressure and vice versa. It might be thought that the upward reduction in pressure which produces a situation whereby high pressure exists at the surface and low pressure at the top of the atmosphere would be conducive to the atmosphere shooting off into space (since air moves from high to low pressure). However, it should be noted that it was the force of gravity which established these differences in pressure in the first place and a mutual balance exists between the force of gravity and the vertical pressure gradient. This keeps the atmosphere attached to the earth and is known as hydrostatic equilibrium. Vertical air motion is thus quite limited by comparison to its horizontal counterpart known as the wind, typically two orders of magnitude less. Exceptions do occur to this, most notably in severe storms.

Measuring pressure has historically been done by measuring the height of a fluid which the air pressure pushing down on a reservoir can support in an evacuated tube sealed at one end. Any fluid will do, although denser fluids are more convenient. A water column of over 10 m would be supported by average surface air pressure at sea level. Obviously this is an unwieldy arrangement and since mercury is 13.6 times more dense than water, it provides a more practical alternative. Evangelisto Torricelli, a student of Galileo, invented the mercury barometer in 1643. Typically air pressure supports a column about 76 cm high, explaining why for a long period pressure was expressed in terms of 'millimetres of mercury'. Aneroid (or liquid free) barometers became widespread in more recent times, having been first constructed by Lucien Vide in 1843. Much more portable than their mercury counterparts, these consist of a partially evacuated metal box which flexes as pressure changes. These flexures are amplified by levers to give a reading on a dial, or in the case of barographs connected to pens which give a trace of pressure changes over time on a revolving drum. Instead of pressure the instrument may be calibrated to produce heights if it is carried aloft into a different pressure environment. This is the basis of the instrument known as an altimeter.

Key ideas

1. Early in its evolution the earth's atmosphere was dominated by lighter gases which gradually escaped to space. Outgassing of water vapour and CO_2 from the earth's interior gradually transformed the composition of the atmosphere and produced conditions conducive for the evolution of life forms.

2. Complex feedback relationships exist between the composition of the atmosphere and the well-being of life forms. The Gaia hypothesis exemplifies one such possible relationship whereby self-regulating mechanisms involving components of the atmosphere, climate change, and the congeniality of the planet for life have been suggested.

3. Altering the composition of the atmosphere may involve the crossing of as yet unknown thresholds from which recovery of the climate system may not readily be achieved.

4. The troposphere is the lowest layer of the atmosphere, extending up to a height of about 16 km over the equator and about half this over the poles. Temperatures generally fall with height in this layer, making it an unstable (turbulent) zone. The troposphere contains most of the mass of the atmosphere and almost all its water vapour. Almost all weather systems are also confined to the troposphere.

5. The stratosphere lies above the troposphere and extends up to 50 km in height. Temperatures generally rise with height, making this a stable (little vertical air movement) layer. The stratosphere is relatively rich in ozone, providing a shield from damaging UV radiation for life forms at the surface of the earth below.

6. Higher levels of the atmosphere extend to about 500 km where air densities become extremely low as the edges of space are approached.

7. Atmospheric pressure at the surface averages about 1,013 hPa, falling to about 700 hPa at a height of about 3 km.

2.2 Mass and air motion

2.2.1 Forces controlling horizontal and vertical movement

(a) Pressure gradient force

Variation in pressure at two locations results in a pressure gradient force existing between them. This is the driving force for all air motions and acts to push air (i.e. as wind) from high to lower pressure. The magnitude of the force, and thus the speed of the wind, is a function of the pressure difference between the two locations concerned. These differences in pressure can be mapped to show the locations that have higher and lower pressures at a particular height. Figure 2.2 shows a chart of air pressure for the surface. The lines connecting places of equal pressure are known as isobars. Tightly spaced isobars are indicative of a steep pressure gradient and therefore strong winds, while widely spaced isobars typify places with gentle pressure gradients and light wind or calm conditions. Areas of enclosed high pressure are known as anticyclones; areas of enclosed low pressure are known as depressions in most middle latitude countries. Unenclosed high or low pressure areas are known as ridges or troughs respectively.

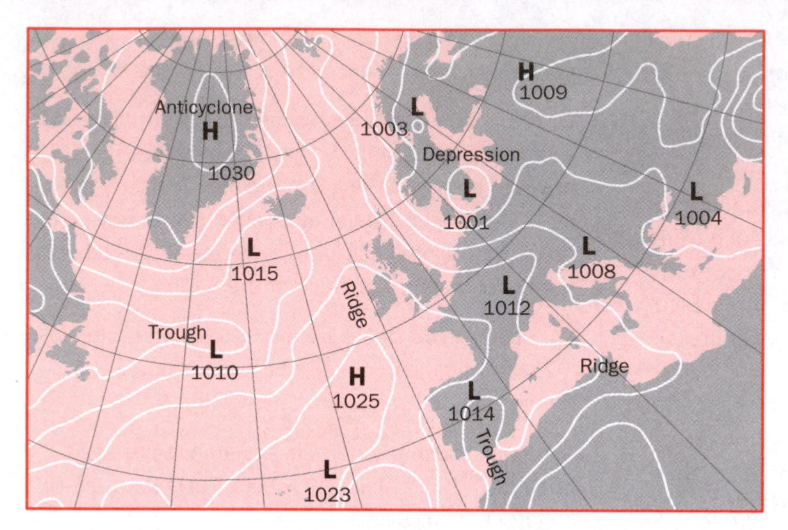

Figure 2.2 Pressure chart for the North Atlantic region showing common pressure features: anticyclones, depressions, ridges and troughs

Charts may also be drawn depicting the height at which a specified pressure value is reached. In this case the lines are contours showing how far an ascent would have to be made before a particular fixed value of pressure is reached. Commonly, values of 850, 700 or 500 hPa are chosen. Figure 2.3 shows part of a 500 hPa chart. Recalling the Gas Law, for the same pressure value (500 hPa in this case), colder air will have to be more dense than warmer air. Accordingly, the height to be climbed through a colder air mass before reaching the 500 hPa surface would be less than through a warmer air mass. Further consideration will reveal that contour lines of low height will indicate regions of lower pressure at the surface and vice versa. Upper air charts can thus provide valuable information on what is happening many hundreds of metres below at the surface. As will be shown in Chapter 4, upper air movements largely control the formation and direction of movement of weather systems at the surface.

(b) Coriolis force

Air, or any object, moving across the earth's surface is moving on a sphere which is itself rotating through 360° in every 24 hours. From the perspective of someone situated on the earth, and not therefore conscious of its rotation, this produces a change in the direction of movement. Consider the view of the earth a satellite stationary over the North Pole would have (Figure 2.4). From that perspective, the earth below would complete one revolution every 24 hours, or 15° rotation

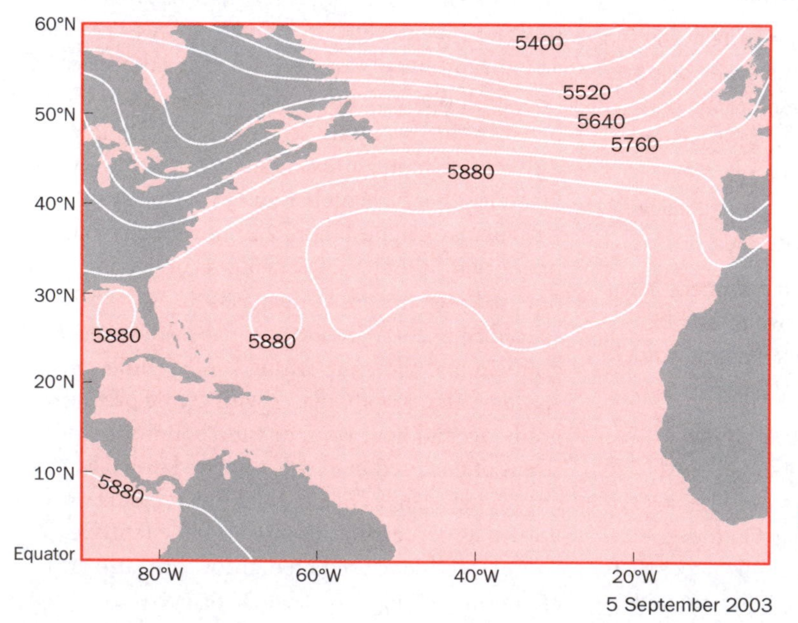

Figure 2.3 A 500 hPa contour chart. The height of the 500 hPa pressure surface is indicated in metres

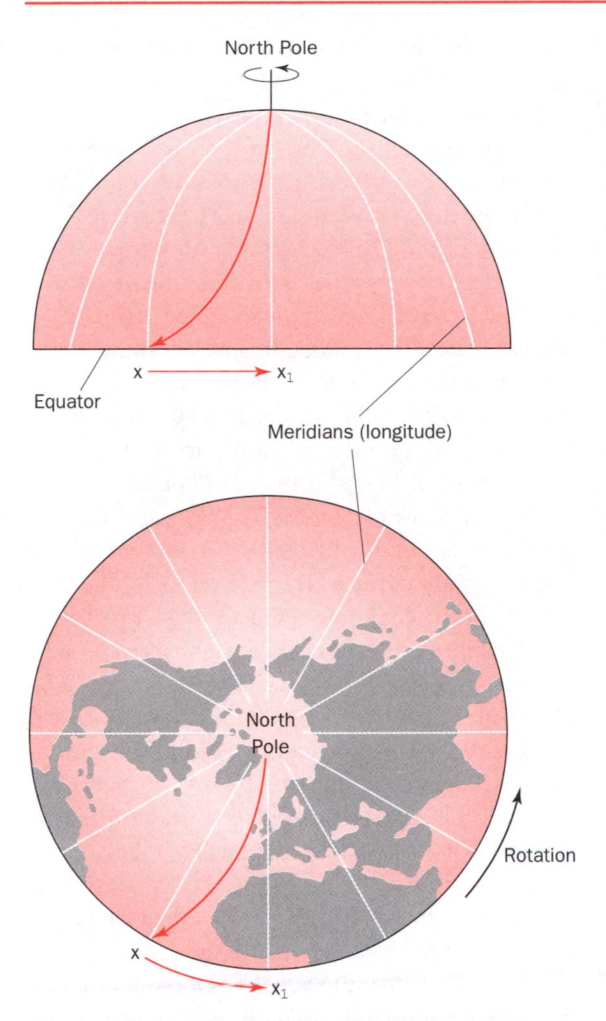

Figure 2.4 Coriolis deflection of southward-moving objects as viewed from the North Pole

per hour. A missile fired from the surface at the North Pole towards the equator would, if it took 2 hours to reach the equator, land 30° to the right of its intended target. Watching this from any point on the earth it would appear that the missile had traced out a curved path to the right. From the satellite however the truth would be apparent – it was the earth that turned and not the missile. Oblivious to earth rotation, it seems to us on the surface that some force must be deflecting the missile to the right. Gaspard Coriolis, a French mathematician, gave his name to this force which deflects any free-moving object to the right of its intended path in the northern hemisphere and to the left in the southern hemisphere (see Box 2.2). This deflection is relatively easy to appreciate for the north–south example illustrated here. Further

BOX 2.2

THINKING FURTHER

The Coriolis force

Coriolis deflection affects ocean currents, rivers, missiles, aircraft – all free-moving objects. A bullet fired at 120 m/s will be deflected 0.5 cm in 2 s. A car travelling at 100 kph would drift 4.5 m in 1.6 km in the absence of friction between the tyres and the road. Mostly though the Coriolis force is slight compared to other forces involved in producing the motion concerned. Mathematically it is expressed as

$$C_f = 2V\Omega \sin \phi$$

where Ω is the angular velocity of spin (for the earth 15°/hour or $2\pi/24$ radians/hour), ϕ is the latitude and V is the velocity of the moving object. The following conclusions can be deduced from the equation and/or by considering the example in section 2.2.1(b):

1. The Coriolis force acts to deflect a moving object to the right of its line of motion in the northern hemisphere and to the left in the southern hemisphere.
2. The Coriolis force is strongest at the poles and is zero at the equator. Only the direction of motion is influenced by the Coriolis force, not the speed.
3. At a given latitude, the Coriolis force increases as the speed of an object increases. (An object which is stationary or slow moving is obviously moving in unison with the earth's rotation and therefore experiencing little or no deflection.)

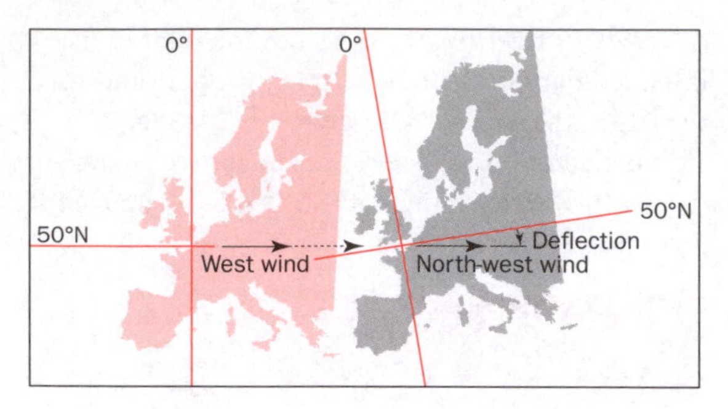

Figure 2.5 Coriolis deflection of a west wind to produce a north-westerly wind

consideration will demonstrate that the deflection also occurs for all directions of movement (Figure 2.5).

(c) Pressure gradient and Coriolis forces combined

Above the influence of surface frictional drag, air movement is governed by the pressure gradient and Coriolis forces alone. Figure 2.6 shows an air parcel at high elevation in the northern hemisphere. The isobars show a pressure gradient from 808 to 800 hPa wanting to drive the air parcel from the higher to the lower pressure. At point 1, once the air parcel begins to move, it becomes subject to Coriolis deflection to the right, swinging it towards point 2. The cumulative effect of the pressure gradient and Coriolis forces is also to speed up the air parcel, so that it is accelerating by the time it reaches point 2. At point 2, the pressure gradient is still acting in the same direction, but the trajectory of the air parcel means that the

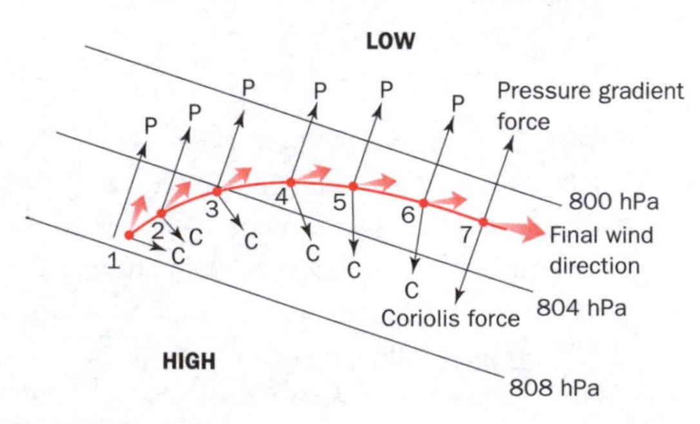

Figure 2.6 The geostrophic wind

Coriolis force has swung around somewhat to the right. So when the air parcel moves past point 2, it is deflected a little more to the right. Since it is also moving faster the Coriolis force is increased. By point 3 the Coriolis force is round to the right even more in response to the slightly changed trajectory of the air parcel, now continuing to accelerate. The parcel has been speeding up all this time because the two forces added together produce a net force pulling it from left to right. By point 6 the air parcel has been deflected round almost to the point of being parallel to the isobars by the ever-strengthening Coriolis force. At point 7, a balance is achieved between the pressure gradient and Coriolis forces. The Coriolis force has now become equal and opposite to the pressure gradient force. They cancel each other out. No net force is now acting on the air parcel which continues to travel at constant speed in the direction it was going when balance was achieved. The result is that the air parcel now travels parallel to the isobars. The speed of travel will depend on the initial pressure gradient. The stronger this was the faster the air parcel would have been travelling before balance was achieved. Equally, if the Coriolis force is weak, such as in the low latitudes, the parcel of air would also have reached a higher speed before the balance was achieved.

The air movement produced as a result of this interaction between the two forces is known as the geostrophic wind. Such winds, blowing parallel to the isobars, are the norm at high altitudes and can be seen in Figure 2.7. The chart also indicates that air blows in a clockwise manner around areas of high pressure and in an anticlockwise manner around areas of low pressure in the northern hemisphere. When there is a curvature in the isobars another force must be taken into account. This is the centrifugal force which acts to push a moving object outwards from a central point. Turning a corner quickly on a bicycle or in a car readily demonstrates the presence of this force. When air is moving around a low pressure centre the centrifugal force is acting in an opposite direction to the pressure gradient force, and the balance which results from the interaction of

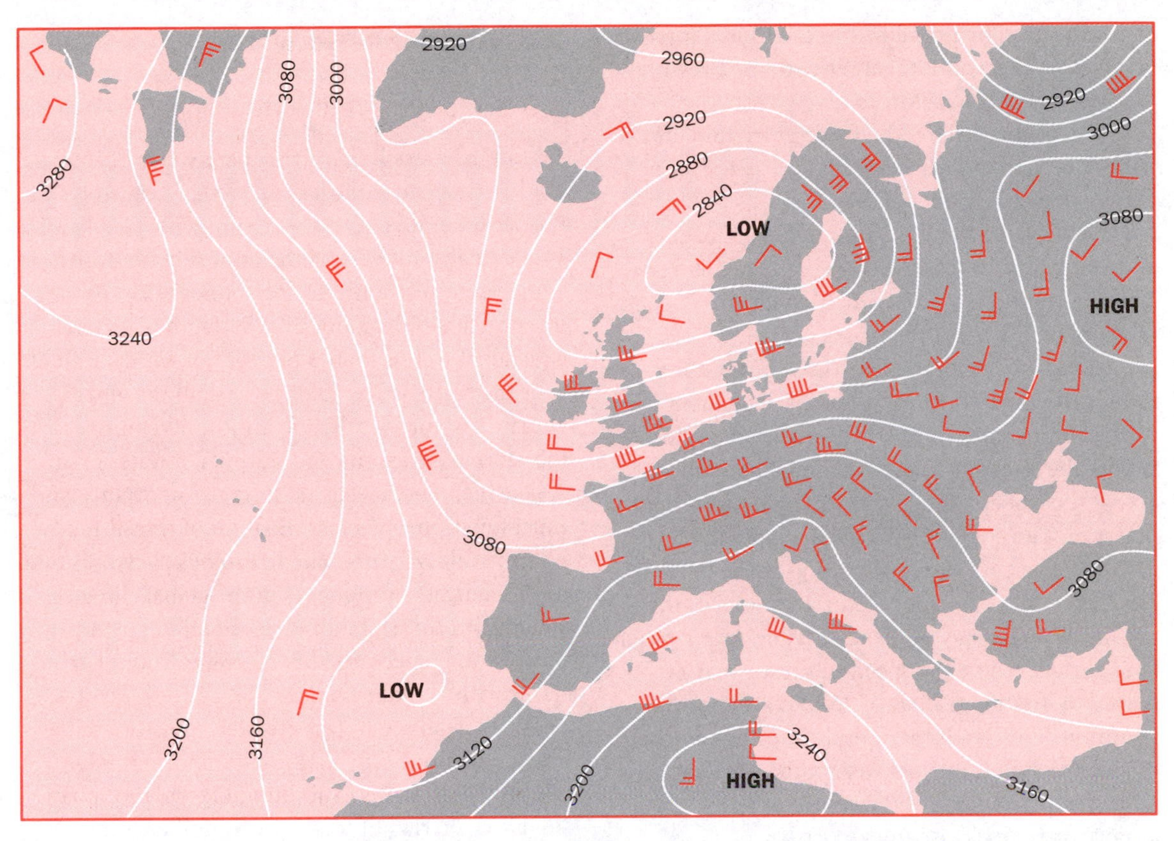

Figure 2.7 A typical upper air chart at the height of the 700 hPa pressure level showing wind flow directions parallel to the isobars

pressure gradient, Coriolis and centrifugal forces is therefore less than if the air was travelling in a straight line. This is what is known as the gradient wind. Like the geostrophic wind it is parallel to the isobars, but of lesser magnitude than would otherwise occur (subgeostrophic). In the case of circulation around a high pressure with a similar pressure gradient, the opposite would apply. The centrifugal force would add to the pressure gradient force, meaning that a stronger Coriolis force would be needed to achieve balance. Wind speeds that are 'supergeostrophic' would result. For practical purposes the concept of the gradient wind is only important when very intense vortices are involved, such as in tornadoes. However, when the directional curvature of air motion changes in the upper atmosphere the airflow may undergo acceleration or deceleration as it adjusts from subgeostrophic to supergeostrophic or vice versa. This may induce convergence or divergence in the airflow aloft and be instrumental in the formation of surface weather systems (see Chapter 4).

(d) Friction

The strongest winds at the surface are experienced over the sea. This is because air moving over the surface is subjected to frictional drag, particularly if the surface is rough and irregular. Large forests or tall buildings can reduce wind speeds close to the surface considerably; 100 m inside a forest the wind speed may be reduced to less than 10 per cent that outside of it. Convective towers of rising air on hot afternoons may also act like obstacles to the regional wind flow. As well as slowing wind speeds, frictional drag may also have the effect of changing wind direction by providing an additional force to be taken into account together with the pressure gradient and Coriolis forces.

Close to the surface, the force of friction affects wind direction as well as wind speed. By slowing the wind speed, frictional drag disrupts the balance between the pressure gradient, Coriolis and centrifugal forces which exists at higher levels. This disruption is achieved by slowing air movement. Such an effect does not affect the pressure gradient force in any way; but it does reduce the Coriolis force (which is dependent on speed of movement). The pressure gradient force is now the stronger of the two and air is pulled across the isobars, moving from higher to lower pressure. The angle at which the wind will blow across the isobars varies according to the friction force exerted. Over smooth surfaces such as the ocean the angle may be 10–20°, while over a rough land surface where the speed of the air may be halved by friction, values of 25–35° are more common. This is why air spirals into the centre of a depression (anticlockwise in the northern hemisphere) and out of the centre of an anticyclone (clockwise in the northern hemisphere). Once again there are implications for weather at the surface due to convergence (and uplift) in a depression and divergence (and subsidence) in an anticyclone.

Key ideas

1. A gradient in pressure between two points creates a force encouraging movement of air from high to low pressure. Subsequent movement is, however, also influenced by a deflective force arising from the rotation of the earth known as the Coriolis force, and by friction which the moving air may encounter with static obstacles.

2. In the higher levels of the atmosphere the pressure gradient and Coriolis forces interact to produce air movement parallel to the isobars known as the geostrophic wind.

2.3 Atmospheric water

2.3.1 Moisture in the atmosphere

The atmosphere holds a tiny percentage of the world's water by comparison to the great stores of the ocean and terrestrial environments (Table 2.2). While in absolute terms the atmospheric store is a huge amount, some 13 teratonnes (13×10^{12} t), it represents only about 2.5 cm depth of water, equivalent to less than 3 per cent of annual rainfall over the earth as a whole. To sustain the global climate system, clearly, the atmosphere must therefore have inputs and outputs of water on a scale which far exceeds its storage importance and must function as a great medium of transit for water in all its forms. The hydrological cycle is one of the central components in the global climate system and understanding its functioning and how it can be perturbed is central to coming to terms with climate change.

As we shall see in this chapter, moisture in the atmosphere not only acts as an important mass constituent in terms of air humidity and precipitation but also in terms of its ability to influence air motion, particularly in the vertical dimension.

(a) Water vapour

(i) Absolute measurements

In the atmosphere, 96 per cent of moisture exists as water vapour, a colourless gas not to be confused

Table 2.2 Global stores of water

Store	%
Oceans	97.2
Ice sheets, glaciers	2.15
Groundwater	0.63
Lakes, freshwater stores	0.009
Saline lakes, inland seas	0.008
Soil moisture	0.005
Atmospheric moisture	0.001
Rivers	0.0001

Source: Strahler and Strahler (1997, p. 88).

with cloud or mist or other forms of liquid water suspended in the air. At any particular time, the proportion of water vapour may range from close to zero to about 7 per cent. This proportion is referred to in a variety of ways. The mass of water vapour that is held by a particular volume of air is defined as its absolute humidity (g/m^3). Another measure used is the specific humidity – the ratio of the mass of water vapour to the total mass of moist air in the same volume (g/kg). A slight variant of this is the mixing ratio that is the ratio of the mass of water vapour to a unit mass of dry air (g/kg). Irrespective of which measure is chosen, the maximum amount of water vapour that can be held by air is determined simply by the temperature of the air. When air is holding the maximum amount of water vapour possible it is said to be *saturated*. Warmer air can hold more water vapour than colder air. At a temperature of 20 °C air may hold 15 g of water vapour, while at a temperature of −10 °C a mere 1.5 g of water vapour is all that can be accommodated.

The amount of water vapour in the air may also be described in terms of the *pressure* it exerts. As a gas, it contributes to the total pressure exerted by an air parcel within which it is contained. Near the surface, this actual vapour pressure exerted by the water vapour would typically be about 10 hPa. The vapour pressure is a good indicator of how many water vapour molecules are present at a particular time. A closely related concept to this is the saturated vapour pressure, that is the pressure that the water vapour molecules would exert if the air were saturated with vapour at a particular temperature. One can imagine the concept of saturation as involving molecules of water vapour escaping from a water surface being exactly matched by a return flow. Warming the water surface would produce a faster escape, and thus to maintain equilibrium an increase in the number of water vapour molecules above the water surface would need to occur. Essentially this demonstrates that more water vapour is needed to saturate warmer air than colder air. Saturation water vapour pressure at 10 °C is about 12 hPa, while at 30 °C it has increased to about 42 hPa. The greatly enhanced water vapour carrying capacity at higher temperatures begins to explain why tropical rainfall rates are so much more intense than in the mid-latitudes.

(ii) Relative measurements

The ratio between actual water vapour pressure and saturated water vapour pressure at a particular temperature is a measure of the saturation of the air. Usually expressed as a percentage, this relative humidity is the most popular way in which atmospheric moisture content is described. It expresses the ratio of the air's actual water vapour holding to the maximum that it could hold at that temperature. It is important to emphasise that the relative humidity of an air parcel may change either by altering its water vapour content, or by changing its temperature. Relative humidity may for example decline steadily, as much as 50 per cent after sunrise as air warms up, only to increase again as the sun goes down. During the interval the absolute humidity may have remained unchanged. Air possessing a relative humidity of 100 per cent is saturated and any further cooling requires that condensation of some of the water vapour load takes place. This is the dew point temperature, defined as the temperature at which an air parcel would become saturated in the absence of a change in pressure or water vapour content. On calm, cool summer nights as the air near the ground loses heat by radiation to the air above, the chilling of the air in contact with the ground may cool the air below its dew point temperature and water vapour will condense in the form of dew. Air forced to ascend may also be cooled below it dew point, producing cloud.

(b) Phase changes of water

Evaporation and condensation involve phase changes for water from liquid to gas. Freezing and melting involve phase changes from solid to liquid. These changes of state may be considered as changes in the molecular arrangement, with ice having molecules tightly packed in a crystal structure, liquid water having bonds which break and reform to allow flow, and water vapour having randomly moving molecules without a bonding structure to confine them. To change from solid to liquid form energy must be applied to break down the bonding pattern. To change from liquid to gaseous state similarly energy input is required. This can be quite a considerable input. To melt a kilogram of ice at 0 °C to produce liquid water at 0 °C requires roughly 335 kilojoules (kJ) of energy. To evaporate water to produce water vapour requires 2,460 kJ/kg. These energy inputs from the environment do not produce changes in temperature – they are merely used to break molecular bonds. Accordingly they are referred to as latent heat, in this case the latent heat of melting and evaporation respectively (Figure 2.8). When the opposite processes occur this latent heat is returned to the environment as sensible heat. This is the heat we can experience with our senses and can be measured by a thermometer. Freezing and condensation thus release energy and warm up the surroundings. On occasion the release of latent heat may occur at distances long removed from where it was absorbed in the first instance. Evaporation from the tropical oceans may result in water vapour which travels to high latitudes before condensing into cloud and releasing the latent heat absorbed during the original evaporation episode. The release of latent heat thus plays an important role in balancing the earth's energy (as well as water) budget by exporting the excess heat of the low latitudes to subsidise the energy-deficient areas of the high latitudes. In addition, such is the magnitude of latent heat release during the condensation process that it provides a major source of energy for atmospheric disturbances such as thunderstorms, depressions and hurricanes.

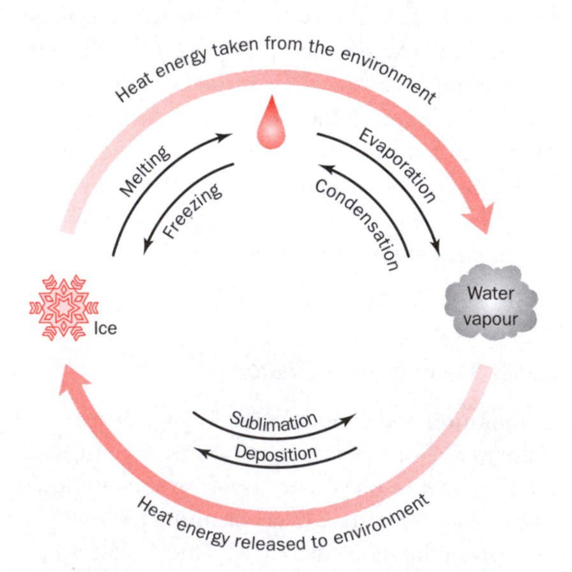

Figure 2.8 Changes of state of water

2.3.2 Atmospheric stability

When an air parcel is displaced upwards, it may respond by moving back to its original position, if it finds itself surrounded by air that is warmer than itself. In this case the colder heavier air in the original ascending parcel will move back down through the atmosphere. Such air is said to be *stable*. A physical analogy for this would be a plastic football sinking to the bottom of a pond when filled with water, i.e. the combined weight of the ball and water is heavier than the surrounding water. An air parcel displaced upwards into colder air may, however, continue to rise until it reaches a level where its temperature, and therefore density, has fallen sufficiently for it to be similar to the surrounding air. This buoyant air is said to be *unstable*. An analogue for unstable air would be an air-filled plastic ball placed at the bottom of a pond. When released the air-filled ball which is highly buoyant will shoot upwards to the surface. Whether an air parcel becomes stable or unstable determines respectively whether subsidence/warming or uplift/cooling is likely to occur. This may have great significance for condensation, the release of latent heat of evaporation and the energising of atmospheric disturbances of all kinds.

(a) Adiabatic processes

A parcel of air which cools and expands, or warms and compresses, without any exchange of heat with its surroundings is said to undergo an adiabatic change in temperature. The word 'adiabatic' simply means 'without a loss or gain of heat'. Two kinds of adiabatic changes in temperature can be envisaged. First, if a parcel of air which is not saturated with water vapour is induced to rise, it cools at about 10 °C/km. This is known as the dry adiabatic lapse rate (DALR). If, however, a parcel of air is saturated, as it rises and cools it must shed some of its water vapour load as liquid moisture. This liberates from the air parcel the latent heat of condensation which cushions its temperature fall. The lower rate of cooling in a moving saturated parcel of air is known as the saturated adiabatic lapse rate (SALR) and is normally about 5 °C/km. Unlike the DALR, the SALR is not constant, but varies with the temperature of the air parcel. Warm saturated air produces more condensation (and hence liberates more latent heat) than cold saturated air. As a result there is normally a reduction in the SALR with height in the atmosphere. To decide whether air is unstable or stable it is necessary therefore to know the temperature changes both of a displaced parcel and of the surrounding ambient air in which it finds itself. The vertical temperature change in the ambient air (i.e. measured by taking a thermometer up through the atmosphere) is referred to as the environmental lapse rate (ELR). Typically this averages a fall in temperature of around 6.5 °C/km, but on occasion the ELR may depart considerably from this value, even to the extent of temperature increasing with height (a temperature inversion).

(b) Types of air stability

(i) Stable air

Figure 2.9(a) shows a situation where a displaced parcel of air would initially cool at the DALR until saturation occurred. Thereafter any further displacement upwards would produce condensation in the form of cloud formation and further cooling would occur at the SALR. The cloud base frequently indicates the dew point temperature height of air displaced upwards in this kind of situation. In this instance the displaced air parcel would find itself at all stages of its upward trajectory cooler (and denser) than the surrounding air (the temperature of which is indicated by the ELR line on the graph). If released from its upward motion it would therefore fall back downwards to its starting level at the surface, warming back up at the DALR as it did so. The air in this case is said to be stable.

(ii) Unstable air

Figure 2.9(b) shows an opposite situation. In this case a parcel of unsaturated air displaced from the surface would cool according to the DALR. At each level it finds itself warmer (and less dense) than

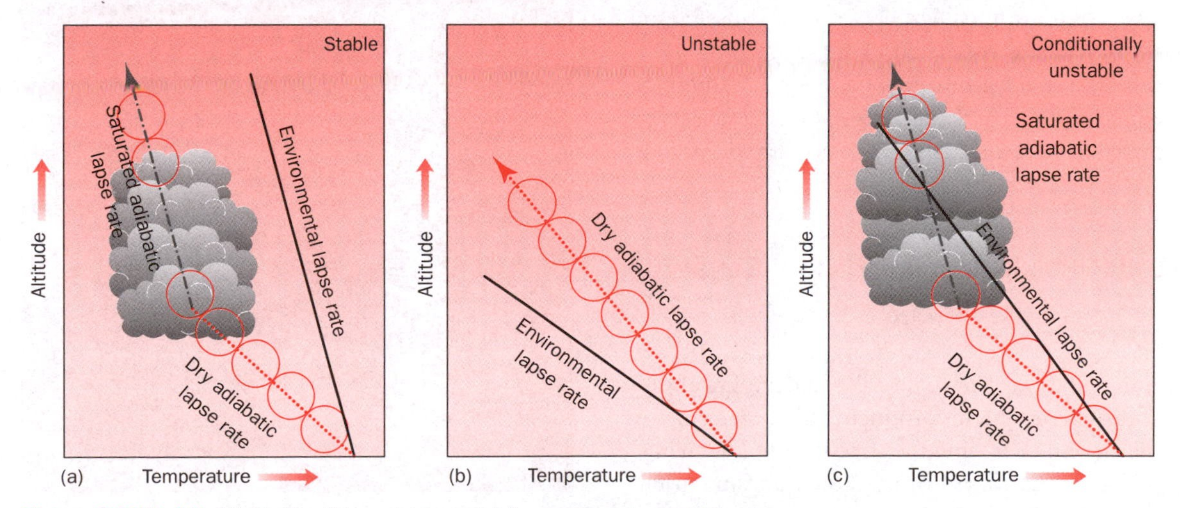

Figure 2.9(a)–(c) (a) Stable, (b) unstable and (c) conditionally unstable air

the surrounding air (the temperature of which is indicated by the ELR line on the graph). Even if the propelling force upwards were to cease, it would continue to rise buoyantly. This air is said to be unstable.

(iii) Conditionally unstable air
A further category of instability known as conditional instability exists. Consider Figure 2.9(c). Air induced to rise from the surface will cool initially at the DALR until the dew point is reached. Further ascent will occur at the SALR, producing condensation, though air parcel temperatures are still less than the ambient temperatures. If, however, air ascent can be maintained (see air uplift, section 2.2.4(b)) past the point at which the SALR and ELR lines cross over, the air parcel temperature becomes warmer than the surroundings and instability takes over. Upward growth of cloud continues as cooling at the SALR proceeds. Thus, only if displacement above a critical height can be achieved is the air unstable, a situation known as conditional instability.

(iv) Air stability and atmospheric pollution
Air pollution tends to be concentrated under stable atmospheric conditions for the same reasons that make the air stable in the first place (Box 2.3). Air parcels containing polluted products will return to the surface under stable air conditions, thus enhancing pollution concentrations in the lower layers of the atmosphere. In contrast, unstable air conditions are associated with good pollution dispersal in the atmosphere and low pollution concentrations at the surface. This is because unstable air is associated with upward air motion and deep vertical air mixing.

2.3.3 Condensation and cloud formation

(a) Fog

Condensation is central to the functioning of the atmospheric system, producing liquid moisture in the air and providing the precursor to life-giving precipitation. Air is brought to saturation most

BOX 2.3

CASE STUDY

Stable air and atmospheric pollution

The temperature inversion

Very stable air can exist under high pressure or anticyclonic conditions, with air gently falling from more elevated levels in the atmosphere. This subsiding air may undergo compressional warming as it falls, leading to a warm layer or layers in the atmosphere where temperatures are higher than either above or below. This warm layer is called an **inversion** layer because it inverts the normal condition where temperatures in the air progressively fall with altitude. Such a situation is not conducive to the dispersal of aerial effluent that moves upwards through the atmosphere (within rising pockets of air) for a time and then encounters air warmer (and therefore lighter) than itself, stopping its further rise. The inversion layer acts as a 'lid' preventing the dispersal of pollution and should the inverted layer fall close to the surface dangerously high ground level concentrations of pollutants may be experienced. Temperature inversion conditions have been implicated in almost all of the world's major air pollution disasters (Table 2.3).

Air pollution episodes

Figure 2.10 shows temperature soundings at various heights in the atmosphere over Valentia in south-west Ireland in January 1982 together with other meteorological data from Dublin, some 300 km further east. An uninterrupted temperature inversion is apparent over a six-day period, sometimes quite close to the surface. The pressure trace suggests that an anticyclone moved across Ireland, with highest air pressures on 12/13 January. -Light winds ▶

▶ and bitterly cold temperatures existed during this winter period. Winter anticyclones bring clear and cold conditions in mid-latitude regions. On 12 January temperatures as low as –8 °C were experienced. As with all cold snaps in winter, this would produce greatly increased fuel demand, in this case from domestic coal burning. Light winds and the presence of the temperature inversion, however, trapped these emissions close to the surface, causing them to build up over a number of days to produce alarmingly high concentrations. A peak daily smoke concentration of 1,812 μ/m^3 is seen to have occurred on the 14th when the inversion probably reached its lowest level in the Dublin area. Figure 2.10 also shows the mortality trend in one of the city hospitals for the month in question. A doubling of deaths in the weeks following the episode can be seen. These were mainly due to respiratory complaints among elderly patients. Advance knowledge of the occurrence of severe temperature inversions can enable preventative measures such as the issuing of public alerts or switching to alternative fuels to be considered. The environmental impact of potential new sources of pollution, such as power stations, can also be modelled in terms of their effluent's ability to penetrate temperature inversions and not contribute unnecessarily to ground-level pollution.

Table 2.3 Major (non-nuclear/non-natural) air pollution disasters

Place	Year	Excess deaths
London	1873	700
London	1880	1 000
London	1911	1 150
Meuse Valley, Belgium	1930	63
Donora, Pennsylvania, USA	1948	20
London	1952	4 700
New York	1953	250
London	1956	480
London	1957	300–800
New York	1962	46
London	1962	340–700
Osaka, Japan	1962	60
New York	1963	200–405
New York	1966	168
Bhopal, India	1984	2 500

commonly by cooling. This can occur when a cold body of air is introduced to a warm mass of moist air. For example, cooling from below may occur as warm, moist air passes over a cold land or sea surface. The layer of air in contact with the cold surface is cooled to the dew point, often producing fog. Horizontal air movement of this kind is known as advection and advection fogs are common occurrences in summer along coastlines where the sea temperatures are cool. A good example of advection fog occurs over the North Sea coast of Britain in summer with a warm easterly flow of air from the continent (Plate 2.2). Chilling of this air over the cool sea surface produces the 'haars' so well known in coastal areas. These dissipate rapidly a few kilometres inland over the warmer land surfaces. On land, on calm, clear nights, outgoing radiation (heat) from the surface may chill the air above the surface. Condensation may also occur in this way, producing a radiation fog. Typically this is associated with the development of a surface-based temperature inversion and very stable conditions. A light wind of 3–10 m/s is however necessary to keep a flow of moist air in contact with the cooling surface and to mix the chilled air with some of the warmer, moist air aloft. A delicate balance between too much and too little turbulence is required since if too much warm air is mixed downwards the temperature may be increased above the dew point and radiation fog will not form. Once a layer of fog has formed, the top of the fog acts like a second 'active surface' able to radiate heat away to the surrounding environment. The fog top thus maintains the temperature inversion and assists with further thickening and deepening of the fog layer.

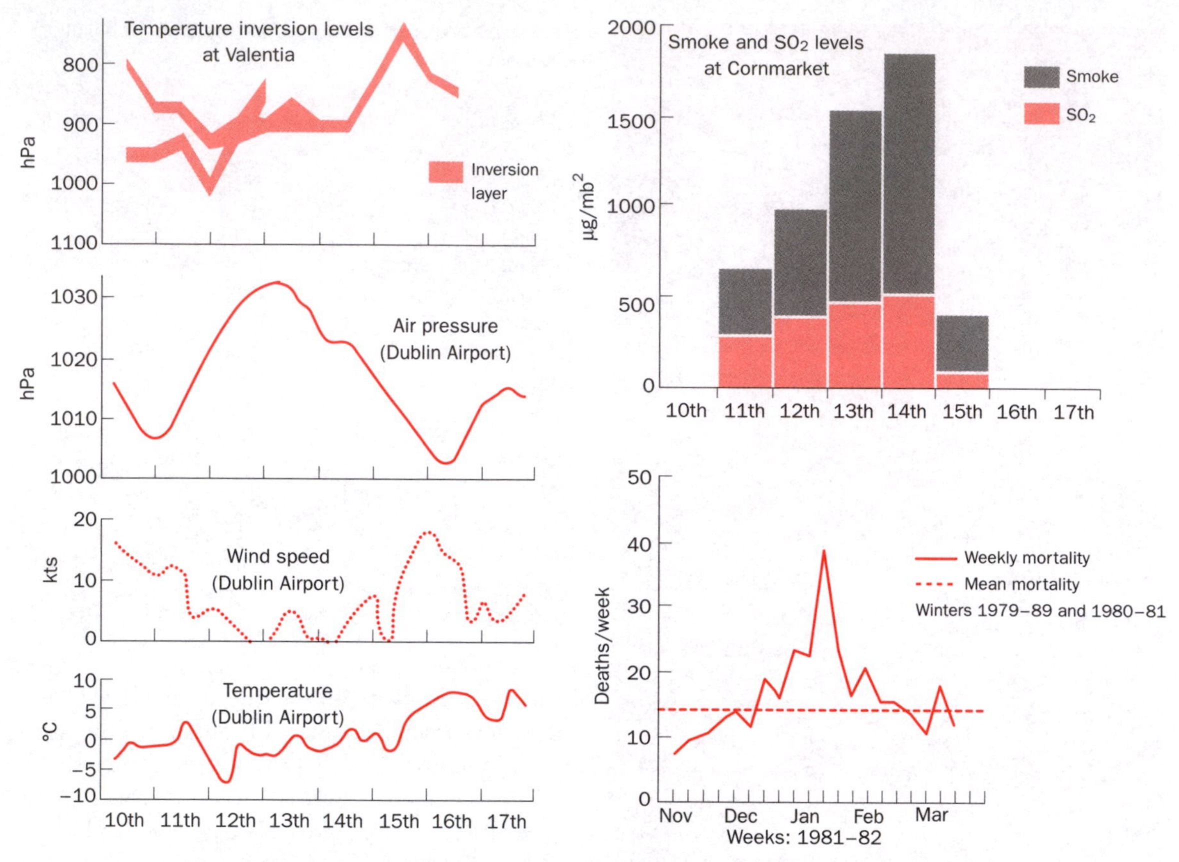

Figure 2.10 An air pollution episode in Dublin in January 1982. The link between weather conditions, especially wind speed, the height of the temperature inversion and smoke concentrations can be clearly seen. The peak in mortality noted by a respiratory surgeon in his records can also be seen after the event

Since cool air is denser than warm air, downslope movement of cold air from hillslopes may pond in valley floors and low-lying hollows where further cooling may produce condensation. Valley fogs may be persistent, effectively resisting evaporation under a morning sun. Valley fogs may also be associated with air pollution problems due to the temperature inversion they typically exhibit. Unless they are very thick, fogs tend not to persist more than a few days. They will eventually dissipate, either (i) by the sun's warming influence as day proceeds causing temperatures to rise above the dew point and causing evaporation of the fog to take place, or (ii) by increased windspeeds mixing the fog with drier and warmer air above it, raising its temperature once again above the dew point and leading to its dissipation.

(b) Air uplift and clouds

(i) Four ways of lifting air

The most effective manner in which saturation and condensation can be produced in a moist air mass, however, is to induce it to rise. Figure 2.11(a) shows the four main ways in which this can be achieved. First, air may be forced to ascend over a physical obstacle such as a mountain. This is orographic lifting and explains why exposed mountainous coasts receive so much rainfall. Second, warm, moist air may be forced to ascend

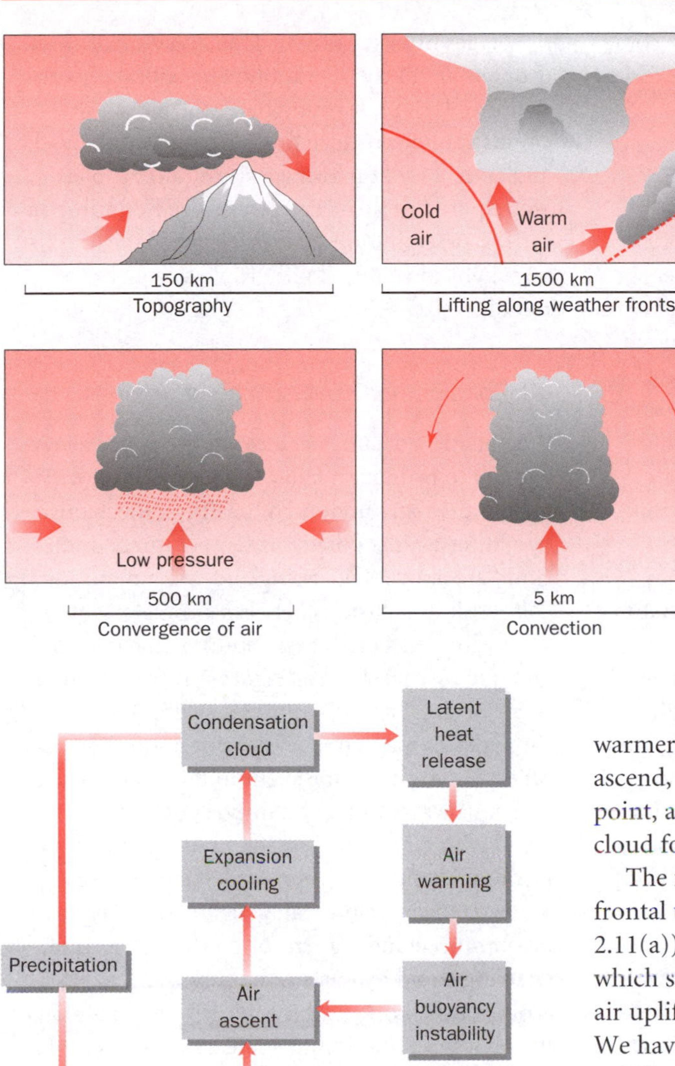

Figure 2.11 (a) Four 'external' ways air can be forced to ascend inducing rainfall; (b) links between external uplift mechanisms and internal buoyancy processes
Source: (a) Ahrens (2000, p. 167)

over an air obstacle, such as a mass of cold air. This upglide causes condensation in the rising warm air body to take place along the boundary zone between the two air masses. This is frontal uplift. A third lifting mechanism occurs when air is drawn into low pressure centres and spirals upwards in the centre of them. This convergence and uplift forces the air to higher, cooler, levels where condensation of water vapour takes place, sometimes as part of major cyclonic disturbances. Fourth, convectional lifting occurs when air is heated from below, such as by conduction from a warm surface on a summer's day. Small convective currents are triggered off which initiate vertical movement. If these ascending bubbles of air find themselves warmer than their surroundings they continue to ascend, even after they have cooled to the dew point, and condensation in the form of cumulus cloud formation has commenced.

The four main processes of orographic ascent, frontal uplift, convergence and convection (Figure 2.11(a)) should be seen as external mechanisms which support the internal adiabatic processes of air uplift through air buoyancy (Figure 2.11(b)). We have already seen in the case of conditional stability, that an initial trigger or forcing agent (i.e. one or more of the named external processes of air uplift) is often needed to elevate air from the surface so that condensation and subsequent air instability can occur. It is also useful to bear in mind that when air is ascending within storms producing copious cloud and precipitation, the external and internal processes of air ascent are working together. For instance, adiabatic buoyancy mechanisms are coupled with surface convection/convergence in thunderstorms, with frontal uplift/convergence within mid-latitude depressions and with strong convergence/convection in hurricanes. The precipitation rate and amount of these storms can be further enhanced if they come up against mountain barriers where they are forced to rise.

(ii) Condensation nuclei

Condensation does not readily take place into the free air. It is facilitated by the presence of microscopic particles around which the water vapour can condense. These *condensation nuclei* have a diameter of less than 0.4 μm and between 10^9 and 10^{10} particles can be present in a litre of air. Over land areas especially, these fine aerosol particles may be derived from windblown soil dust, volcanic ash or carbonaceous or sulphate materials produced from fossil fuel combustion. Over the oceans sulphates may also be formed from dimethyl sulphide (see section 8.5), a gas emitted into the atmosphere from phytoplankton activity; but salt from ocean spray forms the most important constituent. Water vapour is particularly attracted to salt particles and will condense on its surface at relative humidities below 100 per cent, even as low as 75 per cent. Such hygroscopic nuclei are abundant in the lower atmosphere and provide the third requirement for precipitation formation: saturated air, a lifting mechanism and condensation nuclei.

2.3.4 Precipitation mechanisms

When condensation has commenced on a nucleus, a cloud droplet is formed. This is typically around 20 μm in diameter, still 100 times smaller than an average raindrop. If the original cloud condensation nuclei dissolves in the water, the salt ions bind the water molecules more closely, restricting tendencies for them to evaporate. This helps to enlarge the droplet by reducing evaporative losses from it. As the droplet grows in size its curvature reduces. The more this happens the more the droplet finds it easier to retain its moisture. Smaller droplets have more curvature than larger droplets. Curvature acts negatively in terms of encouraging more water vapour to condense and positively in terms of losing water molecules by evaporation. Thus, very small cloud droplets are likely to evaporate or fail to grow to rain droplet size, forming haze. Indeed, most clouds cannot produce precipitation size droplets and since a million average size cloud droplets are required to form an average sized raindrop, a further precipitation-producing mechanism must be sought to enable cloud droplets to grow to raindrop size.

(a) Collision and coalescence of droplets

Precipitation occurs when cloud drops are heavy enough to produce a falling or terminal velocity greater than any updraught or lifting mechanism operating against gravity. If temperatures in the cloud are above 0 °C, rising and sinking motions will result in collisions between droplets, with larger droplets benefiting disproportionately and growing by coalescence (Figure 2.12(a)). A good range of sizes helps the process as droplets of similar sizes may not coalesce but bounce off each other. Equally the time spent in the cloud by the larger growing droplets is important. A thick cloud with a sustained strong updraught is most favourable to the collision/coalescence process. Some tropical regions can generate updraughts in cumulus clouds of tens of metres per second, enabling cloud droplets to be supported to sizes around 1,000 μm. As this raindrop then falls back through the cloud it may increase in diameter to 5 mm. However, this is usually the maximum size that drops arriving at the surface exhibit. Larger ones suffer some attrition from their neighbours on the way down.

(b) Supercooled water and ice crystal growth

In colder clouds, such as in middle and high latitudes, the temperatures may be well below freezing even short distances above the surface. However, even down to temperatures of −40 °C, liquid water droplets can be found in clouds. These supercooled droplets remain liquid in the absence of suitable solid particles around which ice can

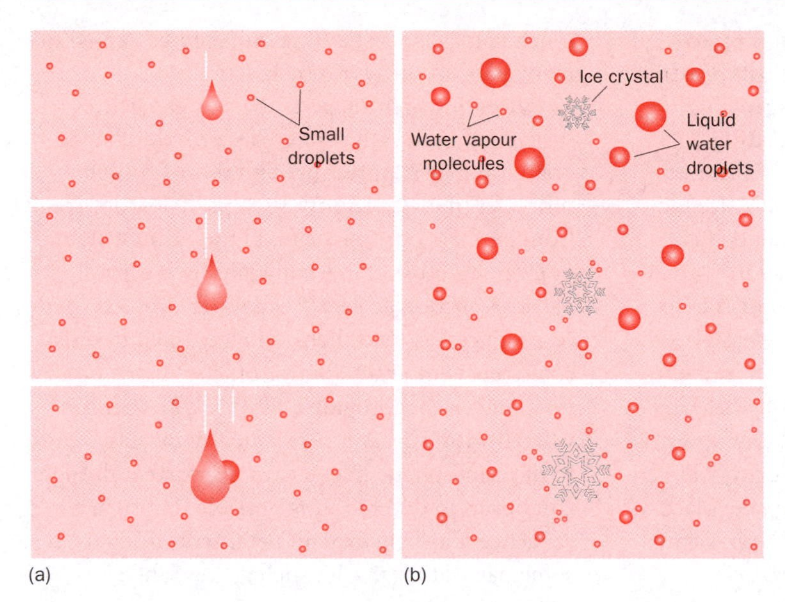

Figure 2.12 Precipitation formation: (a) collision/coalescence; (b) ice crystal growth

form. These ice nuclei can consist of dust or clay particles, or sometimes even biogenic material such as bacteria or fungi. Typically at these temperatures a mixed cloud of ice particles and water droplets exists. Temperatures below about −40 °C produce spontaneous glaciation – freezing occurs even without particular particles being present. These high ice clouds, such as cirrus, are not prolific precipitation providers, however. It is the mixed cloud which requires a mechanism to grow droplets to raindrop size.

In a saturated air body containing both ice crystals and supercooled water droplets the number of molecules of water vapour leaving the supercooled droplet or ice crystal must be the same as the number entering. This is what saturation implies. In the case of the water droplet, however, escape is easier than for the tightly bonded ice. So the number of water vapour molecules in the vicinity of the water droplets is greater than those in the vicinity of the ice particle. Another way of saying this is that the saturation vapour pressure above a water surface is greater than the saturation vapour pressure above an ice surface at the same temperature. This pressure difference between the two surfaces results in the gradual movement of water vapour molecules towards the ice particles where they are absorbed. The water droplets, having lost their balancing water vapour molecules from the vicinity, are no longer in equilibrium with their surroundings and begin to evaporate. The ice particles begin to grow into larger crystals at the expense of any surrounding water droplets. The Swedish meteorologist, Tor Bergeron, and the German meteorologist, Walter Findeisen, are generally credited with this theory of precipitation formation which is often referred to as the Bergeron–Findeisen process. Nevertheless, the distinguished German climatologist Alfred Wegener (the proposer of the theory of continental drift) made significant earlier contributions to understanding the process of selective growth of ice crystals in mixed clouds (Figure 2.12(b)).

Once formed, the ice crystals then begin to collide with each other, perhaps splintering into smaller ice particles that attract a multitude of new supercooled water droplets and instigate a chain reaction. Collisions between the ice crystals also result in their combination by *aggregation* into snowflakes. As these snowflakes fall a similar mechanism of collision/coalescence as applied in warm clouds comes into play to enlarge them. The Bergeron–Findeisen process implies that most of the precipitation outside of the tropics begins as a snowflake.

(c) Forms of precipitation

(i) Rain and snow

The subsequent form precipitation takes largely depends on the conditions it encounters in its fall. In cold conditions when the freezing level is at or near the surface, the snow will fall directly to the ground. Even if the freezing level is up to 300 m above the surface, snowflakes will resist melting. A more extensive passage through warmer air layers will result in the melting out of the snowflake to form a raindrop, typically 0.5–4 mm in diameter. Drops over this size tend to break up during their fall. Large raindrops fall with a terminal velocity of about 9 m/s. If an updraft of this magnitude exists (and of course it was a lifting mechanism which may have been instrumental in inducing condensation in the first instance) then the droplets may be held in the air or evaporate. A weakening of any upward motion, or its reversal to a downdraught, will enable precipitation at the surface to take place. Gentle vertical motions characterise stratiform clouds and these result in longer-duration lower-intensity rainfall than cumuliform clouds where intense, short-lived events are more characteristic.

(ii) Sleet and hail

Falling droplets which encounter a freezing layer close to the surface may reform into ice pellets up to 4 mm in diameter. These are correctly known as sleet though colloquially sleet is also often incorrectly described as a mixture of snow and rain. Sleet is, however, fundamentally different from hail which is also an ice pellet. In the case of hail, the ice pellet is facilitated in its formation by large-scale violent updrafts which carry nuclei of various kinds high into the sub zero zone of a cumulonimbus cloud. Lateral air motions sweep these 'embryos' through the cloud where they grow by accumulating a coating of ice around them. A quite lengthy period of accumulation may take place as a result of the repetition of falling and rising motions. This can result in successive accumulations of ice on the particle which shows a layered structure of opaque and clear ice in cross-section. Hail only falls from cumulonimbus clouds. The ferocity of the updraughts involved can be gauged by the size to which some hailstones grow. Diameters of 14 cm (1.67 kg in weight) have been recorded and it is thus not surprising that so much damage to a standing crop can be caused in many parts of the world by hail.

Recent research suggests that small particles suspended in the atmosphere as a result of air pollution may suppress precipitation-forming processes in their vicinity. Such particles may result from biomass burning or fossil fuel combustion, or even soil erosion mechanisms. Such aerosol particles encourage very small water droplets to be produced around them, small enough to resist coalescence into rain-sized droplets. Such a process may conceivably provide a feedback mechanism whereby drought, and consequently increased dust in the atmosphere, may encourage further drought conditions to develop. It is possible that such a sequence may help explain persistent droughts in some parts of the world, such as the Sahel, in recent decades.

2.3.5 Precipitation systems in action

(a) Orographic precipitation model

One way of synthesising our ideas so far on air motion and moisture is to make use of a simple orographic precipitation model. It can be seen in Figure 2.13 that when mountain barriers intercept rain-bearing winds, rainfall increases with altitude, especially on the windward side, and marked reductions occur in the rain shadow of the lee slope. Unsaturated air, ascending on the windward side of a mountain range, cools first at the DALR (10 °C/km) until condensation and clouds form. As clouds form, latent heat is released within the ascending air parcels, helping to make them warmer and thus lighter than the surrounding air. As a result, the air parcels continue to ascend and cool, forming much deeper clouds in an unstable atmosphere. Updraughts and downdraughts within the ascending air on the upper slopes are now fast enough to allow the small cloud particles to collide and coalesce and grow into rain droplets. Air temperatures above the upper slopes (particularly in mid and northern latitudes) may also be cold enough to allow rain to form by the ice crystal process.

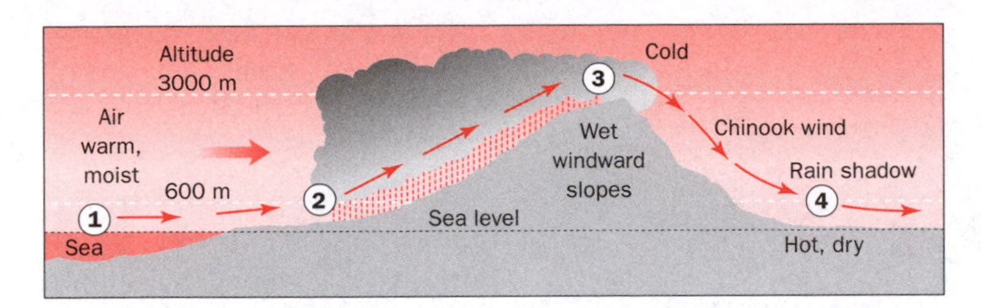

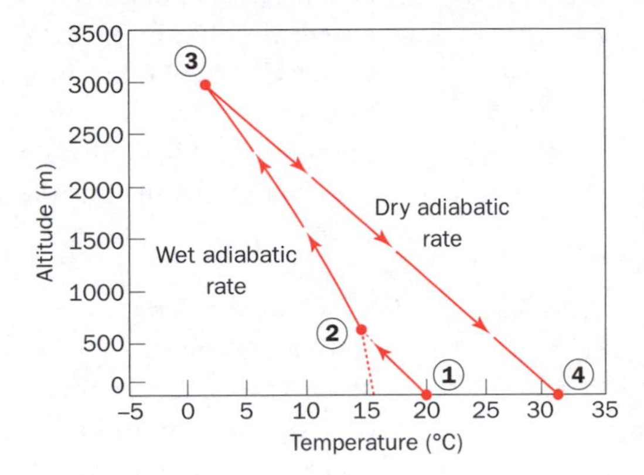

Figure 2.13 Rain shadow diagram and the Chinook effect
Source: Strahler and Strahler (1989, p. 111)

After shedding most of its moisture as precipitation (rain, snow) on the windward and summit slopes, the air that descends the lee slope is not only relatively warm (from latent heat release on the windward slope) but also dry. On descent this air is compressed and warmed at the DALR (10 °C/km) along most of the length of the lee slope. Such processes (i.e. latent heat release and long-fetch adiabatic warming) allow the air at the foot of the lee slope to be much warmer than at the bottom of the windward side. Another result of compressional heating is that any moisture in the air on the lee slope is evaporated, creating good visibility and clear skies.

In the mid and high latitudes, orographic rainfall effects and adiabatic lee-slope warming are particularly pronounced on west coast mountain ranges that lie in the path of west to east moving rain-bearing wind systems. In the Olympic Mountains of Washington State, USA, annual precipitation increases up to 4,000 mm until the air has passed over the summits at an altitude of between 2,300 and 2,700 m. Thereafter, precipitation decreases as the air stream descends onto the leeward side. The warm dry air that descends to the bottom of the lee slope (called the Chinook in the Rockies and the Föhn in the Alps) can be typically 6–8 °C higher than at the foot of the windward slope. As a result of these effects, the windward slopes are often cloaked in dense wet temperate rainforest while the bottom of the lee slopes is covered in dry scrub and grassland. The increase of precipitation with altitude in Britain has also long been recognised. It is an effect that dominates precipitation distribution, not only on annual maps (Figure 2.14), but also on monthly and many daily precipitation maps as well.

(b) Convectional instability: the thunderstorm

As previously discussed, another way of encouraging air to rise up through the atmosphere, where it may subsequently cool, condense and produce precipitation, is through convectional instability. When air is warmed over a *heated land surface*, it expands and becomes less dense than the surrounding air. Being lighter than the surrounding air it will quickly rise up through the atmosphere helping to generate or sustain unstable air conditions. When air is heated in this way under unstable atmospheric conditions (other factors are low pressure, moist air, steep adiabatic lapse rates) severe local storms or thunderstorms can result. Because of the high intensity of sunlight or solar radiation in tropical regions, convectional disturbances that develop into thunderstorms over a land area can occur at any time of the year. In the mid latitudes, however, they generally develop

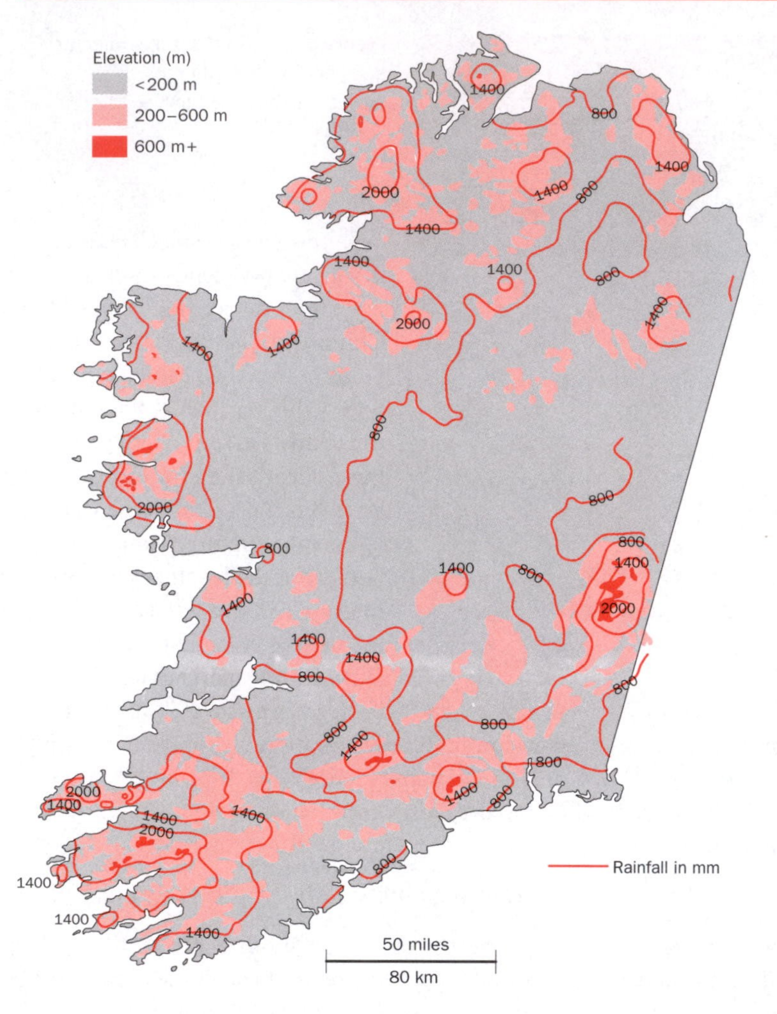

Figure 2.14 The role of altitude in precipitation in Ireland. The Eastern coastline is not included in this diagram

in spring, summer and autumn. The following sections explore the formation and development of the thunderstorm.

(c) Types of thunderstorm

It is generally recognised that there are three types of thunderstorm. The first type is the air mass or supercell thunderstorm. This is a very large mass of unstable air which forms a cumulonimbus cloud, measuring a few tens of kilometres across. Such storms generally have a lifespan of several hours and may produce an hour or more of heavy rainfall (up to 25 mm/hour in most temperate latitude situations, 50–75 mm/hour in tropical regions). Also associated with such storms are hail, thunder, lightning and, if severe enough, tornadoes (see section 10.2). Three stages in the development of the supercell thunderstorm can be recognised: the cumulus (birth), mature (violent) and dissipating (death) stage.

(i) Cumulus stage

During the first stage, or cumulus phase, convectional updraughts of warm moist air from the heated land surface help to form a moderate sized cumulus cloud (Figure 2.15(a)). At this stage there is no precipitation, the cloud being composed mostly of cloud droplets. Cumulus clouds can reach up to 4.5 km and grow outwards to a diameter of about 1.5 km. The convectional updraughts cause convergence at the surface, which adds additional moisture to the growing cloud. The small but continually growing cloud droplets are carried aloft by convectional–convergent updraughts into colder parts of the cloud where they can become supercooled water droplets or small ice particles. There are thus three states of moisture droplet in the growing cumulus cloud – water droplets above 0 °C, supercooled water droplets below 0 °C and ice crystals below 0 °C.

(ii) Mature stage

During this stage, the cloud continues to grow vertically by air updraughts into what is called a cumulonimbus or rain-bearing cloud. The process of cloud growth is greatly assisted in the presence

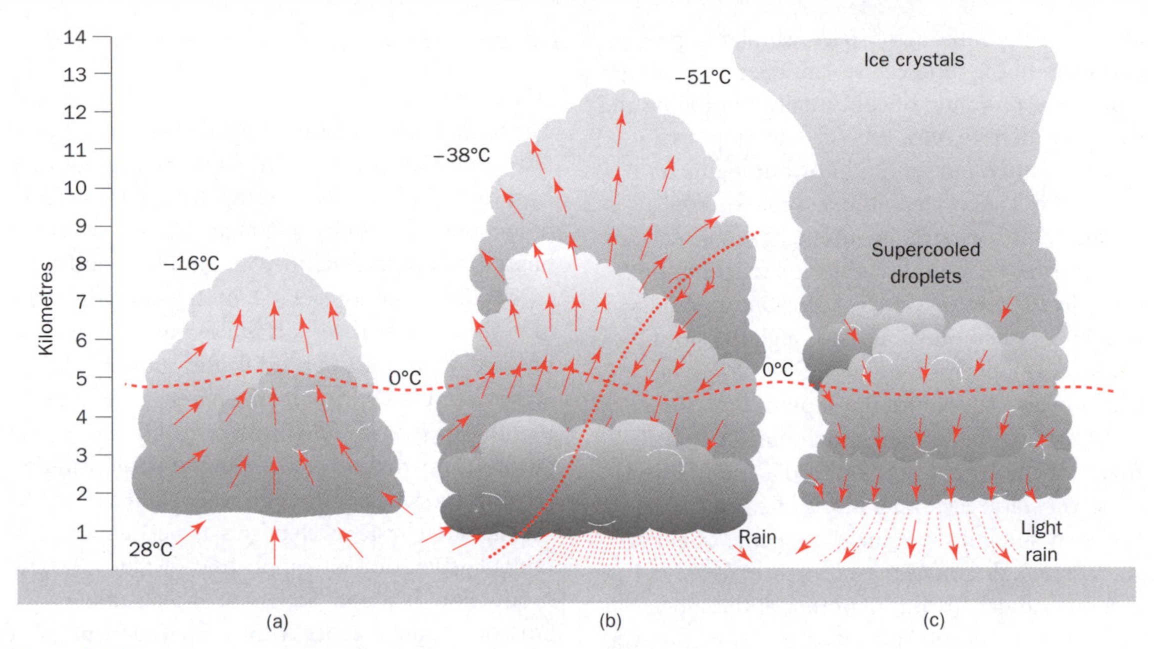

Figure 2.15 Cumulus development and thunderstorm formation
Source: Neiburger *et al.* (1982, p. 152)

of unstable air conditions and continual supplies of moisture available to condense and release latent heat to feed the storm. As the cloud grows, so too do its constituents, the water droplets and ice crystals. When these particles become too heavy to be supported by the rising air (which can reach speeds of 160 km/hour), they fall through the cloud creating downdraughts, which can reach speeds of about half those of the strongest updraughts (80 km/hour). With the existence of both updraughts and downdraughts in the cloud, the mature stage begins (Figure 2.15(b)). As the cold downdraughts become strong enough they pour out of the base of the leading edge of the cloud, resulting in typical gusts of cool air at the front of the storm system. Close behind this wind system is heavy rain as the larger water droplets and ice crystals are pulled down from the uppermost layers of the cloud (with heights up to 10–12 km). Under ideal conditions with the correct balance of updraughts to downdraughts, hail will follow the torrential rainfall. The outpouring of wind, rain and hail also cool the heated ground surface which inhibits updraughts.

(iii) Dissipating stage

When the declining updraughts are less frequent and weaker than the now predominating downdraughts, the dissipating phase of the thunderstorm sets in (Figure 2.15(c)). Just as the cloud enters this stage, an anvil top usually forms at the top of the leading edge of the cloud. This anvil projection helps to exhaust some of the upper air updraughts out of the system. In the absence of highly buoyant air, and with the downdraughts continually depleting the storm of its moisture, condensation processes slow down and the energy released as latent heat is no longer available to fuel the storm. In response, the storm rapidly disintegrates from bottom to top, rainfall diminishes to light rainfall or drizzle and the cloud fades away.

A second group of severe thunderstorms is the multicell storm. As its name suggests, this type embraces a number of adjacent cells of similar size and intensity. Although broadly similar in process and form to the supercell storm, the multicell type comprises a number of continually developing and decaying discrete cells. Each cell develops as a separate entity, but up- and downdraughts from adjacent cells may feed off one another. For instance, the forward downdraught of one cell can be a rear feed and updraught for a neighbouring cell. The lifespan of a single contributory cell is usually less than one hour, although the storm complex will persist for a period similar to or even longer than that of the supercell.

In continental interiors, including central Africa, the northern plains of India and midwest USA, very large agglomerations of convectional cells may sometimes occur. The spatial extent of such mesoscale convective complexes (MCCs) is obviously much greater than that of the supercell storm (up to 50–100 km across, and sometimes, as in central Africa, up to 1,000 km across) and they may persist for over a day. Not unexpectedly, such storms produce high precipitation intensities and amounts over a very large area, and are thus often the main contributors to the total precipitation supply of a district or region.

As mentioned previously, air uplift often takes place in the vicinity of fronts. This leads to the formation of a third type of severe local storm, the squall-line thunderstorm. Although these storms are found in tropical regions (e.g. West Africa), they are most frequently seen as a group of storms aligned along the cold front or 'squall line' of the mid-latitude depression (Chapter 9). Upward air movement within these squall-line storms is assisted by cold frontal air undercutting the storm along the zone or line of the cold front. The complicated but sustained updraughts and downdraughts within such storms often lead to a perpetuation of new storm cells and areas of precipitation along and just in advance of the frontal zone as it moves forward with the depression. One outcome of this is that the total duration time of a squall-line thunderstorm system across an area can be anything from 2–3 hours to as much as half a day.

(d) Blizzard days

(i) Definition

The term 'blizzard' denotes a moderate (5–6 cm) to heavy snowfall (over 13 cm) with wind speeds of at least force 7 (30 knots) causing drifting snow and reductions of visibility to 200 m or less. A severe blizzard is similarly defined and implies wind speeds of at least force 9 (44 knots) and reduction of visibility to near zero. It is also possible for blizzard conditions to prevail after snowfall has ceased, simply by wind increasing to such an extent that the air is again filled with snow lifted from the ground. Snowfall, strong winds and poor visibility are thus the key features of a real blizzard.

The same basic mechanisms associated with rainfall generation also apply to snowfall production, i.e. rapid air ascent as a result of horizontal convergence within surface low pressure cells, orographic uplift, convective instability and frontal uplift. Low temperatures throughout the atmosphere are a prerequisite for snowfall. For snowflakes to occur in the mid latitudes, the freezing level should not be higher than about 600 m above the observer. Higher than this there is too much time for the snowflake to melt. The steeper the environmental lapse rate and fall of temperature with height, the shorter the distance a snowflake will have to travel through any air that happens to be above freezing with less time available for melting.

(ii) UK blizzards

It has been found that the origin and movement of the principal blizzards to affect the UK and Ireland are highly varied. A recent survey of 83 major blizzards which crossed over the region between 1880 and 1989 found the following. Almost half of them (42) moved as mid-latitude Atlantic depressions from off the south-west coast of the UK up through the English Channel and over southern England towards the north and east. A smaller number (8) of mid-Atlantic blizzards moved north directly across Ireland and over the central part of the UK including southern

Scotland. Few blizzards (3) moved as *low pressure systems* (see below) directly from the east towards the UK, but 30 originated from the north, north-west or north-east of Scotland. These northerly blizzards either tracked southwards over the Irish and North Seas or moved eastwards to Scandinavia. As might be expected, 61 per cent of the sampled blizzards (51 out of 83) happened during the winter (Dec., Jan., Feb.), but 31 per cent (26) occurred in spring (March, April, May). Two unusual blizzards took place during the summer (June, July, Aug.) on 11 July 1888 and 2 June 1975, with 4 per cent (5) in autumn (Sept., Oct., Nov.). Some of the worst blizzards to affect the UK arrive on cold winds moving from the east, i.e. on air streams moving anticlockwise around depressions centred to the south of the region. Easterly and northerly slopes on hillsides are almost always associated with rapidly increasing snow depth when subjected to a major blizzard tracking in from the east. In general terms the regions most susceptible to blizzard impacts (Box 2.4) are where

BOX 2.4

CASE STUDY

UK blizzards

18–19 January 1881

The blizzard of 18–19 January 1881 is known as the Great Victorian Blizzard. It was caused by a depression moving from the Bay of Biscay up the English Channel and then turning southwards over France. England found itself in the cold air of the depression to the north over the two-day period. The area of maximum snowfall lay in the south although most of England received snow during the event. An easterly gale which accompanied the heavy snow and whipped it up into blizzard conditions lasted throughout 18 January and continued until noon on the 19th. In London snow depth was 25 cm with drifts of 1–5 m. In Brighton the snow depth was 45 cm, in Exeter 30 cm and on Dartmoor as much as 100 cm. The depth averaged 10 cm over the whole of England. About 100 people lost their lives in the blizzard. The financial cost to the country was enormous, most businesses were halted for a day, Plymouth was deprived of water for a week and it took about a week for road and rail travel to return to normal.

27–29 January 1978

This blizzard crossed the UK and Ireland directly from the west towards Scandinavia caused by a deep depression (980 hPa). In the three days that it took to cross the region (27–29 January 1978), most snowfall and the most intense blizzard conditions occurred in the cold air to the north of the depression, i.e. over the Highlands of Scotland. Here, between 27 and 29 January 1978, snow fell for 50 hours resulting in an average depth of 35–70 cm. In some parts of the Highlands over 1.5 m of level snow fell. Winds up to force 8 on low ground and up to force 10 on high ground produced drifts in excess of 6 m, which totally buried vehicles and isolated large areas for nearly a week.

On 29 January, only two roads were open in the Highlands, as hundreds of cars, buses and even a snow plough became stranded. The biggest airborne rescue operation ever mounted in Europe started on 30 January and 300 people were rescued by helicopter. The death toll was remarkable low, with five people being killed (mostly found dead in their cars). This low figure (compare the 100 dead in the January 1881 blizzard) must be attributed to the magnificent work of the modern rescue services and to people who invited the stranded into their homes.

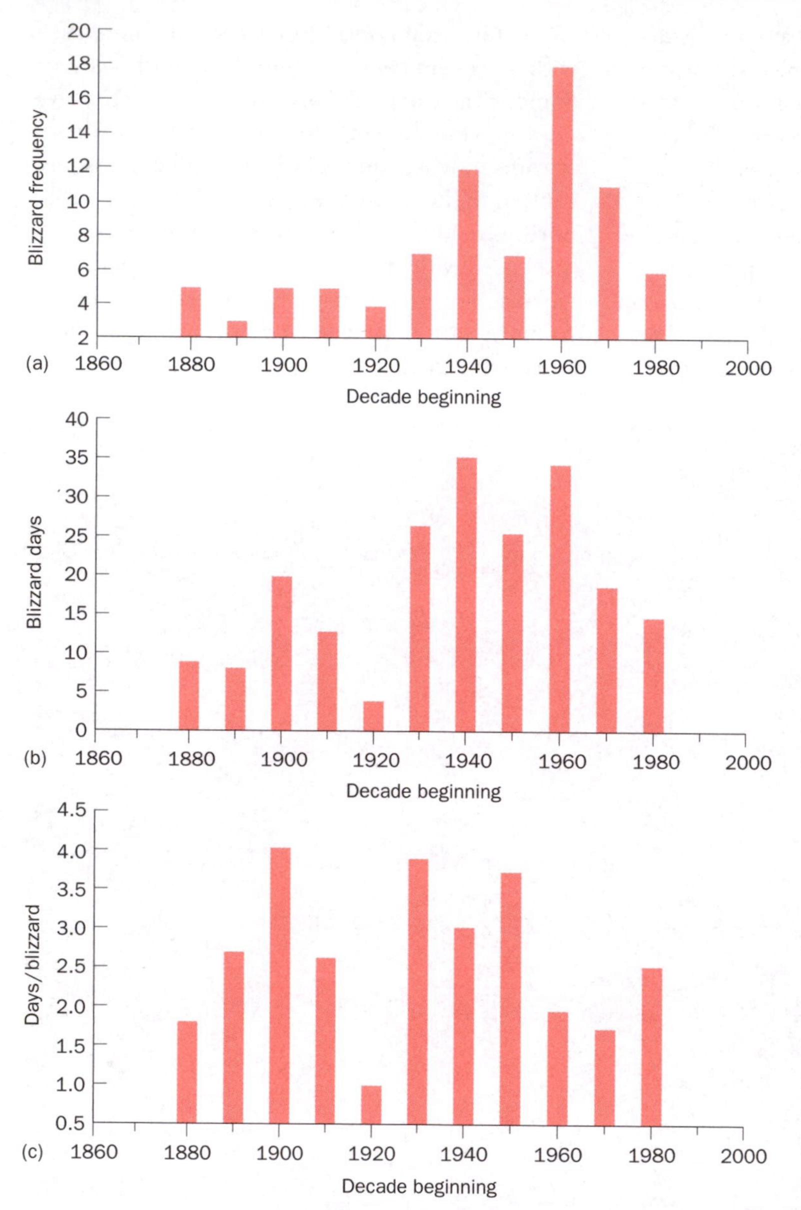

Figure 2.16 Blizzard frequency in Britain, 1880–1989: (a) blizzard frequency by decade; (b) number of blizzard days by decade; (c) average length of blizzards per decade.
Source: Wild *et al.* (1996, p. 88)

easterly winds are forced to rise, i.e. in the vicinity of the Snowdonia range, east Kent, Dartmoor, the eastern Pennines and the Highlands of Scotland.

As blizzards are destructive events causing widespread disruption to life and economy, (Box 2.4) knowledge of their duration and frequency is important. Most UK blizzards are short duration events (one to two days) with low numbers of long duration events. Nevertheless, three 10-day snowstorms occurred during 1880–1989. There is also much variation from year to year and across the decades of the last century. Figure 2.16 shows that while 18 blizzards and 35 blizzard days occurred during the 1960s, only three blizzards were identified during the 1890s, and four blizzard days for the 1920s. Quite high numbers of blizzards were noted during the latter part of the twentieth century, a reflection perhaps of improved data collection and monitoring. The recent decline in blizzard frequency and number of blizzard days by decade since the 1960s may indicate an association with global warming.

Key ideas

1. The maximum amount of water vapour which can be held by air is determined by temperature. Saturated air is said to be at a relative humidity of 100 per cent and the temperature corresponding to this state is defined as the dew point temperature.

2. Freezing and condensation release significant amounts of energy known as latent heat which is a major source of energy for atmospheric disturbances as well as playing a significant role in balancing the earth's energy budget.

3. Air which is not saturated and which is induced to rise will undergo a fall in temperature of about 10 °C/km, known as the dry adiabatic lapse rate. Saturated air cools at a slower rate (due to latent heat release) of about 6 °C/km. These rates may render a displaced rising parcel of air cooler or warmer than its surrounding air.

4. Displaced air which finds itself warmer than its surroundings will continue to rise and is said to be unstable. Displaced air which finds itself cooler than its surroundings will tend to relocate to its original position and is said to be stable.

5. Chilling of air blown across a cold surface may produce condensation leading to advection fog. In calm, stable air conditions, nocturnal cooling may chill the air above the surface producing radiation fog.

6. Uplift of air may occur over a physical obstacle (orographic), over a colder wedge of air (frontal) or by convergent uplift into a low-pressure centre (cyclonic). Instability may also be induced by surface heating (convective) effects. Condensation and cloud formation are the products of such movements where they involve cooling of air below the dew point.

7. Condensation takes place more readily around microscopic particles such as salt particles and condensation may take place at relative humidities below 100 per cent in such circumstances.

8. Precipitation depends on mechanisms to grow cloud droplets, typical diameter sizes about 20 μm, to raindrop sizes typically about 2 mm.

9. Collision and coalescence between cloud droplets subjected to strong updraughts may produce raindrops of up to 5 mm in diameter. In clouds composed both of ice crystals and water droplets the saturation vapour pressure difference between a water surface and an ice surface is responsible for selective growth of the ice crystals by absorption of water droplets. These may fall as snow, or melt during their fall to form raindrops. Sleet and hail may also be formed, depending on the conditions encountered during the descent.

10. Good examples given in this chapter of where external forces of air uplift (convection, frontal, convergence, orographic) combine with internal adiabatic processes of air buoyancy (air instability) include orographic precipitation enhancement on the windward side of mountain ranges, convective thunderstorms and convergence within mid-latitude depressions (blizzards).

2.4 The atmosphere as a staging post

The climate system components described in Chapter 1 consist of five major stores: the atmosphere, the hydrosphere, the cryosphere, the biosphere and the lithosphere. Driven ultimately by flows of solar energy, these components interact constantly and interchange materials between them. Physical, chemical and biological processes occur on a wide range of spatial and temporal scales to sculpt landscapes, form soils, sustain and regulate life. Though, as has been seen with the hydrological cycle, the atmosphere is primarily a medium of transit rather than a major mass store, it is intimately involved in many of these processes and links the various subsystems of the climate system together by means of fluxes of mass and energy. As human impacts on the atmosphere, and on other components of the climate system, increase, it is vital to identify potential destabilising influences. This is important in order to enable preventative measures to be taken before a systems failure occurs, but also to monitor the effectiveness of any actions taken in advance of this to mitigate dislocation of the climate system.

2.4.1 Nutrient stores and recycling

The atmosphere provides an important gaseous reservoir for some elements valuable in ecological systems. Many of the elements concerned constitute essential nutrients for ecosystems. Carbon, nitrogen, sulphur and oxygen use the atmospheric staging post directly, while other elements such as calcium, phosphorus and lead 'piggyback' on atmospheric particles or atmospheric water to make their way to the next phase of their cycle. All elements are dependent on transport by water for at least part of their cyclic journey and so the importance of the hydrological cycle with its atmospheric regulators is once again apparent. Figure 2.17 shows the mean annual fluxes of water and some of the key nutrient ions at various depths beneath a 55-year-old spruce plantation. The losses of water due to evaporation and transpiration, the gains/losses of nutrients due to solution, and the role of vegetation interacting with these mass flows from the atmosphere can be inferred (Box 2.5).

The most common constituent of the atmosphere is nitrogen, and yet nitrogen is often the most limiting element for ecological productivity, both in the terrestrial and marine environments. Achieving a flux of nitrogen from the atmosphere to the soil, or ocean, is of critical importance in sustaining life forms on earth. The chief mechanism for transferring nitrogen from the inorganic atmospheric reservoir to the biological environment is by bacteria and algae. Microbes, such as some forms of bacteria and the blue-green cyanobacteria, possess the enzyme nitrogenase which enables them to break the bonds of nitrogen molecules and release atoms for further reactions. Some of the bacteria work independently in soils; some live within the tissues of host plants such as beans, peas, clover, alfalfa

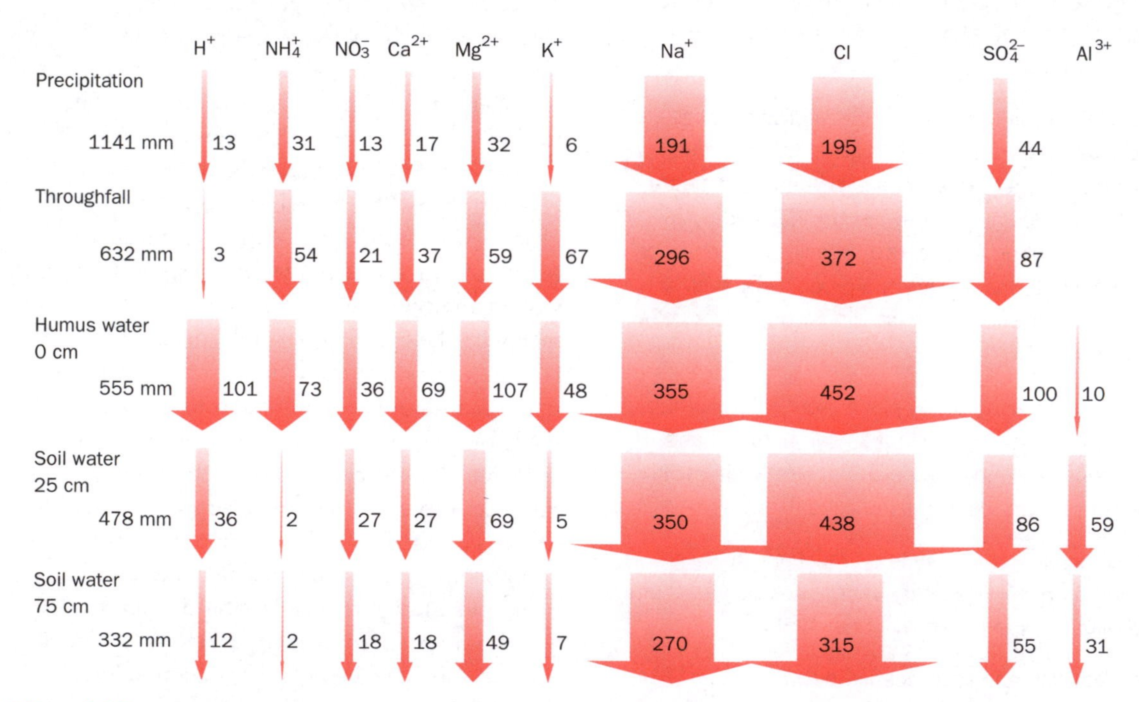

Figure 2.17 Mean annual fluxes of water (mm/year) and ions ($mmol_cM^2$/year) in a mature stand of Norwegian spruce on a podsolised soil in Co. Cork, Ireland.
Source: Sweeney (1997, p. 70)

BOX 2.5

CASE STUDY

Tropical rainforest: the case of Amazonia

Tropical rainforests are the most complex ecosystems on earth, and also among the oldest. The main areas appear to have been in existence for at least the last 40 million years, with some such as the Amazon probably relatively unchanged for the past 100 million years. In more recent times, while regions further polewards were undergoing dramatic geomorphological and vegetational changes during the Quaternary glaciations (2.5 million–10,000 years ago), the tropical rainforests were reduced in size but survived relatively unscathed. They also provided the ecological refugia from which recolonisation of lands released from cold could later proceed. As might be expected from the habitat with neither moisture nor thermal limitations, the tropical rainforest provides the most productive and diverse terrestrial ecosystem, and also one of the most complex; 1 km^2 of virgin rainforest may contain a similar amount of wood to approximately 250 km^2 of temperate forest due to their tall and compact arrangement. Over 50 per cent of the known species of plants and animals, 80 per cent of insect species, 90 per cent of non-human primate species, are found in the rainforest. Amazonia is the most species-rich habitat on earth.

Human interference

Stability should not be equated with invulnerability, however, and increasingly the tropical rainforests are succumbing to human attack. Clearance and burning of forest in Amazonia is not a new phenomenon. There is evidence that it was occurring up to 7,000 years ago. This was small scale and had only minimal impact on the ecosystem. It may even have been beneficial by the creation of a patchwork of varied habitats. During the fifteenth century, English and Dutch migrants saw the forest as a source of charcoal and food. But it was the eighteenth and nineteenth centuries that saw its role extend primarily as a source of tropical plantation crops. However, it was in the 1960s that the roots of massive forest destruction were laid down. Government (e.g. Brazil, Colombia, Peru, Venezuela) policies to build roads stimulated clearance in the vicinity of the highways in the first instance as lands were opened up for cattle ranching on a great scale. Subsidies for agricultural investments in the Amazon reached 100 per cent by 1974 and thousands of new migrants flocked to the area from further east in Brazil. In what was akin to the homesteading idea of the American frontier in the nineteenth century, migrants were given land for free in return for demonstrating an intent to farm. Clearing the forest met these requirements.

Lacking the financial support necessary for larger-scale cattle or other agricultural activities, the peasant farmers who moved in practised slash-and-burn. About 4 ha would be burned, some of the timber sold, and crops grown for 2–3 years. Soil degradation set in very quickly. This is not surprising since a large proportion of nutrients in this ecosystem are stored in the above-ground vegetation. Little is stored in the soils which show high contents of weathered aluminium and iron oxide clays, acidic but poor in plant nutrients. Phosphorus deficiency was a particular problem for cattle which often showed 20 per cent mortality after only 2 years. To render the land fertile required a substantial fertiliser input, costing around $400 per hectare. In contrast, to clear an additional hectare of virgin forest cost around $70.

▶

▶ Tensions between peasant farmers and large cattle ranchers contributed further to deforestation. Land grabbing and the forced eviction of peasant farmers became commonplace events, with 150,000 people being forcibly displaced in a typical year. The end product was that those evicted moved deeper into the forest to start again without long-term security of tenure and therefore without a sense of long-term responsibility for maintaining the productivity of the land they farmed. By the beginning of the twenty-first century forest destruction was averaging almost 20,000 km^2 per annum. This is roughly equivalent to an area 10 per cent the size of the UK being cleared each year.

Forests, if sustainably managed, offer an economic return and have been an important basis for the indigenous peoples. Such is the species diversity of the Amazon, however, that only about 10–15 per cent of the trees are commercially viable to extract and there are typically less than 10 trees worth logging per hectare. Getting at these is expensive and often entails damage to non-commercial adjacent trees. Over 20 trees may be damaged to extract one commercial tree. Selective logging also leads to an increased risk of accidental fire. Such is the slow growth rate of commercially valuable trees, for example mahogany, that the economics of conservation make no sense. A sustainable forest return of 5 per cent p.a. is much less than the return on a bank account or from moving on to clear new areas.

Climate impacts

The dramatic land use changes occurring in Amazonia have impacts on its regional climate, especially precipitation. During the wet season the concentration of cloud condensation nuclei (mostly from mineral dust) is about 300 particles/cm^3. This is quite a low value and the clouds that form are shallow and relatively warm. Average cloud droplet sizes are large, in the range 15–25 μm, and the precipitation mechanisms of collision/coalescence predominate to produce rainfall in copious amounts. During the dry season, when the burning of the forest takes place, the cloud condensation nuclei, mostly from carbon and soot particles, increase enormously to 20,000 particles/cm^3. Because of the availability of the small carbon particles, average cloud droplet sizes are smaller at 5–15 μm. These clouds have to grow above the −10 °C level in order to precipitate, not readily achievable in relatively stable air conditions. Deforestation may thus be contributing to a higher propensity for drought in the region.

Intact Amazonian rainforests play an important role in the global carbon cycle. Their clearance diminishes the photosynthesis mechanisms that remove CO_2 from the atmosphere, and burning the cleared forest exacerbates the situation. A capability to take in up to 6 t/ha carbon per year has been measured, and for the region as a whole, estimates of a total uptake of 0.5 GtC per year have been made. This is a substantial component of the global cycle and to remove this sequestering capability would undoubtedly add further to greenhouse gas loading on the atmosphere. Unfortunately, it is now likely that the size of the carbon sink from forest growth is matched by the size of the carbon source from land use change and that Amazonia has a net zero impact on global atmospheric CO_2 concentrations.

Perhaps the value of the Amazon requires a broader context in which it should be evaluated. Like many ecosystems there are many functions which it fulfils which are not amenable to monetary evaluation (Table 2.4).

Table 2.4 Ecosystem services and examples

Ecosystem service	Ecosystem functions	Examples
Atmospheric gas regulation	Regulation of atmospheric chemical composition	CO_2/O_2 balance, O_3 for UV protection
Climate regulation	Regulation of global temperature, precipitation	Greenhouse gas regulation
Disturbance regulation	Damping of ecosystem response to environmental fluctuation	Storm protection, flood control, drought recovery
Water regulation	Regulation of hydrological flows	Providing water for agricultural, industrial and human uses
Water supply	Storage and retention of water	Provisioning of water by watersheds and aquifers
Erosion control and sediment retention	Retention of soil within an ecosystem	Prevention of soil loss from wind and runoff
Soil formation	Soil formation processes	Weathering of rock and the accumulation of organic matter
Nutrient cycling	Storage, internal cycling, processing of nutrients	Nitrogen fixation, N, P and other nutrient cycles
Waste treatment	Recovery of mobile nutrients and breakdown of excess nutrients	Waste treatment, pollution control, detoxification
Pollination	Movement of pollen	Insects and birds that pollinate crops
Biological control	Trophic-dynamic regulations of populations	Keystone predators, reduction of herbivory by top predators
Refugia	Habitat for resident and transient populations	Overwintering grounds for waterfowl
Food production	Portion of NPP extractable for food	Production of fish, game, crops, nuts, fruits
Raw materials	Portion of NPP used for raw materials	Production of lumber and fuel
Genetic resources	Sources of unique biological materials	Medicines, genes for the resistance of pathogens
Recreation	Providing opportunities for recreation	Ecotourism, sport fishing, other outdoor activities
Cultural	Providing opportunities for non-commercial uses	Aesthetic, artistic, educational, spiritual and scientific value

Source: After Costanza *et al.* (1997, pp. 253–60)

and alder trees. Others work in association with fungi or lichens.

All the time nitrogen is being made available to the ecosystem by biological fixation it is also being lost to the atmosphere. Once again certain bacteria assist with this process of denitrification. This balances the nitrogen budget. Once again, human activities threaten to destabilise the cycle. Denitrification is now augmented considerably by nitrogen compounds released during fossil fuel burning, and by excessive use of nitrogenous fertilisers. Nitrous oxide is damaging both to the stratospheric ozone layer discussed earlier and also is a greenhouse gas.

2.4.2 Climate change and atmospheric fluxes

(a) A case study of the carbon cycle

The carbon cycle is depicted schematically in Figure 2.18 with estimated figures for both the stores in each reservoir and also the annual flux or exchange in carbon between them. The carbon cycle is a good example of how long- and short-term variations in the cycle can spark off climate change.

(i) Carbon fluxes

The vital process for life is photosynthesis whereby atmospheric carbon dioxide (CO_2) is converted to carbohydrate, or stored chemical energy, in the leaves of plants. Green plants are the only organisms that can photosynthesise atmospheric CO_2 and thus are the only organisms that can manufacture their own food. Cells in their leaves, known as chloroplasts, contain a green pigment called chlorophyll and, in the presence of sunlight and water, CO_2 enters the leaf through holes called stomata. This then combines with the stored energy in the chloroplasts to produce a simple sugar or carbohydrate. At later stages, as herbivores consume the plant and carnivores later consume the herbivores, the carbohydrates created are passed up to the higher trophic (feeding) levels. About 120 Gt of carbon is fixed annually on the land surfaces, a quantity known as gross primary productivity, with a comparable figure of 92 GtC for photosynthesis and solution of atmospheric CO_2 in the oceans. Losses of carbon back to the atmosphere primarily occur as a result of respiration by plant tissues and exhalation by animals. Dead organic matter also contributes to the return flow as the detritus food chain operates to decompose biomass back to its inorganic components. The difference between gross primary production and respiration losses yields net primary production, a measure of biomass growth.

(ii) Stores

The carbon cycling through biospheric routes and through the atmosphere represents a very small proportion of the total carbon reservoir. Oceanic carbonate sediments and hydrocarbons laid down in sedimentary rocks constitute the overwhelming bulk of the carbon stock. Over long periods, carbon originating from these stores has been in balance, neither adding to, nor taking from, the atmospheric store. The result is that atmospheric concentrations of CO_2 have been relatively stable over time. It is thought, however, that during major mountain-building episodes, accelerated rates of chemical weathering would have removed large amounts of CO_2 from the atmosphere. This process has been linked with the Alpine orogeny when the Alps, Andes and the Himalayas including Tibet were created. The net removal of CO_2 from the atmosphere at this time (15–20 million years ago) would in turn have led to long-term global cooling (CO_2 is a greenhouse gas) and the ushering in of conditions necessary for ice age formation about 2.5 million years ago (section 8.4). The concentration of CO_2 in the atmosphere during the last glacial period (2.5 million–10,000 years ago) has also been relatively constant, although somewhat higher during warmer interglacial periods when vegetation and biological activity was high, and lower during glacial episodes when biological activity was reduced. Investigations of the composition of bubbles of air trapped in long cores of ice recovered from the polar regions have confirmed this situation. These also showed that in recent centuries a CO_2 concentration in the atmosphere of about 270 ppm applied until the Industrial Revolution. Thereafter a sustained increase has occurred and concentrations today are in the region of 370 ppm. This build-up is largely attributable to increased emissions from fossil fuel burning and land-use changes.

(iii) Anthropogenic loadings

Current estimates of human interference in the carbon cycle from fossil fuel combustion are that there is an additional annual loading on the

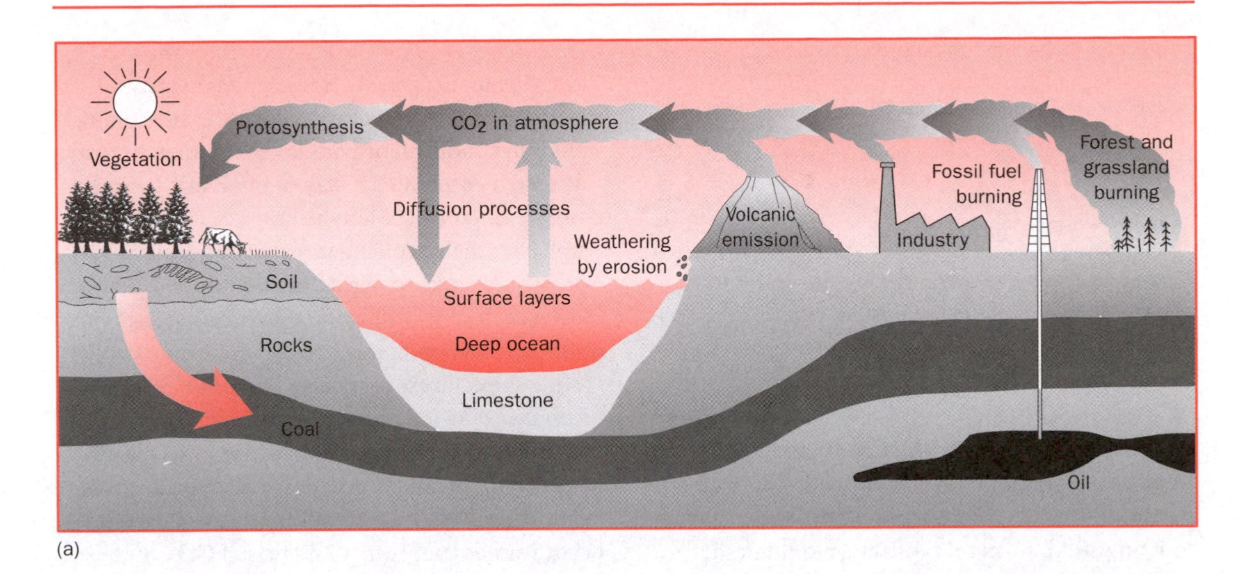

(a)

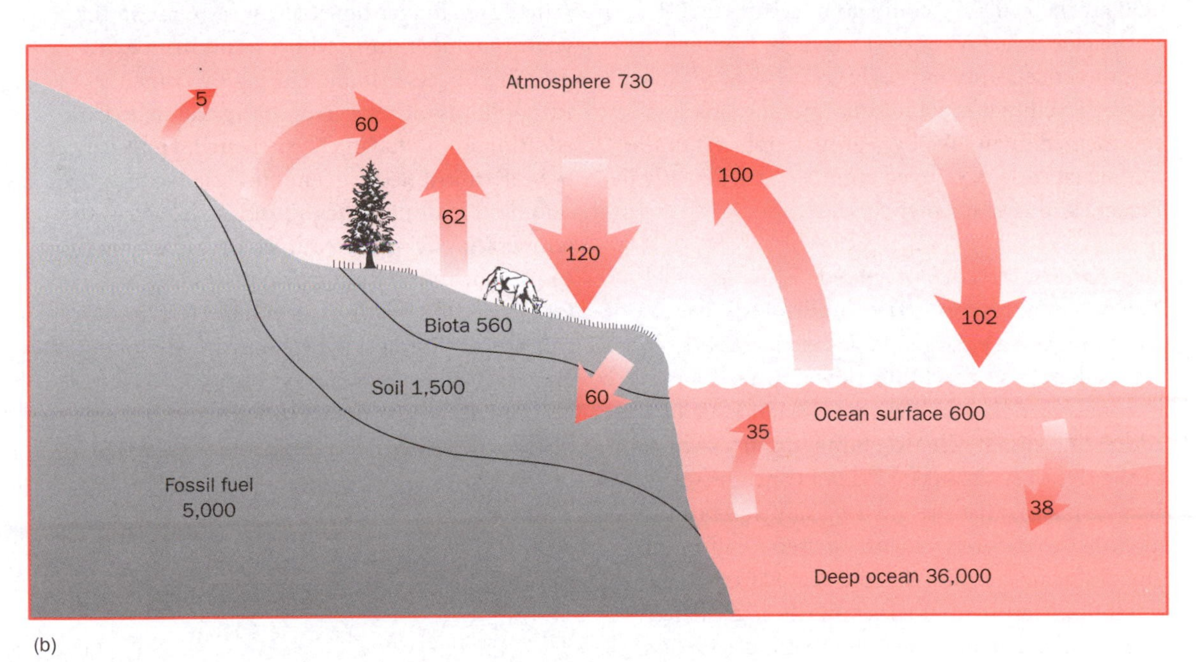

(b)

Figure 2.18 (a) The global carbon cycle; (b) global carbon stores and flows (Gt)
Source: Pickering and Owen (1994, p. 83)

atmosphere amounting to 6.3 GtC. The removal of vegetation, particularly forest cover, is estimated to contribute a further 1.6 GtC annually, though this figure carries considerable uncertainty. Table 2.5 shows estimates of the annual global carbon budget for the 1990s. What is clear is that the 7.9 GtC added annually to the atmosphere does not all build up as extra CO_2. Some 2.3 GtC is removed to the oceans, and the equivalent of 3.3 GtC remains in the atmosphere. This leaves 2.3 GtC

Table 2.5 Global carbon budget

Global carbon fluxes, 1989–98	(GtC/year)
Atmospheric increase	3.3 ± 0.1
Anthropogenic emissions	6.3 ± 0.4
Ocean–atmosphere flux	−2.3 ± 0.5
Land–atmosphere flux	−0.7 ± 0.6
Partitioned into: Land-use change	1.6 ± 0.8
and Residual terrestrial sink	−2.3 ± 1.3

Source: Houghton *et al.* (2001, p. 190)

unaccounted for which must be going to a terrestrial sink, not clearly identified. This is the part of the carbon budget popularly known as the 'missing sink'. It might be strange to think that 2 300 000 000 000 kg could go unaccounted for in the global carbon budget annually, but such was the uncertainty of the fluxes involved. Today it is believed that much of this 'missing sink' is represented by uptake by regrowth of forest in the middle latitude developed countries, particularly in Europe and North America.

(iv) Forests and carbon sinks

The potential for terrestrial sinks (forests, ocean plankton) to be enlarged to absorb increased atmospheric CO_2, and thus slow down the increase in concentrations, has become a major issue in addressing climate change. Similarly, any reduction in the size of such sinks is also of concern (see Box 2.5). Now that the 'missing sink' has been identified, can forest growth be used to offset the global warming problem? In the models used by the Intergovernmental Panel on Climate Change Second Assessment Report in the middle 1990s a very optimistic conclusion was reached. It was believed that the net carbon uptake by terrestrial ecosystems would continue to increase as CO_2 concentrations increased. This 'fertilisation' effect results from the fact that many plants grow more vigorously when extra CO_2 is available.

The idea that forests could be a panacea for global warming gained considerable support during the 1990s and policymakers enshrined the promotion of new forest growth in the Kyoto Protocol, a treaty now signed by most of the world's countries. Tree planting either in an individual country, or paid for in another country, was seen as providing a way of offsetting fossil fuel emission increases. Many countries rushed to emphasise their growing forest inventories as evidence of sinks to negate their sources somewhat. However, recent disturbing evidence suggests this optimism was premature.

Increased forest cover also means increased respiration. As long as the increase in photosynthesis exceeds the increase in respiration, net primary production is positive and carbon is being sequestered. Respiration increases do not occur immediately due to increased CO_2, but rather on a longer timescale in response to the warmer temperatures which result from extra greenhouse gas loading. During this time the extra photosynthesis gives the appearance of an additional sink having been created. However, even if atmospheric greenhouse gas loadings are stabilised at double the pre-industrial levels, the temperature is predicted to go on increasing for a century or more. Respiration losses thus will rise faster over the medium term than 'fertilisation' gains. A declining sink becomes apparent, and after crossover of the two curves (Figure 2.19) forests become net sources of CO_2. Some researchers believe the terrestrial carbon sink will start to

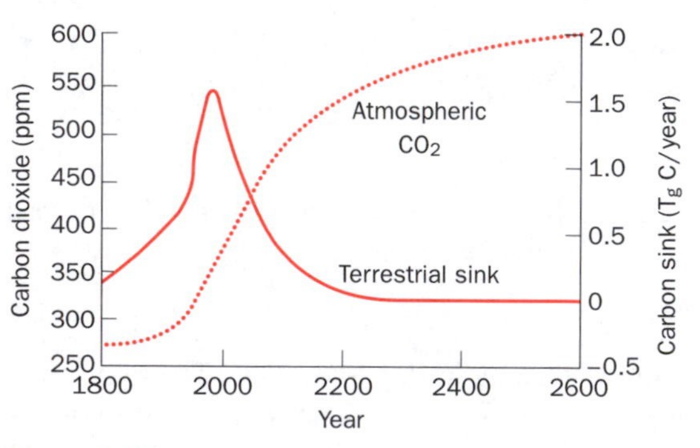

Figure 2.19 A projected declining terrestrial sink for CO_2 in the present century and beyond
Source: Scholes (1999, p. 2)

decline within the next few decades. Ultimately, therefore, only by cutting emissions of greenhouse gases can the spectre of adverse climate change be addressed.

Key ideas

1. The atmosphere acts as a medium of transit for key elements such as carbon, nitrogen, sulphur and oxygen as they move between organic and inorganic stores. Disruption of these pathways by human activities may destabilise the climate system.

2. Human activities are currently adding an additional 7.9 GtC annually to the atmosphere. More than half of this is removed in terrestrial and oceanic sinks, though optimism concerning the role of forests in removing extra CO_2 from the atmosphere is diminishing.

2.5 Observing the system components

2.5.1 Surface networks

Although instrumental observations commenced in some parts of Europe in the seventeenth century it was the Industrial Revolution which provided the impetus for the establishment of the extensive global network which exists today. As the coalfield cities expanded in the nineteenth century, problems of ill health arising from bad sanitation grew ever more pressing. Cholera, typhoid and other epidemics reinforced the need for reliable piped water supplies to improve public health. Reservoirs needed managing and rainfall amounts needed quantifying to assist this. Accordingly it was around the coalfield cities of western Europe that the surface observational network first became established. Temperature (important for evaporation considerations) and rainfall were the primary parameters measured; but wind, cloud cover, sunshine, soil temperatures and humidity followed as instrumentation was developed.

Care needs to be taken in using these early records. Often the exposure of instruments was not standard – some thermometers were hung on north-facing walls for example. Equally, standards of construction varied. Early glass expanded irregularly as temperature increased, distorting the readings that would be obtained. Fahrenheit's claim to fame was not the temperature scale named after him, but rather his skills with glass. It was only after the Stevenson screen became standard that valid comparisons between places could be made. By 1850 western Europe and some of North America had a skeletal meteorological network. Even by 1950 some gaps in global coverage still existed, particularly in the southern hemisphere (Figure 2.20).

Upper air observations were much more difficult to make. Although manned balloon flights were carried out occasionally, these proved dangerous and occasionally fatal. Kites became a favoured alternative. Tethered kites were routinely used to carry an instrument package aloft by the end of the nineteenth century. But the height these could reach was only about 3 km, and observations could only be made in good weather conditions with relatively light winds. In the years before the Second World War, aeroplanes began to displace kites. These could reach higher altitudes, but again could be grounded in poor weather. Their data also could only be analysed some hours after they had been made. Unmanned balloons known as radiosondes ultimately replaced these flights. The radiosonde is a small expendable set of instruments hung from a helium- or hydrogen-filled balloon. Sensors for temperature, pressure and temperature transmit continuously back to a base station as the balloon rises. A radar reflector attached to the balloon enables the ground station to track the height and rate of ascent of the balloon, typically around 300 m per minute, and also the wind direction and speed at various heights. A profile through the atmosphere is provided until the reduction in air pressure results in the balloon expanding to bursting point, or

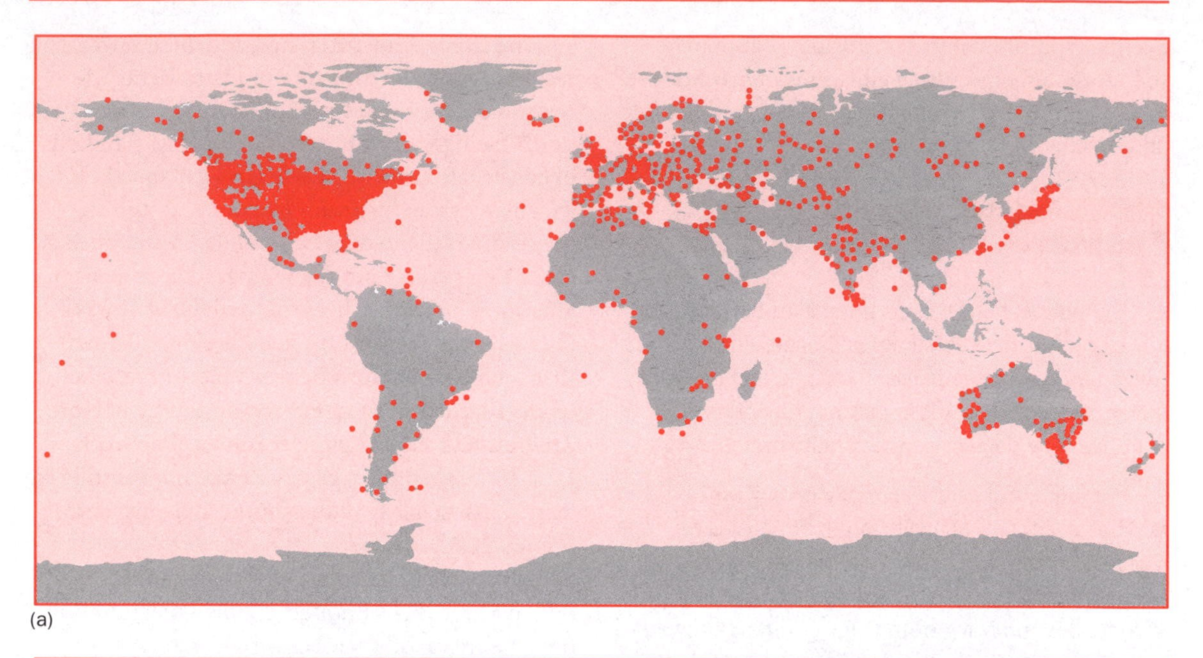

(a)

(b)

Figure 2.20 Growth of the observational network: (a) 1850; (b) 1950

atmospheric conditions cause the ground station to lose contact. Some balloons reach around 40 km before bursting and a parachute allows a gentle descent to occur. Ascents are made at 00.00 and 12.00 hrs GMT at around 900 locations worldwide.

2.5.2 Remote sensing

Since the 1960s, meteorological observations of the atmosphere have increasingly been carried out from orbital platforms. Some of these satellites, such as Meteosat or GOES, are located in geostationary orbits 36,000 km above the surface. As they orbit at the same rate the earth turns, they are therefore capable of imaging the same area below them on a continuing basis. Others, such as the NOAA series, are polar orbiters which image different areas on their pole-to-pole journey as the earth rotates beneath them. Most of these platforms involve passive systems whereby electromagnetic radiation being emitted or reflected from the atmosphere or surface is measured in specific wavebands, particularly in the visible and infrared wavelengths. Gases in the atmosphere selectively absorb both incoming and outgoing radiation at particular wavelengths and by measuring how much radiation is occurring at a particular wavelength, inferences re temperature and atmospheric composition may be made. A second form of sensing depends on active generation of electromagnetic energy at specific wavelengths. By measuring the return flow of pulses of generated energy the satellite can also enable inferences re atmospheric composition to be made.

Passive sensing is particularly useful for cloud analysis due to the high reflectivity (albedo) of clouds. In visible wavelengths, high reflectivity is indicative of high water content. Using the wavebands corresponding to the thermal infrared, inferences regarding cloud temperatures may also be made. Low cloud-top temperatures suggest the presence of high clouds and vice versa. These temperatures may then be compared with those for nearby cloud-free ground or ocean surfaces to enable cloud heights, size and type to be inferred. Of course, snow, mist, fog and sunglint can cause complications to such a classification procedure, but increasingly automated techniques have been developed to enable improved characterisation of clouds (Figure 2.21). As a result, satellite-based rainfall estimates have become increasingly accurate. Today these can be provided on an hourly timescale and at a spatial resolution of less than 5 km. As might be anticipated, satellite estimates are most valuable in poorly monitored areas such as ocean expanses, mountainous areas, or the large tracts of the developing world with only a rudimentary surface observational network. Most people's experience of rainfall estimation, however, comes from ground-based radar sensing as depicted nightly on television weather forecasts. Weather radar operates in the microwave area of the electromagnetic spectrum and is normally operated in conjunction with surface observations to provide ground truthing for the algorithms employed. The UK Meteorological Office currently uses a system known as NIMROD for weather forecasting for much of north-western Europe. This automatically combines ground-based radar, weather satellite and observational data to generate rainfall estimates up to 6 hours ahead. The radar data update rainfall estimates every 15 minutes on a 5 km grid and compares its output to hourly rain gauge data. Meteosat images, both infrared and visible, are employed each half-hour to extend the rainfall analysis out to greater distances than the ground radar operations support.

Satellite sensing of wind speed and direction is also achieved as a by-product of monitoring clouds over time. Watching the movement of clouds in successive images from geostationary satellites enables these parameters to be quantified. Though the resolution of the wind field may not be as good as might be wished, and though it must be remembered that wind at the height of the clouds concerned may not always be symptomatic of surface winds, the technique is valuable in areas where surface monitors are sparse. Over oceanic areas the surface wind may also be estimated from analysis of wave height and orientation. Once again, ground truthing from ocean buoys is often integrated into the procedures.

Key ideas

1. Surface meteorological observations commenced in Europe in the late seventeenth century, though standardisation of equipment and exposure of instruments enabled confident comparisons between places to be made only much later. By the middle of the nineteenth century only Europe and North America had a skeletal observational network.

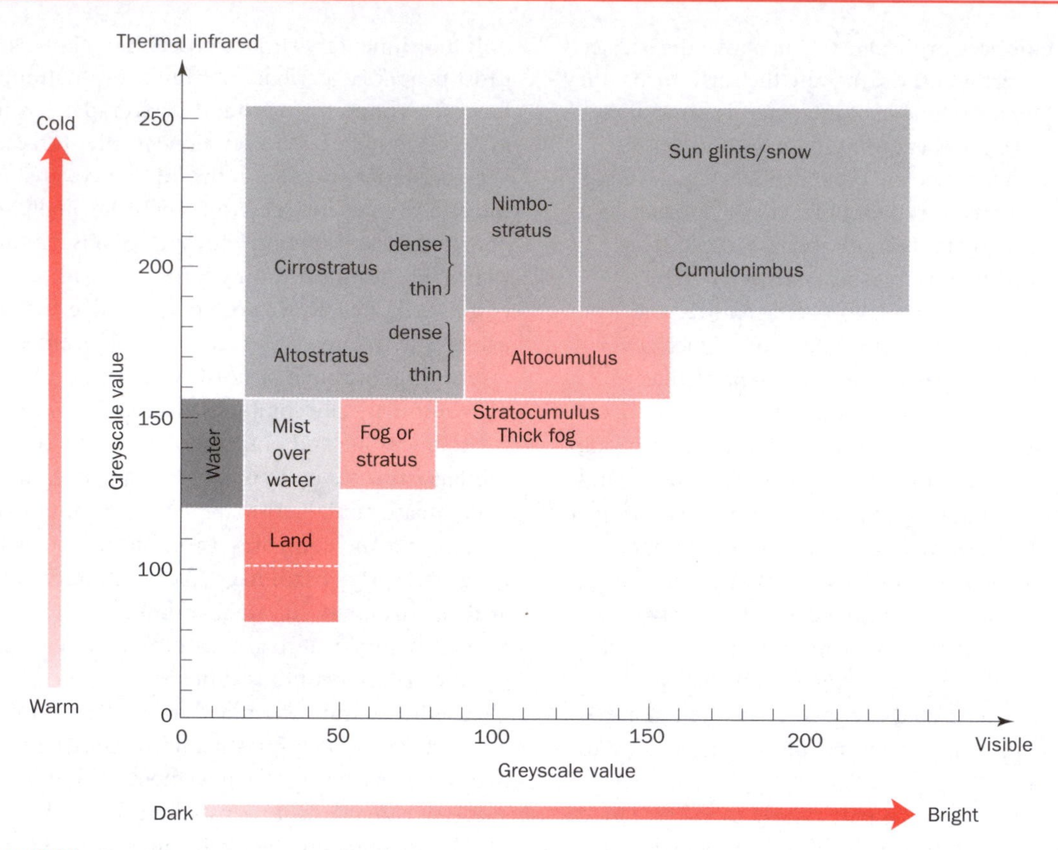

Figure 2.21 Cloud classification from satellites
Source: Gibson and Power (2000, p. 98)

Even by the middle of the twentieth century significant gaps still existed, especially in the southern hemisphere.

2. Upper air observations were revolutionised by the development of the radiosonde after the Second World War. This allowed atmospheric profiling of the troposphere and lower stratosphere.

3. Geostationary and polar orbiting satellites employ passive sensing systems to measure outgoing radiation from which various atmospheric characteristics can be inferred. Cloud characteristics are now routinely derived by this technique. Active systems involve the generation of electromagnetic pulses from the satellites themselves. By analysing the quality of the return signal, inferences regarding atmospheric properties can again be derived.

4. Ground-based radar is particularly useful for precipitation forecasting and is used in conjunction with surface observations and satellite data to extend the forecast capabilities out to greater distances.

Further reading

Ahrens, C. D. (2000) *Meteorology Today*, 6th edn. Brooks/Cole, Pacific Grove, California, 528 pp.

Aguado, E. and Burt, J. (1999) *Understanding Weather and Climate*. Prentice-Hall, Upper Saddle River, New Jersey, 474 pp.

Lachlan-Cope, T. (1999) Back to basics: why does it snow (and can it be too cold to snow)?. *Weather*, 54(1): 16–19.

Lovelock, J. E. (1990) Hands up for the Gaia hypothesis. *Nature*, **344**: 100–2.
Persson, A. (2000) Back to basics: Coriolis: Part 1 – What is the Coriolis Force? *Weather*, **55**(5): 165–70.
Shanklin, J. (2001) Back to basics: the Ozone Hole, *Weather*, **56**(7): 222–30.

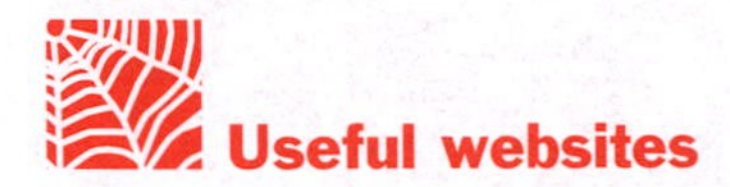

Useful websites

Composition and structure of the atmosphere:
http://www.met-office.gov.uk/education/training/atmosphere.html

Measuring atmospheric pressure:
http://kids.earth.nasa.gov/archive/air_pressure/

Coriolis effect:
http://zebu.uoregon.edu/~js/glossary/coriolis_effect.html

Atmospheric stability:
http://yosemite.epa.gov/oaqps/EOGtrain.nsf/fabbfcfe2fc93dac85256afe00483cc4/1c9d492b7ccef4fe85256b6d0064b4ee/$FILE/si409lesson4.pdf

Precipitation formation theory:
http://www.coolweather.co.uk/htdocsprecipitation.htm

Ozone depletion:
http://www.atm.ch.cam.ac.uk/tour/

Carbon cycling:
http://calspace.ucsd.edu/virtualmuseum/climatechange1/06_2.shtml

CHAPTER 3

Energy components of the climate system

3.1 Nature and role of energy

Earth rotation and energy in the atmosphere are the main factors that drive our weather and climate. Atmospheric motion, including winds and storms, together with the movement of oceanic waters, are primarily generated by the earth's receipt of solar energy and the resulting unequal heating of different parts of the planet. The daily energy received from the sun by the climate system to generate our weather and climate is enormous. For instance, the daily solar energy receipt is 100 times the energy equivalent of a large earthquake, 10,000 times that of a hurricane, 100,000 times that of a nuclear bomb, and represents 100 million times the energy locked up in a typical summer thunderstorm. In addition to heating the planet and driving the world's weather systems, solar energy also performs a number of other important functions.

1. Visible light from the sun is absorbed by plants and used in the process of carbon assimilation or photosynthesis. Because of this, solar energy is crucial to the growth of plants and all animals which are ultimately dependent on them.
2. Solar energy also provides an important source of energy for the world economy. Solar energy and photosynthesis in the geological past have produced vegetation and other living organisms that are the source of present-day fossil fuels.
3. Moreover, light and heat from the sun are used in evaporating and raising water from the land and ocean surface to the atmosphere. The return of this water as rain, and then river flow, to lower elevations is employed as a source of hydroelectric power. In the future, as fossil fuels become scarcer, hydro-power as well as other forms of solar-based energy, for example wind and direct solar heating, will probably become a major and renewable source of energy for humankind.

Energy can be defined as 'the ability to do work'. Energy from the sun is the primary force that is responsible for working the atmosphere and creating all weather and climate. Energy, however, by its very nature, is constantly being changed from one *form* to another, and *transferred* from one area to another within the climate system. For instance, incoming solar energy penetrates our atmosphere to heat the surface of the planet. Some of this surface heat will escape back to space, but most of it will be used to heat the overlying atmosphere. The atmosphere in turn will lose heat to space but much of it will be transferred downwards to reheat the surface of the earth.

One simple classification of energy is into potential energy (i.e. stored or positional) and kinetic energy (the energy of motion). Potential energy is the energy a body possesses by virtue of how that energy is stored. For instance, fossil fuels store and are thus rich in chemical energy. This energy is transformed into heat as its potential is released when such fuels are ignited. Similarly, a body can have potential energy due to its elevation with respect to a datum surface and its weight. A large mass of water in an upland reservoir has considerable potential energy with respect to the valley below it: as water from the high reservoir is released it loses height (i.e. potential energy) by flowing down into the lower valley. However, as it does so its potential energy is converted into kinetic energy or the energy of motion. The power of such moving water of course can be transformed into other types of energy such as hydroelectricity.

3.1.1 Energy and temperature

(a) Sensible heat (you can feel it!)

The kinetic energy of a body can also be observed in motion at the atomic and molecular scale. Temperature is defined as the mean kinetic energy or average speed per molecule of all the molecules in a substance. As energy and heat are applied to a body (e.g. water, air or soil), the average speed or agitation of the molecules within it speeds up, and as a result its temperature will rise. As energy and heat are removed from a substance, the average agitation of its molecules will decline, resulting in a fall in temperature. An example of such direct heating and cooling would be a brick wall warmed by the absorption of solar energy, and the heat emitted by the wall to its surroundings as it cools down during the evening. This type of energy that can be experienced by our senses or measured by a thermometer is called sensible heat (see section 2.3).

Different substances require different amounts of energy or heat to raise/lower their molecular motion and thus their temperature. The heat required to raise the temperature of 1 g mass of a substance by 1 °C is called the specific heat capacity of that substance. Water has the highest specific heat capacity of any substance. The amount of heat needed to raise 1 g of water 1 °C from 14.5 °C to 15.5 °C (conventional range) is termed 1 calorie (4.2 J). The specific heat of water is thus 1 calorie per gram per °C and is labelled as 1.0 (metric units). The specific heat of soils and rocks is much less, being around 0.2–0.3. This means that it takes only about one-fifth to one-third as much energy to heat 1 g of soil by 1 °C as water.

Such variations in specific heat capacity between materials have important repercussions for global and local climate. Because of its low specific heat capacity compared to water, the land surface heats up more quickly than the oceans; but it also cools down more quickly (section 3.3.3). This means that the interiors of continental land masses (e.g. the Sahara Desert, central Canada, Siberia) experience much hotter summers and colder winters than adjoining coastal regions. At coastal locations, thermal differences between the seasons are smoothed out because of the ability of the oceans to act as a large thermal store.

(b) Latent heat (the hidden energy)

We have seen that when energy is applied to a substance, that substance can increase in temperature because the average velocity of the molecules which compose it speed up. Heat can be applied to a substance, however, *without* it necessarily undergoing a change of temperature. As suggested in section 2.3.1, the extra molecular motion produced by the heat is used to change the state of the material (i.e. change it from a solid to a liquid or to a gas) rather than directly changing the temperature of the substance. The heat required to alter the state or phase condition of a material (i.e. from a solid to a liquid or a liquid to a gas) is referred to as latent heat. It is called latent or hidden heat because when it is applied to change the state of a material *no temperature change* in that material takes place.

The nature of latent heat needs fuller explanation and water can be used in illustration. As shown in Figure 2.8, water can exist in three phases or states: as a solid (ice), as a liquid (water) and as a gas (water vapour). We might expect that the kinetic movement of molecules in solid ice is close to zero;

in liquid water relatively slow; and in water vapour relatively fast. In order to change water at the same temperature from its solid (low molecular agitation) to its liquid state (higher molecular agitation) or from its liquid to its gaseous state (very fast molecular agitation), very large amounts of heat are necessary. For instance, to change 1 g of water to 1 g of water vapour, around 590 calories (about 2,470 J) of heat are required. This amount of heat is roughly the same whether the liquid water changes to vapour at relatively low temperatures (i.e. water at 15 °C evaporated from a road surface to water vapour in the air at 15 °C) or high temperatures (i.e. water evaporated and changed to steam at 100 °C). Thus, while it takes roughly 100 calories (419 J) to heat 1 g of water from 0 to 100 °C, it takes almost six times that amount of heat (i.e. 590 calories or 2,470 J) to change 1 g of water to steam. Thus, a more complete definition of specific heat capacity can now be given: it is the heat required to raise the temperature of 1 g of a substance 1 °C *without a change of state.*

3.1.2 Energy transfer

Our weather and climate are conditioned not only because energy can change its form in the atmosphere, but also because it can be transferred from place to place within the climate system. Energy can be transferred in three ways within the earth–atmosphere system, i.e. by conduction, convection and by radiation.

(a) Conduction: agitating the molecules

This is the transfer of sensible heat through a body by molecular collisions without any overall change in position of the molecules. In other words, there is no movement of the body in question. When rock, for instance, is heated by the sun, its surface temperature rises since the molecules which make up its outer layers become highly agitated. These highly agitated molecules will pass on some of their energy by colliding into adjacent but more slowly moving and therefore cooler molecules in the subsurface. By agitating these deeper cooler molecules sensible heat is eventually transferred from the rock surface to the subsurface.

(b) Convection: movement of the substance

This is the process whereby energy is transferred by the movement of the heated substance itself. Unlike conduction, which is a relatively slow process, convection is much faster, and is the main way sensible and latent heat are transferred around the globe by atmospheric motion. The convective transfer of heat is often triggered by *density changes* that take place when a substance is warmed or cooled. The way in which our atmosphere is heated in the vertical dimension provides a good example of thermal convection. On a fine summer's day, as a result of the direct receipt of solar energy by the surface, sensible heat is transferred to the ground which increases in temperature. Air parcels in contact with or close to the heated surface become warmed in turn by that surface (i.e. by conduction). When these air parcels are heated they expand, and as they expand they become less dense than the colder air above. Because of their low density such warmed parcels of air tend to rise up through the atmosphere carrying and distributing their heat with them. An equally important component of such convective energy transfer in the atmosphere occurs when the warmed buoyant parcels of air are replaced in the lower reaches of the atmosphere by colder denser air parcels from higher up.

The convective updraughts and downdraughts which transport sensible heat around the globe also transfer latent heat. When water is evaporated from the oceans to water vapour, latent energy is supplied to the molecules to free them from the strong inter-molecular bonds within liquid water. Convective updraughts help to transfer the water vapour to the higher layers of the atmosphere whereupon cooling the water vapour condenses to form water droplets, usually in clouds. During condensation this latent (or hidden) energy is released from the water droplets to heat the surrounding air. This internal source of heat adds to the cloud's buoyancy, allowing it to penetrate higher into the troposphere with the chance of producing greater rain and storms. Such additions of heat can change pressure gradients within the atmosphere and can help drive large-scale horizontal atmospheric motion.

(c) Radiation: wavy motion

Most direct energy emissions, transfers and absorption within the climate system are by radiation. There are many examples of energy transfer by radiation, including the warmth you feel on your face from direct sunlight (solar radiation), the generally higher temperatures of cloudy as opposed to clear winter nights (caused by the transfer of radiation from clouds to heat the surface), microwave cooking and the sending or receipt of radio signals.

Radiant energy is transferred in the form of waves (Box 3.1). The waves are defined in terms

BOX 3.1

THINKING FURTHER

The nature of radiation (1)

Key features of radiation

1. Radiation is transferred by electromagnetic waves. As the term suggests, these waves have both electrical and magnetic properties. Radiation waves can pass through solids, liquids and gases (solar radiation for instance can pass through glass, water bodies and the atmosphere). In addition, radiation waves do not require a substance to pass through and, unlike say sound waves, can travel through a vacuum. When moving through a vacuum, radiation waves move with the speed of light, i.e. solar radiation moving from the sun to the earth across 93 million miles of space travels at 186,000 miles per second (300,000 km/s).
2. All bodies whose temperature is above the lowest temperature possible, i.e. absolute zero temperature (–273 °C or 0 °Kelvin) emit radiation. This includes every object you can think of, i.e. the air, the land, flowers, trees, people, the stars in the sky.
3. An important characteristic of the radiation emitted by a body is its wavelength. This is measured as the distance between any two peaks or troughs of the electromagnetic wave (Figure 3.1(a)). Electromagnetic waves appear over a wide range of scale from extremely short X-rays to much longer wavelengths of radio and television.
4. The radiant energy emitted by a body occurs not as a single wavelength but as a range of different wavelengths called its electromagnetic spectrum (Figure 3.1(c)).
5. The wavelengths and total amount of radiation that an object emits depends mainly on its temperature (see Box 3.3). For instance, the higher the object's temperature the greater the amount of radiation and the shorter are the waves of the emitted radiation. Most radiation waves from the sun (surface temperature about 5,900 K) range from 0.1 to 100 μm. Solar radiation is dominated, however, by a range of fairly short wavelengths that reach a peak emission level at 0.4–0.7 μm which are the wavelengths of visible light. For this reason solar radiation is often referred to as short-wave radiation. In contrast, the total amount of radiation emitted by a cooler body such as the earth (surface temperature 16 °C or 289 K) is much less, occurring at longer (thermal) wavelengths between 5 and 70 μm, and peaks at about 10 μm. All earth and atmospheric radiation, i.e. the radiation given out by oceans, forests, deserts, clear skies and clouds is described as long-wave radiation. Because long-wave radiation emitted by the earth and atmosphere is in the form of heat only (there is no visible light, etc.) it is often referred to as thermal radiation. Examples of such radiation include heat being emitted or lost from the land and ocean surface to the atmosphere or the heat that is transferred from atmosphere/clouds to the surface of the earth.

of their wavelength, which is the distance between any two adjacent wave crests (Figure 3.1(a)). The wavelengths of radiation cover a wide spectrum from long radio waves to tiny gamma rays (Figure 3.1(b)). Radiation from the sun (called solar radiation) is described as short-wave radiation since the energy wavelengths of which it is composed are generally small. As shown in

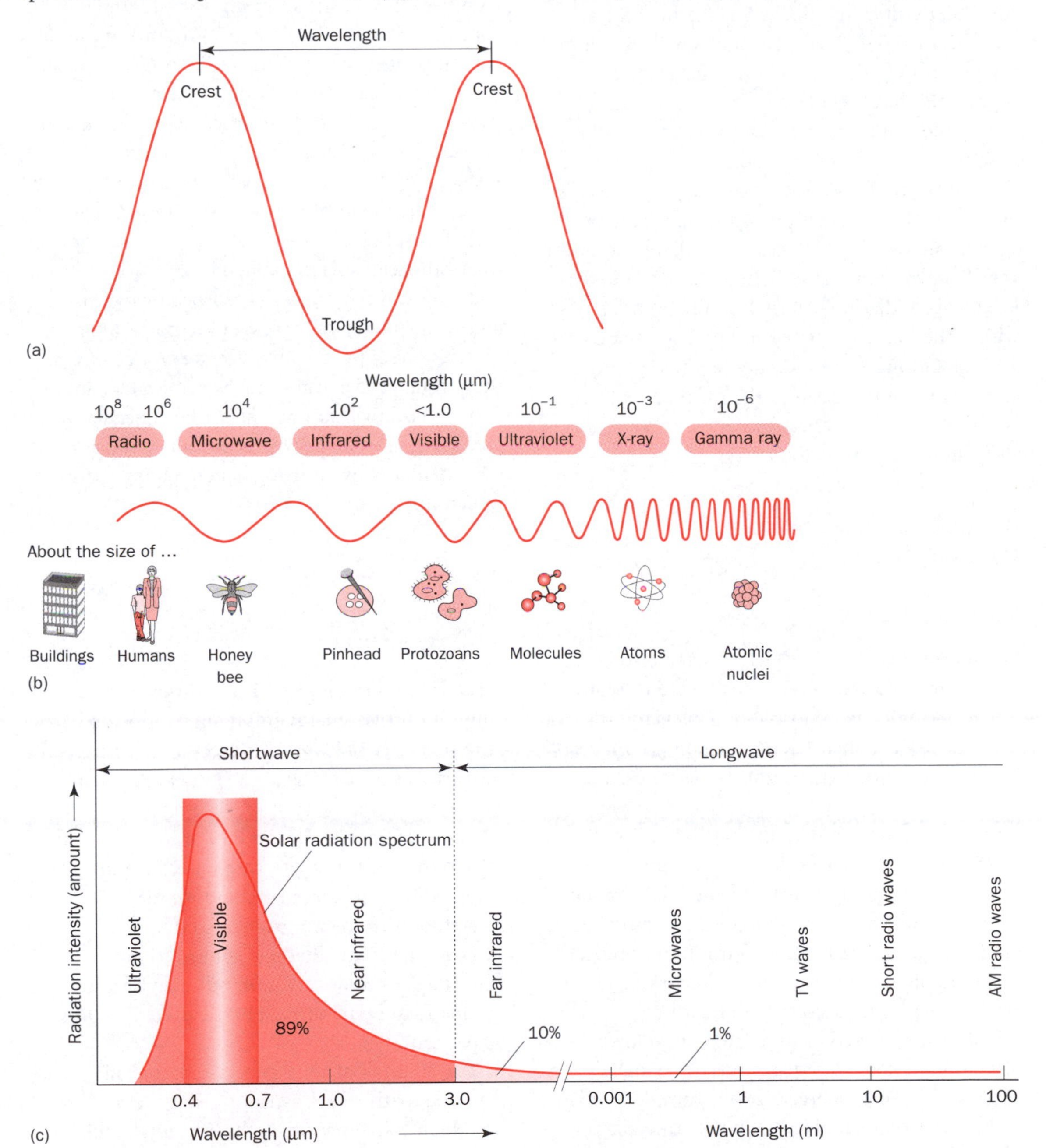

Figure 3.1 The nature of radiation including (a) measurement of wavelength; (b) wavelengths from radio to gamma rays; (c) the electromagnetic spectrum (array of wavelengths) of solar radiation showing the percentage of energy the sun radiates in each part of the spectrum

Source: Various

Figure 3.1(b) and (c), most solar radiation waves have wavelengths in the visible range (0.4–0.7 μm), with the majority of the remainder being less than 3.0 μm (1 μm is one-millionth of a metre). Although visible light including the near infrared (wavelengths 0.4–3 μm) covers a small part of the total solar radiation spectrum, it nevertheless represents about 90 per cent the total energy given out by the sun in the form of the solar radiation spectrum. All *atmospheric* radiation, for example that from clouds, water vapour, oxygen molecules, and all *terrestrial* radiation including that from vegetation, deserts and oceans is radiated at longer wavelengths than that of light and is described as long-wave radiation. All such wavelengths are within the range 3–70 μm and peak at about 10 μm. As long-wave radiation emitted by the earth and atmosphere is in the form of heat (there is no visible light, etc.), it is often referred to as thermal radiation (see Box 3.1).

3.1.3 Energy interaction and transformation

(a) Solar reflection

Because of their long wavelengths, terrestrial and atmospheric radiation are *not* reflected or scattered by the earth or the atmosphere. This is not the case with short-wave solar energy including both visible light (0.4–0.7 μm) and slightly longer non-visible solar radiation (up to 3.0 μm) which can be reflected and scattered by the earth–atmosphere system. Reflection takes place when short-wave solar energy striking a surface at an angle bounces off at the same angle without any heat exchange or change in the light beam. Surfaces that reflect incoming light are many and include mirrors, glass, lakes, clouds, deserts and buildings. Light can also be scattered, i.e. reflected unevenly, in different amounts and directions by particles and gases in the atmosphere. A good example of this is the way oxygen molecules in the atmosphere scatter incoming solar radiation, particularly blue light, giving the planet its blue skies.

(b) Energy absorption: how materials are heated

Natural bodies can also *absorb* short-wave solar (wavelength <3 μm) and long-wave thermal (or heat) energy (wavelength 3–70 μm). If the substance absorbs all of the radiation falling on it, it is called a black body. Most objects in the universe are imperfect absorbers of radiation and are called grey bodies. The earth–atmosphere system and the sun are grey bodies but can absorb about 90 per cent of the energy falling on them. The atmosphere, however, absorbs less energy than the land/ocean surface. When both short-wave and long-wave energy are absorbed by the earth and atmosphere, an increase in temperature takes place. If latent heat is involved, a *phase change* in the substance (i.e. water droplets to water vapour) may also occur. A body that has been heated by absorbing solar and earth–atmosphere radiation may then cool by emitting or reradiating thermal radiation.

3.1.4 Heating and cooling of planet earth

(a) Solar inputs

It can be seen in Figure 3.2 that the average total amount of solar energy impinging at the top of the atmosphere is 342 watts for every square metre of the earth's surface (1 watt (W) = 1 joule/second). This amount can be simplified to represent 100 units of energy. On average around 30 units (107 W/m^2) of this energy are reflected and scattered back to space by clouds or other constituents in the atmosphere and by the surface. The term **albedo** is applied to the percentage of solar radiation reflected by a surface. For the total earth–atmosphere system it is thus 30 per cent.

The 30 units of solar energy reflected and scattered by the earth–atmosphere system plays no role in warming the earth. In contrast, the remaining 70 units (235 W/m^2) of solar energy that are absorbed by the atmosphere, i.e. by clouds

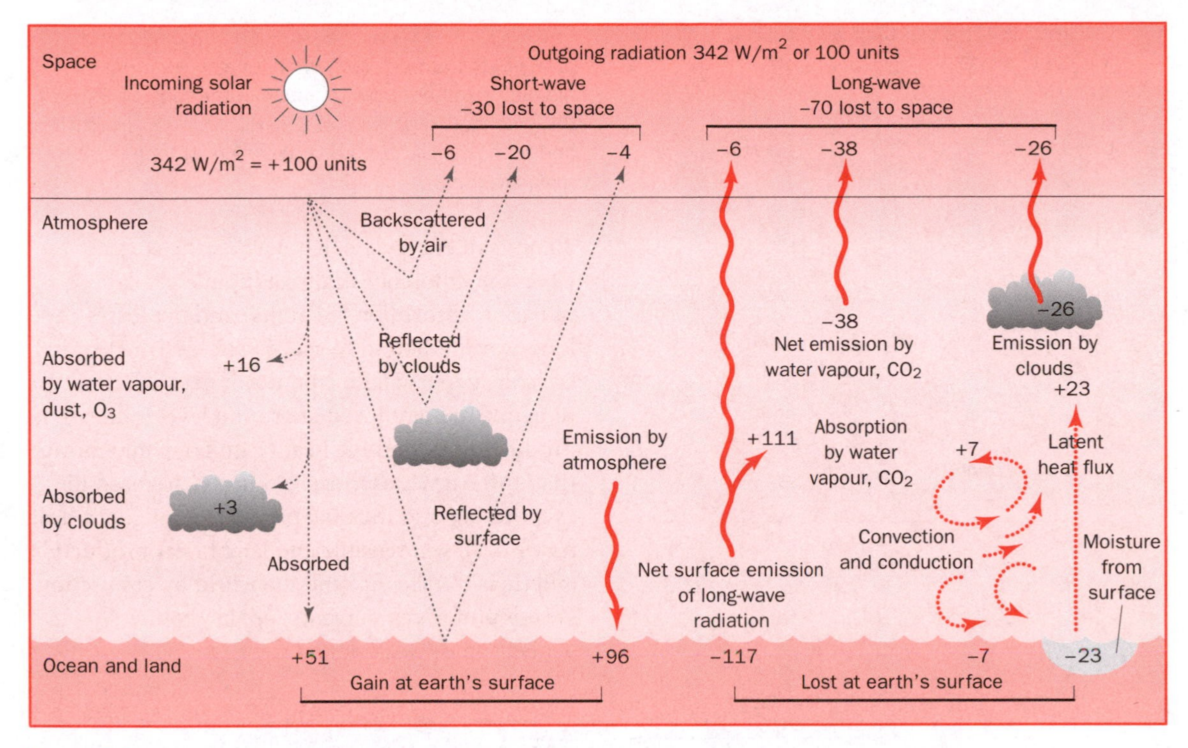

Figure 3.2 The main energy and radiation transfers within the earth–atmospheric system expressed as a percentage of the original solar energy input which is represented as 100 units
Source: Drake (2000)

and other atmospheric constituents (19 units) and particularly by the surface, including both land and sea (51 units), are converted into heat and warm the planet. This absorbed solar energy forms the basis for *all subsequent energy exchanges* within the earth–atmosphere system including (i) the transfer of long-wave thermal or heat radiation and (ii) sensible and latent heat transfers.

(b) Energy exchange within the climate system

The main transfers of long-wave radiation within the climate system include the following: 117 units of thermal radiation (390 W/m^2) are emitted from the surface of the earth as it cools. Of these 117 units, a small proportion (6 units) is lost directly to space through the atmospheric window in the region of 8–14 μm (Figure 3.3). The remaining 111 units are absorbed by the atmosphere, especially by clouds and atmospheric gases such as water vapour and CO_2. This absorbed long-wave radiation is used to heat the atmosphere. The atmosphere also cools by long-wave radiation losses, by emitting heat (a) to the surface (96 units) and (b) to space (64 units). Long-wave heat transfer to space includes 38 units lost by water vapour and CO_2, together with 26 units from clouds.

(c) Not all is radiation: sensible and latent heat transfer

Apart from solar and thermal radiation that dominate energy transfers within the earth–atmosphere system, smaller but critically important exchanges of energy of a non-radiative nature also take place between ground surface and the overlying atmosphere. As shown in Figure 3.2,

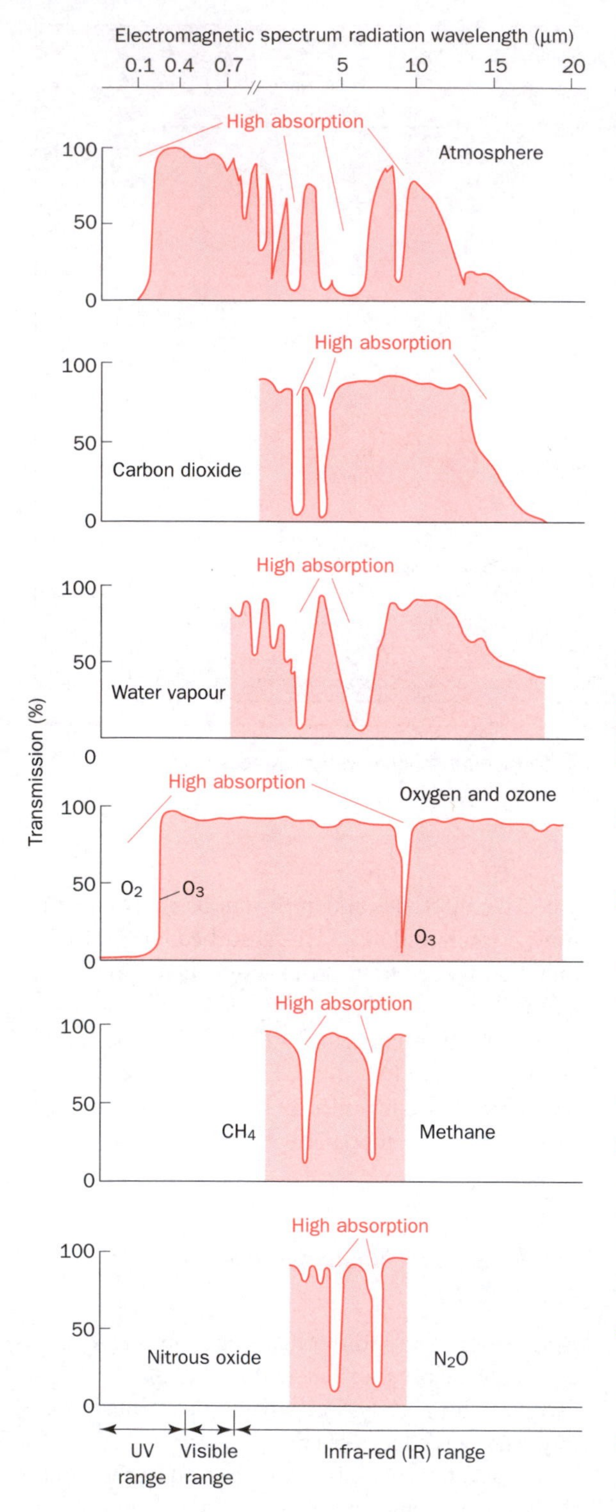

Figure 3.3 Percentage absorption of radiation by gases in the atmosphere. While oxygen and ozone absorb very short-wave radiation in the form of UV light, the greenhouse gases, i.e. water vapour, CO_2, methane and nitrous oxide, are good absorbers of long-wave radiation (heat)

30 units of energy are 'left over' at the surface when the difference in direct inputs of solar radiation absorption (51 units) and net losses of long-wave radiation to space ($117 - 96 = 21$ units) are considered. These 30 units of net surface radiation are employed at ground level in the production of sensible heat (7 units), while more than three times as much (23 units) are used in evaporation and thus the production of latent heat. As we shall see, sensible and latent heat production and their transfer around the world by convection (winds and ocean currents) are key to the generation of weather and climate.

Key ideas

1. Radiation which is a type of energy is invisible and is transferred in the form of waves.

2. There are two kinds of radiation based on the wavelength of the energy waves: (a) radiation from the sun (solar radiation) is composed of, and travels in the form of, short-wave radiation; and (b) heat energy (thermal radiation) from the land surface and atmosphere is composed of and is transferred as long-wave radiation.

3. Inputs of short-wave solar radiation, absorbed by the ground surface and atmosphere, are transformed into heat. Most of this heat (but see sensible and latent heat) is then transferred and exchanged in the climate system in the form of thermal long-wave radiation.

4. Solar energy is the main source of planetary heating and the principal mechanism underpinning the world's weather and climate.

5. Smaller but important amounts of heat (called sensible and latent heat) within the earth–atmosphere system are transferred by

non-radiative processes, i.e. by conduction and particularly by convection.

6. Sensible heat is 'actual' heat that can be experienced by our sensory organs and can be measured by a thermometer as temperature.

7. Latent heat is the heat required to alter the state of a substance from a solid (e.g. ice) to a liquid (e.g. water) to a gas (e.g. water vapour).

8. Latent heat is used to change the movement of molecules within a substance as the phase changes of matter take place. As no observable change in sensible heat or temperature takes place during these energetic phase changes, this type of heat is called latent or hidden heat.

9. The process of evaporation uses large amounts of latent heat from the sun and surrounding environment to convert liquid water to water vapour. By the same token, large amounts of latent heat are released into the environment and atmosphere during the process of condensation which transforms water vapour into water droplets.

10. Convection is the main way sensible and latent heat is transferred around the globe by atmospheric and oceanic motion.

3.2 Heating and cooling balancing acts

Over long periods of time (e.g. several years) and putting global warming aside for one moment, the temperature of the entire earth–atmosphere system is neither increasing or decreasing. This condition applies because all energy inputs (solar radiation input plus incoming long-wave thermal radiation from the atmosphere) and all energy outputs (solar radiation loss plus the outgoing long-wave thermal radiation to space) for the whole earth–atmosphere system balance each other. We can summarise this net radiation budget (or balance) between heating and cooling of the entire earth–atmosphere system by the following simple equation:

$$NRB = (SR_{in} + LR_{in}) - (SR_{out} + LR_{out}) = (\text{net radiation heating} - \text{net radiation cooling}) = 0$$

where NRB is the net radiation budget for the earth–atmosphere, SR = solar radiation and LR = long-wave thermal (heat) radiation. The NRB for the planet is thus the difference between the *absorbed* incoming solar radiation and the *net* loss of outgoing heat or thermal radiation, i.e.

$$NRB = SR_{in(ab)} - LR_{net\ (out)} = 0$$

Moreover, when taken separately, neither the surface nor the atmosphere is changing temperature over time. This must also mean that energy conditions for these two components are also balanced and that the atmospheric and surface net radiation budget is zero (Box 3.2).

3.2.1 Greenhouse balance

(a) The greenhouse model in action

The atmosphere plays a crucial role in the heating and cooling of the climate system because it helps to maintain a balance in surface temperatures higher than they would otherwise be. The raising of global surface temperatures by 'greenhouse' gases in the atmosphere is called the greenhouse effect (cf. the heating of an actual garden greenhouse). The greenhouse effect is usually explained using the following simple model. Short-wave solar energy passes fairly unimpeded through the atmosphere to the earth's surface, and creates a warm surface which then emits heat or thermal radiation. Much of the emitted heat in turn is absorbed by greenhouse gases in the atmosphere (e.g. water vapour, CO_2, methane) only to be reradiated and sent back to the surface. The overall effect is to maintain the surface at a higher temperature than would occur if the atmosphere were as transparent to the loss of outgoing heat as it is to the entry of short-wave solar energy. Thus the atmosphere acts as a glass greenhouse or an insulating blanket over the earth keeping it warm.

The main energy transfer mechanisms of the greenhouse model can be checked using Figure 3.2.

BOX 3.2

THINKING FURTHER

Earth–atmosphere energy (heat) balance

(a) Atmospheric heat balance

When the 111 units of surface energy absorbed by the atmosphere are added to the 19 units of energy absorbed from direct solar absorption, and the 30 units of energy absorbed by the atmosphere by sensible and latent heat transfers, the total amount of atmospheric energy absorption becomes 160 units (Figure 3.3). In order to maintain thermal equilibrium in the atmosphere, the atmosphere will emit exactly 160 units of energy, i.e. 96 units to the surface as back radiation, and 64 units to space (38 units from water vapour, CO_2 and other atmospheric constituents and 26 units from clouds).

(b) Surface heat balance

We can also see from Figure 3.3 that thermal equilibrium is also maintained at the surface when all energy exchanges (radiation together with sensible and latent heat transfers) are considered. This is because energy inputs at the surface, i.e. 51 units from direct solar radiation absorption, plus 96 units from atmospheric back radiation (147 units in total), are exactly balanced by thermal emissions, i.e. 117 units from direct radiation loss and 30 units by sensible and latent heat transfer (total 147).

(c) Heat balance at top of atmosphere

Energy inputs and outputs at the top of the atmosphere complete the overall equilibrium of our planet. In addition to the 30 units of solar radiation reflected from the atmosphere to space, long-wave thermal radiation losses include the 6 units directly radiated from the surface, 38 units from greenhouse gases and 26 units from clouds. When the thermal radiation total of 70 units is added to the solar radiation total of 30 units, i.e. 100 units, it exactly balances the incoming solar radiation.

First, the surface temperature of the earth is much higher than it would be if it did not have an atmosphere. Without an atmosphere the energy characteristics of the planet would be relatively simple. In maintaining a constant surface temperature the earth's *surface* would lose heat from the surface equivalent to the incoming solar energy, i.e. 100 units (342 W/m^2). At this relatively low rate of global energy loss, however, the average surface temperature of the earth's surface would stabilise at −18 °C. It is shown in Box 3.3 that there is a direct relationship between the temperature of a body and the amount of heat it loses. Thus the less radiation that is emitted from the earth the cooler it is. With an atmosphere in place, the energy characteristics of the climate system become very different. For instance, as can be seen in Figure 3.2, the actual amount of heat lost by the *surface* of the earth is not 100 units (342 W/m^2) but 117 units (390 W/m^2). At this rate of radiation emission the temperature of the surface stabilises at around 15–16 °C, i.e. 33 degrees higher than it would be without an atmosphere. The reason why the earth is relatively warm losing a net surface emission of 117 units of energy is because of what is termed the thermal back radiation from the atmosphere to the surface. As shown in Figure 3.2, 111 of the 117 units of outgoing thermal radiation from the surface are absorbed by the atmosphere, mainly by water vapour and other greenhouse gases such as CO_2 and by clouds. A large proportion of this absorbed atmospheric thermal

THINKING FURTHER

BOX 3.3

The nature of radiation (2)

Radiation equations

In physics an ideal radiator is called a **black body**. A black body at a given temperature T emits to its surroundings the maximum amount of radiation possible, i.e. it emits radiation at all wavelengths. It also absorbs all the radiation falling on it at temperature T. A good emitter is also a good absorber of radiation. Most objects like the sun and the earth are not perfect radiators. The sun, however, approximates more closely to a black body than the earth.

The rate or intensity at which energy is emitted (F) from a black body can be calculated using the Stefan–Boltzmann Law:

$$F = \sigma T^4$$

where σ is the Stefan–Boltzmann constant which has a numerical value of 5.670×10^{-8} W m^{-2} K^{-4} and T is the temperature of the surface in degrees Kelvin. This law shows that radiation emission is proportional to the fourth power of the temperature of the object. This means that the higher the temperature, the more radiation is emitted per second. This means that a body like the sun at surface temperature of about 5900 K will emit far more radiation than a cool body like the earth at 289 K.

Another law called Wien's Law allows the calculation of the wavelength at which maximum radiation emission occurs, i.e. the peak of the black body spectrum. Wien's Law states

$$\lambda_{max} = b/T$$

where the temperature (T) is measured in degrees Kelvin and b is a constant with a value of 2.898×10^{-3} m K. Using Wien's Law, the maximum wavelength for the high temperature sun is 0.5 μm while that for the cooler earth maximum emission is around 10 μm. The differences in the wavelength emissions allow the two streams to be treated separately, i.e. solar radiation is described as short-wave radiation, while earth or terrestrial radiation is described as long wave.

Kirchhoff's Law allows us to define how a real body like the earth–atmosphere system, which does not approximate closely to a black body, compares to a black body:

$$\varepsilon = W/W_b$$

where ε is the emissivity or radiation output, W is the real emittance and W_b the black body emittance.

The emissivity of the earth–atmosphere system is only about 0.7. This means that the system emits *and* absorbs only about 70 per cent of the maximum radiation possible. This relatively low figure is partly due to the poor emissivity of the atmosphere itself (less than the land) which is shown in Figure 3.2. The poor emissivity of the atmosphere, i.e. the fact that it is a weak absorber of short-wave solar radiation, allows us to see visible light from the sun so clearly. The atmosphere is better at absorbing ultra-violet radiation (0.1–0.3 μm) and outgoing long-wave terrestrial radiation or heat, especially in the wavelengths around 3 to 8 (Figure 3.3). The relatively high absorption rates of long-wave radiation by the atmosphere's greenhouse gases, especially CO_2 (3, 4 and beyond 15 μm) and water vapour (3 and 4–8 μm), are clearly shown in Figure 3.3. Both emissivity (absorption) ratings and reflectivity (albedo) values are important in terms of energy surpluses and deficits for different surfaces. Polar ice caps, for example, are regions with negative net radiation values, i.e. they lose more energy to space than they receive from the sun. This radiation deficit is a function not just of low radiation inputs from weak polar sunshine but also because of high rates of energy loss due to both high albedos and emissivity ratings.

A common refinement is made to the Stefan–Boltzmann Law to make allowance for the fact that not all bodies act as perfect radiators:

$$F = \varepsilon\sigma T^4$$

radiation, as much as 96 units, are reradiated back to the surface, helping to raise temperatures there.

(b) Developing the model further

A number of considerations can take our understanding of the model further. They are:

1. The atmosphere may be relatively transparent to incoming solar radiation by absorbing only 19 per cent of it, but the earth–atmosphere system reflects a further 30 per cent, leaving only around one-half (51 per cent) of the original solar energy to be absorbed at the surface.
2. The atmosphere is warmed not only by direct radiation from the surface (111 units) but also by direct solar radiation absorption (19 units) and transfers of sensible and latent heat from the surface (30 units).
3. Most importantly, it is only by considering the fact that *temperatures generally decline with height* in the atmosphere that we can fully appreciate the operation of the greenhouse effect. Figure 3.2 shows that the atmosphere cools by emitting not just 96 units of radiation to the surface but also 64 units of radiation to space. Radiation is emitted out to space by greenhouse gases from levels near the top of the atmosphere, typically from between 5 and 10 km high. Here, because of convection processes and falling air pressure, the temperature is much colder (–30 to –50 °C) than the air near the surface. Because the gases at the top of the atmosphere are cold, they emit correspondingly less radiation (64 units) than the gases in the warmer air nearer the surface (96 units). By doing so they help to prevent too much heat loss and maintain the greenhouse effect.
4. In addition, warm air can hold more water in its vapour state than cold air. Data from satellites and radio sonde balloons carrying meteorological instruments are now accumulating which show that there is much more water vapour in the lower warmer atmosphere than in the upper atmosphere where it remains cold. Water vapour is *the* major greenhouse gas, so its high concentration in the lower layers of the atmosphere allows it to trap outgoing heat and send it back to the nearby surface more efficiently. Thus the *vertical distribution and concentration* of greenhouse gases in the atmosphere also serve to reinforce the efficiency of the greenhouse effect.

Key ideas

1. Inputs of solar radiation to the climate system are balanced over long periods of time by losses of heat (thermal radiation) to space, creating a net radiation balance and therefore a constant temperature condition for the earth.
2. The ability of the atmosphere to absorb large amounts of outgoing thermal radiation from the planet's surface, and to reradiate such heat back to the surface, helps to raise and maintain mean global surface temperatures.
3. This process where the atmosphere acts as a blanket retaining the earth's heat is called the 'natural' greenhouse effect.
4. The main gases involved in the natural greenhouse effect include water vapour, CO_2, methane and nitrogen oxide.

3.3 Inequalities in regional energy supply and loss

3.3.1 The net radiation of regions

While the energy or heat balance averaged over several years can be shown to exist for the whole earth–atmosphere system, radiation input/loss conditions for different *regions* of the planet and *time* periods are rarely in equilibrium. Indeed, in order to understand weather and climate patterns, it is necessary to show that major energy or heat *imbalances* (either excesses or deficiencies) for

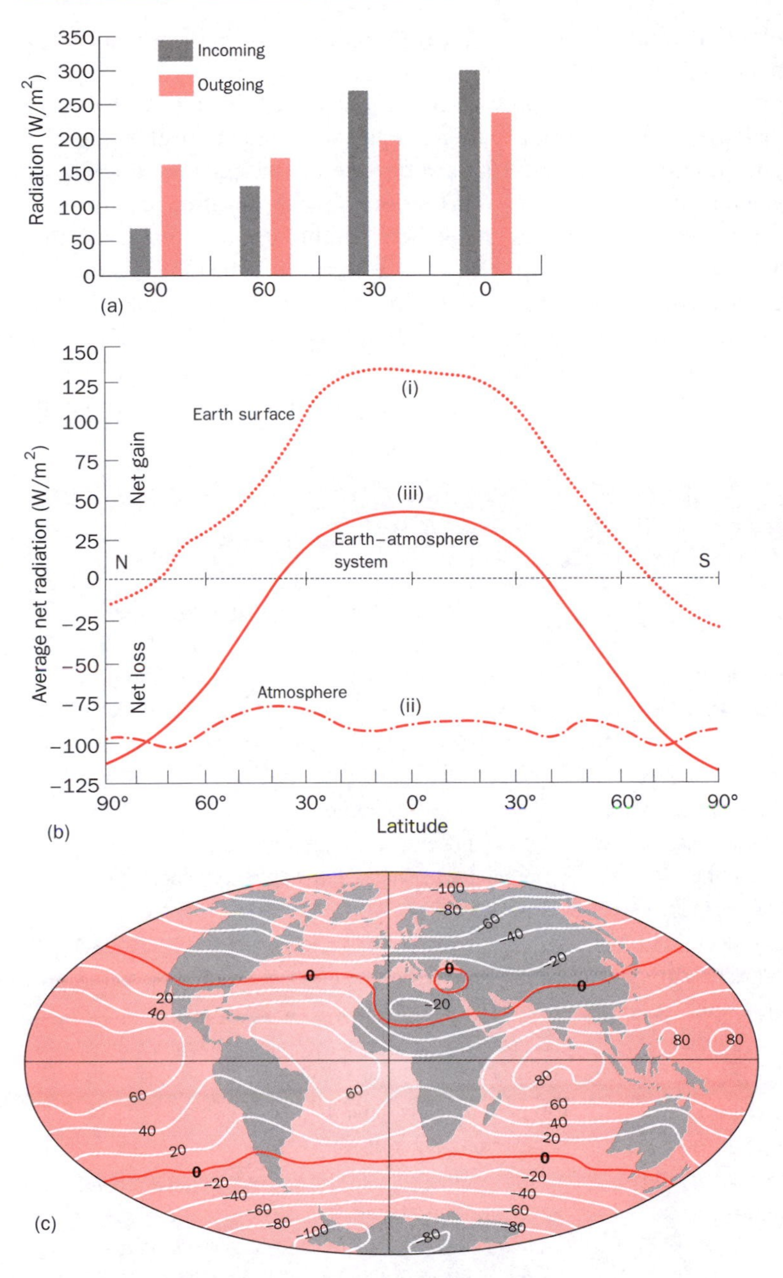

Figure 3.4 (a) Latitudinal variation at 0°, 30°, 60° and 90° of average net radiation (i.e. incoming absorbed heat from the sun minus net outgoing loss of heat to space) for the earth–atmosphere system; (b) latitudinal variation of average net radiation for (i) earth's surface only, (ii) the atmosphere only and (iii) the earth–atmosphere system; (c) global distribution of average net radiation for the earth–atmosphere system. All energy amounts in W/m²
Sources: (b) Sellers (1965).

different regions of the earth at different times are the norm rather than the exception. It can be seen from Figure 3.4(a) that while incoming absorbed solar radiation declines rapidly from equator to poles, the loss of heat or net radiation to space is much more even by latitude. This means that while there is a surplus or net radiation in the low latitudes there is a deficit at higher latitudes. This relationship can be explained further using Figure 3.4(b). It can be observed that when the earth's surface (only) is considered, there is a sharp decline in positive net radiation budgets (at the equator) to negative net radiation budgets at the poles. On the other hand, the atmosphere has a negative net radiation budget across all latitudes. When the whole earth–atmosphere system is taken together, we can see that equatorwards of about 38° latitude, there is a surplus of global net radiation, i.e. in these latitudes inputs of absorbed solar radiation to earth and atmosphere exceed losses of thermal or heat radiation to space. Polewards of about 38° latitude, a mean net energy deficit situation applies, with more thermal radiation or heat being lost to space than ever enters/is absorbed in the form of solar radiation. Figure 3.4(c) shows that the average annual latitudinal decline in global net radiation values can also be identified at the spatial or regional level. Such patterns reflect more than anything else the gradual decrease in solar radiation from the equator to the poles (see next section). Despite high amounts of incoming solar radiation reaching the Sahara under its clear tropical skies, its negative radiation balance stands out however, with its southern margins pushing well into tropical latitudes. The anomalous net radiation deficit of the Saharan region is a result of the tremendous loss of outgoing long-wave radiation to space under cloudless tropical skies together with the large amount of solar radiation loss from the high reflectivity (albedo) of the Saharan desert (Box 3.4).

3.3.2 Inequalities in solar radiation inputs

As previously suggested, the main forcing factor in the pattern of global net radiation shown in Figure 3.4 can be found in the distribution of solar energy. This is because solar radiation absorption varies greatly between the latitudes, i.e. high at the equator and low at the poles, while the loss of long-wave thermal radiation (heat) to space is fairly constant. In the equation $SR_{in(ab)} - LR_{net(out)} =$ net radiation, solar energy absorption is the main driver of net radiation conditions. A confirmation

BOX 3.4

THINKING FURTHER

Albedo

A useful measure is the albedo which is the amount or proportion of light reflected and scattered by a surface (Table 3.1). The earth's different surface albedos can be seen in satellite images where the oceans are the darkest regions. This means that they have the lowest albedos reflecting only around 6–10 per cent in the low latitudes and 15–20 per cent of the incident light near the poles. Ocean albedos increase at high latitudes because at low sun angles water reflects sunlight more effectively. The most reflective and therefore the brightest regions of the globe are the snow-covered Arctic and Antarctic which can reflect as much as 80 per cent of the incident light. The next brightest areas are the major tropical deserts such as the Sahara and the Saudi Arabian Desert which can reflect as much as 40 per cent of the incoming solar radiation back to space. Higher latitude deserts such as the Gobi have lower albedos at between 25 and 30 per cent. The darkest land surfaces with albedos of between 10 and 15 per cent are the tropical rainforests of Amazonia (South America) and the Democratic Republic of Congo (central Africa). Modern satellite images also show that the tops of deep cloud, especially in equatorial and high latitudes in both hemispheres, also have high albedos (80 per cent) appearing white to the observer. The underbelly of some deep clouds viewed from the earth's surface can appear black, however, to the observer. This happens because most of the sun's light is reflected from the top layers of these clouds out to space, creating a dark image underneath.

From this analysis it is clear that if the surface characteristics of the earth–atmosphere system are altered in any major way then the planet's overall albedo and its associated climate will change. If the extent of winter snow and ice cover in polar regions diminishes (this is thought to be happening most effectively today in the Arctic with global warming) it will have a significant warming effect on the planet. This is because with less reflective ice layers and more ice-free land areas, more sunlight will be absorbed and changed into heat at the surface. Conversely, and somewhat surprisingly, any significant expansion or brightening of the major deserts (again this may be happening with the process of desertification) will enhance global albedos and could cause a cooling of the climate.

Table 3.1 Albedo values of various surfaces

	Albedo value
Very high albedo surfaces	
1. Fresh snow	80–95
2. Thick cloud	70–80
(a) cumulonimbus	90–95
(b) thick stratus	60–70
High albedo surfaces	
Thin cloud	25–40
Ice/sea	30–40
Saline deserts	25–50
Hot deserts	25–35
Moderate albedo surfaces	
Savanna grassland	15–25
Tundra	15–20
Crops	15–25
Deciduous forest	15–20
Low albedo surfaces	
Green pasture and meadow	10–15
Dry ploughed fields	10–15
Coniferous forest	10–15
Urban areas	15
Dark soil	5–10
Oceans (average)	7–9
Oceans (high sun)	3–5
Oceans (low sun)	50–80

of the fact that solar radiation does indeed exercise a strong control on the distribution of global net radiation is shown in Figure 3.5. The observed distribution of solar radiation inputs (as with the global net radiation) generally declines from the equator to the poles. This is because for the year as a whole, the angle at which the sun's rays strike the surface becomes more oblique polewards of the equator. Oblique rays deliver less energy at the ground than more vertical rays, when the sun is high in the sky (low latitudes, high latitudes in summer, noonday sun). Figure 3.6 shows that this is because (i) the same amount of solar energy is spread over a larger surface area and (ii) the same solar beam undergoes more severe atmospheric dilution by reflection, scattering and absorption as it has to pass through a thicker layer of atmosphere. Figure 3.5 illustrates that the actual distribution of solar radiation receipt at the earth's surface is strongly influenced by cloud cover. The high solar reflectivity (albedo) of clouds (Table 3.1) sends large amounts of incoming solar radiation back to space, so that the clear sky subtropics (>225 W/m^2) and not the cloudy equatorial regions (<200 W/m^2) stand out as the zone of maximum solar radiation receipt at the surface.

3.3.3 Consequences of uneven heating and cooling

(a) Energy transfers

If, as shown in Figures 3.4 and 3.5, the uneven latitudinal radiation heating and cooling of the earth remained unchecked, the low latitudes (energy surplus) would continually heat up while the higher latitudes (energy deficit) would continually cool down. This does not happen of course (although the equator remains warmer than the poles) because regional energy imbalances are compensated by a net transfer of sensible and latent energy from the low to the high latitudes. Atmospheric winds and ocean currents, for instance, circulate warm air and water towards the poles, and cold air and water towards the equator. The main net poleward exchanges of energy occur as a result of:

1. A transfer by wind of sensible heat that accounts for about 60 per cent of the total transfer. Mobile weather systems in the middle latitudes such as low pressure cyclones and high pressure anticyclones are an important vehicle in transferring heat polewards and cold air equatorwards. Anticlockwise moving air in the northern hemisphere cyclones for instance allow warm air to flow northwards on their eastern flank while cold air from polar latitudes flows south on their western edge. Similarly, clockwise-rotating northern hemisphere high

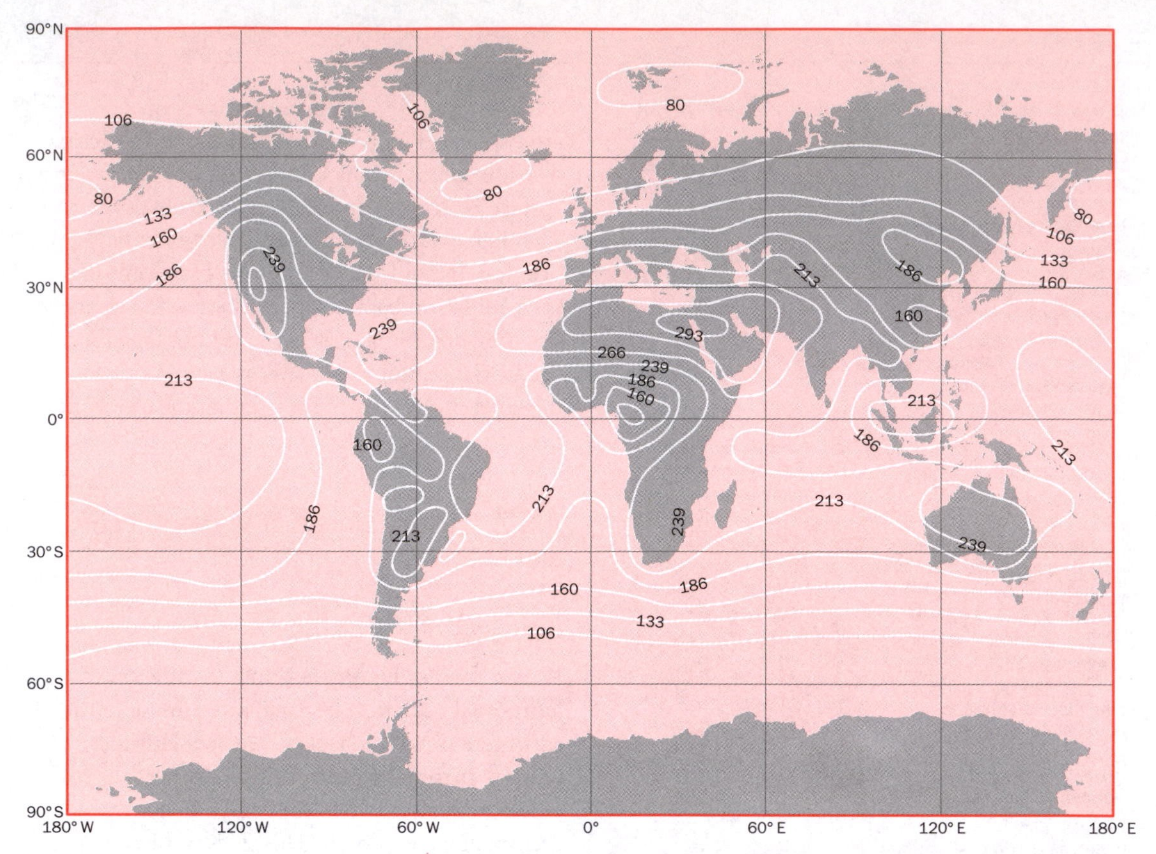

Figure 3.5 Worldwide distribution of annually averaged solar radiation received at the earth's surface (in W/m^2)
Source: Robinson and Henderson-Sellers (1999, p. 29)

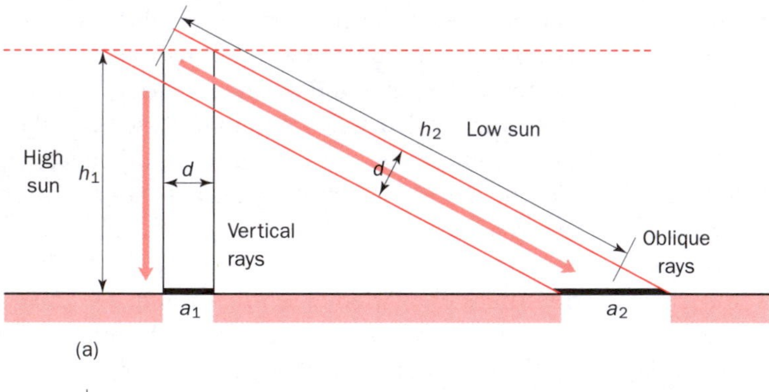

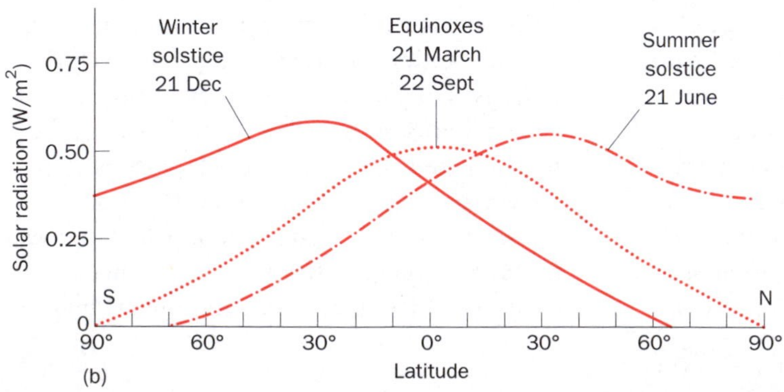

Figure 3.6 (a) Diagrammatic contrast of solar radiation receipt from high-intensity vertical rays and low-intensity oblique rays; (b) latitudinal distribution of solar radiation at ground level for the equinoxes and solstices. It is assumed that the atmosphere reflects 30 per cent of the incoming solar radiation for vertical sun and cloudless skies

pressure anticyclones help to transfer warm air polewards on their western flank, with cold air moving south on their eastern flank.

2. A transfer by wind of latent heat contributing about 15 per cent of the total. When surface ocean water evaporates in tropical regions, it is carried by weather systems into high latitudes where it cools, condenses and releases latent heat.
3. The remaining 25 per cent of the heat exchange between latitudes takes place by the movement of ocean currents (see section 4.4). Good examples here include warm water which is moved poleward in the North Atlantic by the Gulf Stream (North Atlantic Drift), and cold water which is transferred southward towards the equator by a return cold current off the coast of Africa (Canaries current).

(b) Global temperature response

Despite the movement of sensible and latent heat from low to high latitudes, there is still an overall annual decline in mean surface temperature between the equator and the poles. The annual latitudinal decline in surface temperature is modified by *seasonal* changes in solar radiation absorption and resultant net radiation supplies. Before seasonal temperature maps for the world can be described and analysed, we therefore need to look at winter–summer variations in global energy conditions.

(c) Seasonal changes in global solar and net radiation

(i) Seasonal variation in solar radiation

Figure 3.6(b) shows that both latitude and the time of year (season) help to determine the altitude of the sun and the duration of solar radiation across the planet, both of which affect the total amount of solar radiation falling on a region. It can be seen in Figure 3.6(b) that incoming solar radiation values vary between the equator and the poles between the equinoxes (21 March and 22 September) when the sun's noon rays are vertical at the equator. Major variations in solar receipt also occur for the summer (21 June) and winter (21 December) solstices, when the sun's noon rays are vertical at the tropics of Cancer (latitude $23\frac{1}{2}$°N) and Capricorn ($23\frac{1}{2}$°S) respectively. Compared to the situation at the equinoxes there is a substantial poleward shift in solar radiation values in each summer hemisphere. Moreover, each 'summer' hemisphere receives between two and three times as much insolation as the 'winter' hemisphere. In summary therefore, high latitude regions in both hemispheres have low solar radiation inputs at the equinoxes and particularly when the sun is in the 'other' hemisphere (i.e. it is that hemisphere's winter period). By contrast, regions which receive the highest amounts of incoming solar radiation (assuming for the moment cloudless skies) include the low latitudes (equatorial regions) at all seasons, and the high latitudes in their summer season.

(ii) Seasonal variation in earth–atmosphere net radiation

It may be recalled from Figure 3.4(b) and (c) that on an annual basis, regions polewards of about 38° have a negative earth–atmosphere radiation balance while regions equatorwards of this latitude possess a positive radiation balance. During the northern hemisphere summer (July) global earth–atmosphere net radiation values increase and are strongly positive over most of the northern hemisphere (not just south of 38°). At this time, deficit net radiation balances can be found in (a) a small area around the North Pole and (b) over all of the southern hemisphere. This seasonal pattern in the latitudinal distribution of earth–atmosphere net radiation can be explained with a higher altitude sun in the northern hemisphere, and corresponding greater amounts of solar radiation supplied to these latitudes. Because net radiation values increase during summer (July) in the mid and high latitudes of the northern hemisphere, lesser amounts of sensible and latent heat are transferred to these regions from the equator at this time. This is not the case for the southern hemisphere during July which because of its negative winter net radiation balance receives large transfers of heat from equatorial regions. During the northern hemisphere winter (January), the opposite case applies with most of the northern hemisphere in net radiation deficit, and most of the southern hemisphere (except its polar regions) in net radiation surplus. As a result, greater amounts of energy (than the summer or all-year condition) have to be transported from the low latitudes to the northern hemisphere, and much lesser amounts to the southern hemisphere.

This seasonal variation in the latitudinal distribution of earth–atmosphere net radiation is one reason why mobile weather systems (especially low pressure depressions) which help to transfer energy polewards, are more frequent and intense during the northern and southern hemisphere winter than during their respective summer periods.

(d) Seasonal variations in temperature

(i) World distribution of January and July temperatures

We can now explore in some detail the main seasonal patterns in global heating and cooling. The seasonal variation in global solar radiation and net radiation creates very different patterns in hemispherical heating during January and July (Figure 3.7(a) and (b)). Very strong contrasts are evident in the temperature distributions between the two hemispheres, and in the latitudinal or east–west trend of the isotherms (lines of equal temperature). Solar and net radiation values are not the only controls in the seasonal patterns of global temperature, however. Other forcing factors such as the distribution of land and sea and global heat transfers also play a part.

(ii) Land and sea effects

The seasonal global patterns of temperature are also influenced by the distribution of land and sea. This is not surprising since we have already seen that land and sea differ in their specific heat capacities and therefore in their response to solar heating, with land heating up and cooling down much more quickly than the oceans. It will be noticed, first, that there is a more pronounced migration and concentration of isobars over the land than the sea. Second, the annual range of temperature is greater in continental than in coastal locations, with the annual range reaching over 55 °C in north-east Siberia. Third, the large heat storage of the oceans causes them to be warmer on average in winter, but colder in summer than the land in the same latitude.

(iii) Global heat transfers

The seasonal (January–July) control which unequal latitudinal net radiation patterns and global heat transfers have on the distribution of global temperatures is less obvious than that exercised by land and sea. Nevertheless, they remain considerable. For instance, despite the rapid reduction in northern hemisphere temperatures between the equator and the poles in January, the latitudinal reduction in temperature shown is seriously minimised by very large movements of wind-driven sensible and latent heat transfers polewards during this season. More obvious are the effects of heat transfer by ocean currents, especially by warm currents in the winter and cold currents in the summer. The outstanding effect of the warm Gulf Stream and North Atlantic Drift in spreading heat polewards in the north Atlantic in January is clearly identified by the northward shift in the main isotherms. So too is the pronounced equatorward displacement of isotherms along the coast of Peru and Chile in January (southern summer) by the cold Peruvian current.

(iv) Global temperature anomalies

The significance of continentality (section 10.2.2) and ocean current energy transfer can be better observed using the concept of the temperature anomaly. Temperature anomalies reflect east–west, i.e. regional changes in thermal condition rather than north–south (latitudinal) differences. They

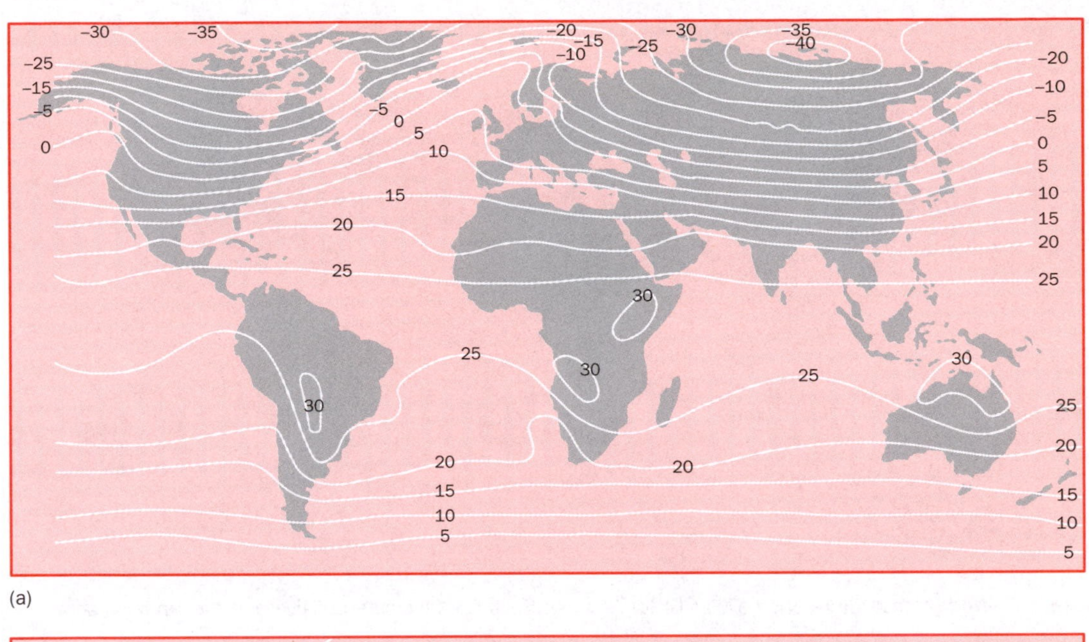

(a)

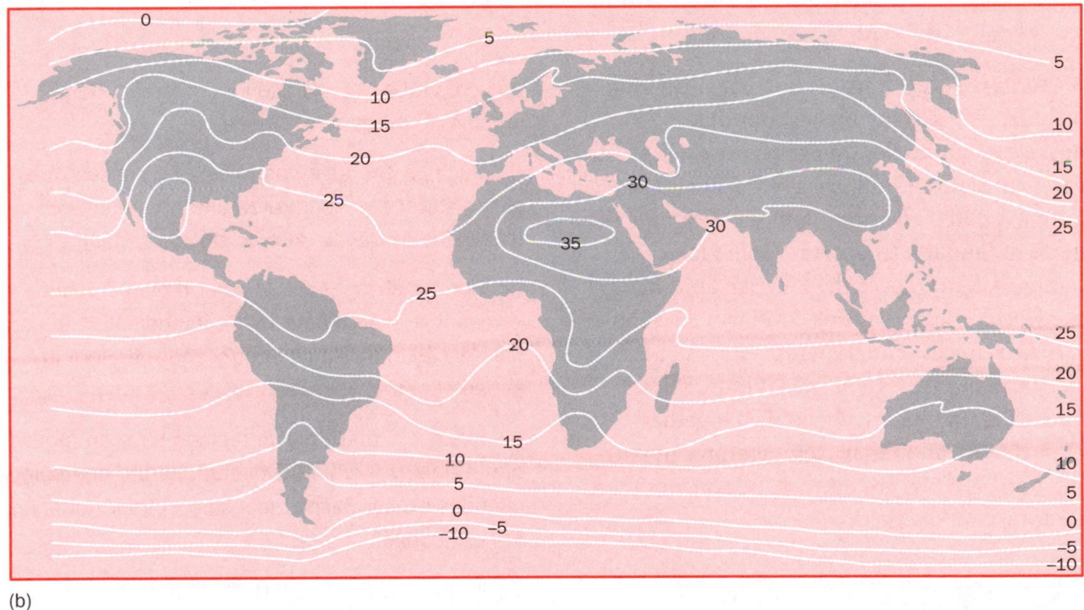

(b)

Figure 3.7 The world distribution of (a) mean January and (b) mean July temperatures (°C) at sea level

are an expression of how a region's temperature differs from the mean condition for its latitude. As shown in Figure 3.8, the largest temperature anomalies are in the northern hemisphere, where continents and oceans alternate in high latitudes, and are seen to best effect in the winter (January) period. At this season positive anomalies are found over the oceans with negative anomalies over the land masses. With the North Atlantic Drift contributing as much heat as solar radiation

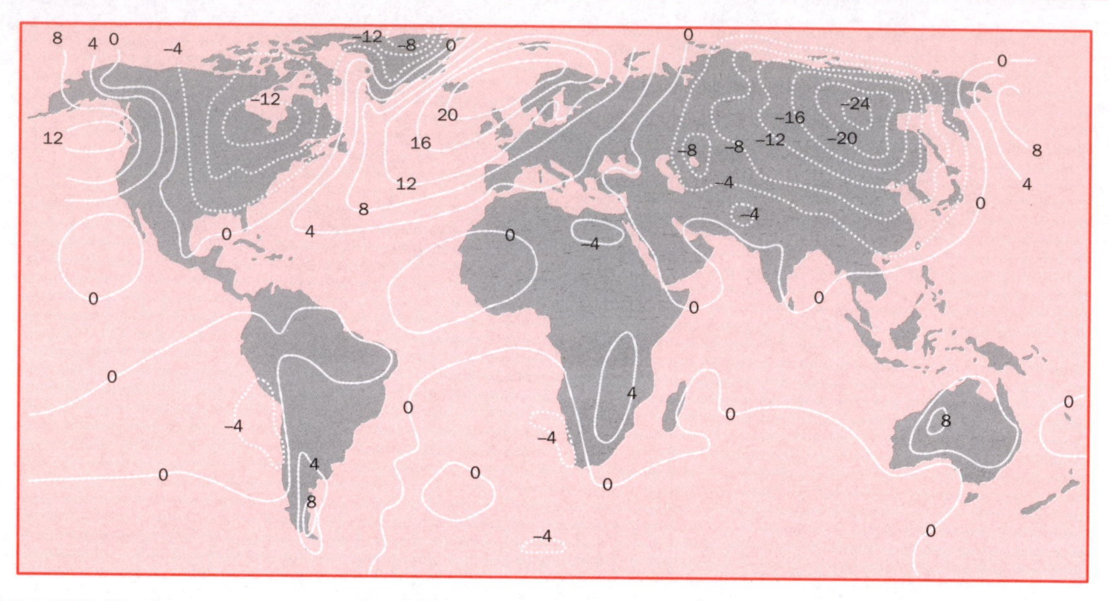

Figure 3.8 World temperature anomalies (°C) for January. Solid lines indicate positive and broken lines show negative values
Source: Various

input to north-west Europe, it is not surprising that this region has the highest positive temperature anomalies on the planet. Coastal Norwegian stations and parts of the northern UK are over 20 °C warmer than the mean for their latitude in January. In contrast, rapid losses of outgoing radiation over the Siberian land mass in the winter mean that this continental region experiences mean January temperatures 26 °C colder than the average for its latitude. These temperature anomalies when taken together suggest that mean January temperatures in parts of the northern UK are up to 45 °C warmer than those found in Siberia.

Key ideas

1. Energy and radiation conditions for different parts of the planet and time periods are rarely in balance. Net energy surpluses and deficits are the norm for different regions and time periods.

2. For instance, the recent massive addition of human-induced greenhouse gases to the atmosphere has created a net radiation increase or energy surplus at the earth's surface.

3. Such greenhouse gas forcing has raised mean global surface temperatures and has thus been implicated in the process of global warming.

4. Horizontal variations in solar radiation input across the globe create major latitudinal differences in available energy and therefore in surface heating.

5. Equatorial and tropical regions have an excess of net energy (heating) while temperate and polar latitudes exhibit deficits (cooling) in their net energy or heat budget.

6. In order to prevent the low latitudes from continually heating up and the high latitudes progressively cooling, major transfers of energy and heat take place between the low and high latitudes.

7. Major latitudinal transfers of heat (both sensible and latent heat) take place by winds and ocean currents, thus helping to keep the earth–atmosphere system in relative heat balance.

8. Nevertheless, global temperature maps show that the low latitudes remain warmer than the higher latitudes.

9. Temperature conditions across the earth are strongly affected by the distribution of land and sea, and energy flow by wind and ocean currents.

10. Planetary temperature conditions are also influenced by the seasonal migration of the sun across the equator between the northern and southern hemisphere.

11. Large seasonal differences in surface temperatures are experienced in winter between land and sea areas especially in the northern hemisphere.

3.4 Altitudinal energy inequalities

3.4.1 Global surface–atmosphere energy imbalance

It has already been shown (Figure 3.2) that a constant net radiation surplus of 30 units can be found at the surface of the earth. To balance this, a permanent net radiation deficit of −30 units of energy can be found in the atmosphere. This deficit can be calculated by measuring the difference between incoming *absorbed* solar radiation and *net* losses of outgoing thermal radiation from the atmosphere, i.e. from Figure 3.2:

$$(SR_{abs}) - (LR_{net/out}) = (16 + 3 = 19) - (111 - (96 + 38 + 26) = -49) = -30$$

3.4.2 Energy transfer in the vertical atmosphere

If the unequal net radiation between the surface and the atmosphere remained unchecked, the surface would continually heat up while the atmosphere would continually cool down. This does not happen because vertical imbalances in the net radiation are continually compensated for by transfers of sensible and latent heat between the surface and the atmosphere. However, despite the vertical movement of sensible and latent heat by convective updraughts and downdraughts, the atmosphere still cools with elevation. This general cooling with height in the atmosphere is called the environmental lapse rate (Figure 3.9). For the planet as a whole the mean vertical temperature decline is about 6 °C for every kilometre of air ascent.

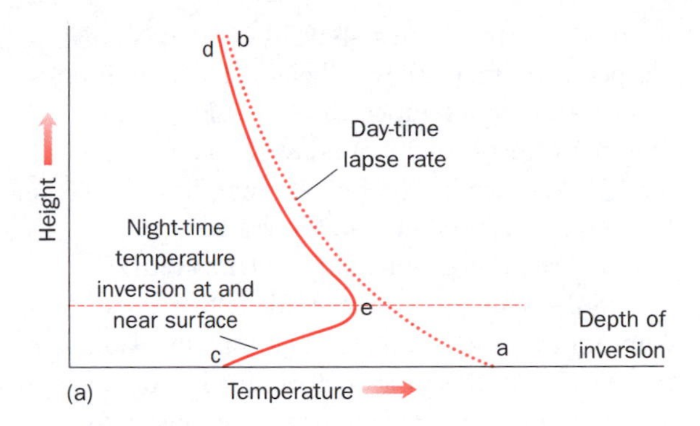

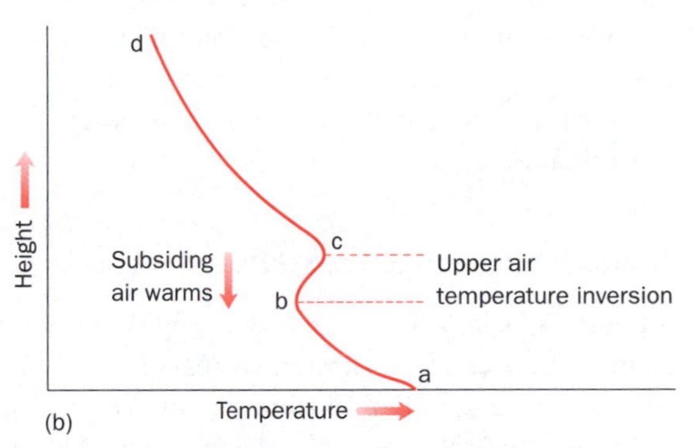

Figure 3.9 (a) Temperature profiles through the atmosphere for daytime (normal lapse rate, a–b) and night-time (temperature inversion at surface, c–e); (b) upper air temperature inversion (b–c)

3.4.3 Daily/local temperature response

(a) Inverted lapse rates

Under certain conditions, the ground surface can lose more energy and heat to its surroundings and cool more quickly than the atmosphere. When this occurs, the air away from the surface can become warmer than the air just above the surface so that temperatures increase with height. This condition is referred to as a temperature inversion, i.e. the inverse of the normal lapse rate (sections 2.3.2–2.3.3). As shown in Figure 3.9(a) and (b), temperature inversions can be created at ground level (i.e. ground surface temperature inversions) usually from ground chilling. Less commonly, atmospheric temperature inversions can develop in the air away from the surface. These atmospheric inversions can be caused, for instance, by middle layers in the atmosphere subsiding, coming under greater air pressure, and then warming more than the atmospheric layers below them.

(b) Ground surface temperature inversions

At night, especially under clear skies and calm weather, there can be a substantive loss of outgoing thermal radiation from the surface. This results in first a marked cooling of the ground surface and then by a marked chilling of the air immediately above. The air chills in contact with the cold ground by (1) conduction (heat is conducted from the warmer air to the colder ground) and (2) by reradiating thermal radiation downwards to the surface which absorbs it. There is thus an increase in temperature with height in the lower atmosphere close to the ground. Such ground-based temperature inversions are responsible for decreasing the daily range of temperature with altitude (Figure 3.9(a)). Five ideal conditions for ground-based inversions exist:

1. Long nights, i.e. in winter when there is sufficient time for a marked loss of net outgoing radiation from the surface.
2. Clear cloudless skies. This is because clouds that are made up of billions of tiny water droplets, are good absorbers of outgoing thermal energy from the surface and can reradiate large amounts of such energy back to the surface. An absence of clouds thus maximises the loss of surface radiation.
3. The presence of relatively dry air increases the amount of outgoing radiation from the surface. This is because moist air containing high amounts of water vapour has a similar radiation response to clouds and would absorb and send significant amounts of heat back to the surface.
4. Calm stable air conditions with an absence of convective or turbulent mixing. Such overturning can increase temperatures in the air at the surface by mixing warmer air from above with the colder air from below.
5. The development of a ground snow cover. This is a response to the fact that snow, especially fresh snow, has many air-filled pockets that prevent heat being conducted through it, i.e. from the surface of the ground to the air above. Snow is also an excellent radiator so that during cloudless nights radiation continues to be lost rapidly from a snow surface thereby chilling it. Moreover, snow is a good reflector of solar radiation. This means that by not absorbing much solar radiation, the surface can remain cold despite the presence of the morning and midday sun. For these reasons, a snow cover cools the air immediately above much more effectively and for a longer period than a snow-free ground surface.

(c) Valley inversions

There is little air movement or circulation where surface temperature inversions occur. This is because the cold air near the ground is denser than the warmer lighter air above resulting in a stable air condition. Low-lying locations such as valleys and depressions can intensify and add further stability to the surface inversion. Air chilled by rapid outgoing thermal radiation at night from surrounding hill land increases in density and sinks down adjacent valley slopes. Such cold air drainage quickly leads to a temperature inversion within the valley, where cold air in the valley bottom is overcapped by warmer air above (section 4.3.2). Some of the overlying warmer air will have been displaced by the cold air ponding in the valley. The speed and degree with which air can be cooled within a valley and the stability of the air under the temperature inversion (reinforced by the confines of the slopes) encourage very intense surface inversions.

3.4.4 Surface energy models

(a) Daily temperature model

Because much of the earth's surface has a positive net radiation balance or an excess energy budget, energy as a whole will tend to be transferred from the surface to the atmosphere or into the ground. Figure 3.10 shows an idealised model of how surface energy budgets, on a daily basis, are used to heat the ground and the atmosphere. During the daytime, as a result of high sun elevation and strong sunlight, incoming solar radiation absorption at the surface is greater than the loss of net outgoing long-wave radiation. The net surface radiation is thus positive during the daylight hours between A and B shown in Figure 3.10. The surplus net radiation at the ground causes temperatures to increase at the surface and then in the air immediately above.

The highest daytime temperature (T_{max}) occurs several hours after the net radiation balance reaches its peak, i.e. at point m in Figure 3.10. This is because air temperature continues to rise as long as there is a surplus of net radiation at the surface, i.e. until B and well into the afternoon. In contrast, during the early morning and evening the sun is low in the sky, producing less intense solar radiation. This is because the sun's rays have to travel through a greater thickness of atmosphere and because such rays are necessarily spread over a wider surface area and diluted when compared to a higher altitude sun. During such times and at night when there is obviously no solar radiation input (points B–C and D–A), the rate of removal of net outgoing thermal radiation (i.e. the loss of heat from the ground) is greater than the rate of energy gain from solar energy absorption. This situation creates a deficit net radiation balance. As long as the surface net radiation remains in deficit, surface temperatures will continue to decline. The minimum temperature is reached just before sunrise (point A), after which time temperatures begin to rise again in response to positive net radiation values (Box 3.5 and Figure 3.11).

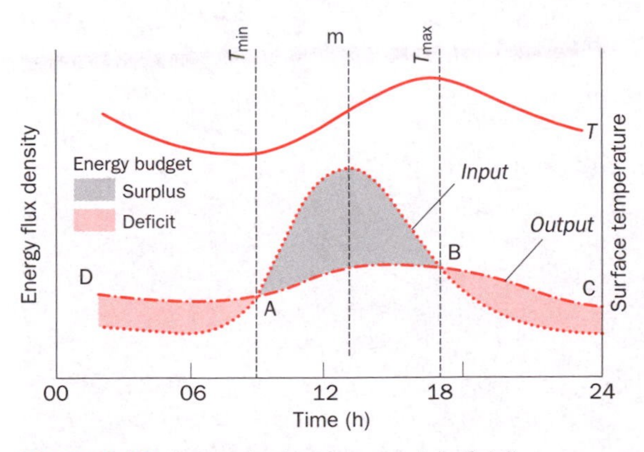

Figure 3.10 A standard surface model of daily atmospheric heating and cooling in relation to radiation gains and losses under clear skies

Key ideas

1. There is also a vertical inequality in net radiation and heating between the earth's surface where there is a surplus (heating) and the top of the atmosphere where there is a deficit (cooling).

2. Because of the surface to atmosphere net radiation gradient, heat energy is continually transferred, by convection and turbulence, from the surface to the top of the atmosphere.

3. Despite vertical heat transfers, the atmosphere in general declines in temperature with altitude. This decline of temperature with height is called the environmental lapse rate.

4. Certain circumstances can lead to a reverse in the normal vertical temperature gradient and produce a temperature inversion where temperatures *increase* with height in the atmosphere.

5. Conditions which favour temperature inversions at the surface of the earth include rapid surface cooling which chills the surface air in contact with it faster than the atmosphere above which retains its heat.

BOX 3.5

CASE STUDY

Net surface radiation and temperature models

The idealised daily surface radiation/temperature model shown in Figure 3.10 represents conditions over a dry surface, so that all heat transfer from the ground to the air is in the form of sensible heat. Because the surface is dry, no latent heat is used in the evaporation of moisture. Moreover, all sensible heat moves into the air: no heat is stored within the system, i.e. within the soil or ground. Figure 3.11 demonstrates a more realistic and complex partitioning in the way available energy is utilised on a daily basis at the surface. The models show (a) the daily variation of surface energy budget components over a bare soil (El Mirage, California) and (b) over a wet grass surface (Hancock, Wisconsin). They show that the available daily net radiation Q^* is utilised not only in sensible heat production and in heating the air directly (*H*) but is also employed in evaporation (using latent heat, *LE*) and in sensible heat storage in the ground (*G*).

At El Mirage, California (Figure 3.11(a)) over bare desert soil, most of the available surplus daytime energy (Q^*) is used in heating the soil and the air. Sensible heat flows first into the soil (*G* becomes positive just after sunrise at about 5.30 a.m.) and then into the air after the surface has begun to warm (*H* does not become positive until 7 a.m.). By comparing the daytime increases in *H* and *G* above the 0.0 horizontal line, about twice as much energy is used in heating the air as warming the soil. Because most available energy at this site is used in sensible heating of the atmosphere and very little in latent heat production (*LE*), air temperatures rise very quickly to about 28 °C by around 2.30 p.m. After this time with continual declines in the available energy (Q^*) a number of other thermal trends can be identified. These include (a) by 4.00 p.m. a rapidly cooling soil (temperatures not shown) begins to lose heat to the surface and the air above. This is shown by the negative trend of the *G* graph; (b) at about 6.00 p.m. a small amount of heat from the air begins to pass into the soil which as suggested began to cool much earlier than the air. Again a negative-trending *H* graph indicates the air to soil heat flow; (c) a negative energy budget develops in the evening and during the night between 5.00 p.m. and 6.00 a.m. and (d) air temperatures continue to decline in response to negative energy budgets until a minimum is reached (15 °C) just before sunrise.

Energy partitioning over the wet grass site at Hancock, Wisconsin (Figure 3.11(b)) is generally similar to that over El Mirage. At both sites the sensible heat flux is upward from the surface during the day and downwards to the surface at night. Air temperatures over Hancock reach a maximum during the early afternoon. However, there is one major difference. Much larger amounts of surface energy are used in evaporation (i.e. not in heating the air directly) at Hancock than at El Mirage. At the Hancock site most energy is used in this way with lesser amounts employed to heat the air and soil. An important meteorological consequence of this energy partitioning is that maximum air temperatures at Hancock are only about 15 °C, i.e. one-half that at El Mirage. This is despite total available energy budgets being comparable at both sites (maximum Q^* at Hancock is 550 W/m^2, and 600 W/m^2 at El Mirage).

These two examples demonstrate some wider meteorological principles. They help explain why moist coastal sites have smaller temperature ranges than continental locations. It is partly to do with the relative division of *LE* and *H* in the energy budget equation. At moist coastal sites much of the net radiation is used in evaporation and not in heating the air. In contrast, in dry desert areas much of the net radiation becomes sensible heat resulting in high temperatures.

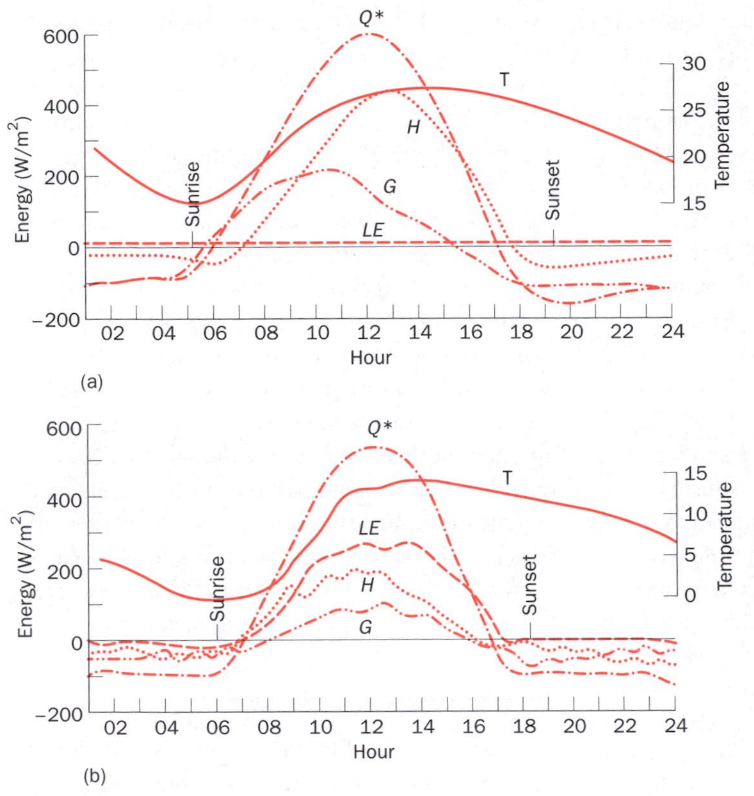

Figure 3.11 Diurnal variation of surface energy transfers and temperature over: (a) dry lake soil at El Mirage, California (9–11 June 1950); (b) irrigated alfalfa grass, Hancock, Wisconsin (27 Sept. 1957). Q^* is net radiation, H and LE are sensible and latent heat transfer, and G is the heat flow into the surface. T is the diurnal temperature response
Source: Sellers (1965)

3.5 Energy conditions and climate change

3.5.1 Energy at the surface

Climate change occurs when the climate moves beyond certain previously defined boundaries or limits. These limits that include calculations of mean weather conditions in temperature, rainfall, wind, etc. together with their extremes, have usually been based on 30-year periods, i.e. 1931–60, 1961–90. The timescale of climate change is therefore mid to long term, involving changes over decades, centuries and millennia. Climate change can occur, however, at any spatial scale affecting the world as a whole or impacting at local and regional scales. It has been shown that the net radiation condition or available energy at the surface of the earth plays an important role in determining sensible and latent heat transfers and therefore weather and climate. As a result, any change or forcing in the surface net radiation will have repercussions on climate change. Radiation forcing at the earth's surface can shift in relation to both external and internal factors. For instance, changes in the earth's receipt of solar radiation is normally classified as an *external* factor while all other forcing agents which cause climate change and variability, for example changes in atmospheric composition, land surface change and alterations in ocean currents, are classified as *internal* factors. Radiation forcing at the earth's surface can shift in five different ways to create climate change.

(a) Changes in solar output

These include known variations in the sun's internal energy output as well as solar variations due to the changing orbit of the earth around the sun. Both of these influence the amount of solar radiation reaching the earth, and thus the energy balance at the surface.

(b) Modifications to air chemistry

This involves both natural and human changes in atmospheric composition. A large explosive volcano that ejects millions of tonnes of particulate and gaseous matter into the atmosphere is an example of the former, while the accumulation of human-induced greenhouse gases in the air is an illustration of the latter. Volcanoes tend to lower energy budgets at the surface and cool the climate by increasing solar reflection from the top of the atmosphere. On the other hand, human-induced greenhouse gases are associated with the reverse process, i.e. increasing energy budgets at the surface and warming the planet by enhancing back radiation from the atmosphere to the ground.

(c) Land surface change

Land surface change (including changes in ice cover) can be the result of natural long-term processes such as continental drift and mountain-building episodes, factors that for example have been implicated in the formation of cold ice age conditions. Changes in land surface character can also be induced by human activity. Changes in recent times in the vegetation cover (i.e. deforestation) leading to farmland or desert land is known to increase solar reflectance at the surface and to reduce energy budgets there. The result has been regional and even global cooling. One region that probably has had its climate altered by vegetation removal and desertification is the Sahel region of sub-Saharan Africa. Enhanced desert albedos may have cooled the surface and the lower atmosphere sufficiently to cause other weather changes including reductions in rainfall. Human constructions such as cities also alter the background albedo and lead to the phenomenon of the urban heat island (see section 3.6.2).

(d) Biogeochemical cycling

This is the circulation of materials (nutrients and gases) from living organisms (plants, animals, microbes) to rocks, soils, oceans and atmosphere and then back again to living organisms. By changing the concentrations of gases and other materials in the atmosphere these cycles can both cool and warm the planet over time. Biogeochemical cycling has been responsible for reducing CO_2 concentration in the atmosphere over geological time by long-term weathering activity (e.g. carbonation). The gradual removal of CO_2 from the air is thought to have coincided with equally slow increases in the sun's energy output over time. This fortunate combination of increasing solar output and declining greenhouse gases (CO_2) over time is thought to have preserved temperature ranges on earth within the boundaries suitable to life. More recently biogeochemical cycling has been implicated in accelerated global warming. It is more than likely that many cold organic-rich soils in the northern latitudes will heat up under anthropogenic greenhouse gas global warming. With such warming and the accelerated organic decomposition it might produce, there are fears that these carbon-rich soils might release large quantities of greenhouse gases in the atmosphere (thus increasing global warming further).

(e) Internal energy variations

It is now recognised that the earth's climate is not and never has been constant or stable. This is because the planet itself possesses many internal energy mechanisms that are in constant flux and adjustment. By continually adjusting, these internal structures ensure continuous global climate change. The best example of an internal planetary mechanism which can change the climate are those related to the natural swings and cycles in the ocean–atmosphere circulation. Large-scale periodic variations in energy conditions at the surface of the oceans such as those attributed to El Niño events or the North Atlantic Oscillation can change regional and global climates.

Key ideas

1. Climate change occurs when the net radiation condition at the surface of the earth alters in a measurable way.

2. The net radiation can alter in five different ways to allow climate shift: changes in solar output, modifications to air chemistry, land and ice surface change, biogeochemical cycling and internal energy variation.

3.6 Case studies in climate change

Two case studies have been selected here to explore further the notion that climate change is fundamentally driven by changes in energy conditions at the ground surface and in the atmosphere above.

3.6.1 Changes in direct solar energy output

(a) Sunspots

Many studies have suggested that variations in the energy output of the sun have been implicated in climate change. There are many types of solar variations and disturbances that can influence energy output (solar irradiance). Perhaps the best known solar disturbances are *sunspots* which are large dark cooler patches (4,000 °C) on the surface (6,000 °C) of the sun. The number of sunspots, which is a measure of sunspot activity, varies a great deal. There can be periods with virtually no sunspots (e.g. 1650–1700) and others when there are as many as 200 (1950s). The numbers of sunspots seem, however, to vary in a cyclical mode with 11-year, 22-year and 80–90-year and longer cycles being identified. The length of individual sunspot cycles can also vary, so that for instance the 11-year cycle can take anything from 10 to 12 years. The presence of sunspots infers a cool sun but the opposite is actually the case. When they are present so too are many whiter (but invisible) hotter spots called *faculae* whose overall effect is to increase solar irradiance.

(b) Sunspots and climate

Sunspot activity is said to increase when there are high numbers of sunspots and/or when the length of the sunspot cycle shortens in time. This latter point means that high numbers of sunspots occur or are cycled at a faster rate. Periods with low sunspot activity cover the reverse process, i.e. when there are few sunspots and/or when the sunspot cycle time increases. Figure 3.12 shows that variations in solar output or irradiance as measured by changing sunspot *numbers* can be matched with twentieth-century surface temperature changes in the northern hemisphere (see Figure 1.1 for global surface temperature rise). The association between sunspot number and hemisphere temperatures is good, i.e. as sunspot number increases so too does the temperature plot. Nevertheless, solar activity according to this measure does not peak until some 15 years after the temperature trend. This plot tells us that solar output might not be the only factor forcing climate change. A closer match between temperature and solar conditions can be found when the sunspot cycle is used. There is almost a perfect association between northern hemisphere surface temperatures and variations in the solar cycle length (Figure 3.12(b)). As the cycle length increases towards 12 years, i.e. 1890s, 1960s and 1970s, hemisphere temperatures decline: as it decreases towards 10 years (1870–1940, 1965–90) temperatures rise. Using evidence from Figure 3.12(b), it would seem that most global warming during the twentieth century could be the direct result of variations in solar heating (see next section).

A number of other solar sunspot cycles have been linked to changes in climate (Table 3.2). One of the best statistical links is Midwestern drought in the USA which is well correlated with the 22-year cycle. Other links have been made between the 80–90-year 'Gleissberg' cycle and regional temperature and precipitation changes. Some of these connections include Icelandic ice concentration varying on an 80-year cycle, Central England Temperatures (76 years), winter severity in Europe (80 years), Beijing rainfall (84 years) and Nile floods (77 years). Solar–climate connections have also been made over a longer historic record.

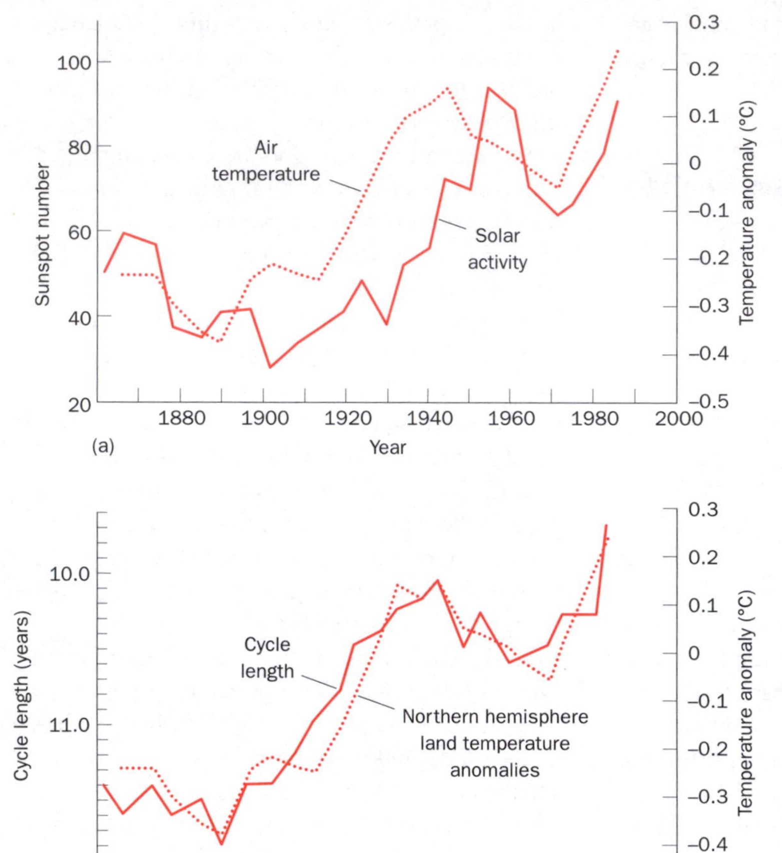

Figure 3.12 Northern hemisphere air temperature anomalies (°C, departures from the long-term mean) and levels of solar activity as measured by (a) sunspot number, and (b) variations in solar cycle length

Source: Friss-Christensen and Lassen (1991, pp. 698–700)

These include the likelihood of high solar activity being present during a particularly warm episode in medieval times (the Medieval Warm Period, AD 900–1100), and the occurrence of periodic low solar activity during a cold phase between about 1430 and 1850 known as the Little Ice Age. Between 1645 and 1715, sunspot activity was largely non-existent. During this period known as the Maunder sunspot minimum, some of the most severe winters of the Little Ice Age occurred, particularly during 1674 to 1682 and 1695–98. In the 1660s and 1670s there was frequent freezing over of the Thames River.

(c) Reassessing theory

While there are good statistical links between solar activity and climate change, part of the long-standing weakness of solar theory (the idea that variations in solar energy could affect climate was postulated in the nineteenth century) has been in terms of a causal mechanism. Some estimates of solar output during the twentieth century have shown that it has only increased by about 0.1 per cent of the solar constant, an amount judged by many scientists to be too low to be able to raise the world's temperature by 0.5–0.6 °C (see Figure 1.1).

To sidestep the problem of weak solar forcing, a number of scientists have postulated an indirect but largely unverified mechanism to boost the strength of the solar signal. They affirm that changes in the intensity of solar radiation alter the strength of the solar wind (a spray of subatomic particles from the sun). Changes in the solar wind could affect the amount of cosmic rays entering the earth's atmosphere, which could in turn influence cloud formation. More solar wind from enhanced solar radiation (this century) equals less cosmic rays, equals less cloud, and as clouds reflect sunlight, higher global surface air temperatures could be the result.

More realistically, it has been suggested that with a 0.1 per cent increase in energy output, solar

Table 3.2 Weather events and their association with cyclic variations close to the 80–90-year Gleissberg cycle

Temperature related	
Central England Temperature	76 years
Prague climate	80 years
Winter severity in Europe	80–83 years
Icelandic ice concentration	80 years
Oxygen isotopes (Greenland)	78 years
Precipitation related	
Beijing rainfall	84 years
Low level of Nile	83 years
Nile floods	77 years
Midwestern US drought	90 years
Tree ring cyclical evidence	
Lapland tree rings	90 years
Sequoia growth per decade	83 years
California tree rings	80 years

Source: Hoyt and Schatten (1997)

forcing could be responsible for about one-quarter to one-third of the observed global temperature rise over the last century, i.e. by about 0.15–0.2 °C. The importance of solar forcing, however, varies on a decadal basis (see Figure 7.14). It has been most effective in the early part of the century from 1890–1950 when it accounted for most of the temperature rise recorded, but comparatively less since the 1950s and especially since the 1970s when greenhouse gas emissions are estimated to have dominated the temperature forcing.

3.6.2 Land surface change: human agency

(a) Local and regional effects: the urban heat island

Many urban areas are 0.5–1.0 °C warmer on an annual average basis and can have daily winter maxima as much as 3–6 °C higher than their rural surroundings. The dome of heat which hovers over many cities, especially those in high latitudes, has been termed the **urban heat island** (UHI). A simple model of the thermal structure of an urban heat island is shown in Figure 3.13. The general contrast between the relatively low temperatures found in the countryside and those in the urban area can be seen. So too can the cliff-like temperature rise as the city is approached from the countryside. Other features include the uniformly high temperature field of the suburban zone, the slightly lower temperatures recorded in open areas within the suburbs, such as parkland, and high peak temperatures of the central city.

The heat island effect and wider climate response of cities can be explained by the following.

(i) Size and shape

Cities have quite a different aerodynamic shape compared to the country for instance. The tall size of city buildings and sharp angular surfaces can act as an obstacle and reduce wind speed, although strong turbulence is not uncommon in windy cities like Chicago.

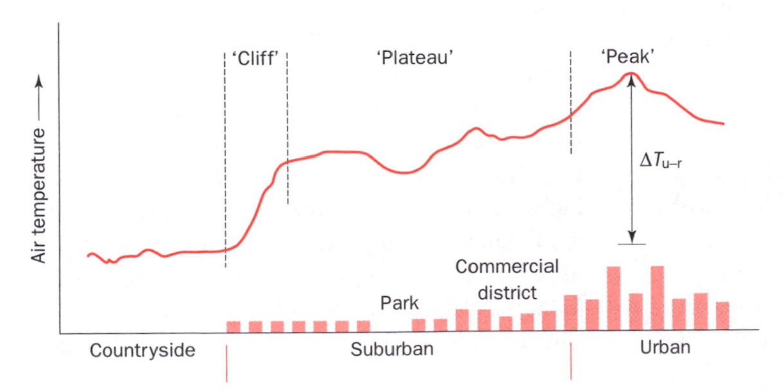

Figure 3.13 Model of the urban heat island showing temperature variations in relation to rural countryside, edge of the urban area, suburbs, parkland and the city centre
Source: Oke (1990)

(ii) Concrete jungle/urban deserts

Cities are virtually land deserts with large ground areas made up of concrete and tarmac. These materials are virtually impermeable and with lightning fast runoff in cities remain virtually dry. Because there are few wet surfaces in the city, energy is not expended in evaporation and latent heat production which would eventually be lost and removed from the city by later winds. Most energy available at the urban surface is therefore used directly in the production of sensible heat warming the city. In contrast, much more heat is used in evaporation in the countryside with their much damper surfaces of soil and vegetation.

(iii) Urban canyons

Urban canyons formed by tall buildings trap radiant energy in their walls. This applies to both incoming solar radiation and outgoing thermal radiation. Comparisons of the UHI of European cities (low canyon effect) and North American cities (high canyon effect) suggest that the density and height of buildings are significant factors affecting the rate at which the urban heat island forms. In other words the denser and taller the buildings, the more rapidly will the UHI effect develop.

(iv) Humidity question

Although the surface of the city can be likened to a desert, cities are not much different from the countryside in terms of the actual amount of water they retain in their atmospheres (absolute humidity effect). However, the higher urban temperatures serve to lower the relative amount of water that city atmospheres can hold (warm air can hold more water than cold air). This effect is expressed as a drop in urban relative humidities compared with the countryside.

(v) Urban haze

The haze that hangs over many cities is composed of air pollution (i.e. dust, smoke, sulphur dioxide, nitrous oxides, ozone). This pollution envelope increases atmospheric albedos, thus reducing sunlight at the surface. Nevertheless, a more important consequence of such pollution is to return more outgoing thermal radiation back to the urban surface, thus enhancing urban temperatures compared with those in the countryside. This back radiation effect together with other heat-enhancing effects is best seen during calm winter nights. At these times the countryside can lose large amounts of radiation to space through relatively clear skies (section 3.4.3(b)) while urban areas receive much back radiation.

(vi) Heat release effect

The residual heat release from fossil fuel burning in cities can also raise city temperatures compared to the countryside. This is particularly the case in high latitude cities in the winter period when heat release is greater and solar radiation less than during the summer. The anthropogenic energy emission over the 60 km^2 of the Manhattan district of New York is about four times the insolation falling on the area in winter.

(vii) Cloud and rain makers

The high concentrations of tiny pollutant particles or aerosols in city atmospheres may act as a surface for condensation (condensation nuclei) and thus serve to encourage cloud formation and even rainfall. Many of the millions of sulphate aerosols found in city atmospheres formed when emitted sulphur dioxide gas dissolves in water may act in this way. Increased cloud layers can drift downwind of the city and upwards in the atmosphere to produce rain in the vicinity of urban areas.

(viii) The Sunday effect

Urban areas have strong weekly air pollution cycles which in turn produce their own meteorological consequences. There are noticeable weekday highs/weekend lows for the loss of bright sunlight, the frequency of fog days and the concentrations of particulate, ozone and sulphur dioxide pollution in cities. Weekly cycles in urban air quality – characterised by high end of week accumulations of low-level ozone compared to smaller concentrations in the early part of the week – have been nicknamed the 'Sunday effect'. From this analysis it can be seen that the urban heat island will be more effective during the week than at the weekend. One corroboration of this is that in many cities, Sunday rainfall totals are often 10–20 per cent less than the average for all the days of the week, and that the effect is most pronounced in the winter.

As suggested in the introduction, the UHI effect is at a maximum during the night-time when urban temperatures can be 3–6 °C higher than the adjacent countryside. Calm anticyclonic weather conditions are also essential in generating a large UHI since windy weather diminishes the temperature contrast between urban and rural regions. There is some debate, however, in relation to the season of greatest UHI effect. Most literature on the topic suggests that *winter* is the season of maximum temperature difference. Examples of winter night-time maxima in the UHI stress the importance of winter space heating (e.g. Manhattan) and greater winter pollution over the city trapping heat at the surface. During the longer winter evenings there is also a greater length of time for the urban–rural temperature differentials to develop. Some recent work, in contrast, has underscored the importance of the *summer* as the season of greatest temperature difference between city and countryside. This work focuses on cities in dry, sunny climates where large amounts of solar radiation are stored in urban structures during summer days to be released slowly at night, keeping temperatures in the city warm compared to the countryside. It should be noted however that many cities located in hot dry climates are actually constructed to avoid heat build-up during the day. These include, for instance, the Islamic cities of North Africa and the Middle East where white-painted three- and four-storey buildings are built close to one another along narrow streets to cut out sunlight and heat retention. These cities are often cooler during the day than in the surrounding countryside.

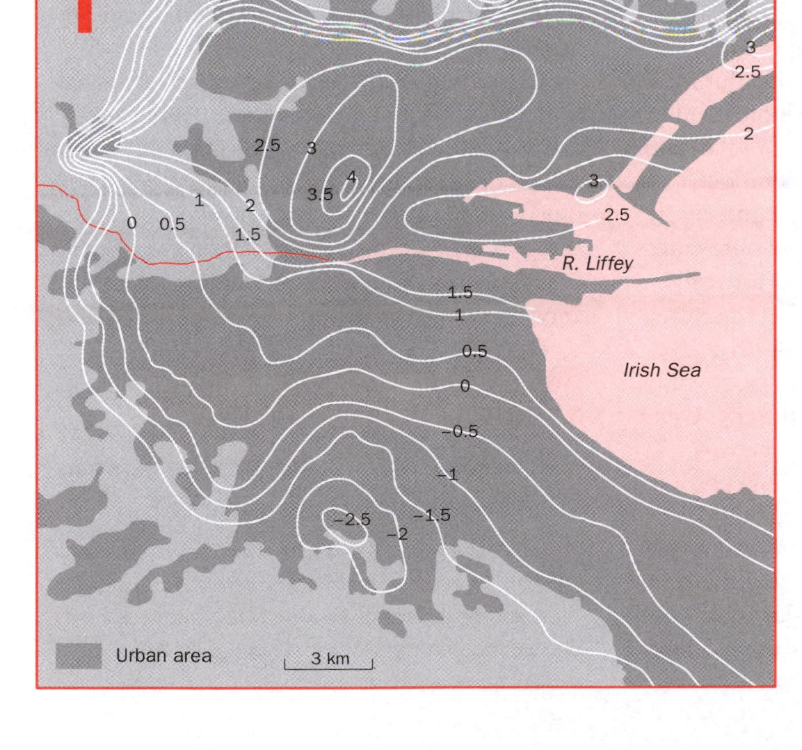

Figure 3.14 The Dublin urban heat island, for 20.00 to 01.00 GMT, 22–23 November 1983 (isotherms in °C)

(b) Case studies of the urban heat island

(i) Dublin's heat island

A field example of the UHI, measured during winter nights, can be seen for the city of Dublin in Figure 3.14. A closed island-like system of isotherms, embracing warmer temperatures, separates the city from the cooler temperature fields in the surrounding rural areas. The actual urban temperature pattern is modified and slightly 'displaced' to the north of the urban area due to the action of cold southerly breezes blowing on to the city from the Wicklow Mountains to the south. This effect can be seen in a more gradual temperature gradient over the southern half of the city and a much steeper temperature 'cliff' along the northern and north-western edge of Dublin. Peak city temperatures of 4 °C are recorded just to the north of Connell Street near Dublin's central business district. Winds advected from a relatively warm Irish Sea also help to maintain higher temperatures over the eastern dockland areas of the city. In contrast, notable cellular areas of lower temperature can be observed along the southern edge (–2.5 °C) and northern sections (–1.5 °C) of the city where more open suburbs are found.

(ii) London's heat island

In common with Dublin, observed temperature data have shown that London's heat island is most intense near the centre of the city, during night-time and under stable anticyclonic weather. The effect diminishes with distance from the centre, during daytime and under windier conditions. One recent survey of London's heat island effect, however, has shown that summer and not winter is the season of maximum urban heat intensity.

Figure 3.15(a) shows the UHI effect in central London (St James's Park) compared with conditions at Wisley, a representative rural station located 32 km to the south-west of the city. Using a 40-year temperature record, an estimate of the night-time UHI effect was calculated by subtracting the daily minimum temperature at Wisley from the daily minimum temperature at St James's Park. Similarly by subtracting the maximum daily temperature at Wisley from the maximum at St James's Park, a measure of the daytime heat island effect was calculated. Average values of the resulting night-time and daytime temperature gradients (UHIs) were calculated as shown for each calendar month.

Figure 3.15(a) demonstrates that London's heat island effect is much greater during the night than during the day, and that maximum night-time temperature contrasts of up to 2 °C occur in the month of August with minimal differences of about 1 °C in January. Long-term trends in the differences between the intensity of the night-time UHI effect and that of the daytime UHI effect from 1959 to 1998 are shown in Figure 3.15(b). Despite some variation in annual temperatures from year to year, the strength of the night-time heat island is consistently greater than that of the daytime, and this difference has increased in all seasons except winter. It can be seen that the all-time highest night-time HI effect occurred in the summer of 1976 where it reached 3.5 °C. In contrast, the lowest night-time HI effect was recorded at 0.7 °C in the winter of 1983.

One of the main reasons for London's summer peak UHI effect lies in its near continental climate (compare Dublin's possible winter maximum with its more western oceanic climate). Located in south-east England, the city is relatively sheltered from the influence of mid-latitude depressions, and as a consequence enjoys one of the driest and sunniest parts of mainland Britain. London's UHI is maximised in summer because of relatively large amounts of incoming solar radiation absorbed by the city's urban structures and their release of heat at night.

Key ideas

1. There is a good relationship between temporal variations in solar radiation input to the climate system, sunspot number/sunspot cycles and climate change.

2. A good example of climate change induced by land surface change is the raising of local temperatures by cities in 'urban heat islands'.

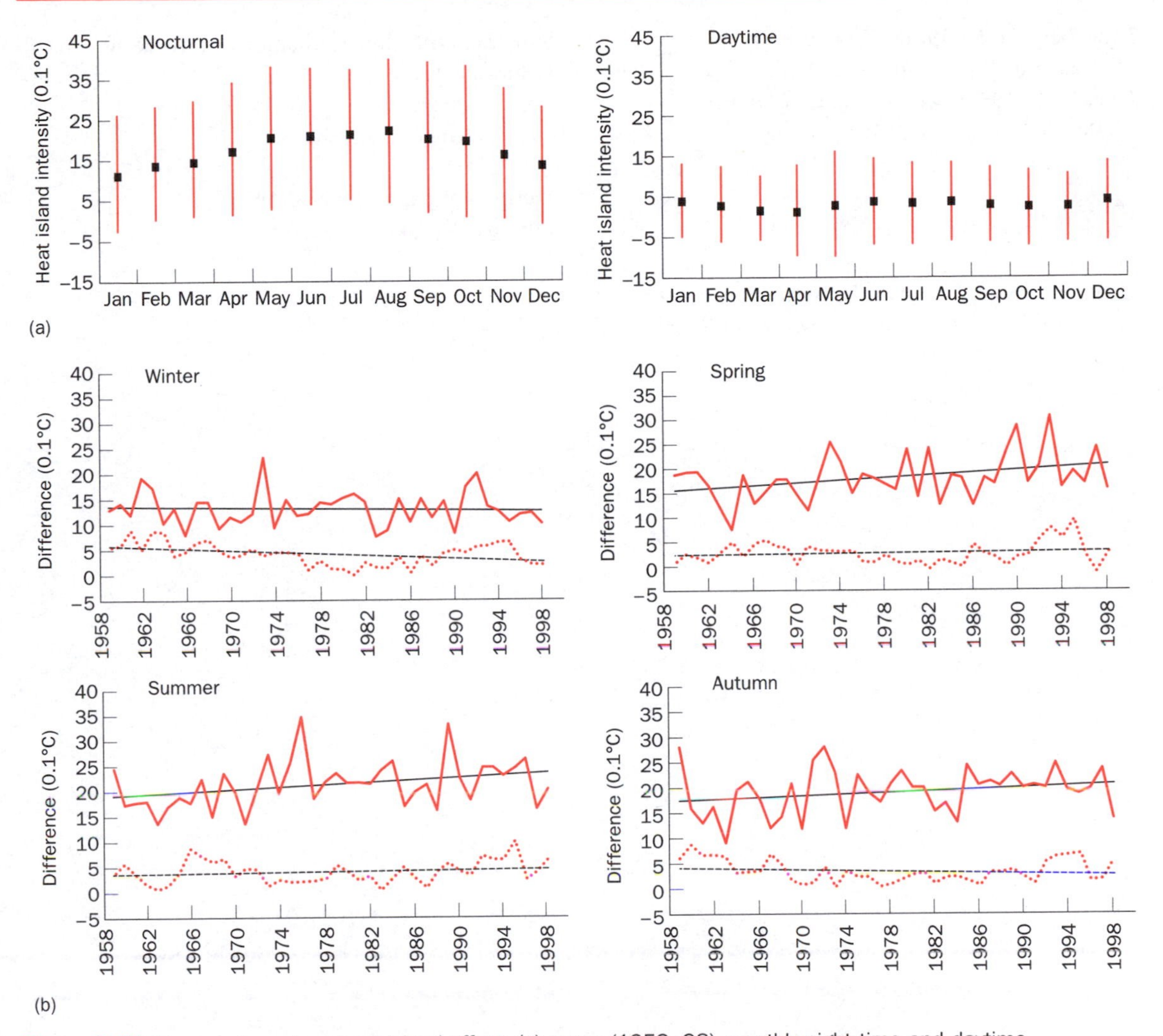

Figure 3.15 The London urban heat island effect: (a) mean (1959–98) monthly night-time and daytime temperature differences between central London (St James's Park) and the rural countryside (Wisley); (b) seasonal and year-to-year variations in the London night-time (solid line) and daytime UHI (dotted line), 1959–98
Source: R. Wilby

3. Urban heat islands are most intense near the city centre, during night-time and under stable anticyclonic weather.

4. There is evidence to show that for many northern towns with cool summer periods, winter is the season of maximum urban heat intensity, while southern towns with warmer summers often show summer to be the season of maximum heat island effect.

Further reading

Friss-Christensen, E. and Lassen, K. (1991) Length of the solar cycle: an indicator of solar activity closely associated with climate. *Science*, **254**: 698–700.

Hoyt, D. V. and Schatten, K. H. (1997) *The Role of the Sun in Climate Change*. OUP, Oxford, 279 pp.

Landsberg, H. E. (1981) *The Urban Climate*. Academic Press, London, 277 pp.
Oke, T. R. (1987) *Boundary Layer Climates*. Routledge, London, 435 pp.

Useful websites

Urban heat islands: Tulane University:
http://www.toronto.ca/cleanairpartnership/pdf/uhis_sailor.pdf

Sun cycles and climate change: University of California, San Diego:
http://calspace.ucsd.edu/virtualmuseum/climatechange2/06_3.shtml

Sunspots and climate: NASA:
http://earthobservatory.nasa.gov/Library/SORCE/

CHAPTER 4

Motions of the climate system

4.1 General circulation

4.1.1 Function and processes

When local and regional wind directions and velocities are averaged over a long period, a consistent pattern emerges at a global scale. This is what is referred to as the general circulation of the atmosphere, and reflects global-scale controlling influences. The most fundamental of these global influences is the unequal heating of the earth's surface discussed in section 3.3. There it was apparent that, although averaged over the whole globe, incoming radiation to the earth–atmosphere system was approximately equal to outgoing radiation from it, this situation did not apply for individual latitudinal zones. A net gain in radiation energy was apparent for the latitudes equatorwards of about 38°, while a net loss existed polewards of these latitudes. Since temperature is simply a reflection of incoming and outgoing energy fluxes, it is apparent that if such a situation prevailed without rebalancing, a very different planet would have resulted. If an unchecked surplus of energy were repeated year after year, the high latitudes would have become progressively colder, while the low latitudes would become progressively hotter as they lost each year more energy than they gained from the sun. Only a small zone in the mid latitudes would be habitable, sandwiched between these two hostile zones. This is not the case, implying some planetary-scale mechanism exists to export the surplus radiation energy of the low latitudes to subsidise the deficit radiation balance of the high latitudes.

The general circulation of the atmosphere is the response to this latitudinal heat imbalance, meaning that the temperature of the planet is regulated by the movement of air and oceans. As already outlined in section 3.3, a planetary-scale wind system moves 75 per cent of the heat (60 per cent as sensible heat and 15 per cent as latent heat) needing to be relocated from the zone of surplus to the zone of deficit, with a further 25 per cent moved by the wind-driven ocean currents. The profound importance of the general circulation as a means of exporting the heat surplus of the low latitudes to the heat deficit zone of the high latitudes is apparent. Indeed, in the absence of the general circulation of the atmosphere the temperature of a vertically averaged column of air in the northern mid latitudes would be approximately 100 °C colder in winter than it is at present (Figure 4.1).

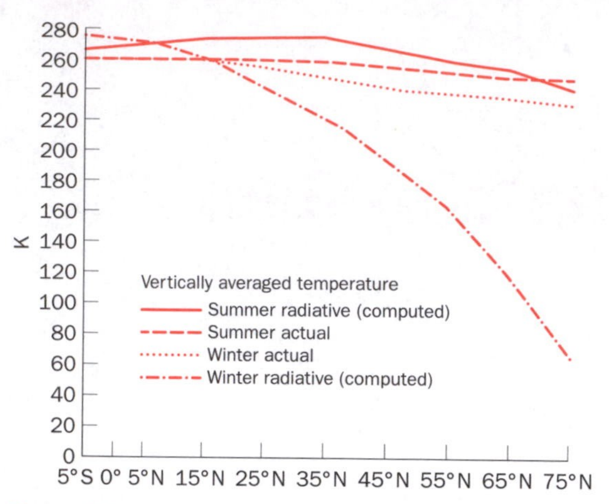

Figure 4.1 Actual and theoretical temperatures by latitude for summer and winter. The theoretical temperatures are those which would be likely in the absence of heat transport from the low to the high latitudes and show the extremely low temperatures that might otherwise occur

Source: Robinson and Henderson-Sellers (1999, p. 108)

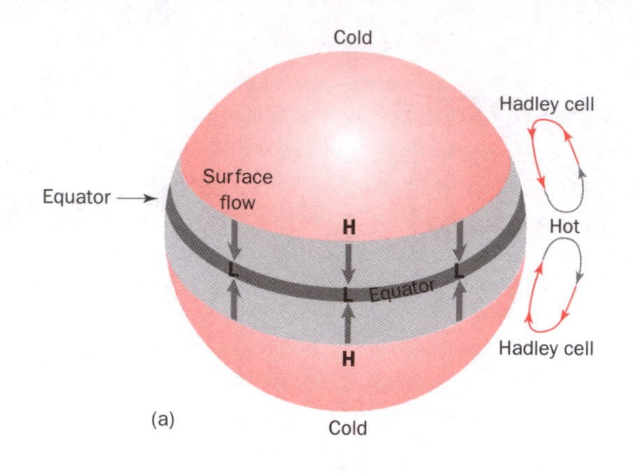

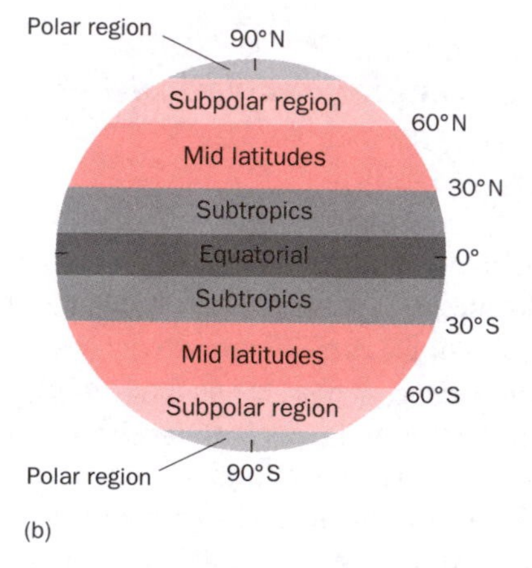

Figure 4.2 Hadley's unicellular model of the global wind circulation: (a) single cell Hadley circulation; (b) main global climate zones

4.1.2 Vertical mixing and the cellular models

(a) Unicellular model

A heat source in low latitudes and a heat sink in high latitudes prompted many early attempts to conceptualise the planetary wind system as a series of convective cells involving the transport of heat out of the equatorial zone. Air rising over the warm low latitudes, moving polewards at higher levels in the atmosphere, and descending over the cool high latitudes was the principal mechanism suggested by Edmund Halley in 1686. His vision was of a stationary earth and a slight modification to this model was suggested by George Hadley in 1735 when he pointed out that the rotation of the earth would result in deflection of the airflows. For the surface components, the Coriolis deflection would give rise to north-easterly winds in the northern hemisphere and south-easterly winds in the southern hemisphere. Hadley further suggested that the descending limb of this large convective cell would take place around the subtropics as radiative cooling stripped the pole-bound flow of its energy and buoyancy (Figure 4.2). The Coriolis deflection prevents this upper airflow making progress towards the poles, causing it to bank up and seek an outlet downwards. Thus, as the piling up (or convergence) of air aloft occurs at these latitudes, subsidence takes place and downward movement of air generates the great subtropical high pressure systems which ring the earth around 30° N and S of the equator.

At first sight a general circulation based on one convective cell in each hemisphere driven by rising air over the equator agrees well with observational evidence. The unstable, rising air columns of the equatorial zone produce the hot, wet climates which support the tropical rainforests. Water is generally abundant, characteristically being delivered by the mid afternoon thunderstorm. Areas close to the equator may experience two wet periods in the year as the overhead sun makes its annual journey to and from the tropics. By contrast, the stable descending limb of what became known as the Hadley cell produces the cloudless skies and desert climates of the subtropical anticyclones. Water is always in short supply. Cloudless skies in summer make these areas the hottest on the planet, though nights can be cold, as energy escapes freely without hindrance from water vapour or clouds. Between these two regions of predominantly vertical air motion in the subtropics and equatorial zones, a reliable zone of surface winds known as the trade winds returns the air to the Intertropical Convergence Zone (ITCZ) to begin the cycle again. In the age of sail, mariners were very familiar with this threefold zonation of the climates of the low latitudes. Typically on a voyage south from Europe they first encountered calm conditions around 30° N where they were forced to jettison (or eat) non-essential cargo, such as livestock. These latitudes thus became known as the Horse Latitudes. The trade wind belt then provided reliable north-easterlies which disappeared as the equator was approached. Again calm conditions prevailed – the Doldrums – until the circulation system of the southern hemisphere was picked up. Of course all these zones migrated north and south with the seasonal shift in the earth's attitude towards the sun, producing seasonal climatic variation. Some regions will have wet 'equatorial' climates for part of the year and dry trade winds for the other, producing a seasonal wet period. This may be short-lived on the desert margins but quite reliable on the equatorial side. Others will be so entrenched in their respective circulation zones that abundant rainfall or aridity may exist all year round. As a result, the wet equatorial rainforests, the wet and dry tropical grasslands, the less wet and much drier semi-arid grasslands (the Sahel), and the permanently arid or desert are the four biomes which result from this zonation in many parts of the tropics.

With any conceptual model there is a constant process of fault finding, rejection and replacement. The unicellular model had a major flaw. In both hemispheres the only hypothesised surface winds have an easterly component. On a rotating earth there is inevitably frictional coupling between moving air and the surface, which in this case rotates from west to east. Over time, air dragging along the surface in the opposite direction to the direction of rotation of the earth would exert an unbalanced force which would slow the earth's rotation and extend the length of day beyond 24 hours. On the other side of the coin, a rotating world in which only easterly surface winds existed would have brought the atmosphere to rest within a couple of weeks. Clearly this is not occurring, so easterly winds must be balanced somehow by winds from the opposite direction in the general circulation.

(b) Tricellular model

William Ferrell in 1856 suggested a three-cell model which retained a direct (Hadley) cell at low latitudes but introduced an indirect cell in the mid latitudes. This cell was termed indirect because it was not driven directly by heating and cooling, but rather as a consequence of the energy supplied by the descending limb of the Hadley cell. Not all of the air spiralling out at the surface from the downward limb of Hadley's cell, he argued, would move equatorwards as the trade winds. Some of the descending air would generate surface flow towards the poles (giving a south-westerly wind at mid latitudes in the northern hemisphere) and ascending air when it encountered the cold polar air masses along the contested zone which would

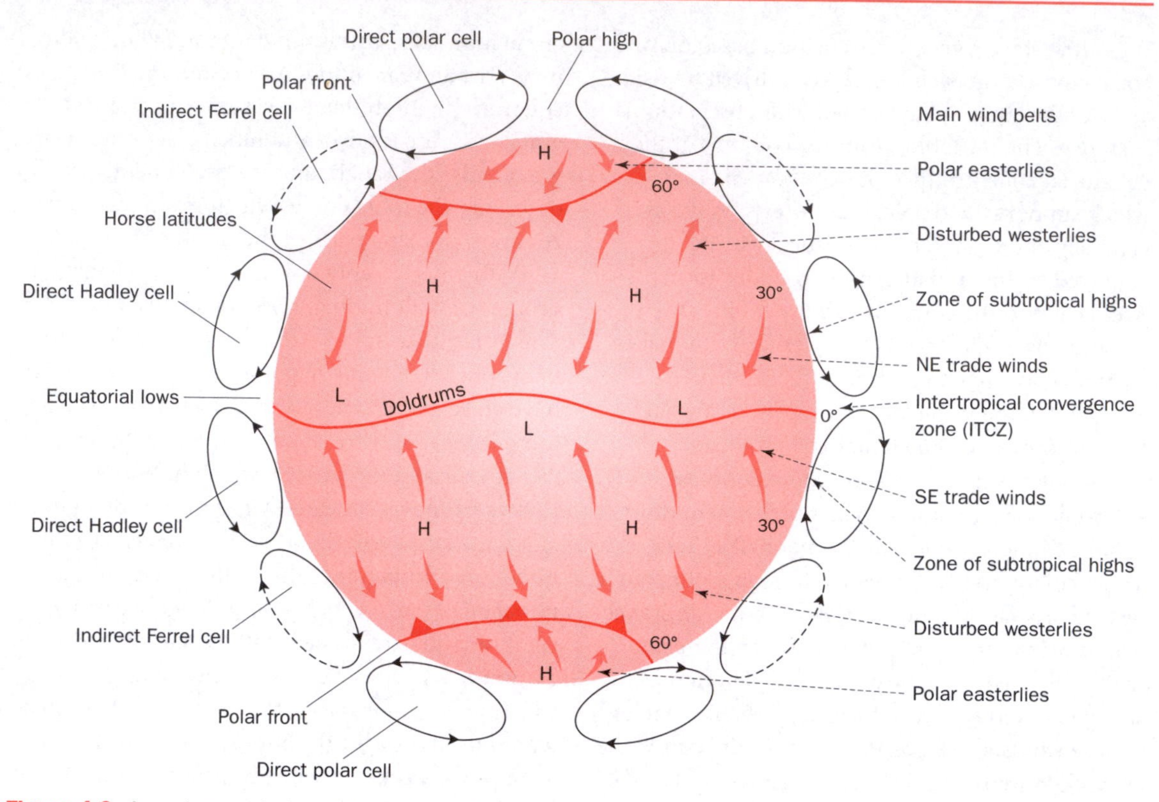

Figure 4.3 A modern evolution of Ferrel/Rossby's tricellular model of the global wind circulation

be later known as the Polar Front. A return flow aloft would complete this indirect cell. In addition, a third small direct cell driven by the outflow of cold dense air from the poles would supply polar air to the Polar Front Zone. Here it would encounter, and be mixed with, tropical air, thus helping to achieve the exchange of heat required to rebalance the earth–atmosphere's energy budget. For almost a century the tricellular model was thought to be the answer to understanding the general circulation of the atmosphere, particularly as it seemed to explain well the tripartite tropical climatic zonation (wet, seasonally wet/dry, dry). It also offered a reason for the prevailing westerlies of the mid latitudes and the easterlies of the high Arctic and Antarctic (Figure 4.3). But it too had a fatal flaw which led to it joining the earlier versions on the scrap heap.

Simple reasoning should suggest that the high-level return flow of Ferrel's indirect cell back towards the equator should experience Coriolis deflection to the right in the northern hemisphere and to the left in the southern hemisphere. This would produce north-easterlies and south-easterlies respectively in the higher reaches of the troposphere. However, by the 1940s, upper air motions were being regularly investigated by radiosonde balloons. These revealed the existence of strong westerly winds throughout the mid latitudes. This was in stark contrast to the easterlies the indirect cell was suggesting. Although Rossby in 1945 tried to explain this as a consequence of the influence of westerly flow aloft in the two adjacent direct cells, the explanation was unconvincing. More radical reassessments were necessary.

When a model is seen to be defective, it is time to reappraise the fundamental underpinnings on which it was based. Reality should never be confused with model output. If a model fails it is flawed in some way and requires further consideration. Maybe the flaw lies in the basic assumptions? In the case of the unicellular and tricellular models the underlying assumption was that vertical overturning of air in huge convective cells restored global energy imbalances. Could an alternative means of heat transfer be involved?

4.1.3 Horizontal mixing and the wave models

To try and simulate the general circulation experimentally, a series of laboratory experiments were set up involving a flat circular pan filled with water several centimetres deep. This dishpan was placed on a turntable. The axis of rotation could be considered analogous to the pole, while around the edges of the pan an electric element was positioned to heat the water at the outer extremities. The contrast between this 'equator' and the 'pole' was further emphasised by the insertion of a refrigerating agent, a cooling cylinder, in the middle of the rotating fluid. Aluminium powder was added to provide a visual tracer to show the manner in which the water moved (Figure 4.4).

With a slow rate of rotation, a simple 'Hadley' convective cell was seen to develop, with warm heated water rising to the surface at the edges and moving slowly to the 'poles'. However as the speed of rotation was increased to levels comparable with the speed of rotation of the earth, the vertical overturning cell was replaced by a series of waves and eddies. At times, the amplitude of these extended a substantial proportion of the distance between the 'equator' and the 'pole' and the waveforms themselves moved as distinctive entities. A predominantly 'west-to-east' flow of water occurred as the rotation increased, particularly in the 'mid latitudes'. Although carried out in a laboratory, the looping waves shown by the dishpan experiments seemed to replicate motion in the atmosphere as indicated by the tracking of weather balloons at high altitude in some mid-latitude regions. From a combination of observational and experimental evidence it became clear that wave motions and westerlies are central components of the mid-latitude general circulation.

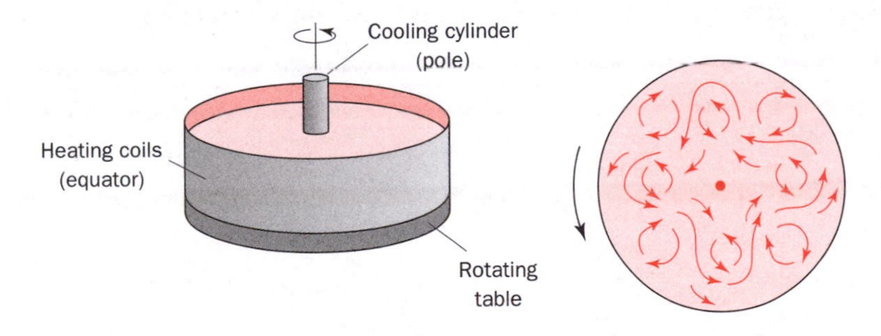

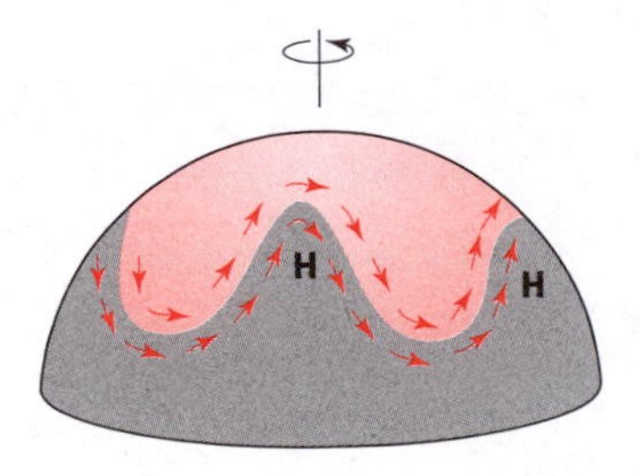

Figure 4.4 The dishpan experiment: the outside is heated to simulate the equator and the centre is refrigerated to simulate the poles. Rotation at a speed similar to the rotation of the earth produces west-to-east wave motions
Source: Ahrens (2000, p. 284)

4.1.4 The upper westerlies

(a) Why do they exist?

The upper westerlies are caused by the unequal heating of the earth. This produces a latitudinal temperature gradient which causes pressure surfaces to slope polewards. This can best be visualised by considering two columns of air, one at the equator, the other at the pole. The column of air at the equator is warmed and has expanded. The tropopause is up to 16 km above the surface at the equator. By contrast, the cold dense air at the pole produces a shrunken column. The tropopause may be only 8 km above the surface. To ascend from the surface to a given pressure level, e.g. 700 hPa, would require a much lengthier ascent over the equator than to reach the same pressure level over the poles. At any comparable stage in this ascent, the pressure would be higher for the equatorial ascent than for the same altitude for the polar ascent. A polewards-directed pressure gradient thus exists (Figure 4.5). Air moving along it however would be subject to the Coriolis force, deflecting it to the right in the northern hemisphere (and left in the southern hemisphere) to produce a westerly wind. The greater the equator–pole temperature difference, the greater the pressure gradient and the stronger the westerly winds will be. This is why the upper westerlies are stronger in winter. It is also apparent that the pressure gradient increases with height. Westerly wind speeds therefore generally increase with altitude.

(b) Wave motion explained

The reason for wave motions in the westerlies relates to the changing magnitude of the Coriolis force at different latitudes and also to the effects of the earth's rotation. In their movement pattern, westerlies conform to a principle known as the conservation of absolute vorticity (Box 4.1). A consequence of this is that as the Coriolis force increases, the cyclonic vorticity diminishes and vice versa. On a poleward-bound limb of a westerly wave, therefore, as the Coriolis force increases with latitude the airflow curvature becomes less cyclonic and more anticyclonic, bending the flow equatorwards. On an equatorward-bound limb, as the Coriolis force decreases, the cyclonic vorticity increases, bending the flow back polewards (Figure 4.6).

Viewed on any particular day, a collection of these long waves or Rossby waves may be

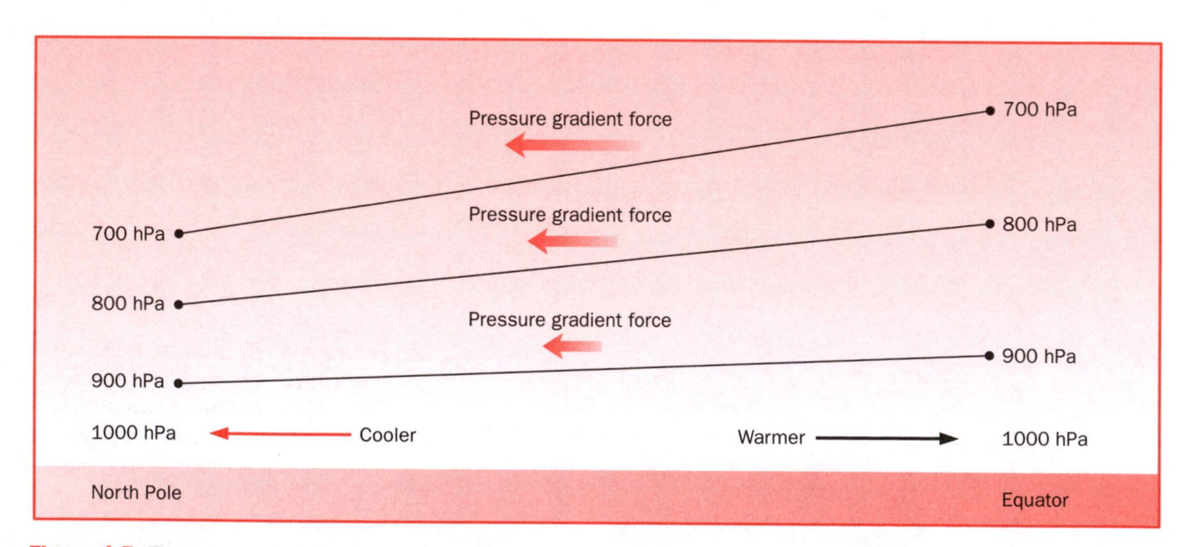

Figure 4.5 The upper westerlies in cross-section

THINKING FURTHER

BOX 4.1

Conservation of absolute vorticity

Vorticity is a measure of the rotation of a fluid. For air moving relative to the surface of the earth its rotation has two components. First, there is the turning of the air relative to the surface, termed its *relative* vorticity. Second, there is rotation by virtue of the fact that any motion is taking place on a rotating earth. This latter form of vorticity is termed 'earth vorticity', a quantity which increases towards the poles and decreases towards the equator. The sum of these two kinds of vorticity is termed 'absolute vorticity' and is a property which tends to be conserved, i.e. it persists or remains constant for a parcel of air not undergoing divergence or convergence. If an air parcel moving for example in a Rossby wave undergoes an increase in earth vorticity it therefore experiences a corresponding decrease in relative vorticity and vice versa. A constant trade-off between the two occurs.

The concept of the conservation of absolute vorticity can be used to understand why the westerly flow in the upper atmosphere tends to form waves. Consider the air moving south-eastwards towards position 1 in the Rossby wave shown in Figure 4.6. As it moves south it is encountering decreasing earth vorticity, and so must increase its relative vorticity to compensate. This manifests itself as a cyclonic curvature which swings the air north-eastwards towards position 2. As the air parcel moves into higher latitudes the earth vorticity again increases and the relative vorticity decreases. An anticyclonic curvature ultimately develops which turns the parcel back towards the south-east towards position 3 and in this way a series of long waves results, ultimately girdling the earth.

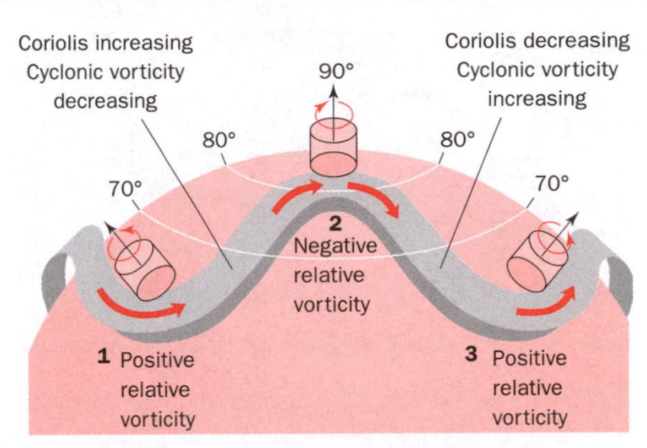

Figure 4.6 Rossby waves. As the air moves polewards the Coriolis force (earth vorticity) increases. Cyclonic vorticity therefore decreases, eventually turning the air around into an equatorward trajectory. As the air moves equatorwards, the Coriolis force decreases. Cyclonic vorticity increases, eventually turning the air around into a poleward trajectory

seen in the upper air circulation moving around a circumpolar vortex. The most common arrangements entail between two and five long waves which typically travel slowly from west to east. The waves may however become stationary or even retreat from east to west for a time. Despite the behaviour of the long waves, ripple-like shorter waves may propagate through them at a much faster rate (Figure 4.7). These short waves tend to deepen when they approach a trough in the long wave and also tend to deepen the long wave trough itself. Such a disturbance to the upper airflow may initiate or intensify the development of depressions and so they are keenly watched for on upper air charts. The long waves tend to be the main steering mechanisms, however, and thus their orientation and behaviour exert a major influence on surface weather systems.

(c) Global heat transfer

The westerly waves (sometimes called Rossby waves) are important for the general circulation because they provide a means of mixing warm and

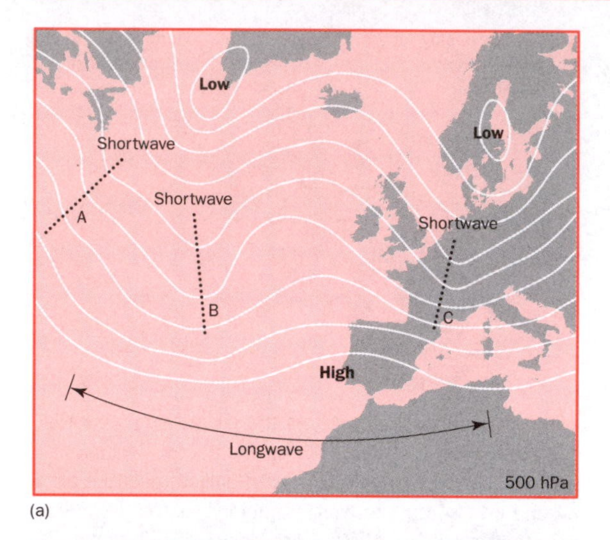

(a)

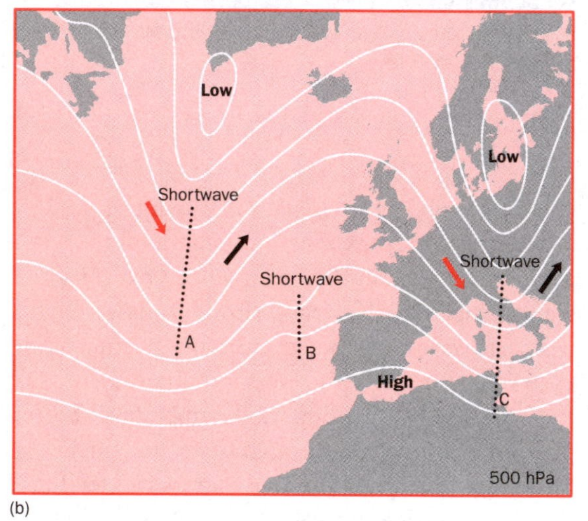

(b)

Figure 4.7 Short waves embedded within long waves in the westerlies: (a) day 1; (b) day 2 (24 hours later)

cold air masses in a horizontal fashion. On some occasions the westerly waves are poorly developed. Amplitudes are low and airflow is predominantly west to east. Rapid movement of weather systems from west to east at the surface is facilitated. A string of depressions may affect surface locations in such a situation, sometimes characterised as a High Zonal Index. At other times the westerly waves exhibit extreme undulations of large amplitude. Large cells of air may be detached from the main flow and may become stranded in very different air bodies. These huge eddies thus provide the horizontal mixing method so important in rebalancing the earth's energy budget (Figure 4.8). This was the mixing method which earlier attempts to model the general circulation failed to consider. Often these Low Zonal Index situations may be associated with weather anomalies because of the way the anticyclones they encourage to form act to 'block' the normal progression of weather systems from west to east at the surface. Persistent blocking of this kind is invariably associated with seasonal droughts, such as in the summers of 1976 and 1995 for example in Britain and Ireland. Of course it equally may steer depressions into areas where they do not normally venture, bringing other kinds of weather anomalies to such areas.

(d) Favoured tracks

It might be thought of as reasonable that if a large number of maps showing the westerly waves on individual days were averaged, the result would show no waves. Troughs and ridges might be expected to average out at a particular location over time producing a concentric circle pattern around the circumpolar vortex. That this does not occur on long-term average charts means that there must be some locations where ridges and troughs are more frequent or persistent. On the January 500 hPa mean chart two troughs are particularly prominent (Figure 4.9), one over North America and one over eastern Siberia. In both cases, the features are downwind of major mountain barriers which lie at right angles to the westerly flow: the Rockies and the Tibetan Plateau. Both of these mountain barriers extend over 3 km into the atmosphere and clearly can disrupt the westerly flow. Indeed the troughs they create are further reinforced by the coldness of these continental interiors in winter. (Bitterly cold columns of air result in pressure levels being lower at comparable heights than in warmed 'expanded'

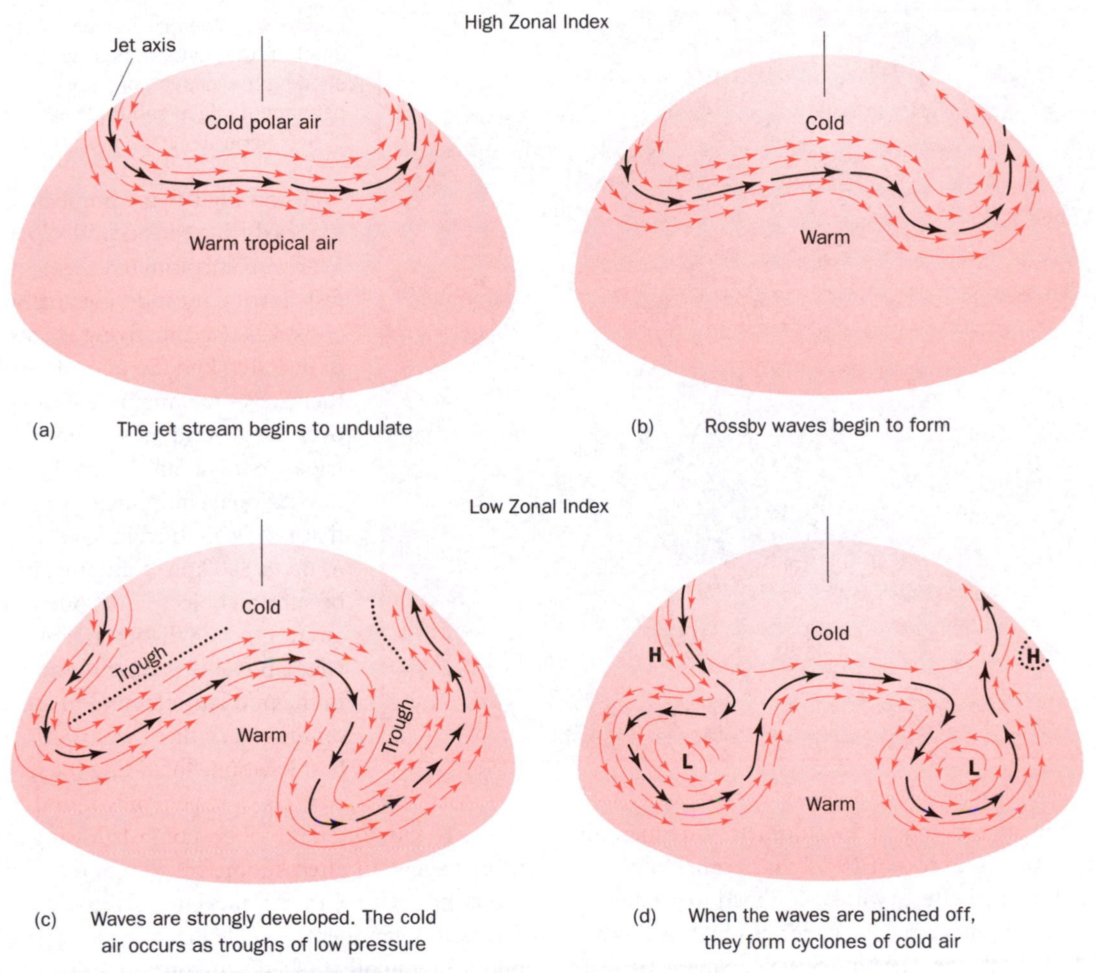

Figure 4.8 The development of different modes of Rossby waves
Source: Strahler and Strahler (1997, p. 134)

columns.) In contrast, weak ridges are apparent over the relatively warm eastern Atlantic and eastern Pacific oceans at the same time. These seemingly preferred positions for long waves in the westerlies thus reflect both thermal and dynamical influences which are apparently capable of anchoring moving waves more readily when they reach key locations such as these. This is an important control on regional climates given the role of ridges and troughs in influencing cyclogenesis, a topic dealt with later in this chapter.

4.1.5 The jet streams

(a) The polar jet

The logic of an equator–pole temperature gradient producing westerly winds in the upper atmosphere can also be employed to explain fast-flowing westerlies produced by steep gradients over shorter distances. The most marked thermal gradient in the atmosphere occurs abruptly where tropical air and polar air come into contact in the middle latitudes. This battleground zone is known as the

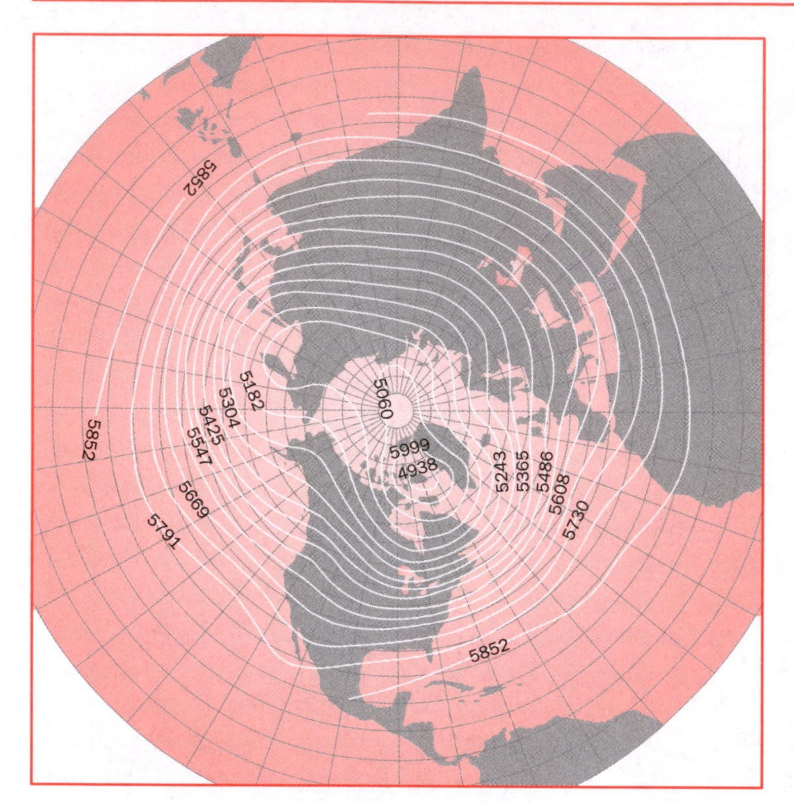

Figure 4.9 Average January 500 hPa chart for the northern hemisphere showing the location of semi-permanent ridges and troughs
Source: Neiburger (1982, p. 217)

It was high-flying bomber planes during the Second World War who encountered these fast-flowing meandering airflows (Box 4.2). Despite flying at speeds of over 400 kph the pilots found themselves making little headway over the ground on westbound trips. On eastbound trips their navigational equipment indicated that they were travelling at speeds of up to 800 kph. Naturally, it became an objective of commercial aircraft to ride the jet stream where possible on eastbound trips in the mid latitudes and to avoid it on the return flight westwards. The best locations for riding the jet streams were generally found to be above 10 km. Often the ribbon of fast-flowing air is only a few hundred kilometres wide and a few thousand metres in vertical extent. Skilled aviation forecasters can identify likely locations for their pilots. Frequently the jet is discontinuous or bifurcates into two branches. It thus does not stand out on charts of the mean circulation and must be pinpointed on individual charts by trained meteorologists for aviation purposes.

polar front and, over a few kilometres, temperature differences in excess of 10 °C may occur. Very fast-flowing westerly winds, confined to a small latitudinal range, may be generated by this thermal difference, particularly in winter (Figure 4.10). Near the top of the troposphere the temperature gradient between polar and tropical air is greatest, and this is where the polar front jet stream is to be found.

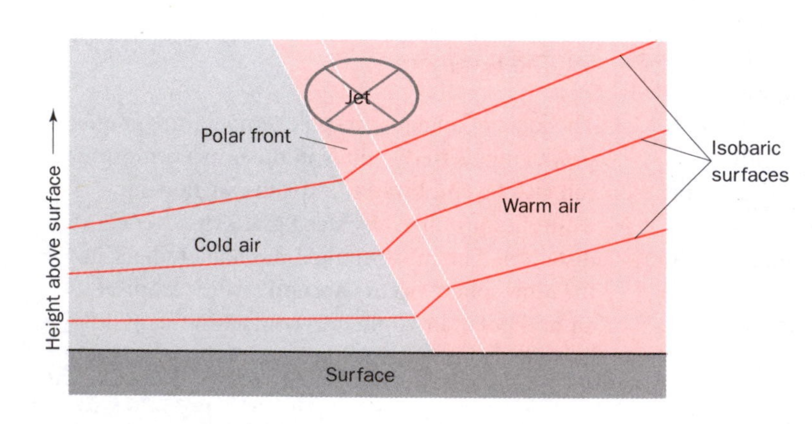

Figure 4.10 The jet stream forms as a result of a large temperature discontinuity over a very short distance, e.g. above the polar front

BOX 4.2

CASE STUDY

The jet streams

On 1 November 1944 the crew of the first American reconnaissance flight over Tokyo during the latter phases of the Second World War found themselves encountering surprisingly strong headwinds. This reduced their speed considerably and made them very vulnerable to being shot down. The pilot noted:

> I found myself over Tokyo with a ground speed of about 70 mph (115 kph). This was quite a shock, particularly since we were under attack from anti-aircraft guns and were a sitting duck for them. Obviously the headwind was about 175 mph (300 kph).

The Japanese military, however, were well aware of the jet stream as a result of research carried out before the Second World War. They had already developed high altitude bombs and incendiary-carrying balloons designed to float eastwards on the jet stream across the Pacific and set about launching these. Approximately 9,000 were launched of which about 1,000 reached America – some as far east as Michigan and Texas. To avoid panic in the civilian population, particularly regarding the possibility of the balloons carrying chemical or germ warfare weapons, media reports of damage were heavily censored.

(b) The subtropical jet

A less pronounced jet stream, the Arctic front jet, can sometimes be detected where polar and Arctic air masses interact. However, a more persistent feature is the subtropical jet which exists in conjunction with the maximum poleward extent of the upper limb of the tropical Hadley cell. The subtropical jet can be thought of partly as a response to the much higher altitude and therefore colder troposphere the tropical air is encountering at this latitude. It can also be explained as a result of the faster over-the-ground speed that it possesses by the time it reaches 30° N or S. Air at the equator moving with the earth's rotation has a speed of 490 m/s, while air moving with the earth's rotation at 30° N has a speed of 425 m/s. Moving air from the equator to 30° N or S therefore gives it a relative westerly speed of 65 m/s even before any other considerations are taken account of. Using this analysis, the subtropical jet as with the polar jet, flows in a westerly direction (i.e. moving from west to east). However, as if to prove the importance of the thermal gradient in generating airflows, an easterly (moving from east to west) tropical jet exists over India and Africa in summer. This is produced when the desert zones are warmer than the cloudy moister areas nearer the equator. As a consequence, high pressure is found at altitude over the deserts and low pressure at the same elevation over the equator, thus reversing the normal latitudinal gradients on the globe.

The jet streams provide additional dynamism for the mixing processes of the atmosphere. Since the polar jet in particular is closely related to the mixing zone between polar and tropical air masses, the meandering of the jet axis is an effective agency for achieving the interpenetration of energy-rich and energy-poor air bodies. However, the configuration of the westerly waves which the jet stream gives expression to has an even more profound influence on surface weather features. They are the chief determinants of how and where depressions will be formed, and how and where they will travel to.

Key ideas

1. The global circulation of winds and ocean currents is necessary to export the heat surplus of the low latitudes to the heat-deficit zone of the high latitudes. Without this mechanism only a relatively small part of the earth would be habitable.

2. Early models of the planetary wind circulation based on one convective cell in each hemisphere

were flawed. Tricellular models were also found to poorly explain the mid-latitude upper westerlies.

3. Horizontal mixing of warm and cold air via waves in the westerlies provided an alternative perspective. These westerlies were driven by thermal differences between high and low latitudes and the waves they exhibit, Rossby waves, are instrumental in the formation and guiding of weather systems.

4. Westerly waves show preferred positions due to interaction with the underlying topography, particularly the Rockies and Tibetan Plateau.

5. Fast-flowing ribbons of air within the westerlies, known as the jet streams, occur close to the temperature discontinuity of the polar front. The large temperature gradient over a short distance is responsible for wind speeds of about 400 kph at heights of 10–12 km.

4.2 Synoptic systems: upper air and surface weather

4.2.1 The concept of vorticity

The tendency for airflow in a westerly wave to turn right and left in response to changing values of the Coriolis force was discussed earlier. These changes in direction were shown to relate to a principle known as the conservation of absolute vorticity (see Box 4.1). The overall rotation of a fluid, such as air, has two components to it. First, there is rotation relative to the earth's surface, such as in turning left or right in wave motion. This is termed its relative vorticity. Second, there is rotation due to the fact that the air is moving in an atmosphere which is itself moving with the rotation of the earth. This is termed 'earth vorticity' and increases in magnitude the higher the latitude the air is moving at. One way of thinking about earth vorticity is to imagine a column of air at the equator over a 24-hour period. It essentially does a somersault on itself, with no rotational forces at right angles to the column. At the North Pole over a similar time, the air column is twisted round on itself as the earth spins. Maximum earth vorticity thus occurs at the poles and minimum earth vorticity is at the equator.

If air is flowing in the same direction as the earth is turning (anticlockwise in the northern hemisphere) both relative and earth vorticity are reinforcing each other and the two forms of vorticity act to increase absolute vorticity. In the northern hemisphere, therefore, air rotating cyclonically (anticlockwise) is described as having positive vorticity. The opposite applies to anticyclonic (clockwise) rotation, which is said to have negative vorticity.

4.2.2 Jet streams, vorticity and surface weather

The vorticity changes exhibited by airflow through a Rossby wave have important ramifications for the encouragement or discouragement of weather systems many thousands of metres below at the surface. Note from the conservation of angular momentum (Box 4.3) that increasing spin is achieved when the turning skater pulls in her arms (or converges in on herself). Increasing absolute vorticity is thus associated with upper air convergence. Vice versa, decreasing absolute vorticity is associated with upper air divergence. These concepts of convergence and divergence are important characteristics of airflow. A simple way of conceptualising them is to think in terms of air arriving at a point faster than it is leaving (convergence) or leaving faster than it is being replaced (divergence).

The fastest-flowing part of the jet streams are usually in the trough of a Rossby wave. But as the air exits the trough, where the isobars are tightly packed, the pressure gradient reduces slightly and the strengthening Coriolis force exceeds it. This causes divergence to occur. A zone of divergence can thus be identified just downwind of an upper level trough. This divergence draws air upwards from the surface to 'feed' itself and thus encourages the initiation of a low pressure centre or depression at the surface. Inspection of the upper air wave pattern may thus enable forecasters to identify places favourable for depression

BOX 4.3

THINKING FURTHER

Conservation of angular momentum

An object moving in a straight line possesses a property called 'linear' momentum which is measured by the product of its mass times its velocity. Similarly an object following a curved path has 'angular' momentum, defined as the product of its mass, velocity and the radius of curvature of the arc it is following (*mvr*). At the equator the rotation of the earth means that air is carried eastwards at a speed of 1,764 km/hr. If the earth was to suddenly stop rotating this air would have sufficient momentum to keep moving until friction brought it to a stop.

Angular momentum is conserved in a system unless external twisting forces are applied to it. This means that the product *mvr* at one location must equal the product *mvr* if the air mass concerned is relocated to another location. For example, if air is moved polewards from the equator to 30° N the smaller radius of curvature at that location means that the air must move faster than the surface is rotating to conserve its angular momentum. This manifests itself as a strong westerly wind of 234 km/hr.

The conservation of angular momentum is demonstrated in many areas where spinning actions are involved. Perhaps the most easily visualised is that of the spinning ice skater. With arms fully stretched out she spins quite slowly. As she reduces her radius of curvature by drawing in her arms, the speed of spin increases dramatically. To slow up she extends her arms once again.

formation or cyclogenesis. The opposite state of affairs applies to areas immediately downwind of an upper level ridge. Here convergence associated with decreasing absolute vorticity promotes subsidence, inducing pressure to rise at the surface and anticyclone formation to occur. The behaviour pattern of the upper air thus provides strong controls on the surface pattern by both enabling excavation of rising air from surface lows to proceed easily or not, and also by producing conditions conducive to their formation or otherwise (Figure 4.11).

Mid-latitude cyclones

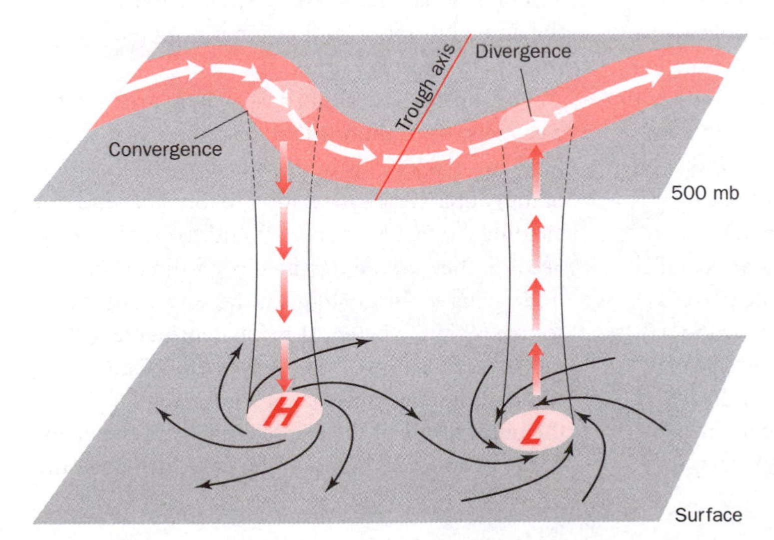

Figure 4.11 Upper air convergence/divergence and surface weather system formation

Key ideas

1. Relative vorticity, due to local rotation, and earth vorticity, due to the turning of the earth, may act to reinforce or negate each other. Reinforcement occurs when air in the northern hemisphere is rotating anticlockwise, producing a state known as positive vorticity, and vice versa.

2. Positive vorticity is associated with diverging air motions aloft. Negative vorticity is associated with converging air motions aloft.

3. A zone of divergence downwind of an upper air trough creates favourable conditions for uplift and formation of a depression at the surface. A zone of convergence downwind of an upper air ridge encourages subsidence and anticyclone formation at the surface.

4.3 Regional and local scale motions

4.3.1 Air masses

Air masses are large parcels of air bringing distinctive weather conditions to a region or country. An air mass is defined as a mobile body of air whose physical properties, such as temperature, moisture content, lapse rate, stability, visibility and general weather, are relatively homogeneous over hundreds of square kilometres. The weather features of an air mass change more abruptly with height in the atmosphere but they do so at a more or less constant rate. Thus under ideal barotrophic conditions (i.e. there is a continuous fall in pressure and temperature with height), the isotherms and isobars of the air mass are parallel with height in the troposphere. The degree of uniformity of an air mass is governed by the character of the underlying surface, the age of the air mass, and the effects of different heat and moisture surfaces encountered along its track. An air mass can be separated from another air mass having different weather features by a transitional zone or boundary called a front. The frontal separation of contrasting air masses and the related topic of mid-latitude frontal depressions are discussed in section 9.2.2.

(a) Source regions

A source region is a large area of the earth's surface with relatively uniform physical characteristics, for example ocean surfaces, ice-covered areas, deserts, large plains. These are effective areas for forming air masses if the air above them is slow moving or gently subsiding. This gives time for the transmission to the overlying air of surface-related characteristics. Thus, provided an air mass remains stationary over a given geographical area for a period of 3–7 days, the body of air and the surface will gradually attain a degree of equilibrium through radiation exchanges and vertical mixing, with the air mass reflecting the nature of the surface. Such conditions are generally restricted to regions of slowly divergent flow from extensive thermal and dynamic high-pressure cells. As shown in Figure 4.12, these conditions are found in polar and subtropical regions where there are large semi-permanent anticyclones (high-pressure zones). In order for the air mass to become a significant feature of the general circulation, however, the source area must be of the order of a million square kilometres. These criteria restrict the annual duration and number of potential sources to a few key areas.

For most practical purposes, there are only two basic sets of conditions that describe an air mass. First, air masses may be classified by the location of the source region in relation to *thermal* conditions, yielding Arctic (A), polar (P) and tropical (T) air. Second, they can be classified according to the underlying *moisture status* of the surface of the source region, whether dry, i.e. continental (c) or moist, i.e. maritime (m). As shown in Figure 4.12, some of the most important source regions lie at the outer limits of the Hadley cell, as in the vicinity of the subtropical highs. Others are formed at the

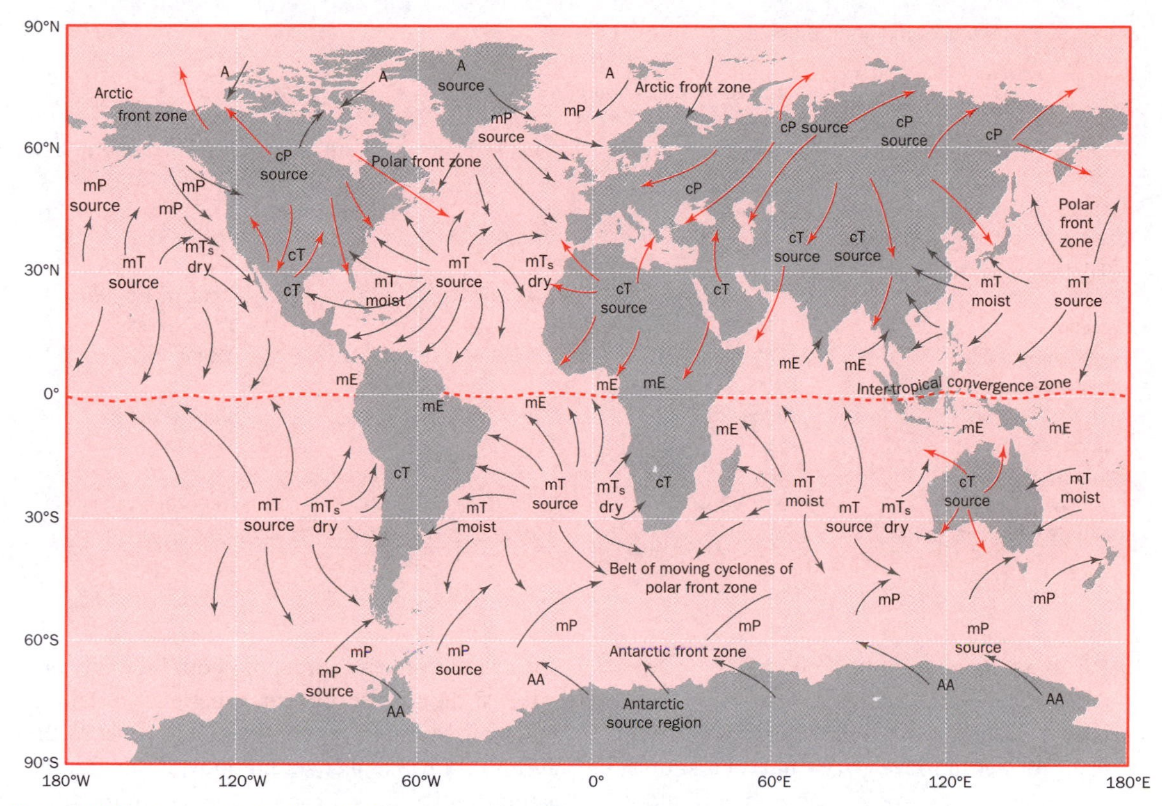

Figure 4.12 Global distribution of air mass source regions and main air mass trajectories. Key: mE maritime equatorial, mT maritime tropical, mT_s maritropical subsiding, cT continental tropical, cP continental polar, mP maritime polar, A arctic, AA antarctic

Source: Robinson and Henderson-Sellers (1999, p. 161)

edges of the polar continental margins, as in the continental anticyclones of northern Canada or Siberia and the Arctic Basin. Air masses forming over snow-covered regions are characterised by marked cooling and stability near the surface and are described as continental polar (cP) and continental arctic (cA) air respectively (Table 4.1). Further radiative cooling and subsidence from upper atmospheric levels under high-pressure conditions typically result in marked temperature inversions from the surface to the 850 hPa pressure level (about 800 m elevation). Cloud formation is limited in both the cP and cA types due to their extreme dryness, low temperatures and stable atmospheres. These air types are less extensive in summer than winter because continental heating over Siberia and northern Canada extinguishes the sources of cold air, while weakening the Arctic Basin source.

Other common source areas in the northern hemisphere include the oceanic subtropical high-pressure cells (mT) of the Atlantic and Pacific and, in summer, the warm surface air masses of the continental interiors (cT) of central Asia and North America. Maritime tropical air masses are characterised by high temperatures, air subsidence, stable conditions and high humidity in the lower layers over the oceans (Table 4.1). The winter continental type (cT) is also a warm and stable air mass, but dry by virtue of the North African source area (Figure 4.12). Because of the low humidity, development of cloud and precipitation is unlikely despite some instability generated by summer surface warming.

Table 4.1 Properties of major air masses

Air mass	Characteristics at source: Temperature (°C)	Characteristics at source: Specific humidity (g/kg)	Typical properties
Maritime tropical (mT)			
summer	22–30	15–20	Mild, moist
Maritime polar (mP)			
winter	0–10	3–8	Cool, moist
summer	2–14	5–10	Cool, moist
Continental tropical (cT)	30–42	5–10	Warm, dry
Continental polar (cP)			
winter	–35 to –20	0.2–0.6	Cold, dry
summer	5–15	4–9	Cold, dry
Continental Arctic (cA)			
winter	–55 to –35	0.05–0.2	Very cold, very dry
Maritime equatorial (mE)	~27	~19	Warm, very moist

(b) Air mass movement and change

Variations in the atmospheric general circulation cause air masses to move away from their source regions. As air masses move they gradually adopt the features of the surfaces over which they move. For instance, as polar air masses move southward in the northern hemisphere, they encounter increasingly warm surfaces and undergo heating from below. They may also increase in humidity if they move over ocean surfaces. When cP air from northern Canada comes into contact with the North Atlantic Drift its humidity is typically increased through evaporation into the lower layers. The associated instability caused by surface heating (i.e. increase in lapse rates) together with rises in humidity can spread through a considerable thickness of the air mass causing some cloud formation. The weather of southward-moving cP air is characterised by squally showers and bright interludes, with changeable cloud cover dominated by cumulus and cumulonimbus. The instability promotes gusty, turbulent winds leading to good visibility and air pollutant dispersal. If the cP air reaches the central Atlantic it will have been transformed into a cool, moist maritime polar (mP) type. This air mass is often of neutral stability or stable at the surface, but can experience instability following heating over continental Europe. As the air mass moves as a coherent body, it can also assume anticyclonic properties and even modify the pattern of Rossby waves as it encounters the main path of the westerlies.

In contrast, tropical air masses tend to become more stable with lower lapse rates as they move from warmer southern locations to cooler northern ones. In this case, surface cooling may produce a temperature inversion that acts as a lid to further cooling through the vertical profile of the air mass. Air quality is also reduced under conditions of limited vertical mixing within relatively stagnant air resulting in poor visibility and higher pollutant concentrations (see Box 9.1). If the mT air has high moisture content, surface cooling may result in advection fogs or low stratus cloud as commonly experienced in the English Channel during spring and early summer. However, rainfall formation in mT air depends on the forced ascent of the air by the underlying terrain, or on vertical mixing due to high wind speeds. In comparison, cT air is generally hot and dry, but potential instability and light rain can be triggered if moisture is acquired over the Mediterranean, en route from North Africa.

Finally, as an air mass ages, there is a progressive loss of its original identity, a decline in the rate of energy exchanges and dissipation of the associated weather phenomena. Eventually the air will assume the properties of the surrounding environment, particularly at lower levels.

(c) *Air mass frequency and weather*

Air mass changes thus depend on first, the trajectory of the air with respect to the source area, second, the degree of heat and moisture exchanges with contrasting underlying surfaces, and third, the extent of mechanical mixing. When considering the effect that air masses have on specific weather and climate, another important factor in addition to air mass modification, concerns the frequency with which different air masses invade an area. Year-long hot and dry conditions prevail in the Sahara Desert because this region is continuously dominated by continental tropical (cT) air for 12 months of the year. In the UK, however, at least five or six major air masses affect the region, bringing different assemblages of weather at different seasons. Thus the number and balance of contrasting air masses at different times of the year need to be included in any regional climate analysis. The influence which different air masses and their frequencies have on the climate of the UK is discussed in Box 4.4.

4.3.2 The heating of land and sea

Air masses that originate over land tend not only to be drier than their maritime counterparts but also warmer (in summer) and colder (in winter). This is because different substances do not absorb incoming radiation uniformly (section 3.3). This is particularly the case with land and water surfaces as anyone walking from hot beach sand to cool sea water will testify to. Uneven absorption of radiation energy by water and land surfaces arises for a number of reasons. First, water is transparent, allowing radiation to penetrate the surface and dissipate heat over a greater volume than is the case with land, where heat builds quickly at the surface. Second, water may move when heated, setting up convection currents that

BOX 4.4

CASE STUDY

Air masses and UK weather

The following case study demonstrates the role of air masses in the weather and climate of the UK. At least five major air masses influence the climate of the region, including as shown in Figure 4.12, maritime Arctic (mA), continental polar (cP), continental tropical (cT), maritime tropical (mT) and maritime polar (mP). A distinction is often made between colder maritime polar (kmP) air which arrives over the UK directly from the north-west and warmer maritime polar (wmP) air which has a longer western track over the North Atlantic before arriving over the UK from the west.

1. Maritime tropical (warm and cloudy)

Maritime tropical air usually approaches the UK from the south-west, its source region being the subtropical Atlantic Ocean (Azores high-pressure cell) or even the Caribbean. The air usually reaches the UK on south-westerly winds, commonly in the warm sector of a depression (see section 9.2.2) or around the periphery of an anticyclone located over Europe (Figure 4.13, no. 4). During its passage across the Atlantic the air is cooled from below as the air passes over a progressively cooler ocean, and therefore becomes more stable. While it cools down little ▶

of its moisture is lost, so it reaches south-west England or western Ireland almost saturated. This gives dull, warm, clammy and overcast (i.e. with stratus cloud) weather. On the coasts, sea fog is common. Bodmin Moor, Dartmoor, Dyfed, western Ireland and Scotland can be shrouded in mild damp conditions whether it be summer or winter. Inland in summer the low stratus may be heated sufficiently for evaporation to remove the clouds. This is especially the case with adiabatic warming in the lee of hills or mountain ranges where the clouds sometimes break up giving a lot of sunshine. Favoured locations like north Somerset, North Wales, Northumberland and the Moray Firth can bask in spring-type weather even during midwinter.

2. Maritime polar (cloudy and showery)

Polar maritime air is the most common type of air mass affecting the UK and Ireland (Table 4.2). It flows towards the UK on north-westerly winds from the Arctic regions of Greenland and northern Canada. As shown in Figure 4.13, no. 2, it reaches the UK from the west or north-west having swung around the western side of a depression. As the cold air travels over the relatively warm sea it is warmed from below and becomes unstable. Unstable air as shown in section 2.3.2 tends to produce convection and so cumulus and cumulonimbus clouds and showers are likely in maritime polar air. As this air mass is typically unstable it is associated with deep vertical mixing in the lower layers of the atmosphere resulting in pollution dispersal and good visibility.

Maritime polar air brings cool conditions in winter but especially in summer. In winter convection is initiated over the Atlantic and showers affect the coasts by day and night, spreading inland if winds are strong. The Scottish and Welsh mountains often shelter the eastern side of Britain which lies in the rain shadow of this westward-moving air mass. Breaches in the mountain and hill ranges can let maritime polar weather through to interior and eastern regions. When the flow is westerly, winter showers can cross Glasgow and central Scotland to reach Edinburgh and Fife; when north-westerly, some showers push through the Cheshire Gap to reach Birmingham and sometimes London. In spring and summer showers and short-lived thunderstorms can be set off in the maritime polar air over inland areas by surface heating.

3. Maritime Arctic (cold and showery)

After a low has crossed eastwards over the UK, winds veer to a northerly direction and true Arctic air may reach the UK. As shown in Figure 4.13, no. 1, when high pressure is situated to the north-west of the UK and low pressure to the north-east, maritime Arctic air can be drawn south over the UK. This air mass is similar to mP air but tends to be more unstable, colder and drier. It releases showers of rain, snow, sleet or hail especially on northern coasts and over high ground. The Highlands of Scotland usually take the brunt of a 'screaming northerly' with blizzards on low and high ground (see Box 2.4). Elsewhere there tends to be clear skies and good visibility.

4. Returning maritime polar (mild and cloudy)

The classic returning maritime polar airflow occurs when a large depression is situated to the north-west of the UK (Figure 4.13, no. 3). With this situation, air reaches the UK after travelling around the southern edge of the depression and the winds are between south and south-west. This air mass brings cool and mainly fair weather to the UK, but showers can occur in western districts. This happens when the already unstable air (remember it has been heated on its journey southward over the Atlantic) is heated over inland surfaces.

5. Continental polar (dry and cold in winter, warm in summer)

When there is a high pressure area north or north-east of the UK, clockwise-moving winds can bring cold continental polar air masses from Scandinavia or Russia over the UK (Figure 4.13, no. 6). Bitterly cold easterly winds on the southern limb of the high pressure cell can drop temperatures below the average in winter. Higher temperatures occur with this air mass to the west of the UK, i.e. on the lee slopes where adiabatic warming occurs (see section 2.3.2). In summer, conditions are warm and dry. As the moisture content is low the weather is usually fine and sunny. This is often the case when the air mass has made a short journey over the English Channel from Calais to Dover. However, if the air has made a longer sea track across the North Sea from Denmark to Scotland, surface cooling can result in clouds over the sea (see the 'haars' in section 2.3.3 and Plate 2.2) and can make eastern districts cloudy with perhaps drizzle or snow flurries. Visibilities for this air mass vary. They are good when the air has come from Scandinavia but moderate to poor when the air originates in the industrial regions of central and eastern Europe (see Box 9.1). Table 4.2 shows that, as a direct airflow, this air mass is an infrequent visitor to the UK. However, this air is often entrained within anticyclones and can thus affect the climate of the UK over significant periods during the summer and winter.

6. Continental tropical (warm/hot and dry)

Continental tropical air is also a low frequency air mass across the UK (Table 4.2) and does not appear during the winter months. On the small number of days that it does invade, it usually comes with south-easterly or southerly air streams. As shown in Figure 4.13, no. 5, this airflow occurs when high pressure lies directly to the east and low pressure directly to the west of the UK. As the air originates from North Africa and often travels over Spain and France before reaching the UK, there is little surprise that it is hot and dry. In summer, easterly winds from central Europe and the Ukraine could be included in this category in view of the high temperatures there in summer. The heatwaves and long dry sunny weather produced by cT air can be broken by thunderstorms. This happens when moisture is picked up and trapped at medium heights in the atmosphere on its way north over the Mediterranean. If there is sufficient instability in the atmosphere convection can trigger deep cloud development, i.e. altocumulous castellanus with high-level turrets. These are often the precursor of intense thunderstorms which can occur during the day or night.

Air mass frequency

Air mass analysis can offer a better description or interpretation of a region's climate if air mass frequencies are included. Table 4.2 shows us that the frequency of air mass invasion at Stornoway in north-western Scotland is very different from that at Kew Gardens (London). Stornoway has a much wetter climate than London because this station receives most of its weather from the moist North Atlantic in the form of Arctic, mP and mP-returning air masses. The drier climate of London is seen in its higher frequency of continental air masses including cP and cT. London is also drier because it is affected by many more anticyclonic days which bring descending air flows, stable weather and dry conditions. It is more difficult to decide which station has the colder climate from the air mass data shown. No seasonal data are shown, but in summer, Stornoway will certainly experience cooler conditions than Kew because of the prevalence of mP air and the low frequency of warm cP and cT air masses compared to London. Nevertheless, London is often colder than Stornoway in winter because of its more continental/less maritime location and the resultant higher frequency of cold continental air masses (cP), which are often entrained within anticyclones.

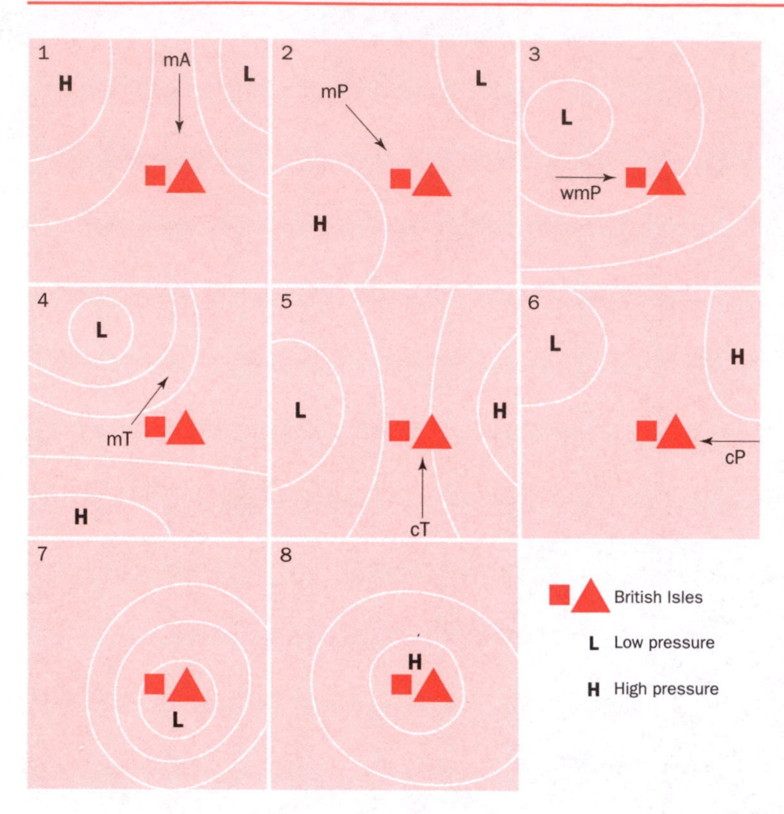

Figure 4.13 Schematic illustration of the main air masses affecting the UK and Ireland. The main pressure fields and circulations associated with air mass movement are also shown
Source: Musk (1988)

mix the heat well throughout its volume. On land this cannot occur; the heat is confined close to the surface. A moist surface also takes radiation energy preferentially for evaporation, leaving less behind for building heat than does a land surface. Finally, as a material, water has also a higher specific heat, meaning that more energy is required to produce the same temperature change as would a land surface. What this means is that land surfaces, and continental areas, warm quickly in spring and summer, and cool quickly in autumn and winter. Oceanic surfaces, and maritime land areas, display a more sluggish temperature response.

Table 4.2 Percentage frequency of air masses affecting Kew (London) and Stornoway (north-west Scotland)

	Kew	Stornoway
Maritime Arctic (mA)	6.5	11.3
Maritime polar (mP)	24.7	31.5
Maritime polar returning (wmP)	10.0	16.0
Continental polar (cP)	1.4	0.7
Maritime tropical (mT)	9.5	8.7
Continental tropical (cT)	4.7	1.3
Air masses entrained within anticyclones, particularly cP	24.3	13.8
Air masses in vicinity of fronts (depressions), particularly mT and mP	11.3	11.8

Source: Belasco (1952)

(a) Monsoon winds: an introduction

Unequal heating of land and sea can be partially invoked to explain the greatest wind reversal in the world, the monsoon wind reversal which occurs over large parts of the lower latitudes including West and East Africa, the Middle East, India, South-East Asia and China. This is a dramatic reversal which even in north China produces 60 per cent of the winds from the west, north-west and north in January and 60 per cent of the winds from the south, south-east or south-west in July, a remarkable contrast in wind regime. Though the monsoon reversal has many other linked causes (section 11.2), it is also a consequence of the heating up of the landmass more quickly during the spring and summer than the surrounding ocean. Mean temperatures in Delhi, for example, in May rise to 33 °C, and warm air masses (cT) with ascending air flows begin thereafter to form a low-pressure centre over much of southern Asia. Cooler, but still relatively warm and moist air masses (mE) from the southern tropical ocean (see Figure 4.12), dramatically invade the continent in June as the monsoon breaks, bringing the life-giving rains. By October the land–sea temperature contrast has begun to reverse. The landmasses have begun to cool rapidly in the interior of the continent, with high pressure reasserting itself over much of Asia by November. The sea is now warmer than the land and wind reversal occurs, bringing the outblowing winds from the interior of the continent and dry conditions (Figure 4.14).

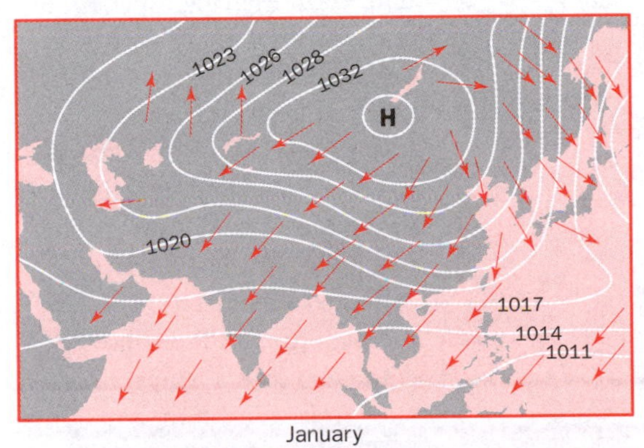

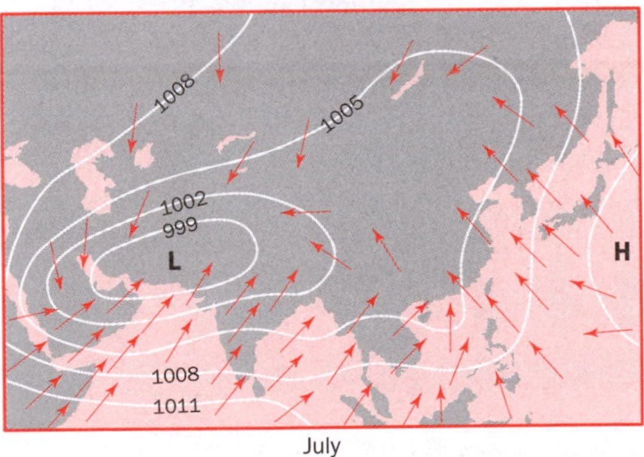

Figure 4.14 The Asian monsoon as a response to land–sea heating and consequent pressure contrasts
Source: Strahler and Strahler (1989, p. 90)

(b) Land and sea breezes

The principle that uneven heating can produce air movement also applies at a smaller scale. In coastal areas the warming of the land during the day produces a thermal low over the land, while the cooler marine air remains at a higher pressure. A pressure gradient directed towards the land exists and typically in the early afternoon a flow of onshore air commences. At night the thermally induced pressure gradient is reversed and an offshore flow may develop (Figure 4.15). This land and sea breeze is not an everyday occurrence, however, and depends on slack regional pressure gradients and thus calm conditions existing. When these enable local heating inequalities to dominate air movement, penetration of the marine layer inland may be considerable and is often marked by a linear cloud pattern, sometimes referred to as the sea-breeze front.

(c) Valley winds: anabatic and katabatic airflows

Local winds may also develop as a result of unequal heating and cooling of slopes. In valleys, especially sides facing the sun, intense heating can occur. This causes rising air motions, well known by birds and gliders as elevators to get them to higher altitudes. To replace the air rising in these thermals, air is drawn uphill from lower down in the valley. An upslope wind, termed an anabatic wind, is created. During the night, however, the hillslopes cool more quickly than the more sheltered valley floors initially. This cooler, denser air begins to flow down the slope to collect in the valley floor, especially if the slope profile of the valley is concave which discourages air from resting midslope. This cold air may then drain downslope as a katabatic wind. Speeds are quite slow, of the order of 1 m/s, and quite shallow, of the order of a few centimetres in depth. The latter means the flow may become ponded behind obstacles and may build up to some depth before breaking out as mini avalanches of cold air (Figure 4.16) heading

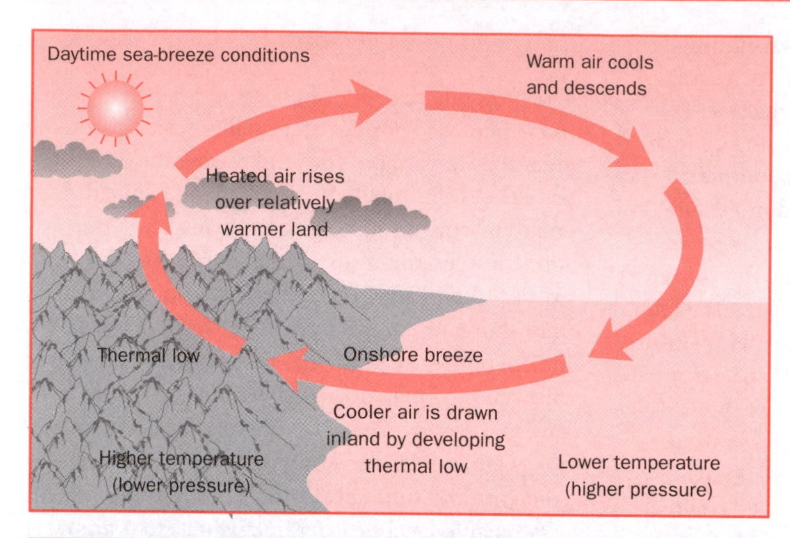

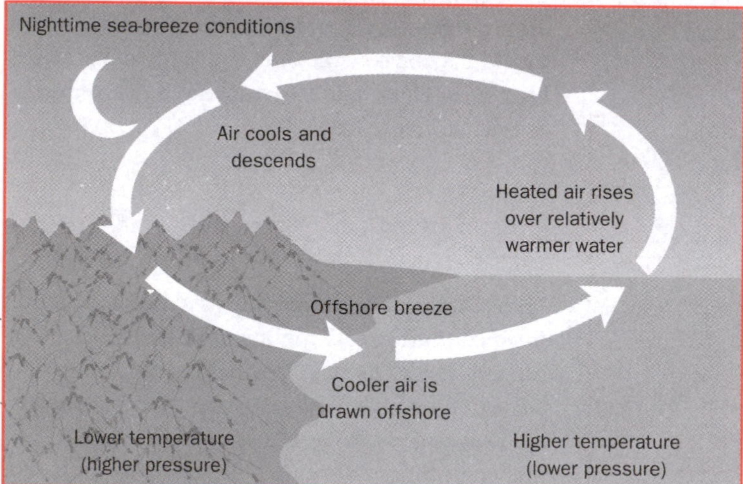

Figure 4.15 (a) Sea breeze; (b) land breeze
Source: Christopher (1995)

downslope. This katabatic wind may result in the formation of ponded chilled air in hollows producing fog and frost on cold, calm, clear nights and may also be implicated in air pollution episodes. In some places where there is no outlet for the valley the downslope flow of katabatic air may result in extremely low temperatures being experienced and a reversal of the normal vegetation zonation with altitude. One of the most dramatic instances of this can be observed in an enclosed limestone hollow in the Austrian Alps which on one night early last century exhibited a temperature of −25 °C in the bottom of the hollow as compared with a temperature close to freezing at the top of the hollow (Figure 4.17).

Katabatic air movement also explains why the slopes of a hill are often warmer than the valley floor, a phenomenon well known to fruit growers and viticulturalists in frost-prone areas. Assistance to downslope winds can occur following the passage of a depression. Such a situation draws cold polar air out of the mountains along valleys

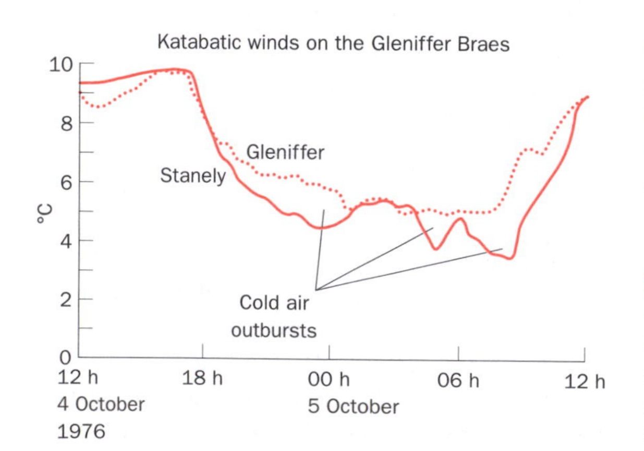

Figure 4.16 Cold air drainage into the Clyde Valley, central Scotland. The two temperature traces are located near the valley floor (Stanely) and on the overlying plateau (Gleniffer). Several bursts of cold air can be seen to arrive during the night at the lower site as katabatic flow events occur

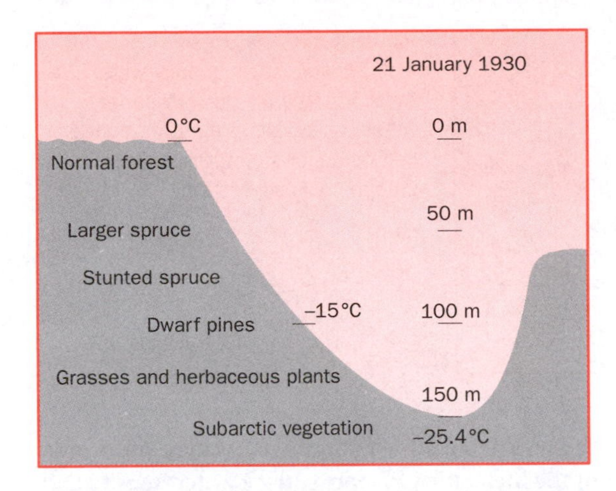

Figure 4.17 The Gstettner-Alm frost hollow on the night of 21 January 1930 as measured by Rudolph Geiger

aligned to the regional airflow. In southern France the mistral of the Rhone Valley is such a powerful, and potentially damaging, cold, dry, drainage wind, as is the tramontana of the Catalan coast of Spain or the bora of the northern Adriatic coast. As might be expected, the strongest katabatic winds exist in the Antarctic where they may occur as frequently as 100 days per year. Flows of up to 300 m depth occur as cold air surges downslope from the central plateau. Indeed, such is the outflow that subsidence of air occurs in their wake over the polar region, a component of the direct cell referred to above.

Key ideas

1. An air mass is a large mobile body of air with uniform characteristics in terms of its basic weather features, i.e. pressure, temperature, humidity, wind, rainfall and cloudiness, visibility, stability.

2. The basic or original weather features of air masses are modified when they move out of their areas of origin (source region) to invade other regions.

3. The original weather features (source region) and modified weather conditions (area of invasion) of air masses together with their frequency (dominance) can be used to describe the climate of regions.

4. Land and water surfaces respond differently to solar heating, with land surfaces warming more quickly and cooling more quickly than their oceanic counterparts.

5. Unequal heating and cooling rates can induce air movements at local, regional and global scales.

4.4 Ocean mass/energy fluxes

The importance of the links between the oceans and the atmosphere is so important in our understanding of weather, climate and climate change that a separate chapter of this book is devoted to it (Chapter 5). Here, however, a number of general points can be made about the significance of ocean movement especially at the global level.

4.4.1 The oceans as heat stores and heat transporters

The ocean, covering 71 per cent of the earth's surface, offers heat storage capacities greatly in excess of the atmosphere. Even considering the upper 70 m of the ocean, where most vertical mixing by wind action occurs, the storage capacity is approximately 30 times that of the atmosphere. In addition, the ocean and atmosphere interact in complex ways to change the physical and chemical characteristics of each other. They exchange heat, they exchange fresh water and water vapour and they act to alter the densities of each other. Over a number of different timescales they interact to provide feedback mechanisms which may induce or mitigate climate change. The main volume of the ocean involved in these processes is obviously the top few metres in contact with the atmosphere. This is where the most important energy interchanges take place. Indeed, half the solar radiation input to the ocean is absorbed in the first 6 cm. The thermally mixed layer extends

Figure 4.18 The oceanic conveyor belt. Sinking, cooling, salty water over the North Atlantic draws warmer waters polewards. When the conveyor belt falters or stops, the climate in Europe cools significantly

on average less than 100 m down, perhaps to a maximum of 400 m in the lower mid latitudes. Below this well-mixed warm layer close to the surface is the thermocline. Here a rapid rate of temperature decrease begins with depth, and the density of the sea water also increases significantly. At depths below 1 km cold, dense, ocean water exists where movement occurs primarily due to differences in salinity. This is the deep ocean thermohaline circulation. It is now known that this exhibits a distinctive worldwide pattern and is commonly referred to as the global thermohaline conveyor or the global oceanic conveyor belt (Figure 4.18 and Box 4.5). Also, as shown in Chapter 8, it is likely that disturbances to this deep-water circulation, for example, by the influx of large volumes of light, fresh water from melting ice sheets have been responsible for some abrupt climate changes in the past. For the same reasons, an understanding of the workings of the ocean conveyor is important in considering the future of global climate.

(a) The North Atlantic conveyor

The North Atlantic is very much the key ingredient in the global thermohaline circulation. In the North Atlantic, as surface waters head north, salinity increases due to evaporation and chilling by cold Arctic air occurs. This salty, dense water begins to sink (known as deep water) and a major driving force for the deep ocean circulation mechanism is provided. As the northward limb of the system encounters the cooler waters and winds of the high latitude zone, surface water density and salinity increases. A bottom water flow back south commences which leads through the southern ocean and into the Pacific before a warmer less salty return flow resumes. The formation of dense, deep water most commonly occurs in the southern oceans around the Antarctic as sea ice forms and formerly dissolved salts enter the layers of ocean below the surface, rendering it denser and with a tendency to subside. Deep water is formed which may spread out on the ocean floor to resurface perhaps millennia later in zones of upwelling. However, the positive aspect of this process is that CO_2 in waters close to the surface may be brought down to great depth, and taken out of circulation, or sequestered, and thus not be destined to add to the problems of global warming in the medium term. This is an important mechanism to understand since its functioning is crucial to anticipating the rate of build-up of CO_2 in the atmosphere.

(b) The switch mechanism

A mechanism exists to switch off this ocean conveyor belt. An influx of cool fresh water from melting ice can induce the sinking to occur much earlier in more southern latitudes of the North Atlantic. When the conveyor belt weakens or collapses to more southern latitudes, winters in Europe become much colder. If the aberration is prolonged enough, ice masses may advance. Such

BOX 4.5

THINKING FURTHER

The thermohaline conveyor

The thermohaline conveyor, sometimes called the global ocean conveyor belt, is the transport of ocean waters around the world by surface and deep ocean currents. The main features of the global ocean conveyor belt, both its deep water (Figure 4.18) and surface components (Figure 4.19), are shown in sections 4.4.1 and 4.4.2. Thus, in order to explain the thermohaline conveyor, we need to take into account not only surface wind patterns which fuel the surface ocean currents but also *density differences* produced in sea water by temperature and salinity changes which drive the deep water circulation.

Like the atmosphere, the ocean is warmed by solar radiation impinging at the surface. But whereas the surface of the earth is at the bottom of the atmosphere, it is at the top of the ocean. So the water that is warmed by the sun in the tropics is already at the top of the ocean, and cannot rise by convection. This warm tropical water tends to sit quasi-permanently on the surface of the ocean and overlies deeper, colder, denser water. A zone called a thermocline separates the lighter, warmer, top water from the deeper colder water. This stable situation is not found everywhere, however. There are regions and circumstances where surface waters become denser than the deeper water. The instability set up causes ocean turbulence with denser top waters sinking to great depths and deeper waters rising to the surface.

Antarctic deep water

Two main regions can be identified where deep water is formed, i.e. where dense water sinks down to the deep ocean, namely in the North Atlantic Ocean between Scandinavia and Greenland and in the region of Antarctica called the Weddell Sea (Figure 4.18). The densest water forms in the Antarctic. There, salt water freezes to form ice at a temperature of around –2 °C. The ice is fresh water so the sea water left behind is more salty and therefore more dense. It flows down the continental shelf to form a great deep water current known as the Antarctic Bottom Water. This cold dense current sweeps around Antarctica at a depth of 4 km and spreads branches out into the deep basins of the Atlantic, Indian and Pacific oceans. These deep cold ocean currents then rise slowly and link to a network of warmer surface currents that carry water in great volumes around the world.

North Atlantic deep water

There are two reasons why surface waters in the North Atlantic should be saltier and therefore heavier than those in the Pacific (where water upwells from below to the surface) and sink to the bottom of the ocean as deep water. First, the Gulf of Mexico acts like an enclosed simmering pond where high rates of evaporation induce the formation of warm saline water. This warm salty and therefore slightly heavier water flows north-eastwards into the Atlantic to form the Gulf Stream which eventually becomes the North Atlantic Drift in the North Atlantic. A second and even more important reason is that there is a net transfer of fresh water from the Atlantic Ocean Basin to the Pacific and Indian oceans that results in the enrichment of salt in the surface layers of the Atlantic. Figure 4.20 shows this net loss of fresh water (as water vapour) to adjoining basins. Moist temperate westerly winds crossing the Rocky Mountains and the Andes shed much of their water as rainfall on the western leeward slopes of these mountains. This moisture is returned via drainage systems back to the Pacific Basin. ▶

The result is that fairly dry air enters the Atlantic from the northern and southern Pacific oceans. Moreover, westerly winds crossing Europe, Asia and Africa from the Atlantic do not encounter a similar north–south barrier of mountain ranges. Thus the moisture that is evaporated from the temperate Atlantic Ocean tends to be shifted eastwards into the Indian and Pacific basins. Similarly, in the tropics the passage of easterly trade winds across the narrow isthmus of central America transports Atlantic-derived moisture into the Pacific, while the extensive mountains of Equatorial Africa exert a drying action on the returning easterly trades. There is therefore, as shown in Figure 4.20, a major net loss of fresh water (as water vapour) from the Atlantic to the Indian and Pacific basins, a fact which makes the surface waters of the Atlantic more salty than other ocean regions.

Salt export in the North Atlantic

The Atlantic Ocean therefore needs to export salt, and it does this through the action of surface currents and deep circulation within the ocean. First, as mentioned previously, warm salty surface waters flow north-eastwards into the Atlantic to form the Gulf Stream and eventually the North Atlantic Drift (NAD). Much of the heat stored in the surface waters of the NAD is given up via evaporation to cold westerly winds blowing across the Atlantic from Canada. Eastward-moving winds which are warmed as they cross the Atlantic bring substantial heat particularly in winter to the UK and Europe. The amount of heat they convey across Europe is substantial and their overall effect can be seen in global temperature anomaly maps (see Figure 3.8). For every cubic centimetre of water transported by the NAD, there is a release of approximately 8 calories (33.5 J) of energy to the atmosphere. This is equal to nearly one-third of the region's annual solar radiation (equivalent to 5×10^{21} calories or 21×10^{21} J).

As the dense North Atlantic salty water cools (to about 2–3 °C) it becomes even denser and begins to sink, contributing to the formation of deep water. Moreover, surface waters which move into the far North Atlantic (the Norwegian and Greenland seas) freeze to produce sea ice. As sea ice formation takes place brine is expelled from within the sea ice, making the surface waters beneath even saltier and heavier. This water also sinks to become deep water. As the relatively dense surface water sinks to the floor of the ocean, it completes the link in the North Atlantic conveyor belt that allows surface waters to sink and return southwards along the bed of the North Atlantic Ocean thus exporting the salt excess to the surrounding oceans.

An important link exists between deep water circulation and the hydrological cycle. Model simulations have shown that North Atlantic deep water (NADW) formation is extremely sensitive to lighter freshwater inputs, whether these come from melting icebergs as during the glacial period, or in a future globally warmed greenhouse world, from higher rates of precipitation in the North Atlantic. Model simulations also indicate that the resumption of the conveyor belt circulation after a shutdown is a rapid process (see Chapter 8).

an episode occurred 11,000–12,000 years ago when retreating ice sheets in North America began to send meltwater out into the Atlantic through the Gulf of St Lawrence, instead of through the Gulf of Mexico. Within a matter of decades the North Atlantic conveyor belt shut down and failed to recover for over a millennium (from 12,700 to about 11,500 years ago), during which severely cold conditions became established in much of northern Europe. Pollen records indicate a return to tundra vegetation in many parts of the continent at this time. This extremely cold event, known as the Younger Dryas (see Box 8.4), also demonstrates how volatile climate can be in some

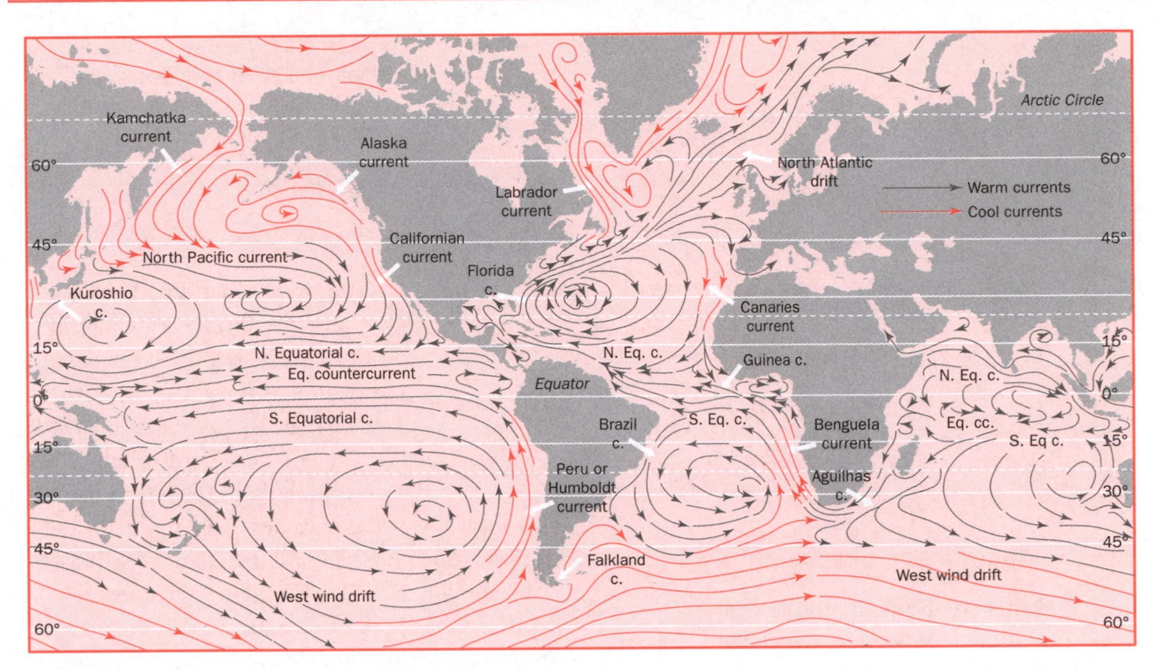

Figure 4.19 Surface ocean currents (January)
Source: Strahler and Strahler (1997, p. 137)

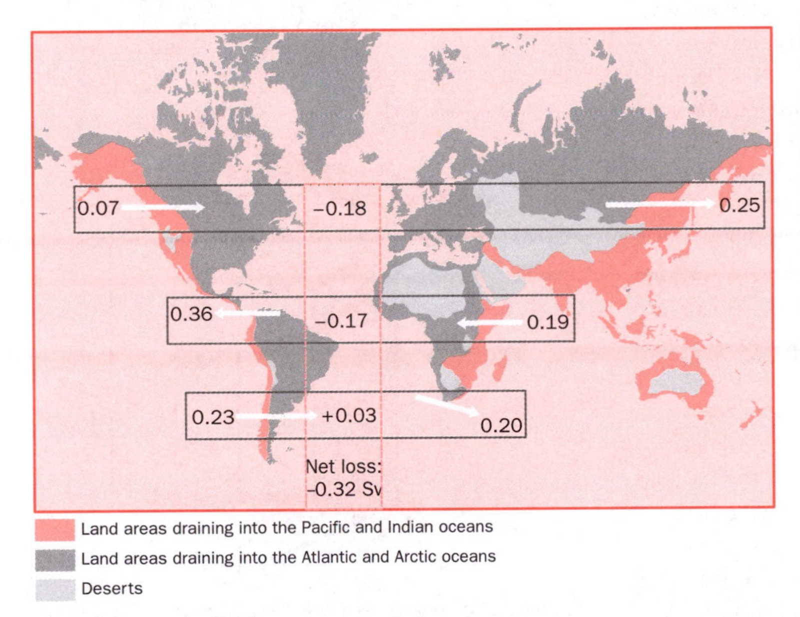

Figure 4.20 Moisture fluxes in the Atlantic and other oceans. More water vapour is transported out of the Atlantic Basin by winds than is brought into it, showing why the Atlantic is more salty than other oceans. Transport units in sverdrups (Sv); 1 sverdrup = 10^6 m^3/s (i.e. more than the flow rate of three Amazon rivers)
Source: Broecker (1995)

regions. Concerns exist that a similar influx of fresh water into the North Atlantic could occur as global warming melts the ice at high latitudes and increased rainfall reduces the salinity of the ocean. This could produce a repeat performance and prolonged cooling in much of Europe while elsewhere increasing warmth might be observed. It is apparent therefore that even the deep ocean may affect or be affected by climate change. Such is the holistic nature of the climate system.

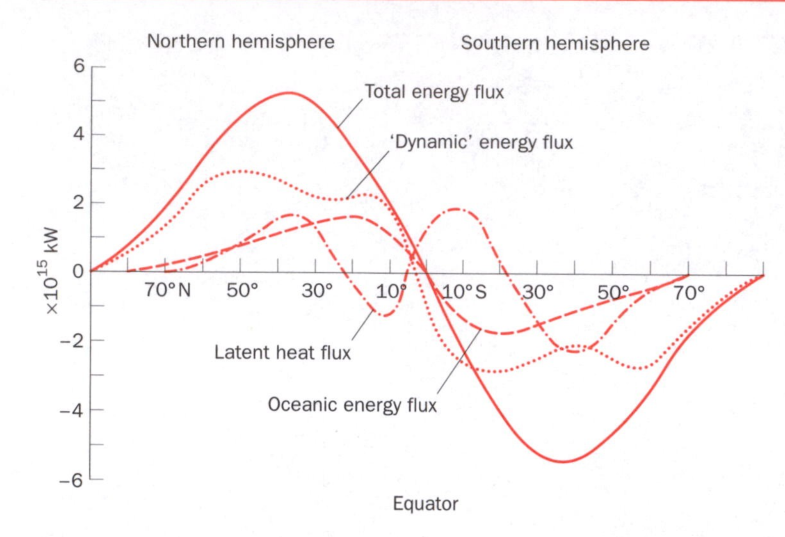

Figure 4.21 Northward-directed global energy flows: oceanic, latent and sensible heat annual mean amounts by latitude
Source: Robinson and Henderson-Sellers (1999, p. 108)

4.4.2 Surface ocean circulation

Ocean currents are driven by the general circulation of the atmosphere, with heat transport predominantly involving the polewards movement of warm surface waters. The heat moved is considerable, with annual mean poleward energy fluxes of approximately 2×10^{15} W typical over much of the world's ocean (Figure 4.21). Ocean current speed ranges from a few kilometres per day to a few kilometres per hour. In the case of the North Atlantic Drift, for example, the average speed is about 25 km/day, meaning that a bottle thrown overboard off the Florida coast of the USA would take approximately 8 months to be washed up on the Kerry coastline of south-west Ireland. The spatial distribution of ocean currents shown in Figure 4.19 indicates the controlling role of the subtropical highs in driving the ocean currents. Two great gyres are seen in each of the Atlantic and Pacific oceans and another one in the Indian Ocean. These are essentially the product of the descending limb of the Hadley cells described above. These great subsiding air cells cause a clockwise current rotation to prevail in the northern hemisphere and an anticlockwise one in the southern hemisphere. It can also be noted that although the subtropical highs are associated with outward-blowing air masses at the surface, the oceanic circulation appears to exhibit a tighter, more circular pattern. This is because the surface waters tend not to move exactly in the same direction as the wind, but rather at an angle of 20–45° to the right in the northern hemisphere and similarly to the left in the southern hemisphere due to the Coriolis force. As the frictional effect of layers of water moving against each other increases at depth, so the speed of the current slows. As a result, the Coriolis force deflects water movement increasingly with depth. This produces what has become known as the Ekman spiral (Figure 4.22). At depths of about 100 m the deflection is about

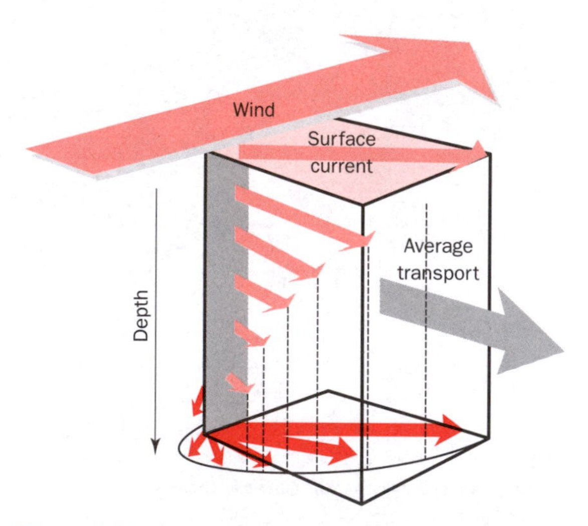

Figure 4.22 The Ekman spiral. As the wind blows over the surface water the layers below move more slowly and the Coriolis force deflects them to a greater extent. The net transport direction of the surface layers of water is to the right of the wind direction
Source: Adapted from Sverdrup, Johnson and Fleming (1970, p. 493)

opposite to the surface direction and the current is not detectable much below this. The Ekman spiral is a factor in delivering cooler waters south-eastwards from the North Atlantic Drift in summer to the west coasts of Britain and Ireland.

Key ideas

1. Below the well-mixed upper layers of the ocean, colder, denser water exists which circulates on a global scale driven primarily by differences in salinity.

2. The principal sinking limb of this oceanic conveyor belt occurs in the North Atlantic. This also provides an important mechanism for sequestering CO_2. Concerns exist that shutting down of this oceanic conveyor belt, as has happened in the past, could occur due to an influx of fresh water from melting ice at high latitudes.

3. Surface ocean currents in the low latitudes are closely matched to the air pattern associated with the subtropical anticyclones.

4.5 Observing motion

4.5.1 Surface networks

Surface obstructions and roughness intimately affect air motion close to the surface. Convective towers of rising air may also act to disrupt the flow regime and enhance turbulence. This fluctuation is reflected in a more gusty form of motion so that observations of wind speed are often averaged over 10 minutes or longer to remove short-lived events. Gust strength may, however, also be measured as indicative of damage potential in high wind conditions. Wind observations are also typically made at a height of 10 m above the surface for similar reasons. Where obstacles exist in the vicinity of the instrumentation, complicated rules are used to calculate the *effective* height of the wind-measuring apparatus or anemometer. The most widely used instrument historically has been the Robinson cup anemometer which was invented by Thomas Robinson of Armagh in 1846. Until quite recently one of his original anemometers was located on Dun Laoghaire pier near Dublin. A counting mechanism to log how many times the cups rotate is needed and the derivation of a relationship between revolutions and wind speed forms the basis of this instrument. More recent developments of the cup anemometer suggested that a three-cone cup arrangement works best. Similarly, though the instrument tends to overrun when wind speed drops, and though very light winds may be unable to turn the spindle at all, errors of about 1 per cent in the measurement of wind speed overall are typical. The rotation of the shaft may also be used to drive a generator which enables automation of data collection.

Pressure tubes, such as the Dines pressure tube, whereby the wind colliding with a fluid under pressure exerts a force which can be converted to speed, have increasingly displaced the cup anemometer at many observing stations. These instruments have been proven to be very robust and to require little maintenance and have thus become very popular. Other types also exist. Many roadside wind monitors employ an instrument based on a propeller-type arrangement. These have to face constantly into the wind, and this is achieved by means of a vane, which also enables direction to be recorded.

Up to 100 m in height a logarithmic increase in speed occurs as the effects of surface friction are gradually lost. Normally measured in metres per second, wind speed is also frequently expressed in nautical miles per hour (1 m/s = 1.94 knots). Historically, the scale devised by Admiral Sir Francis Beaufort in the early nineteenth century is still widely used to provide qualitative/semi-quantitative estimates of wind speed. Though most commonly used for marine purposes, the scale also has a terrestrial version (Table 4.3).

Table 4.3 Estimating wind speed from surface observations

Beaufort number	Description	Wind speed	
		Knots	km/hr
0	Calm	0–1	0–2
1	Light air	1–3	2–6
2	Slight breeze	4–6	7–11
3	Gentle breeze	7–10	12–19
4	Moderate breeze	11–16	20–29
5	Fresh breeze	17–21	30–39
6	Strong breeze	22–27	40–50
7	High wind	28–33	51–61
8	Gale	34–40	62–74
9	Strong gale	41–47	75–87
10	Whole gale	48–55	88–101
11	Storm	56–64	102–119
12	Hurricane	65	120

4.5.2 Remote sensing of motion

Although instruments such as the Dines pressure tube are widely used on land, a dearth of wind information has historically existed for oceanic areas. In some instances this absence has proven to be a serious deficiency in forecasting storm development. The devastating storm of 15/16 October 1987 which passed over southern England had gusts of over 90 knots and blew down 15 million trees. Its explosive development took place mainly in the Bay of Biscay, an area where the density of wind-monitoring instrumentation is relatively poor. Much better prediction of the storm characteristics and earlier warning of its fury would have been possible had a better network of data points been available to forecasters and to the meteorological models existing at the time. Although many parts of the world are now busy deploying wind-measuring instruments on buoys and ships to fill such monitoring gaps, their coverage is still insufficient to provide a comprehensive global wind field map.

Satellite-based sensors are increasingly being used to provide global coverage of wind speed and direction, and also of the movement of the oceans (Plate 4.1). Operating at microwave frequencies, both active and passive sensors are employed in the measurement of wind speed and direction. Passive sensors work by measuring the intensity of emitted infrared radiation. Studies show that sharp variations in temperature are frequently accompanied by substantial variations in wind speed, and as a result formulae which relate the variations in emitted radiation to wind speed have been successfully deployed to convert detected radiation changes to wind speeds. In addition to these passive systems, active systems termed 'scatterometers' are increasingly being used over extensive oceanic areas to measure wind speed and direction at all times of the day and night, and under all cloud conditions. Wind scatterometers operate by recording the change in radar reflectivity of the sea due to small ripples created by the surface. Since the energy of these small wind-formed ripples increases as wind velocity increases, so does the backscatter. The backscatter measurements may thus be used to calculate surface wind speed and direction very efficiently over large areas of the ocean where surface observations may be sparse. Some of the modern sensors may target a band on the earth's surface up to 1,800 km wide, yielding approximately 400,000 measurements per day. This approximates to 90 per cent coverage of the global oceans every day and yields wind speed measurements which comparisons with actual data suggest are accurate to 2 m/s. As well as ocean winds, scatterometer data are also useful for distinguishing vegetation and ice coverage (Plate 4.2).

Key ideas

1. Wind observations have traditionally been made at a height of 10 m using a cup anemometer or more recently a pressure tube.

2. A logarithmic increase in wind speed typically occurs from the surface up to 100 m as the effect of surface friction diminishes.

3. Satellite-based wind measurements are based on both active and passive systems. Passive systems interpret marked spatial changes in emitted radiation as air movement. Active methods employ scatterometers to interpret changes in radar reflectivity of water bodies as waves whose characteristics can be linked to wind speed.

Further reading

Aguado, E. and Burt, J. (1999) *Understanding Weather and Climate*. Prentice-Hall, Upper Saddle River, New Jersey, 474 pp.

Ahrens, C. D. (2000) *Meteorology Today*, 6th edn. Brooks/Cole, Pacific Grove, California, 528 pp.

Barry, R. G. and Chorley, R. J. (2003) *Atmosphere, Weather and Climate*, 8th edn, Routledge, London.

Lutgens, F. and Tarbuck, E. (1995) *The Atmosphere*. Prentice-Hall, New Jersey, Ch. 3.

Robinson, P. and Henderson-Sellers, A. (1999) *Contemporary Climatology*, 2nd edn. Longman, London.

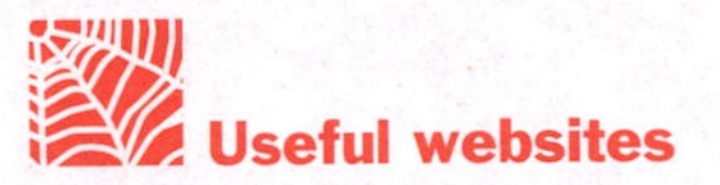

Useful websites

Measurement of wind:
http://oceanworld.tamu.edu/resources/ocng_textbook/chapter04/chapter04_04.htm

Planetary wind circulation models:
http://atschool.eduweb.co.uk/kingworc/departments/geography/nottingham/atmosphere/pages/pressureandwindsalevel.html

Upper air motions:
http://homepage.ntlworld.com/booty.weather/metinfo/uppair.htm

Local winds:
http://www.bbc.co.uk/weather/weatherwise/factfiles/thebasics/wind/localwinds.shtml

Ocean currents and climate change:
http://www.pik-potsdam.de/~stefan/Lectures/ocean_currents.html

Remote sensing of winds:
http://winds.jpl.nasa.gov/

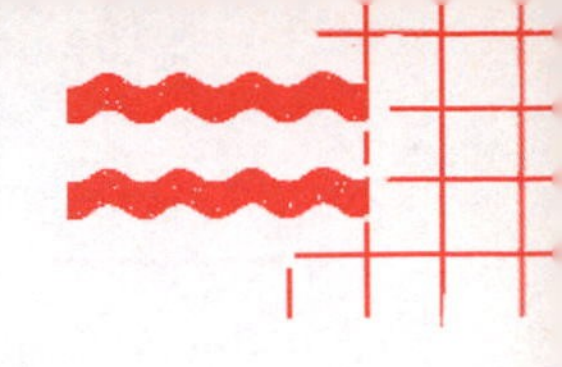

CHAPTER 5

Ocean–atmosphere interactions

5.1 Energy exchange

In the previous chapter, the importance of the oceans as a powerful heat store and transfer mechanism was shown. As a direct result of the massive capacity of the oceans to both sequester and transport heat, some of the most powerful and complex circulations controlling weather and climate involve the interaction between the atmosphere and the oceans. The atmosphere and oceans continuously interact with one another, particularly in the way they transfer and exchange energy. Warm ocean water is able to exchange large amounts of energy vertically with the atmosphere during evaporation. This energy is eventually released in the atmosphere as latent heat during air ascent, cooling and condensation (cloud formation). The clouds that are formed in turn reflect incoming solar radiation and can thus temper the warming of ocean waters. Most energy exchange involving evaporation and condensation over the oceans takes place across a wide area with clouds, for instance, being a frequent and extensive feature of the main ocean basins. Certain ocean–atmosphere exchanges can be highly localised and intense. The tropical hurricane, examined in this chapter, represents a dramatic example of localised ocean–atmosphere energy exchange where evaporation over the oceans provides the energy base for the growing tropical storm.

Winds play an important role in transferring ocean–atmosphere energy exchanges across the planet, displacing them in space and time. They can move tropical hurricanes and their energy-releasing mechanisms far from their oceanic region of origin to other parts of the ocean or to where they make contact over land. Winds also move rain-bearing atmospheres from the oceans over the land as seen in the monsoons of the tropics and along the western coastal margins in the mid latitudes. Winds also drive surface water over the oceans forming the main ocean currents (see Figure 4.19). The speed of these surface currents is relatively fast so that they can shift warm water polewards and cold water equatorwards over relatively short time periods. The surface currents are tethered to, and interact with, a much deeper, more extensive, and slower-moving current of water called the thermohaline conveyor (see Box 4.5). As we shall see, the behaviour of the deep ocean water can strongly affect the performance of the surface currents and vice versa. These interactions between the surface and deep water are important because the surface ocean currents can strongly modify weather and climate patterns in their vicinity (like the Gulf Stream and Europe).

As water reacts more slowly than the air to energy exchange, most energy transfers are slightly out of phase or equilibrium. With conditions in the atmosphere changing much more rapidly than those in the ocean, the atmosphere and ocean continually shift in response to each other. Some of the most complex ocean–atmosphere responses are those that suddenly flip or oscillate from one phase or mode to another over predictable but irregular time periods. A number of complex ocean–atmosphere oscillations between the oceans and atmosphere existed during cold or glacial periods. There is much evidence to show that during the last glacial period, from about 90,000 to 10,000 years ago, the climate of the North Atlantic frequently switched from a cold 'stadial' phase to much warmer 'interstadial' phase over relatively short time periods. It is thought that such rapid changes in climate during the ice age are related to the 'switching' on and off of the Gulf Stream (sections 4.4.1 and 8.6.1). Evidence of major ocean–atmosphere oscillations in today's post-glacial climate can also be found. In this chapter, examples of five major contemporary ocean–atmosphere oscillations that alter local, regional and global climate are examined. They include the North Atlantic Oscillation (NAO), the Arctic Oscillation (AO), the Antarctic Polar Wave or Oscillation (APW or AAO) and the Pacific Decadal Oscillation (PDO). The main discussion, however, is reserved for the El Niño–Southern Oscillation (ENSO) in the Pacific.

Key ideas

1. Ocean–atmosphere interaction is one of the most powerful controls on weather and climate.

2. Mass and energy interactions between ocean and atmosphere can be relatively localised and direct as in the development and formation of the hurricane, or regional and time lapsed as in certain large-scale ocean–atmosphere oscillations.

5.2 Local ocean–atmosphere interaction: the hurricane

In this section, the origin, development and distribution of tropical storms, especially those with hurricane wind speeds (mean wind speed >33 m/s or >74 mph) are explored. These great storms are of interest for their ocean–atmosphere interactions (this chapter) and because these disturbances cause more fatalities on average than any other natural disaster (Chapter 12).

5.2.1 Location and frequency

Tropical hurricanes are found in seven main ocean basins (Figure 5.1). They include the North Atlantic and the Caribbean (Basin 1), the Arabian Sea and Bay of Bengal (Basin 2), the north-eastern Pacific Ocean (Basin 3), the north-western Pacific Ocean (Basin 4), the southern Indian Ocean (Basin 5) and the south-west Pacific Ocean (Basin 6). Hurricanes occur most frequently in the north-west Pacific and north-east Pacific regions where on average 20–30 storms can form each year. Although well covered by the US media, relatively few hurricanes (4–5) develop each year in the North Atlantic and Caribbean. Ocean basins such as the Bay of Bengal show intermediate to high hurricane frequencies of about 12–15 tropical storms per year on average. From this analysis it is clear that only about 80–90 hurricanes develop globally each year compared to thousands of extra-tropical cyclones (mid-latitude depressions) which develop annually along the polar front (sections 4.1 and 9.2.2). To take our understanding of the distribution and frequency/intensity of hurricanes further we need to explore the factors necessary for hurricane formation and intensification.

5.2.2 Formation and intensification

Six conditions are usually stipulated for the initiation and intensification of tropical hurricanes. These are detailed below.

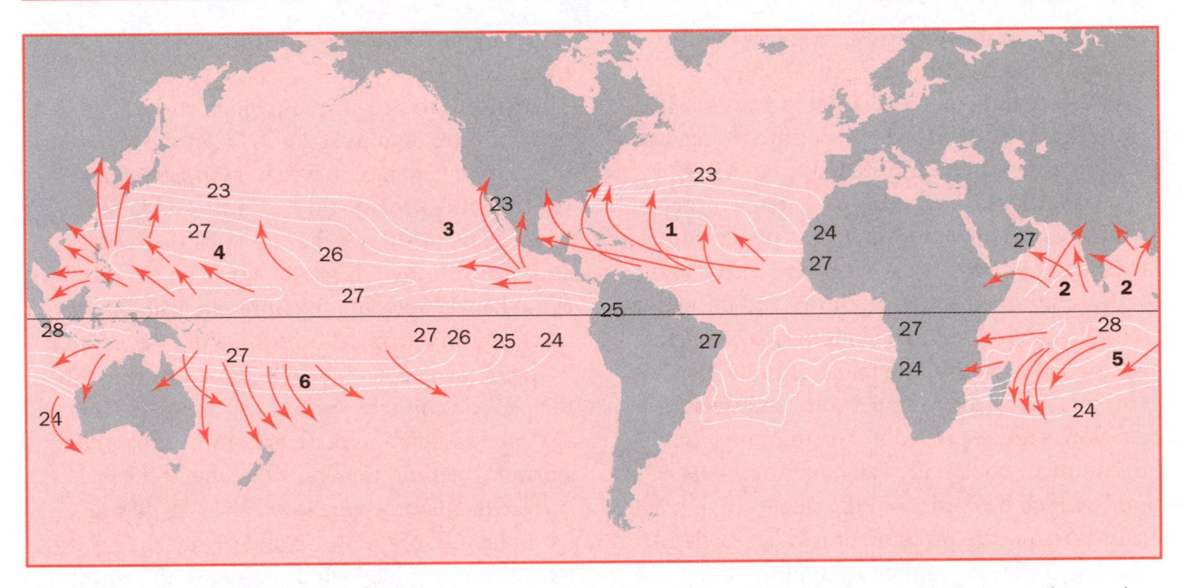

Figure 5.1 Map of global hurricane distribution showing main ocean basins where they originate (start of arrows) and main hurricane tracks (arrow directions). Note the link between the oceanic basins of origin and ocean temperatures above 26–27 °C. See text for main ocean basins and hurricane frequencies

(i) Low-pressure areas (easterly waves)

The first requirement is for some initial small-scale low pressure disturbance in the tropical oceans to attract ground-level wind convergence. On occasion, when the correct meteorological conditions apply, low pressure regions known as easterly waves can form within the normally high-pressure trade wind belt between the subtropics and the ITCZ. These low pressure waves slowly move westwards under the influence of the surface trade winds. They are frequently observed as shown in Figure 5.2, in the western tropical Atlantic, originating off the coast of Africa and moving westwards towards the Caribbean. They also form in the south China Sea

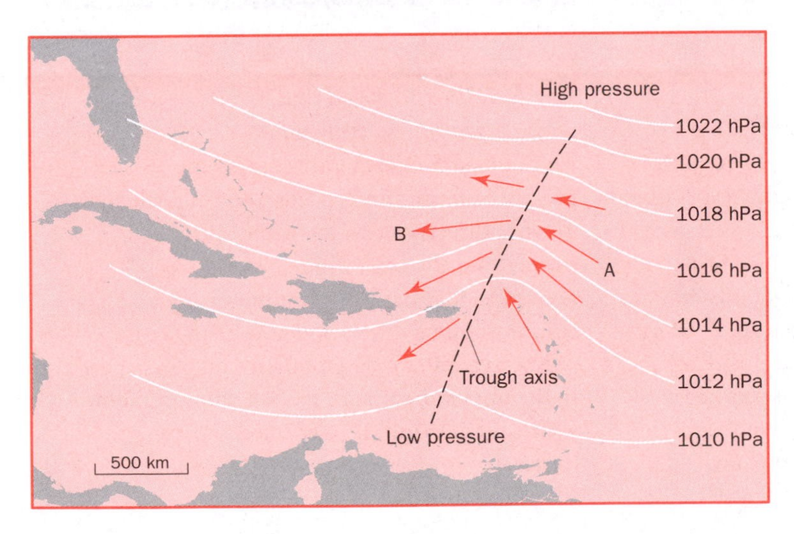

Figure 5.2 Horizontal view of an easterly wave in the southern North Atlantic. The dashed central line represents the axis of the low-pressure trough near the surface. At A, the surface winds converge towards the trough axis producing rainfall: at B the winds diverge and subside, creating clear skies and dry conditions

Source: Robinson and Henderson-Sellers (1999) from Critchfield (1983)

where they can move westwards across the Malayan Peninsula into the Bay of Bengal.

As shown in Figure 5.2, surface winds situated east or downwind of the low-pressure trough axis at A, are warmed and moistened from below by the warm Atlantic becoming potentially unstable. As these winds move together or converge towards the trough axis, they are forced to rise, cool and condense out their water vapour (cloud formation). During condensation and cloud formation, latent heat is released within the rising saturated parcels of air increasing their temperature and thus their buoyancy with respect to the surrounding (unsaturated) cooler air. Tall cumulonimbus clouds punch upwards into the stable air aloft and the disturbance becomes organised into a cloud system with intense rain and thundery showers. In the zone of air divergence at B (where the wind arrows are spreading out), air subsidence and fair weather are found. When suitable atmospheric conditions are in place (see following conditions ii–vi), easterly waves can develop into severe tropical storms with hurricane wind speeds. Suitable conditions are not always in place since only about 10 per cent of wave disturbances intensify into the more violent rotating storms so feared in many parts of the tropics.

(ii) A warm moist tropical atmosphere that is conducive to convective overturning and that leads to the development of deep cumulonimbus clouds

Air ascent created by low-level wind convergence will continue if the air remains unstable (see Figure 2.11(b)). This means that there should be a *moist* atmosphere with a large enough temperature decrease with height (strong lapse rate) such that the atmosphere overturns and becomes turbulent when the air becomes saturated and clouds begin to form. With condensation and the release of latent heat, the saturated air is warmer and positively buoyant with respect to adjacent unsaturated cooler air, and begins to bubble up like the bubbling of water when heated from below.

(iii) Ocean surface temperatures greater than 26–27 °C

There needs to be sufficiently high surface temperatures of at least 26–27 °C in the top 50 m of ocean to generate sufficient evaporation and subsequently provide the moisture and latent heat release to sustain the energy of the storm, and the continued development of cumulonimbus cloud banks. Such temperatures are usually reached in the tropical oceans during the *autumn* in the respective hemispheres as oceans reach their maximum temperatures at this time. There is a close association between temperature and storm formation. Generally the higher the sea surface temperature the lower the central sea level pressure of the storm and higher the maximum sustained wind speeds. With respect to Figure 5.1, some of the warmest surface ocean waters and highest hurricane frequencies are found in the two northern Pacific basins. Moreover, the largest and most powerful hurricanes usually form in the tropical north-west Pacific region. In contrast, hurricanes do not form in tropical oceanic areas off the west coast of South America and central Africa. This is because such areas are cooled below 26 °C by the action of the cold currents moving from the Antarctic towards the equator along the western coast of South America (Humboldt current) and Africa respectively (Benguela current).

(iv) Small wind speed and direction changes with height in the atmosphere

An important consideration is for the presence of low wind speeds and small changes in wind direction (i.e. low wind shear) throughout the atmosphere. These are needed to maintain maximum heating and cumulonimbus development over the region of lowest pressure. High horizontal wind speeds in the upper air disrupt the tall cumulonimbus which literally topple over the developing hurricane.

(v) A sufficiently strong Coriolis force is required to spin the system and create the hurricane vortex

The effect of the rotation of the earth gives the rising air a 'twist' so that the whole system can begin to revolve. The organisation of clouds into spiral bands is a critical stage in this process, transferring energy from individual rising packages of air into a more coherent vortex. The Coriolis force helps to spiral winds inwards cyclonically (anticlockwise in the northern hemisphere, clockwise in the southern hemisphere) at low levels towards the low pressure, and outward anticyclonically (clockwise in the northern hemisphere, anticlockwise in the southern hemisphere) at upper levels away from the system. As indicated in section 2.2.1, the Coriolis force varies with latitude, being zero at the equator and maximum at the poles. This means that hurricanes will not form equatorwards of about 5° (see Figure 5.1) where the Coriolis factor exerts too small a force to rotate the tropical hurricane. The various wind mechanisms of a hurricane can also produce a series of individual small-scale but very violent vortices (similar to tornadoes), usually in the front right quadrant of the storm where wind speeds are greatest. On occasion these tornado-like circulations can result in wind speeds well over 89 m/s (200 mph) and can create localised but nevertheless high losses to lives and property.

(vi) The development of a strong anticyclone in the upper troposphere over the surface low

Finally, there needs to be a strong upper-level anticyclone to lower surface pressure by allowing winds to evacuate the system at the top of the atmosphere faster than they are drawn in at the surface where surface friction slows the winds slightly. Generally, surface pressure does not fall very far because the air which diverges aloft sinks at the periphery of the cumulonimbus cloud system and is recycled into the storm. In addition, the subsiding air warms and dries the region surrounding the cloud system, providing an atmosphere that is less conducive to vigorous deep cumulus convection. A powerful upper-air anticyclone that is able to evacuate air and moisture away from the storm will continue to lower surface pressure and the hurricane will continue to intensify.

When *all* of the above conditions are satisfied, it may take only 2–3 days for a weather system to intensify from a tropical depression with sustained wind speeds of less than 17 m/s (about 38 mph) to full hurricane strength with sustained wind speeds of over 34 m/s or 74 mph.

5.2.3 The mature hurricane

(a) Structure and function

Mature, fully developed hurricanes range in diameter from 300–400 km to 1,000 km. Central pressure is usually below 970 millibars (Figure 5.3), exceptionally below 880 millibars, and around this vortex the full fury of the storm is experienced. The mature hurricane is also vertically extensive. Towering walls of cumulonimbus cloud, highest and widest near the centre of the storm (Figure 5.3) often occupy the whole of the lower atmosphere from the ocean surface to heights of 19–22 km (12–14 miles). These great central cloud banks are packed most densely around the centre of the storm where air ascent, cooling, condensation and latent heat release are greatest. Since the stratosphere inhibits vertical motion due to a strong temperature inversion, the air and clouds transported upwards spread out horizontally at the top of the troposphere. It will be noticed in Figure 5.3 and Plate 5.1 that the cloud banks become smaller and more separated by regions of clear air (i.e. where the air is falling) with distance from the centre of the storm.

As shown in Plate 5.1, at the heart of the cloud spiral is a small circular cloud-free zone known as the 'eye' of the storm. The central 'eye', commonly 8–25 km in diameter, stretches from

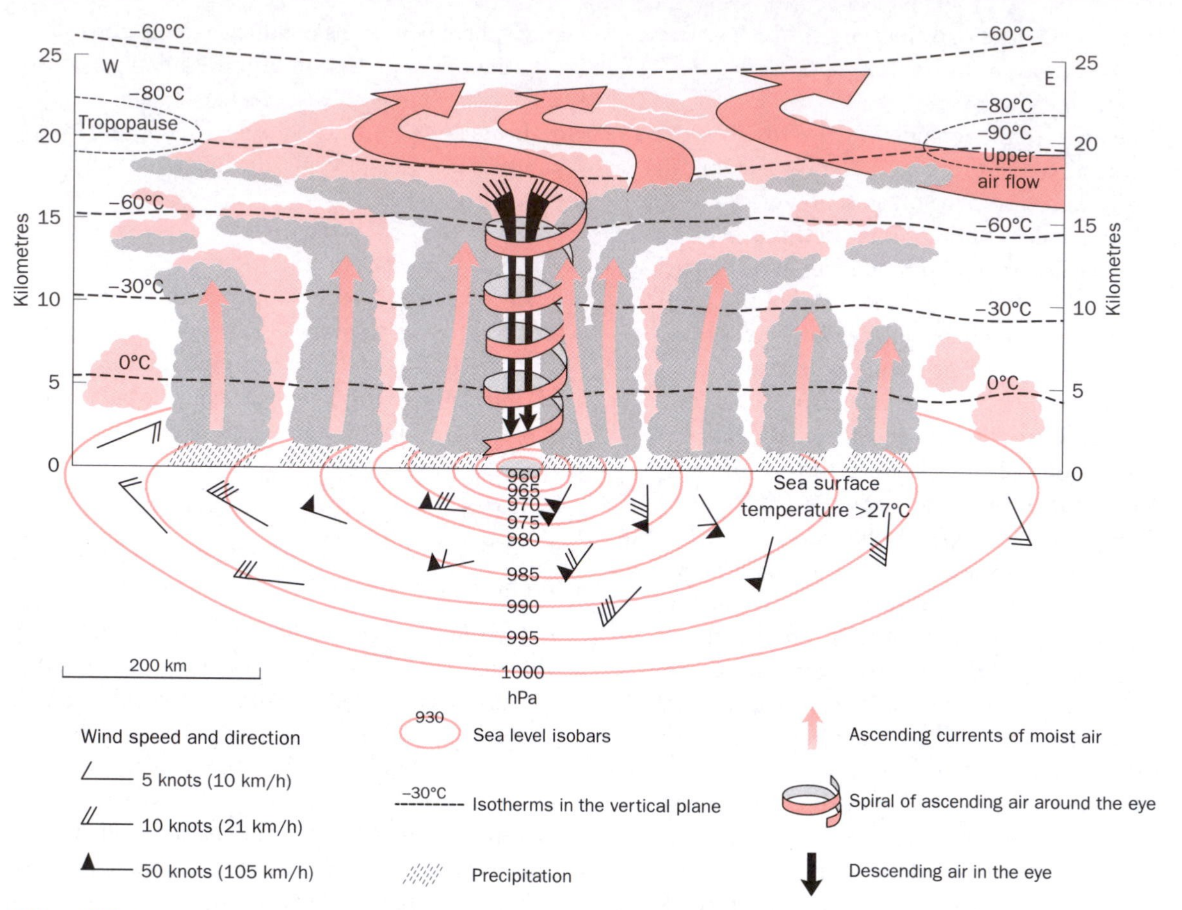

Figure 5.3 The anatomy of the mature hurricane, illustrating overall dimensions, main pressure regions, temperatures, airflows, clouds and rainfall
Source: Various, including Buckle (1996)

the ocean surface to the top of the troposphere (Figure 5.3). In this zone with clear skies and light winds, air is subsiding from the top of the storm to near its base. Here air is forced to descend and is warmed by compression. Cloud particles and droplets located in the eye are evaporated by the warmed descending air so that clear skies result. Its importance is twofold. First, the central eye signifies strong divergence aloft, a vital aspect in removing the rapidly rising air of the hurricane and permitting convection to proceed at the surface. Second, the warming of the descending air in the 'eye' itself induces instability at the surface and stimulates the storm's intensity.

(b) Hurricane impacts

A detailed account of the impact of hurricanes on society and environment is given in section 12.2. Here, in this section, the primary meteorological agents of hurricane destruction are previewed, namely the system's violent winds, excessive rainfall and high tidal waves.

(i) High wind speeds

Despite the calm of the central 'eye' itself, hurricanes are known for their great wind speeds and the structural damage they cause. Horizontal wind speeds are actually greatest near the centre of

Table 5.1 Wind speed classification for the most powerful hurricanes

Category	Wind speed km/h	Storm surge Metres	Damage
1	117–153 (74–95 mph)	1.2–1.5 (4–5 ft)	No real damage to buildings. Damage to unanchored mobile homes, shrubbery and trees. Minor pier damage
2	154–177 (96–110 mph)	1.6–2.4 (6–8 ft)	Some roofing material and window damage. Some trees blown down. Considerable damage to mobile homes and piers
3	178–209 111–130 mph	2.5–3.6 (8–12 ft)	Some structural damage to small homes. Large trees blown down. Mobile homes destroyed. Terrain lower than 1.5 m (5 ft) above m.s.l. may be flooded 13 km (8 miles) inland. Evacuation of low-lying residences
4	210–249 (131–155 mph)	3.7–5.4 (13–18 ft)	Complete roof structure failures. All signs blown down. Complete destruction of mobile homes. Terrain lower than 3 m (10 ft) above m.s.l. may be flooded. Massive evacuation of homes as far as 10 km (6 miles) inland
5	>249 >155 mph	>5.5 >18 ft	Complete roof failure on many residences and industrial buildings. Complete destruction to mobile homes. Severe damage to constructions, windows and doors on properties less than 4.5 m (15 ft) above m.s.l. and within 500 m of coastline. Massive evacuation within 8–16 km (5–10 miles) of shoreline

the hurricane (in the cloud wall surrounding the eye) where pressure is lowest, and decrease with rising pressure outwards away from the central vortex. Hurricanes are officially defined as tropical storms with sustained wind speeds in excess of 34 m/s (74 mph) for one minute or more. Since the most powerful hurricanes can register mean wind speeds of 80–89 m/s (180–200 mph) hurricanes are classified into five categories based on their mean wind speed and associated destructive capabilities (Table 5.1).

A related aspect of wind damage by hurricanes is tornadoes. These rapidly rotating small-scale vortices are produced in squalls usually where winds are highest, i.e. in the front right quadrant of the storm. As the hurricane makes landfall, the lower winds are retarded by ground friction. Frictional drag at the surface can provide the circulation or spin needed for tornadogenesis. While these tornadoes are not as severe as the major tornadoes that are associated with continental convective thunderstorms (section 10.2.1), loss of life and property does occur as a result of them.

(ii) Heavy rainfall

In addition to high wind speeds, hurricanes are also well known for their heavy rainfall and associated flood damage. Up to 15×10^9 t/day of water vapour passes through the hurricane system, half of which may fall as rain. Consequently, excessive rainfall often occurs when a tropical hurricane makes landfall, particularly if the cumulonimbus cloud system is forced up and over mountain barriers. Over several days, many hurricanes can deposit 30–50 cm (12–20 inches) or more of rainfall over a wide area and cause severe flooding. In the developing world such rains can quickly fill upland reservoirs to maximum capacity. When these reservoirs are poorly constructed, the water pressure in the reservoirs can easily cause their dams and surrounding walls to burst. The resultant flash flooding downstream can wash away villages and produce high death tolls. In lowland areas, flooding associated with hurricane rains causes enormous damage to standing crops, for example paddy rice.

(iii) Storm surge

Despite the considerable structural damage and property loss that hurricanes can generate because of their high winds and heavy rainfall, most hurricane deaths are a product of the storm surge. These are the tidal waves pushed up in front of the hurricane as it makes landfall. Such tidal waves are generated by three different but interrelated processes. First, the hurricane's low sea surface pressures allow the sea surface to push up under the storm thus elevating the surface level of the ocean. This effect can raise sea levels 80 cm to about 1 m. Second, and more importantly, frictional drag created at the surface of the elevated ocean by the storm's very high wind speeds, pushes up a wall of water typically 80–160 km (50–100 miles) wide in front of the hurricane as it makes landfall. Third, the generated tidal waves are elevated further by water converging or 'piling up' along the increasingly shallow ocean shelves over which the hurricane travels as it makes landfall. Along the coast of the Gulf of Mexico and eastern India, storm surge heights are typically in the region of 2–5 m. Exceptionally, with the largest and most violent hurricanes, they can attain 6–7.5 m in the Gulf of Mexico and as much as 12–13 m at the head of the Bay of Bengal in Bangladesh.

5.2.4 Hurricane decline and energy dissipation

Once a tropical cyclone reaches hurricane intensity, it will not weaken unless:

1. Its source of heat and moisture is reduced as a result of passage over land, or relatively cold water.
2. Dry cool air is transported into the system. This will disrupt deep cumulonimbus convection.
3. The anticyclone aloft is replaced by a low-pressure cyclonic system which adds winds and mass to, rather than subtracting them from, the developing system (hurricane heat engine).

Energy dissipation over land or sea is the most commonly observed way in which hurricanes lose strength. Passage over land, in particular, quickly robs a hurricane of its source of energy, i.e. evaporation over the oceans, so that hurricanes making landfall can dissipate into much weaker storms in a matter of hours. Atlantic hurricanes initiated by easterly waves off the west coast of North Africa and driven west by easterly trade winds often intensify over the warm waters of the Caribbean. Such hurricanes also regularly dissipate in strength when they make landfall over the coasts of Central America and south-east USA. Caribbean hurricanes that are pushed north-eastwards by westerly winds over colder Atlantic waters dissipate more slowly into less energetic storms before they reach France and the UK.

5.2.5 Main tracks

Hurricanes are synoptic weather systems and like the mid-latitude depression are influenced and driven by larger atmospheric circulations. The most favoured mean storm tracks (see Figure 5.1) that hurricanes adopt are determined by global wind direction, although the actual day-to-day movement of hurricanes and eventual landfall positions are difficult to predict precisely (see Box 5.1). Hurricanes developing in the northern hemisphere reach their peak in the autumn when sea surface temperatures (SSTs) are warmest (August, September, October and even November). Most storms here are pushed westwards by the north-east trade winds. In the North Atlantic and Pacific basins some longer-lasting hurricanes (i.e. more than about 5–7 days) that move northwards are eventually picked up by westerly winds and driven north-eastwards over colder waters where they eventually dissipate respectively into less severe storms over the UK

BOX 5.1

THINKING FURTHER

Hurricane track predictions

Because of the great loss of life and damage to property attributed to hurricanes, knowing exactly where and when a hurricane of a certain strength will make landfall is important. Accurate prediction of the timing and location of hurricane landfalls enables individuals and communities particularly in the more developed countries to prepare (evacuation measures, concrete shelters, house structural reinforcement) in advance for the hurricane's approach. A number of statistical models are used by the US National Hurricane Centre to forecast the movement of tropical storms and hurricanes in the Caribbean (see Figure 5.1). Statistical analysis of prior tropical hurricane tracks permits the development of expected storm movement which is based on a given set of current and forecast synoptic-scale meteorological flows. The models used relate storm motion to large-scale observed and predicted atmospheric conditions. These include the positioning of surrounding low-pressure cells (that attract hurricanes) and high-pressure centres (which deflect hurricane movement).

Because hurricanes represent a system with fluid motion (the movement of air and moisture) whose condition is difficult to predict and which *also* moves according to the nature of other surrounding fluid motions (low- and high-pressure systems) whose behaviour is also not easy to forecast, hurricane prediction is not yet a precise science. Despite great advances in the monitoring of hurricanes in the Caribbean (with radar, satellite, reconnaissance aircraft) and in computer power to process and analyse data, the precise landfall location of many forecasts is still fairly hit and miss. The average error of 24-hour forecasts of the positions of hurricane landfall along the south-east coastlands of the USA is still around 130–160 km (80–100 miles).

and Japan. In the southern hemisphere, SSTs are highest and hurricane frequency greatest in the autumn period, especially during the months of March, April and May. In the south Indian and south Pacific basins, south-east trade winds divert most hurricanes westward along favoured storm tracks. Some of these are eventually diverted south by westerly winds and continental land areas over Australia and along the coast of East Africa.

5.2.6 Hurricane futures

There are no modelling studies suggesting that hurricanes will increase in frequency in a globally warmed world. Observed climate variability over the last 100 years shows that there is no appreciable long-term variation in the total number of hurricanes in the various global ocean basins (Figure 5.4(a) and (b)). Despite evidence of increases in tropical ocean surface temperatures in recent years, we have seen that ocean temperature is only one of six possible factors that govern the development of hurricanes. Working against higher hurricane frequencies forced by increases in ocean temperature is wind shear, which might also increase in a globally warmed world. As we have seen, wind shear can prove very destructive to a developing tropical storm.

On the other hand, increases in the persistence and frequency of the most *intense* hurricanes are not discounted under climate change, a fact which will have serious implications for future economies and societies in the tropical regions of the world. However, there is no statistically sound evidence to date for such a speculation. A recent weak trend in hurricane intensity has been observed in several ocean basins (North Atlantic, north-west Pacific and south-west Pacific). Figure 5.4(c), for example, shows that despite declines in the frequency of the most intense hurricanes since the 1950s (category 3, 4 and 5), hurricane strength has increased in the Atlantic since the mid-1990s. This indicates possibly that while factors such as wind shear might modulate total hurricane frequency, other factors like ocean heat might be more important in regulating their intensity.

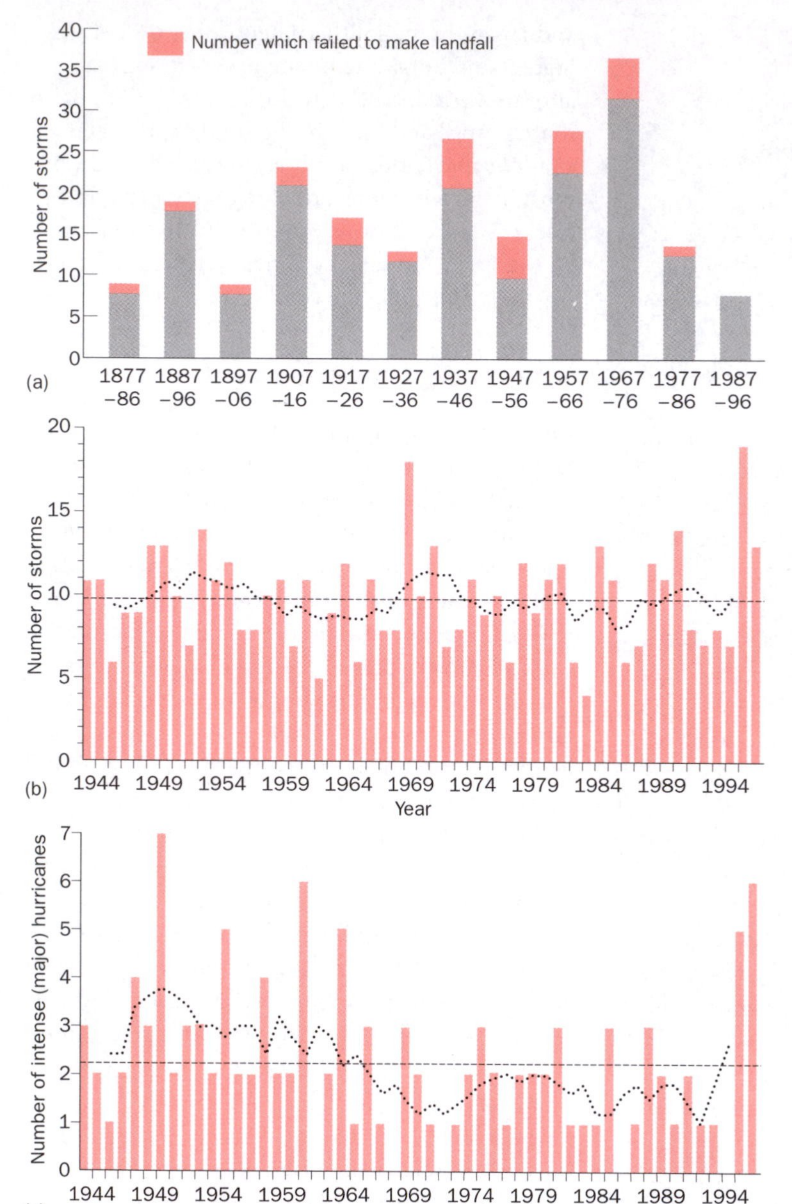

Figure 5.4 Tropical storm frequency and strength in selected ocean basins: (a) the decadal number of severe tropical storms (28–33 m/s) and hurricanes (>33 m/s) in the Bay of Bengal, 1877–1996; (b) the annual number of severe tropical storms and hurricanes in the Atlantic Basin, 1944–94; (c) the annual number of major (category 3, 4, 5) hurricanes (i.e. wind strength >49 m/s) in the Atlantic Basin, 1944–96
Source: (b) and (c) NOAA

3. Tropical hurricanes also have the following distinctive features: low sea surface pressure, high pressure at the top of the storm, steep environmental lapse rates, spiralling upward wind movement, deep cumulonimbus clouds, high and prolonged rainfall, and huge storm surges.

4. At least six factors have been identified for the initiation, formation and intensification of hurricanes.

5. Warm ocean temperatures are essential to the formation of hurricanes since they produce sufficient evaporation to fuel the storm during later air ascent, cooling, cloud formation and the release of latent heat.

6. Equally important, however, are low and persistent sea surface pressure, high pressure at the top of the storm, sufficient Coriolis force to spin the system, low wind shear levels to stop the storm from collapsing, and a moist unstable atmosphere to ensure rapid air ascent.

7. Hurricanes decline when their chief source of fuel, evaporation over a warm sea, is removed.

8. The most damaging effects of hurricanes are related to high wind speed, prolonged and heavy rainfall/flooding and high tidal waves.

Key ideas

1. The hurricane is defined as a tropical cyclone with wind speeds exceeding 34 m/s (74 mph).

2. The well-known characteristics of a hurricane are its high wind speed and an 'eye' at its centre.

5.3 Global ocean–atmosphere oscillations

In contrast to the tropical hurricane which may be regarded as an example of a fairly localised and direct interaction between the oceans and the atmosphere, there are other much larger weather systems which represent a more complex and indirect coupling between the atmosphere and ocean. We have already seen in section 4.4.1 and Box 4.5, the tremendous capacity of the oceans to store and transport heat and moisture across the planet. Because energy transfer within and between the ocean and atmosphere are different and time lagged, the ocean and atmosphere are always slightly out of balance with each other. In many of the world's oceans, the Atlantic, the Arctic, Antarctic and Pacific, regional scale ocean–atmosphere couplings can be found which adjust their energy balances very quickly by switching mode over a matter of days, months or several years. The most dramatic example of a dual phase ocean–atmosphere system, is the El Niño Southern Oscillation (ENSO) in the tropical Pacific. Other important oscillations exist, however, and include the North Atlantic Oscillation (NAO), the Arctic Oscillation (AO), the Antarctic Polar Wave or Oscillation (APW or AAO) and the Pacific Decadal Oscillation (PDO). All ocean–atmosphere couplings have been shown to strongly influence weather and climate patterns in their vicinity.

5.3.1 The Atlantic region

In the North Atlantic, the strength of the North Atlantic Drift (NAD) and westerly winds blowing over the UK and Europe are influenced by pressure differences between semi-permanent low pressure over Iceland (Icelandic Low) and semi-permanent high pressure off the coast of Portugal and Spain (Azores High). This pressure gradient has been shown to vary or oscillate every two to three decades over the last 150 years between two distinct phases: when in positive mode (between 1900 and 1930, and strongly since the 1980s) pressure differences are high and both westerly winds and Gulf Stream advection are strong. When the cycle switches to negative mode (between 1930 and 1960), pressure variations, westerly winds and ocean advection are much weaker. As discussed in section 9.3.3, this decadal scale cycle in surface pressure, winds and ocean currents has been called the North Atlantic Oscillation (NAO). When in (warm) positive mode, westerly winds and frontal activity strengthen and shift northwards over the Atlantic bringing relatively mild, but wet weather to Britain, especially western Scotland, and Scandinavia. Warm temperatures have lengthened growing seasons but damaged the ski industry in Scotland, Scandinavia and other parts of northern Eurasia. The northward shift of rain-bearing winds over the Atlantic during a warm phase NAO, however, reduces rain further south in Europe, the Middle East and even Africa. Reduced rainfall loadings have caused serious drought since the 1970s in parts of Spain, the Mediterranean and in the Sahel of Africa (Ethiopia, Sudan, Somalia). When in the opposite (cold) negative phase, the NAO encourages much colder drier weather over north-west Europe and wetter conditions in southern Europe.

There is evidence that the NAO may be subsumed by a larger more fundamental flip-flop circulation – the Arctic Oscillation (AO) – that influences climate over the entire northern hemisphere. As discussed in section 8.1.3, the strength of the AO is directly related to the intensity of the upper westerlies (Circumpolar Vortex). When the vortex strengthens as it has done for the last 20 or so years, the AO matches and reinforces the effect of a warm phase NAO. In this mode, higher temperatures but lower pressure in the upper atmosphere increase the flow of the westerly jet streams and thus the surface winds that bring warm wet Atlantic air eastward over Europe and Siberia. When the Circumpolar Vortex loses strength (relatively cold temperatures and high pressure in the upper air), weaker upper-level westerly jet streams prevail allowing more cold air to penetrate Europe from the north. There is evidence that the NAO and AO could become phase locked over long periods of time. This possibility with a prolonged and combined negative phase lock, might help explain the Little Ice Age, an intensely cold period in the UK and Europe from 1350 to 1880.

5.3.2 The Antarctic and Pacific regions

There are also ocean–atmosphere circulations that strongly influence weather patterns in the Antarctic such as the Antarctic Circumpolar Wave (ACPW). Like the Arctic, fast-moving 'wavy' upper air currents move eastwards (i.e. clockwise in the southern hemisphere) around the poles. These winds have recently been discovered to shift warm and cold pools of water across the Antarctic Ocean. Every 4 years or so, a cold or warm pool passes over the southern ocean. When a warm pool nears Australia, for instance, winds moving off the ocean advect warm moist air over the continent and produce warmer and wetter than average winters. As expected, colder pools induce opposite weather over Australia, i.e. colder and drier winters.

It has been observed that the vast northern Pacific Ocean switches in 20–30-year cycles between 'warm' and 'cold' phases in what has been called the Pacific Decadal Oscillation (PDO). When this circulation is in warm mode (since 1977 and 1925–46) central interior Pacific waters remain cold while northern coastal waters around Alaska, Canada and western USA become warm. These warm coastal waters modify air circulation in their vicinity and produce warmer if drier winters (i.e. lower water supplies and less snow for skiing). They also carry less nutrients than when cold and damage local fish catches. During a cold phase (1947–76 and 1890–1924) there is warm central interior water and colder coastal waters. While these colder coastal waters chill winter winds and bring colder winters, they are nutrient rich and support bumper fish harvests.

The Pacific region also lays claim to the most widely known and most dramatic example of a weather-changing, dual phase ocean–atmosphere circulation. This is the El Niño–Southern Oscillation or ENSO. In the 1920s, Sir Gordon Walker drew our attention to the dynamic role that the oceans play in atmospheric circulation. He noted that ocean–atmospheric circulation in the equatorial Pacific switched mode or direction periodically, radically altering weather and climate patterns in its vicinity. This phenomenon which became known as the El Niño–Southern Oscillation (ENSO) also led Walker to recognise (without knowing how) the possibility of atmospheric 'teleconnection', i.e. that air circulation in one part of the world (the Pacific Basin) can influence air circulation and thus weather and climate in another part of the world (the South Asian summer monsoon). ENSO events, and the impact they produce, have been the focus of a great deal of study in the last quarter of the twentieth century, not least because they seem to have increased in frequency and intensity during the last 30 years or so.

Key ideas

1. In contrast to the hurricane, many ocean–atmosphere mass and energy interactions are large scale and out of phase with each other, both in terms of time and location.

2. In order to regulate their mass–energy balances, many out-of-phase ocean–atmosphere interactions suddenly switch mode, oscillating from one energy phase to another.

3. Five major ocean–atmosphere oscillations that strongly influence the world's weather and climate have been identified, comprising (1) the North Atlantic Oscillation (NAO), (2) the Arctic Oscillation (AO), (3) the Antarctic Oscillation (AAO) or Antarctic Circumpolar Wave (ACPW), (4) the Pacific Decadal Oscillation (PDO) and the most well known, (5) the El Niño–Southern Oscillation (ENSO).

Plate 2.1 The Antarctic ozone hole on 11 September 2003 extended over 28.2 million km^2. The dark blue hues represent areas where major depletion of ozone existed. In some of these areas concentrations as low as 40 per cent of normal were measured

Source: http://www.gsfc.nasa.gov/topstory/2003/0925ozonehole.html

Plate 2.2 Banks of sea fog can be seen here over large parts of the North Sea and extending into the approaches to the Baltic on 26 March 2003. The land is warm enough to prevent condensation and the fog dissipates once it crosses onto the coast

Source: MODIS Rapid Response Team (http://www.redtailcanyon.com/items/14478.aspx)

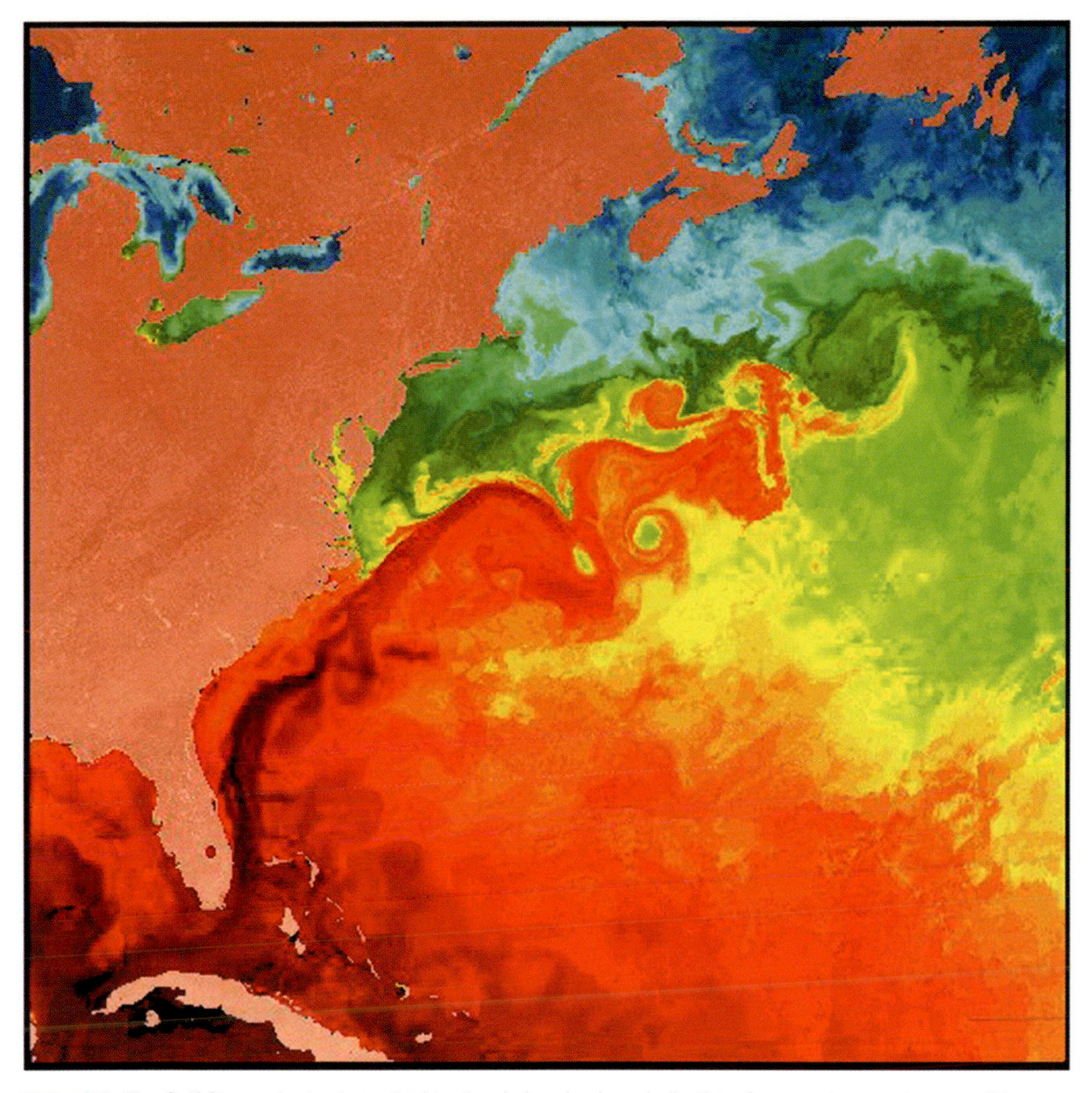

Plate 4.1 The Gulf Stream is clearly marked by the dark red colours indicative of warmer temperatures on this satellite image of the Atlantic Ocean off the eastern seaboard of North America

Source: NASA (http://seawifs.gsfc.nasa.gov/SEAWIFS/IMAGES/eastcoast.gif)

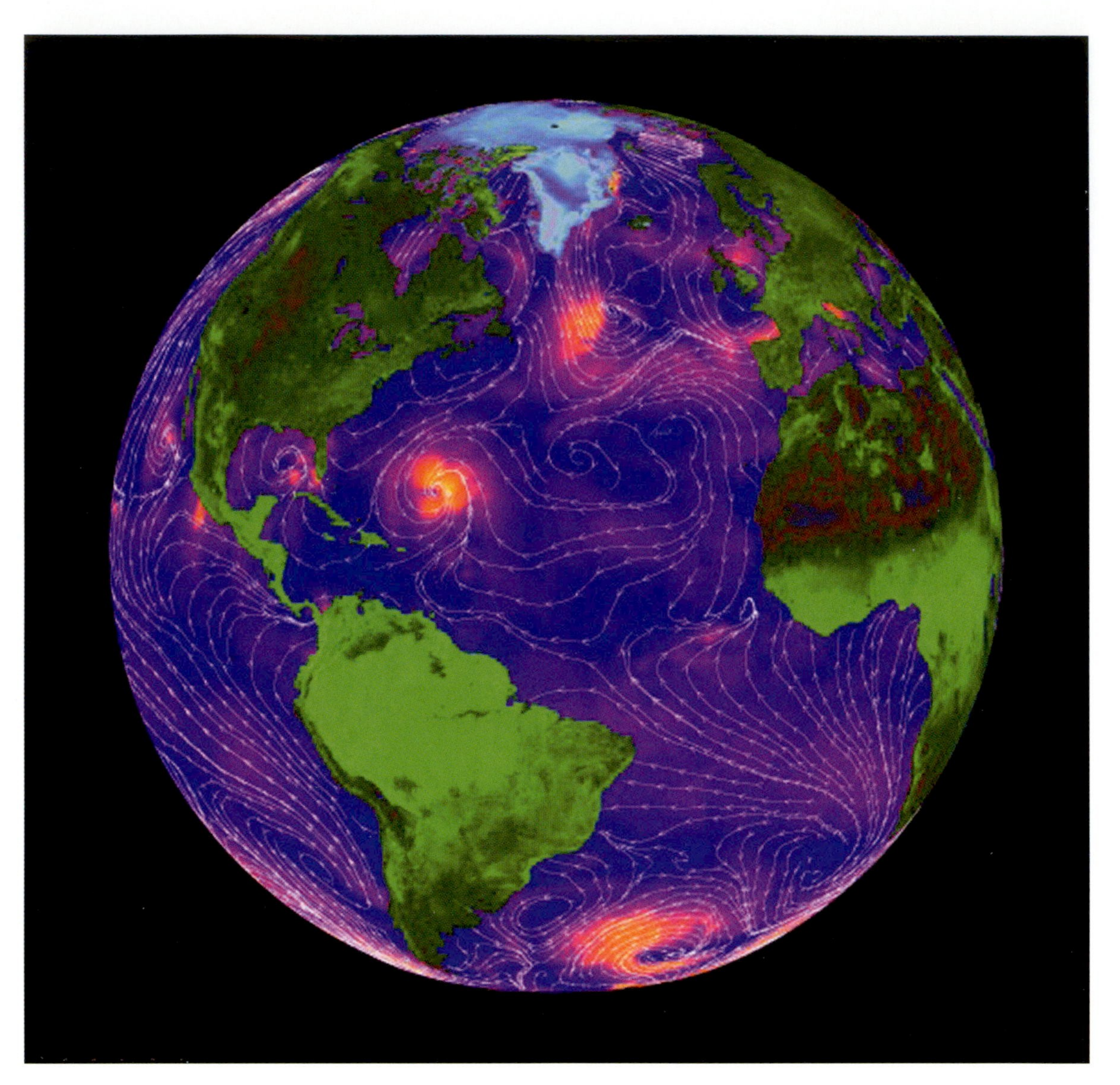

Plate 4.2 This false-colour image, based on scatterometer data, shows wind speeds over the ocean, with orange as the fastest wind speeds and blue as the slowest. White streamlines indicate the wind direction. A hurricane is present in the western Atlantic. The light green areas of land correspond to the greatest radar backscatter. The bright Amazon and Congo rainforests contrast with the dark Sahara Desert. Cities also appear as bright spots, while rivers such as the Amazon can be seen as dark lines

Source: NASA (http://winds.jpl.nasa.gov/news/pacific_global_winds_5.html)

Plate 5.1 Satellite image of Hurricane Mitch making landfall over Nicaragua and Honduras on 26 October 1998
Source: NOAA

Plate 5.2 Warming of the eastern Pacific during the El Niño event of 1997. The white areas show sea surface warming of 6–8 °C above normal.
Source: NASA

Plate 6.1 (right) Annual ice accumulation layers in the Quelccaya ice cap in the Andes
Source: National Oceanic and Atmospheric Administration, Paleoclimatology Program/Department of Commerce
Photo: Lonnie Thompson, Byrd Polar Research Center, Ohio State University

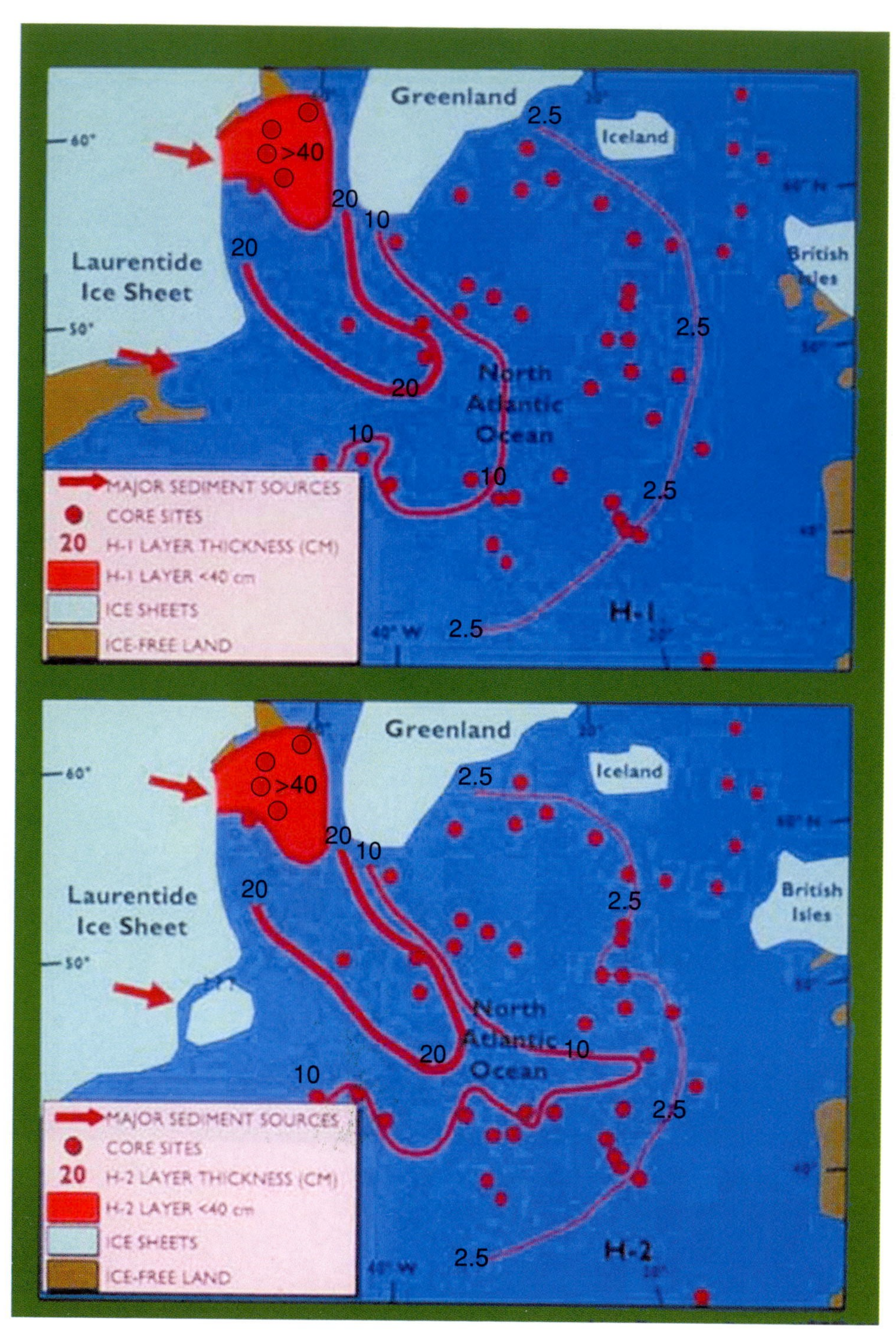

Plate 8.2 Thickness and geographical distribution of Heinrich deposits H1 and H2 from North Atlantic sediment cores. Origin and diffusion of the ice-rafted debris (IRD) from the Laurentide (Canadian) ice sheet is shown

Source: NOAA Paleoclimatology Program and INSTAAR, University of Colorado, Boulder, USA.

Photo: Bedford Institute of Oceanography

5.4 The El Niño–Southern Oscillation

5.4.1 What is it?

Originally, the term 'El Niño' (Spanish for little boy) was used to describe a warm, seasonal ocean current that arrives along the coastline of Ecuador and Peru around Christmas – hence the celebration of El Niño, as the Christ child. However, today the term is generally used as a description of a *dual* phenomenon that has a closely linked oceanic and atmospheric element. Hence, most analyses of El Niño describe first the links between the dramatic warm phase increases in sea surface temperature over the central and eastern equatorial Pacific region and colder waters in the western Pacific. Second, they examine the associated periods of heavy rainfall in the central Pacific (Tahiti), western South America (Ecuador and Peru) and California, with corresponding droughts in Australia, Indonesia and New Guinea.

However, an El Niño is only one combined element (warm period) of a larger dual phase oscillating ocean–atmospheric system. When the system switches to a condition termed a cold phase, Pacific waters return to a cool state with cold water off the coast of Ecuador and Peru and warm water in the western Pacific in the vicinity of Indonesia and eastern Australia. As a result, dry conditions return to the central and eastern Pacific (Peru and Ecuador) with wet conditions in the western Pacific (Indonesia and Australia). Sometimes, in an El Niño's wake, eastern Pacific waters not only return to a cool state, but become unusually cold. When this happens a 'La Niña' (Spanish for little girl) occurs which causes among other climate effects, extreme drought in coastal Peru and Ecuador and very heavy rains in the western Pacific. The entire ocean–atmospheric system so described incorporating warm phase El Niños and colder phase events often with La Niñas is called the El Niño–Southern Oscillation or ENSO. It was originally described as a southern oscillation because most countries in the Pacific Basin where its effects were most immediately felt (southern Ecuador, Peru, Tahiti, Australia, southern Indonesia) are in the southern hemisphere.

5.4.2 Analysis: El Niño and La Niña events

(a) The 'normal' condition

In a 'normal' year, in the southern Pacific Ocean, high-pressure anticlockwise circulation (Figure 4.19) is associated along the west coast of South America with a cool northward-flowing ocean current known as the Peru current. The anticlockwise winds related to this current tend to move warm water away from the coast towards the central Pacific, allowing cold nutrient-rich water to upwell near the surface (Figure 5.5(a)) with the thermocline, a zone which divides warmer lighter top water from deeper colder denser water, rising close to the surface. The cold ocean surface water in the eastern Pacific positively feedbacks into the system and reinforces the high-pressure anticyclonic system which generated it in the first place. This is a further illustration of the closely coupled relationship between the ocean and atmosphere. Anticyclonic development in the lower latitudes of the Pacific Ocean on either side of the equator results in a belt of trade winds that blow westwards along the margin of the equatorward side of the anticyclones. These high-pressure-driven winds generally blow across the eastern Pacific Ocean towards a major cell of low pressure located over Indonesia (Figure 5.5(a)). The consequence of these trade winds is a corresponding flow of surface water in the same direction. As the water is transferred westwards, it is subject to progressive heating from the atmosphere with the consequence that the surface ocean waters along this equatorial zone are warmer (by up to 8 °C) in the west than in the east. The warm surface water created in the western Pacific heats the air immediately above it, encouraging atmospheric instability and rising air currents that

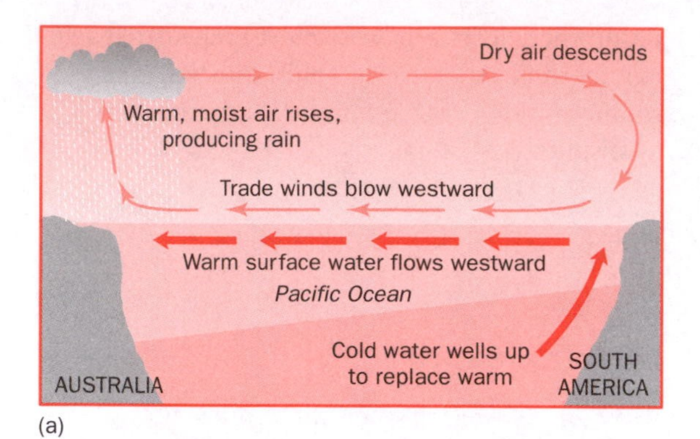

(a)

Dry air descends, droughts occur

Warm moist air rises, producing rain and flooding

Trade winds weaken or head east

Pacific Ocean

Warm water pools in east

AUSTRALIA

SOUTH AMERICA

(b)

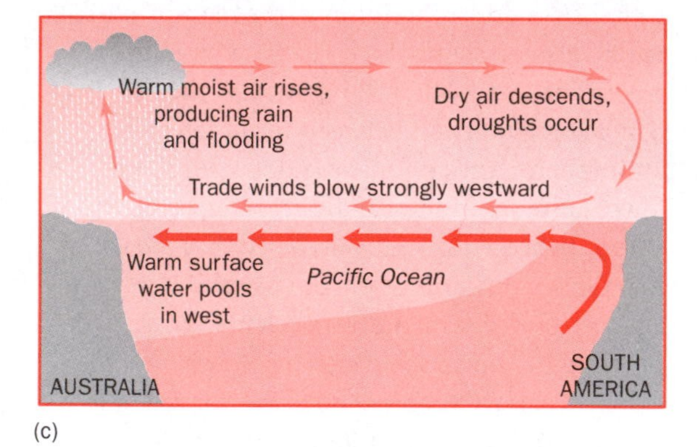

(c)

Figure 5.5 An ENSO event showing three stages in ocean–atmosphere coupling: (a) standard condition; (b) warm phase El Niño; (c) cold phase La Niña
Source: Dawson and O'Hare (2000) from Lippsett (2000)

in turn reinforce the surface low pressure (Figure 5.5(a)). The transfer of warm surface water also results in a slight increase in the level of the sea surface in the western Pacific Ocean than in the eastern Pacific.

The westward flow of surface trade winds on either side of the equator and rising air currents over the low-pressure cell in the western Pacific are compensated by eastward-moving winds aloft and cold descending air over the high-pressure cell in the eastern Pacific. The low-level (about 5–6 km altitude) thermal cell so produced generates its own classic weather at the surface. As suggested previously, cloud and high rainfall are associated with rising cooling and condensing air over the western low-pressure region, while clear skies and dry conditions are found in association with the descending and warming air of the eastern high-pressure cell.

(b) The El Niño condition

During a major El Niño event (Figure 5.5(b)), there is a reversal of the prevailing pattern of ocean surface temperatures and pressures, thermal circulation in the atmosphere, and weather conditions already described for the normal situation (Figure 5.5(a)). There is first of all an increase in surface air pressure across Indonesia and the western Pacific and a fall in average air pressure across the central and eastern Pacific Ocean. A key effect of such a change is a slackening of the trade winds, and indeed during strong El Niño events, the surface trade winds are occasionally replaced by westerly winds. Under these conditions there is a return flow of warm surface water across the tropical Pacific towards South America (Plate 5.2). This ebb return is in the form of a Kelvin (stationary) wave that is typically no more than 15 cm deep but covers a vast area of ocean surface. The creation of a warm pool of surface ocean water towards the east lowers the thermocline by as much as 100 m. This warm pool also creates low pressure, the convective ascent of unstable air and heavy rains in the central and eastern Pacific. Westward-moving winds are created in the upper atmosphere and air descends over the western Pacific. These ocean and atmospheric changes lead in turn to cool waters and high pressure in the vicinity of Indonesia and Australia and are associated with clear skies and reduced rainfall amounts.

(c) La Niña condition

The end of a major El Niño event is accompanied by a see-saw reversal in the distribution of surface air pressure such that there is a progressive return to high pressure over the eastern Pacific Ocean and the strengthening of a large cell of low pressure across Indonesia and the western Pacific Ocean. A La Niña event occurs (Figure 5.5(c)) when the switch back from El Niño conditions to a normal condition overshoots the mark so that anomalously cold conditions and high pressure develop in the central and eastern Pacific with anomalously warm conditions and low pressure in the western Pacific. With very cold water in the eastern equatorial Pacific and extra warm water in the western Pacific, La Niña occurrences are linked to severe droughts in coastal Peru and Ecuador, and higher than average rainfall with flooding and landslides in Indonesia and north-east Australia.

Key ideas

1. An ENSO event is a large-scale ocean–atmosphere system in the Pacific that switches mode periodically from a warm phase El Niño episode to a cold phase La Niña episode.
2. Under 'normal' conditions in the Pacific, cold water, high pressure and low rainfall are found in the central and eastern regions, while warm water, low pressure and wet conditions prevail in the western regions.
3. El Niño events change the 'normal' condition and are recognised by the development of anomalously warm water over the central and eastern Pacific.
4. During an El Niño, warm water and low pressure over the central and eastern part of the Pacific Basin bring wet conditions to countries like Tahiti, Peru and Ecuador, while cool water and high pressure over the western part of the basin encourage low rainfall in Australia and Indonesia.
5. A La Niña is the converse of an El Niño with very cold water, strong high pressure and very dry conditions in the eastern Pacific and very warm water, severe low pressure and very wet conditions in the western Pacific.
6. ENSO events involving the switch from El Niño to La Niña conditions alter not only local and regional climate (droughts, intense rainfall) but also climate patterns at the global scale.

5.5 Monitoring and predicting ENSO events

5.5.1 Measurement

ENSO events can be measured using different parameters, each giving a different perspective on the cycle of El Niños and La Niñas in the Pacific Basin. Changes in the average air-pressure difference across the equatorial Pacific between Tahiti (central Pacific) where pressure is normally high and Darwin, northern Australia, where it is usually low, have been a traditional way of measuring ENSO events. With a lowering of the mean Tahiti–Darwin air-pressure gradient, the onset and development of an El Niño are suggested, while an increase identifies a La Niña. The recording of anomalous sea surface temperatures (SSTs), i.e. departures from average conditions in the eastern Pacific, is another single factor method of ENSO analysis. When this indicator is used the recent El Niño of 1997/98 shows up as the strongest El Niño that century. Multivariate analyses, i.e. using more than one factor at various locations across the whole Pacific Basin, is thought to be more reliable and a better method for predicting the onset, development and decline of an ENSO event. A multivariate method using the weighted average of six ENSO features, including sea level pressure, SSTs, surface air temperature, surface wind speed and direction, and total cloud cover is shown in Figure 5.6. Positive departures above the zero line indicate the development and progress of an El Niño, while negative departures indicate the same for a La Niña. When this type of indicator is employed,

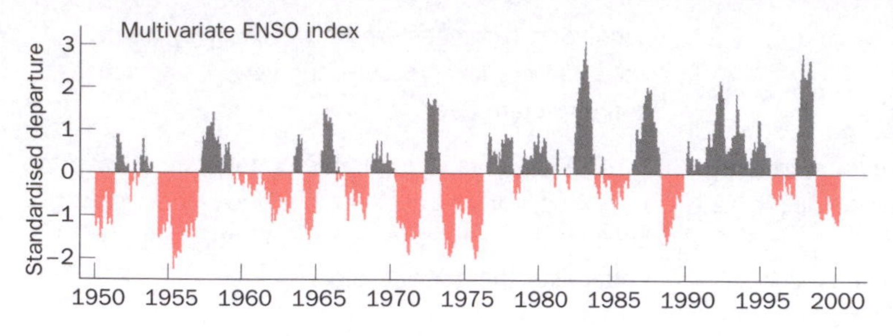

Figure 5.6 A multivariate ENSO index for the period 1950–2000, showing warm phase El Niño episodes (positive values above the zero line), and cold phase La Niña episodes (negative values below the zero line). The larger the index (positive and negative) the more intense the El Niño and La Niña
Source: Wolter and Timlin (1998, pp. 315–36)

warm El Niño events appear to occur on average every 4 years, but the cycle is irregular having a periodicity of anywhere between 3 and 8 years. El Niños usually last from several months to about a year, but sometimes, as during 1991–93, the phenomenon can last for 2 years. Figure 5.7 shows in more detail how early El Niño events during 1965/66 and 1972/73 all featured an early warming in the far eastern Pacific and reached their peak development before the end of the first year. In contrast, the more recent El Niños (1982/83, 1986/87 and 1991/92) took longer to mature, typically reaching their peaks in the spring of the following year. The El Niño of 1997/98 featured two peaks, one in July/August 1997 and the other in February/March 1998. In relation to peak conditions, the El Niños of 1982/83 and 1997/98 have been the strongest events.

5.5.2 Atmospheric teleconnections and weather impacts

(a) El Niño warm phase

Some of the main atmospheric and environmental impacts of the 1982/83 and 1997/98 El Niño events can be seen in Table 5.2 and Figure 5.8 respectively. It can be seen that these events were responsible for a good deal of anomalous climate behaviour within the central Pacific Basin, for example flooding in Ecuador and northern Peru, serious drought in Indonesia and Australia (Dec.–Feb.). What is surprising is the extensive climate effects in locations and places far removed from the core Pacific area, i.e. drought in southern Africa and flooding in the US Gulf states (Dec.–Feb.) and drought in India, with storms in the US mountain and Pacific states (June–August).

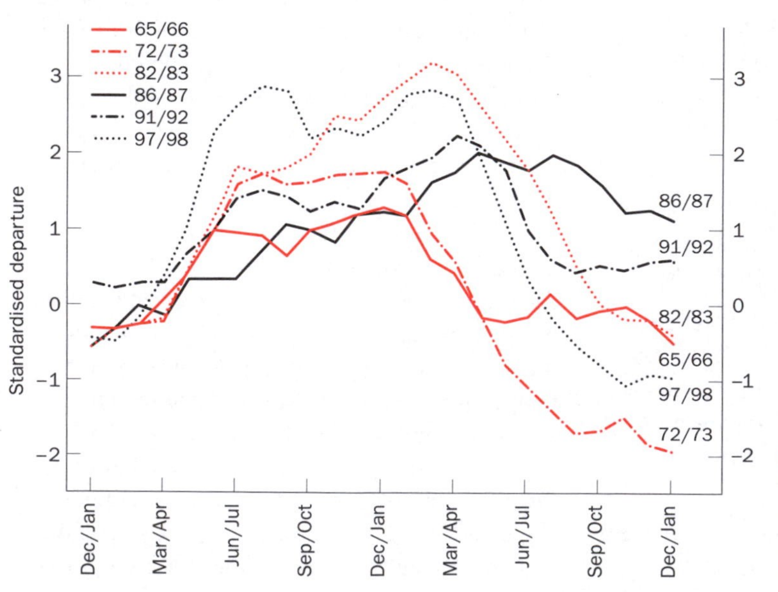

Figure 5.7 The development of six El Niño events from 1965 to 1998 using the multivariate ENSO index
Source: Wolter and Timlin (1998, pp. 315–36)

Table 5.2 Global weather impacts and costs attributed to the 1982/83 El Niño event

Location	Anomaly	Major social impacts	Costs (US$)
US mountain and Pacific states	Storms	45 dead	1.1 billion
US Gulf states	Flooding	50 dead	1.1 billion
Hawaii	Hurricane	1 dead	230 million
North-eastern USA	Storms	66 dead	N/A
Cuba	Flooding	15 dead	170 million
Mexico and Central America	Drought	N/A	600 million
Ecuador and northern Peru	Flooding	600 dead	650 million
Southern Peru and western Bolivia	Drought	N/A	240 million
Southern Brazil and northern Argentina	Flooding	170 dead 600,000 evacuated	3 billion
Bolivia	Flooding	50 dead, 26,000 homeless	300 million
Tahiti	Hurricane	1 dead	50 million
Australia	Drought, fires	71 dead, 8,000 homeless	2.5 billion
Indonesia	Drought	340 dead	500 million
Philippines	Drought	N/A	450 million
Southern China	Wet weather	600 dead	600 million
Southern India	Drought	N/A	150 million
Middle East, Lebanon	Cold, snow	65 dead	50 million
Southern Africa	Drought	Disease, starvation	1 billion
Iberian Peninsula and northern Africa	Drought	N/A	200 million
Western Europe	Flooding	25 dead	200 million

Source: *The New York Times*, 2 August 1983

It would seem that El Niño events are capable of influencing not just local climates and environments but also distant, regional and even global ones (Box 5.2).

(b) La Niña cold phase

Although cold phase (cold in the eastern Pacific) La Niña events are less disruptive than their warm phase El Niño counterparts, they still pack their own set of climate and environmental effects (Box 5.3). Like an El Niño, a La Niña event exerts not only local impacts, i.e. severe drought in Peru, excessive rainfall and landslides in Indonesia, but also has wider regional and global repercussions including cool conditions in the north-east Pacific and dry conditions in southern Brazil (Figure 5.9). By increasing normally high surface temperatures in the western Pacific and subsequently in the neighbouring Indian Ocean, La Niña events raise overlying air temperatures and moisture in these regions. La Niña episodes thus correlate strongly with high rainfall monsoons (June–August) with bumper rice harvests in India and excessive flooding in Bangladesh (see Box 5.3). Extra warm temperatures in the western Pacific and Indian oceans during La Niña events are also implicated in the bleaching and destruction of extensive areas of coral reef (e.g. the Great Barrier Reef, Maldives).

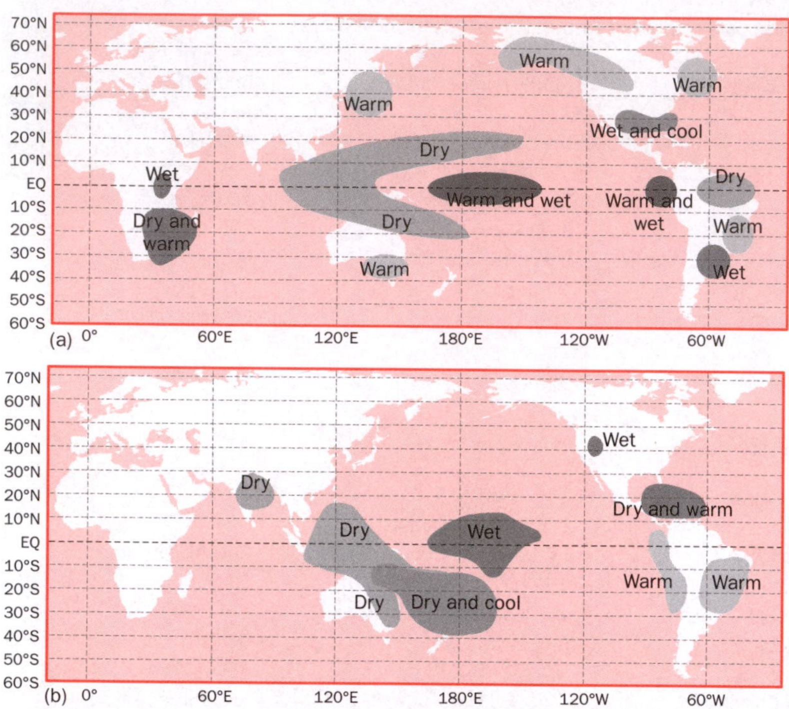

Figure 5.8 Geographical impact of an El Niño warm event during (a) December–February and (b) June–August

BOX 5.2

The 1982/83 El Niño event

As illustrated in Figures 5.6 and 5.7, the 1982–83 El Niño was the strongest of the century, perhaps in recorded history. During the powerful 1982–83 ENSO event there was an abnormally strong eastward movement of surface water across the tropical Pacific that resulted in an increase in relative sea level across the eastern Pacific Ocean. As the water was progressively warmed due to solar insolation, the surface water temperatures reached more than 7 °C higher than usual off the coast of Peru. The arrival of this warm surface water caused the thermocline off the coast of South America to drop more than 150 m (500 feet) and almost completely stopped the process of oceanic upwelling.

The 1982/83 El Niño produced major changes not only to local, but to global weather patterns (Figure 5.9 and Table 5.2). The development of an anticyclone over the western Pacific led to devastating droughts, dust storms and bush fires in Australia, Indonesia and southern Africa. In the eastern Pacific intense low pressure produced the heaviest rainfall ever recorded in Peru – 3.8 m (11 feet) in places where 2.5 cm (6 in.) was the norm. Some rivers carried 1,000 times their normal flow and produced devastating floods in the

▶ country. There was also heavy rainfall and flooding in Ecuador, Central America and the Gulf states, and severe storms rearranged the beaches of California. Further afield, the 1982/83 El Niño caused droughts across southern Africa and in parts of India. In these two examples, there is a good relationship between the occurrence of El Niño events, low summer rainfall and crop failure.

There were also a host of more indirect environmental problems associated with the 1982/83 El Niño event. The replacement of cold nutrient-rich sea water off the western coast of South America by warm nutrient-deficient conditions resulted in the widespread mortality of fish and sea birds. With the virtual collapse of Peru's fishing industry thousands of poor fishermen lost their jobs, many migrating to find work in the squatter settlements of the capital Lima. There was also a rise in bubonic plague in New Mexico where cool wet weather provided favourable conditions for flea-ridden rodents to spread the disease to humans. Altogether, the 1982/83 El Niño event has been blamed for between 1,300 and 2,000 deaths. The global total insurance damage attributable to its associated floods, wind storms, drought and other economic catastrophies is generally believed to have exceeded $13 billion.

The 1997/98 El Niño

It can be seen from Figures 5.6 and 5.7 that another severe El Niño event developed during 1997–98, proving comparable in strength and impact with the 1982–83 episode. This El Niño event developed very rapidly and culminated during December 1997–January 1998. By autumn 1997, there was already a 50 per cent decrease in monsoon rainfall across Indonesia, with drought conditions exacerbated by extensive forest fires developing in many areas. There were many media reports of damage such as choking smog in Jakarta that persisted throughout the last 4 months of 1997. Similarly, severe drought ravaged parts of Australia, particularly in Queensland and locally in south-east Australia where forest fires were common. Drought conditions also affected large areas of New Zealand as well as extensive areas of southern Africa, particularly in Namibia and Zimbabwe. Similarly, the disruption of upwelling along the west coast of South America led to very heavy rain during the 1997–98 winter in Peru together with floods and landslides. Severe flooding also took place in East Africa, particularly in Kenya, while storms and heavy rainfall also occurred in California and north-west Mexico. The exceptionally high ocean temperatures in the eastern Pacific Ocean may also have been an important factor leading to the development of numerous tropical storms which were particularly destructive in Mexico.

One of the most extreme demonstrations of the power of the 1997/98 El Niño to alter weather on a global scale is the great Canadian *ice storm* in the winter of 1998. It began as a warm belt of rain over Texas. This warm moist air was pushed by high pressure induced by the El Niño much further north than usual over the eastern coastal states of the USA into Canada. When this northward-moving air stream met a much colder air mass from the Arctic moving south, the conditions were set for exceptionally severe weather. For over two weeks, heavy but very cold rain fell over the city of Montreal. As the rain fell it froze at ground level creating a thick blanket of surface ice over the city, damaging power lines and property and bringing transport and normal work patterns to a halt. The 1998 ice deluge struck with such severity that the city came within hours of being completely abandoned. In all, 27 people died in the province of Quebec, 17,800 were made homeless and 3.2 million were without electricity for 4–5 days. The total cost of the Montreal ice storm has been put at over $4.5 billion.

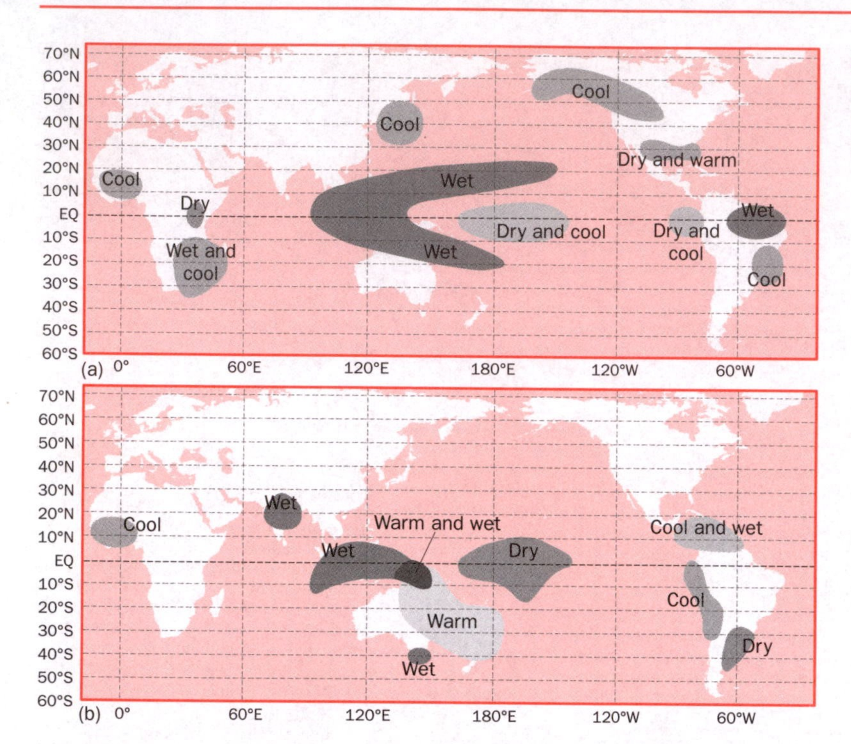

Figure 5.9 Geographical impact of a La Niña cold event during (a) December–February and (b) June–August

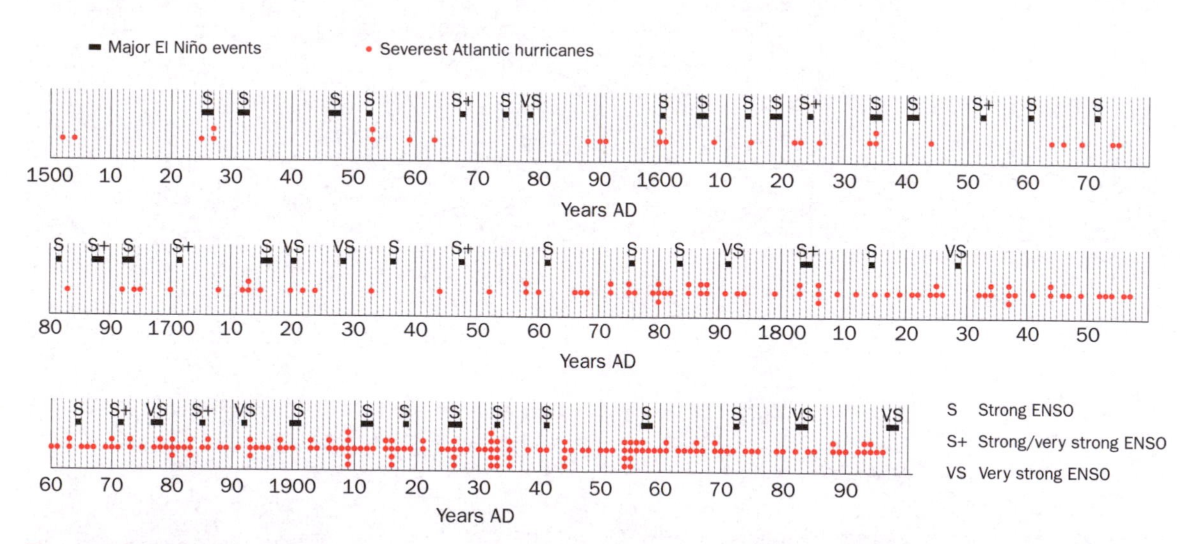

Figure 5.10 The occurrence of different intensity El Niño events and the pattern of hurricane impacts in the southern Atlantic since AD 1500

Source: Dawson and O'Hare (2000, p. 205)

BOX 5.3

CASE STUDY

Case study: La Niña impacts

The particularly intense La Niña that followed immediately in the wake of the 1997/98 El Niño produced some dramatic weather incidences worldwide, including unprecedented heavy and prolonged rain in the summer of 1998 in central China, India and Bangladesh. Bangladesh floods every year from heavy monsoon rains (the annual average for the country is over 4,000 mm or 160 in.), but those of 1998 were the highest on record. The rains started in early July. After two months of heavy rain, 30 per cent of the capital city of Dacca (8.5 million population) was under water. As the water stagnated and the sewerage system broke down, severe outbreaks of water-borne diseases, including malaria, cerebral malaria, dysentery and diarrhoea took many lives, in addition to those who had drowned. During 3 months of flooding, 75 per cent of the country was under water, 30 million people were rendered homeless and over 1,000 people died.

The La Niña of 1998 was not only unusually intense but also lasted for over 2 years, only diminishing during the spring of the year 2000. By February 2000, high temperature anomalies were still present in the south-western Indian Ocean. Warm moist air over the seas off the coast of Mozambique was drawn inland, subsequently cooling and producing extensive cloud and rain over the country's interior mountain ranges. The massive slow-rise flooding that later developed across the country's southern river systems and which paralysed many districts, was well recorded by the world's media. Despite foreign assistance (too little too late) many hundreds of people were drowned and over 100,000 were made homeless. In all, the great La Niña-induced Mozambique flood of February 2000 set the country's development efforts back at least 10 years.

(c) Atmospheric teleconnections

Climate scientists refer to the term 'atmospheric teleconnection' to describe the global-wide climate impacts that ENSO events can produce. There are a number of ways such teleconnections can be delivered. First, more energy is exchanged by evaporation between the earth's surface and the atmosphere over the tropical oceans than anywhere else on the planet, and of the world's oceans, the Pacific is the most important source of energy supply. When the Pacific Ocean warms during an El Niño event, additional large amounts of heat and moisture are added to the atmosphere. Since this extra energy and moisture can be transported thousands of miles by wind, it can modify distant weather patterns, changing and often intensifying local storms including tropical cyclones, thunderstorms and even mid-latitude depressions. Second, pressure changes induced in the lower atmosphere over the Pacific during ENSO events modify neighbouring and even distant surface-level pressure systems and thus the weather in their vicinity. Third, it has been suggested that ENSO events are capable of causing weather and climate changes in different parts of the globe because they also affect upper air circulations first in the Pacific and then in adjacent regions.

5.5.3 ENSO periodicity: a brief history

A question in relation to ENSO events is whether or not such events have taken place with any regular periodicity during recent centuries. Such an analysis would be particularly valuable since it would assist climate scientists attempting to understand why ENSO events took place during certain periods and not others.

Detailed reconstructions of ENSO events and their climate impact (wet periods, drought) have been traced back several thousand years. This has been carried out using tree ring analysis, while oxygen analysis of coral growth rates in the oceans (see section 6.3.3) offer a further proxy source to take our understanding of ENSO incidence back further into prehistory. Figure 5.10 shows, using documentary sources, detailed constructions of ENSO events and their relative strengths dating from *c.* AD 1500. Inspection of the timing of strong, strong/very strong and very strong El Niño events shows some significant trends. For example, there is a very noticeable gap between 1830 and 1865 during which time no major ENSO events appear to have taken place. Similarly, between 1942 and 1974 only one major ENSO event occurred in the category S, i.e. strong or higher. Comparison between the timing of major El Niño events and the most destructive tropical Atlantic hurricanes is also instructive. It would appear that strong El Niño events limit hurricane development and the destruction they cause (measured in terms of loss of life) in the Atlantic. This may be because SSTs decline in the Atlantic (though they increase in the Pacific) when there is an El Niño event. It may also be due to evidence which suggests that wind shear in the upper atmosphere over the Atlantic increases during El Niño episodes. Nevertheless, these results are striking and point to a set of processes in the earth's climate system demonstrating a 'see-saw' link between ocean–atmosphere processes in the tropical Pacific and those in the tropical Atlantic oceans. The fact that these 'see-saw' patterns may extend back to *c.* AD 1500 (and surely beyond) indicates that such processes were in operation prior to the Industrial Revolution, and that these may truly be considered part of a 'natural' global climate process.

5.5.4 ENSO futures

(a) Disentangling global warming and ENSO events

One of the most perplexing problems associated with understanding the earth's climate system is how to distinguish natural factors such as ENSO events which may induce short-term or cyclical climate changes, from longer-term climate trends caused by human-induced global warming. The climate change effects of global warming and ENSO events are closely interwoven with regard not only to long-term averages (both have been increasing in the last 30 years), but also to extreme weather events. Records of warming SSTs and rising sea levels are given as proof positive for global warming. Nevertheless, major ENSO events are extremely important contributors to the natural variability of SSTs and base level changes in the tropical oceans and may influence global sea temperatures in their own right. Moreover, one of the tenets of greenhouse-induced global warming is that as the world warms, there will be a higher frequency and intensity of extreme weather events such as wind, storm, flood and drought. But these changes are precisely those that are attributable to ENSO events. For instance, episodes of drought in Australia and excessive (higher than normal) rainfall with slow-rise floods in Bangladesh in the late 1990s could be blamed on climate change processes linked to global warming, but they could also be placed at the doorstep of ENSO – to El Niño and La Niña respectively.

(b) ENSO uncertainty

It is not possible to be confident about future changes in the behaviour of ENSO events. If ENSO events are being forced by global warming then we should expect El Niño and La Niña episodes to increase in the future. Recent evidence suggests that this may be the case. During an apparent 'shift' in temperature of the tropical Pacific to warmer conditions between 1976 and 1998, ENSO events were more frequent, intense and persistent (see Figure 5.6). The persistence of this warm state, however, remains unclear, as there was a long cold La Niña from late 1998 until early 2001. A future with more frequent, strong, prolonged and disruptive warm phase El Niño events therefore looks highly probable; but the Pacific Basin in particular and the climate system in general is so complex that other ENSO futures are possible.

Key ideas

1. Evidence shows that ENSO events have become more intense and more frequent in the last 30 years or so, and it is probable that such a trend will continue in the future.

2. It is difficult to separate out the individual effects of El Niño events and those attributed to global warming since many of their impacts are similar (i.e. global heating, greater storminess).

3. An understanding of historic ENSO events together with climate modelling of the system may allow us to produce more accurate scenarios of ENSO frequency and intensity in the future.

Further reading

Bigg, G. R. (1996) *The Oceans and Climate*. CUP, Cambridge.

Brugge, R. (ed.) (1998) El Nino–Southern Oscillation. *Weather* (Special Issue), **53**(9): 270–335.

Dawson, A. and O'Hare, G. (2000) Ocean–atmosphere circulation and global climate. *Geography*, **85**(3): 193–208.

Diaz, H. F. and Pulwarty, R. S. (eds) (1997) *Hurricanes: Climate and socio-economic impacts*. Springer-Verlag, Berlin.

Lippsett, L. (2000) Beyond El Nino. *Scientific American Presents – Weather*, **11**(1): 77–83

Pielke, R. A. Jr and Pielke, R. A. Sr (1997) *Hurricanes: Their nature and impacts on society*. John Wiley & Sons, London, 279 pp.

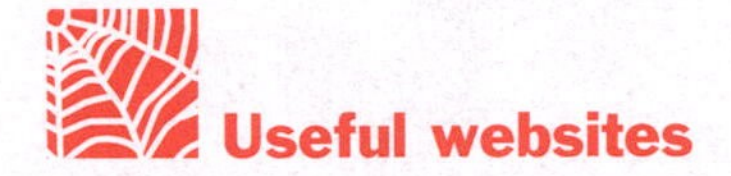

Useful websites

The Pacific Decadal Oscillation:
http://tao.atmos.washington.edu/pdo/

Hurricanes: video/audio representations:
http://www.hurricanecity.com/

Hurricane climatology:
http://earthobservatory.nasa.gov/Library/Hurricanes/hurricanes_1.html

The most intense US hurricanes, 1900–2000:
http://www.aoml.noaa.gov/hrd/Landsea/deadly/

Impacts and prediction of El Niño events:
www.pmel.noaa.gov/tao/elnino/impacts.html

Current and past ENSO events:
http://www.cdc.noaa.gov/ENSO/

CHAPTER 6

Changes in the climate system

6.1 Components and interactions in the climate system

6.1.1 Introduction

The public in many parts of the world have become highly sensitised to climate change in recent years as individual extreme events have forced a reappraisal of what is meant by 'normal'. In the past, it was common for a few decades of meteorological observations to be taken as indicative of the likely range of conditions which would be experienced at a place. Using statistical means, such climatic fingerprints, typically established over a 30-year reference period, were extended to provide estimates of the likelihood of specified extremes occurring. The flood defences necessary to cope with the once-in-century rainfall, or the bridge strength necessary to cope with the once-in-500-years gust of wind could be estimated. Climate extremes could be expressed in probability terms, and engineers and architects could design structures which provided an acceptable trade-off between construction cost and public safety. But all of this inherently assumed a stable climate, one for which a baseline established on data from the past could be relied upon to provide safe guidance for the future. If the climate is changing, such an approach is fundamentally flawed.

It is now clear that climatic change is an inherent characteristic of climate. It is, for example, now evident that both the rate of warming and overall warmth of the twentieth century are likely to have been unsurpassed during the past 1,000 years. Globally, the 1990s was the warmest decade of the twentieth century, and 1998 the warmest year of the instrumental record. Yet this is the century considered to be 'normal' for the purposes of designing structures and for environmental management and protection generally. Clearly, what is normal may turn out to be very abnormal when a longer time perspective is employed. This implies a need for establishing the range of climate conditions existing in the past. But it also emphasises that the climate of future decades and centuries will most probably also hold surprises in store. A prudent approach to climate change issues is required.

6.1.2 Components of the climate system

Spatial variations of climate are a response to local considerations of latitude, distance from the sea, topography, land cover and a host of other factors. Temporal variations are equally a response to changes in the relative importance of particular controls. Seasonal changes in the radiation balance or surface pressure, diurnal changes in air movements – these may be just some of the influences producing changes in mass, momentum and energy. The climatic system is constantly seeking to readjust itself to accommodate changes in the mix of external and internal forcing agents.

(a) *Structural components*

The *atmosphere* is only one of six major components instrumental in determining the status and directional trend of climate at a particular time (see Figure 1.2). The water bodies of the world, or *hydrosphere*, store and relocate the sun's energy and respond also to changes in the composition of the atmosphere. The ice bodies, or *cryosphere*, influence the reflectivity of the surface to solar radiation and also the operation of deep ocean circulations. The *lithosphere* moves to rearrange places over long timescales and provides mechanisms, such as volcanoes, that change the composition of the atmosphere. The marine and terrestrial *biosphere* also achieves atmospheric transformations through the processes of photosynthesis and respiration, and by the emission of gases. Finally, modern interpretations of the climate system include *human activity* as a sixth component, justified by the ability of human industrial society to alter present climate by greenhouse gas global warming. Each of these six subsystems is itself operational on different timescales and is responding to forcing exerted on it by other earth systems. Some responses are slow and their present operation may reflect perturbations occurring decades or even centuries ago. If solar energy input were suddenly to stop, the atmosphere would have ceased to move after about two weeks; the oceans would continue to circulate for a century or more (Table 6.1). The difficulties of addressing present climate change or anticipating future climate change are thus obvious: how can these differing timescales be reconciled for a selected period?

Table 6.1 Adjustment time for components of the climate system to accommodate forcing

Climate component	Adjustment time	
	Days	Years
Atmosphere – free	~11	
Atmosphere – boundary layer	~1	
Hydrosphere – upper ocean		~7–8
Hydrosphere – deep ocean		~300
Hydrosphere – terrestrial	~11	
Cryosphere – mountain glaciers		~300
Cryosphere – ice-sheets		~3,000
Cryosphere – sea ice	Days to centuries	
Cryosphere – surface snow and ice	~1	
Biosphere – soils/vegetation	~11	
Lithosphere – mantle		~30,000,000

Source: After Henderson-Sellers and McGuffie (1987)

(b) *Feedback interactions*

The extent of climate change in response to changes in forcing also depends on feedback effects. These involve internal interactions between the changing components which either enhance (positive) or dampen (negative) the initial perturbation. Among the most significant of these are the ice-albedo, water vapour and cloud feedback effects. The ice-albedo effect is a positive feedback which occurs because ice/snow coverage changes enhance themselves. For example, a warming which results in a reduction of snow/ice cover produces a greater area of darker, more heat-absorbing surfaces. These warm more quickly to produce more melting and enhance the initial changes. Positive feedback also occurs when water vapour changes in the atmosphere occur. For example, a warming which produces more evaporation produces more water vapour in the atmosphere. But water vapour is a highly effective greenhouse gas which traps outgoing terrestrial long-wave radiation. The atmosphere warms, then the surface, resulting in more evaporation, more warming, etc. Cloud feedback is more complex and exhibits both positive and negative feedback characteristics. Cloud is reflective, so more evaporation of surface water and its eventual condensation could be expected to produce a greater reflectivity for incoming solar energy, cooling the surface. This would be an example of negative feedback where the initial perturbation is dampened. However, higher clouds, which are fairly transparent to incoming short-wave radiation, nevertheless contain sufficient water vapour to trap outgoing long-wave radiation, thus producing a warming tendency. Deciding which process is dominant is not easy, particularly as cloud cover changes may not correspond with cloud volume changes. Stratiform clouds may be horizontally extensive but vertically shallow and cumuliform clouds may be the opposite. Further consideration of the feedback effect is provided in Chapter 7.

Key ideas

1. Calculations of 'return periods' for climatic extremes based on historic data can no longer be relied upon. Risk management in a wide range of areas now requires to be 'climate change proofed'.

2. Forcing of climate change can occur due to changes in the relative importance of either external or internal controls. Components of the climate system respond to such forcing on a variety of timescales.

3. Uncertainties in predicting climate change are magnified by complex positive and negative feedback effects which become operative as the climate system responds to changes in forcing.

6.2 Concepts of climate change

In the previous chapter, it was suggested that climate change occurs when the climate moves beyond certain previously defined boundaries or limits, such as long-term means of temperature, precipitation, wind, etc., together with their

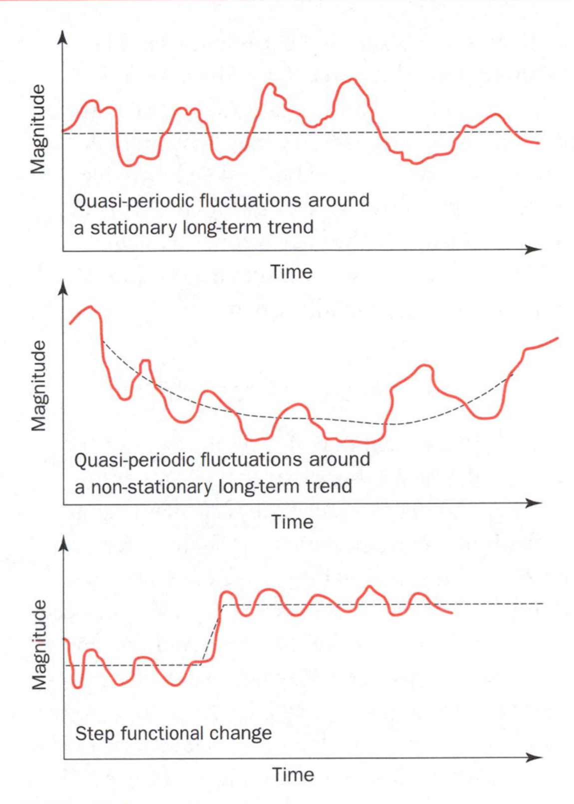

Figure 6.1 Types of change in climate time series

extremes. By doing so it is possible to define 'climate change' in a wide variety of ways. These may range from what might be more appropriately described as *short-term* fluctuations to more *long-term* fundamental shifts in climate. Some of these types of variation are displayed in Figure 6.1. Intuitive concepts of climate change might centre on *step-functional* change as the most obvious response to a change in boundary conditions, for example solar energy receipt, while *quasi-periodic* variation might be more easily understood as resulting from internal readjustments, for example oceanic fluctuations such as El Niño.

The energy content and mass of the climate system are difficult to comprehend. Solar energy input is about 1.81×10^{14} kW or about 342 watts for every square metre of the earth's surface, and almost 500,000 km^3 of rain falls each year. It might be thought that such a vast system is impregnable to change as a result of what might seem like relatively puny human action. However, it is important to remember that systems may have pressure points where they may be vulnerable. The climate system may be rendered unstable when key pathways for energy and moisture are interfered with. Rather like a cone balancing on its apex, seemingly small forces may be capable of producing large readjustments.

6.2.1 Three types of climate change

The climate system may not respond in a measured, predictable manner to changes in forcing, making the causes of particular changes difficult to ascertain. Figure 6.2 shows three possible kinds of climate change outcomes to a change in forcing. A transitive response may occur with a gradual transition to a new equilibrium. An intransitive system may not respond to a change in forcing and may have more than one mode of operation with a particular set of controls. A quasi-intransitive response might involve little or no response until some threshold was crossed and then a 'flip' to a new state occurs. For climate scientists this is the most worrying mode, since it suggests climate change may be rather like a coiled spring, somewhat unpredictable as to when it will produce a surprise. Unfortunately the concept also suggests that recovery may be equally

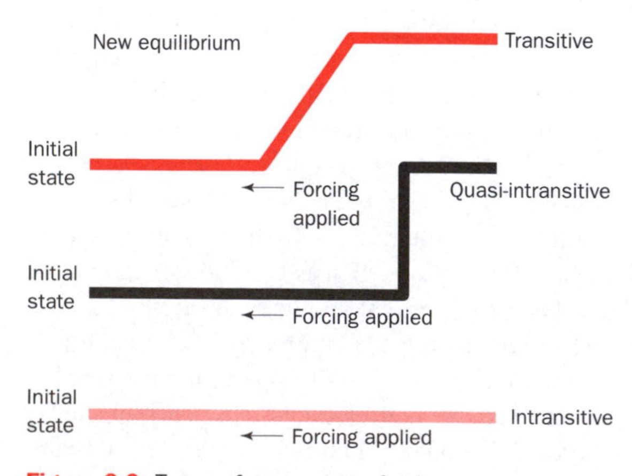

Figure 6.2 Types of response to forcing

unpredictable. There are obvious implications here for assessing how human impacts on the atmosphere should be considered and in measuring how effective control measures such as the Kyoto Protocol might be for mitigating future climate changes.

Key ideas

1. Destabilisation of the climate system may occur as a result of relatively small interference in key pathways. Critical thresholds may also exist beyond which recovery mechanisms may not succeed.

6.3 Techniques for reconstructing past climates

There are basically two major techniques that can be used to reconstruct former climates. These include the use of meteorological instruments (thermometer, barometer, satellites) that measure the weather relatively *directly* and the application of methods that estimate the weather or climate *indirectly*, i.e. using documentary sources (ships' logs, estate reports) or environmental sources (tree rings, pollen, sea and lake sediments, ice cores).

6.3.1 Instrumental observations

The most satisfactory means of comparing past with the present climate is by using instruments exposed under standard conditions. The availability of such data is today taken very much for granted. Worldwide communications and the Internet mean that conditions existing at a particular instant in many parts of the world can be called up and used to understand present and predict future weather changes. Archived data sets from professionally maintained observing stations can be interrogated and climate change patterns and trends established. This provides, however, only a relatively restricted insight into the past.

(a) Early history

Although elementary instruments such as the rain gauge can be traced back to the fifteenth century in Korea, the development of modern instrumentation can be mostly sourced to Renaissance Italy. Galileo is credited with the invention of the thermometer in 1590 and Torricelli with the barometer in 1643. Problems with the quality of early instruments, and with the manner in which they were deployed, render their results unreliable however until much later. The thermometer was particularly problematical. Two fixed points could be established with some ease: the freezing point and boiling point of water. But how many subdivisions should be made between these? Seventy-seven different scales were reportedly in use during the nineteenth century, and the quality of the glass used posed further problems. G. Daniel Fahrenheit (remembered for his scale using 180 divisions between freezing and boiling point) is perhaps more appropriately remembered as a thermometer maker who used the best glass. Equally, exposures varied considerably. Sometimes the thermometer was hung indoors, or on a north-facing wall. England's first rain gauge in 1676 was located on the roof of Richard Townley's house, with a pipe leading down to his bedroom where he could make his measurements in comfort!

(b) Problem of homogeneity

Amateur observers usually consisted of the better-off members of society with time to devote to their hobby, and little coordination of readings is apparent until the end of the eighteenth century. By then the Industrial Revolution was leading to an influx of people into the coalfield cities of Europe. As these urban agglomerations grew, they began to experience major problems of public health and sanitation. Installing a public water supply and sewerage network began to be important considerations for municipal authorities. This necessitated a quantitative assessment of rainfall and river resources as sources of water supply and waste dispersal. The beginnings of many meteorological networks can be traced to this time (see section 9.3). The coming of the railways in the early 1800s in the United Kingdom necessitated a system of standard time, and thus meteorological observations began to be organised in a more rigorous temporal manner. The UK Meteorological Office dates from 1854, the USA Weather Bureau from 1870 and the World Meteorological Organisation from 1873. The European colonial powers also extended their observational networks into their overseas territories such as India and West Africa during the nineteenth century, although a coastal bias was evident in the coverage which presented a misleading picture of climates in many tropical areas. Yet even by the middle of the twentieth century there were large tracts of the world with poor or non-standardised instrumental coverage and only with the advent of satellites can a truly global coverage be claimed (see Figure 2.20).

The longest reliable (i.e. homogeneous) instrumental temperature record dates from 1659 and is considered representative of a typical site in central England (Figure 6.3 and section 9.3). This Central England Temperature (CET) record has been painstakingly assembled by comparing overlapping records and allowing for changes of site and exposure from a variety of locations. Even over this relatively short period, climatic fluctuations of up to 2 °C in annual and seasonal means are apparent. Although this may not appear to represent major change, a small fluctuation in a primary climatic parameter such as mean temperature may manifest itself much more dramatically in other aspects of climate, such as storminess or snowfall, as will be demonstrated later in this chapter.

6.3.2 Documentary sources

(a) Human proxy data

Long before instrumentation was available, individuals made written comments about the weather in diaries and journals. In some countries such as Iceland, China and Ireland some ancient

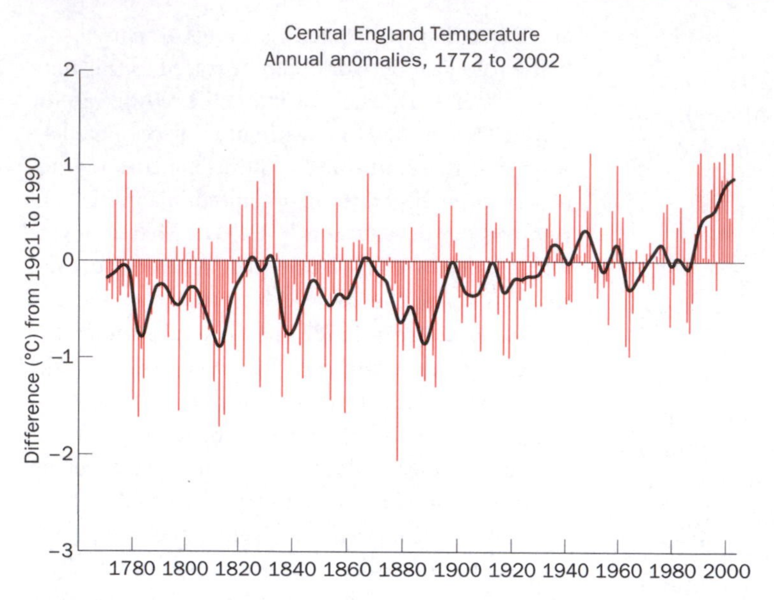

Figure 6.3 Central England Temperature series
Source: Met. Office, UK (Hadley Centre): http://www.met-office.gov.uk/research/hadleycentre/CR_data/Annual/cet.gif

manuscripts provide snippets of information about particular events which provide fragmentary insights into climatic conditions at the time (Figure 6.4). Sometimes these were subjective, relating to aspects which might be important for their agricultural activities, such as the occurrence of frost or rainfall. Sometimes the meticulous tabulation of harvest dates or grain prices enables rudimentary reconstructions of seasonal climate for these past years to be attempted. The latter is known as proxy data since it is indirectly a function of climatic conditions prevailing at the time. In the wine-growing areas of northern France, for example, a collective decision is often made by the village elders each year as to when to harvest the grapes. Experience gained over a lifetime has been transmitted down the generations from as far back as the fourteenth century. Logic suggests if the date chosen in recent years is the same as that chosen centuries ago, it is probably reasonable to assume the growing season conditions were similar. The conditions at a nearby meteorological observatory can then be examined with reference to this present date, applied by implication to the past one, and thus rudimentary

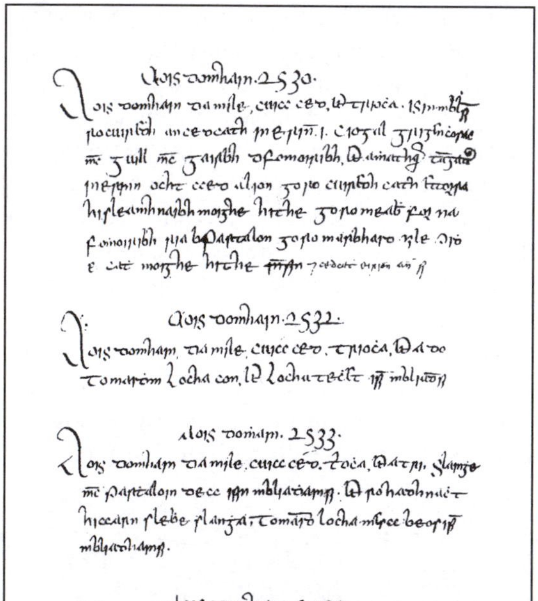

Figure 6.4 A storm on Lough Conn in 4668 BP. In this ancient Irish manuscript, reference is made to a storm on Lough Conn in Co. Mayo which is alleged to have occurred more than two millennia before the birth of Christ. Although the veracity of such sources is questionable, they can occasionally enable inferences on past climatic conditions to be made
Source: Royal Irish Academy

reconstruction of past conditions attempted. Such an approach is also possible with records of dates of flowering or fruiting, providing what is known as a phenological record. This may facilitate climatic reconstructions long before instrumental data became available. In Japan, for example, it was traditional that the date the emperor travelled to the ancient capital of Kyoto each spring to view the new cherry blossoms be recorded, providing a source of proxy data.

Observations and comments on the weather can be found in documents such as early newspapers. These tend to focus on extremes, and as today, frequently exaggerate their severity. Corroboration of events is desirable, and it also must be remembered that the target audience for early newspapers may not have been the populace at large, but rather a small privileged elite. So reporting is often selective and care has to be exercised in identifying the severity of events. Nonetheless, event frequencies can be reconstructed to some extent (Figure 6.5).

Ships' logs are a particularly valuable documentary source for climatic reconstruction. During the era of sailing ships, matters such as the wind and sea conditions were of crucial importance for the safety of the ship, particularly during times of war. On board ships of the British Navy a logbook was required to be kept by the master. Unlike the captain, the master often stayed with the same ship throughout his career. Using his own personal equipment, such as a quadrant or sextant, a range of observations were marked on a slate every hour when at sea. These typically included the wind direction and strength, cloudiness, visibility and precipitation. Rainfall was noted partly because it affected the routine on board of when the decks would be scrubbed. At noon each day the master made the main observations. A sighting of the sun would be made to record the position of the ship and the principal characteristics of the weather would be noted. All the earlier details from the slate would then also be entered in the ship's log. The log was also updated three times daily when the ship was in port. Given a sufficient distribution of ships, it is thus possible to reconstruct rudimentary weather charts for particular periods and to trace the movement

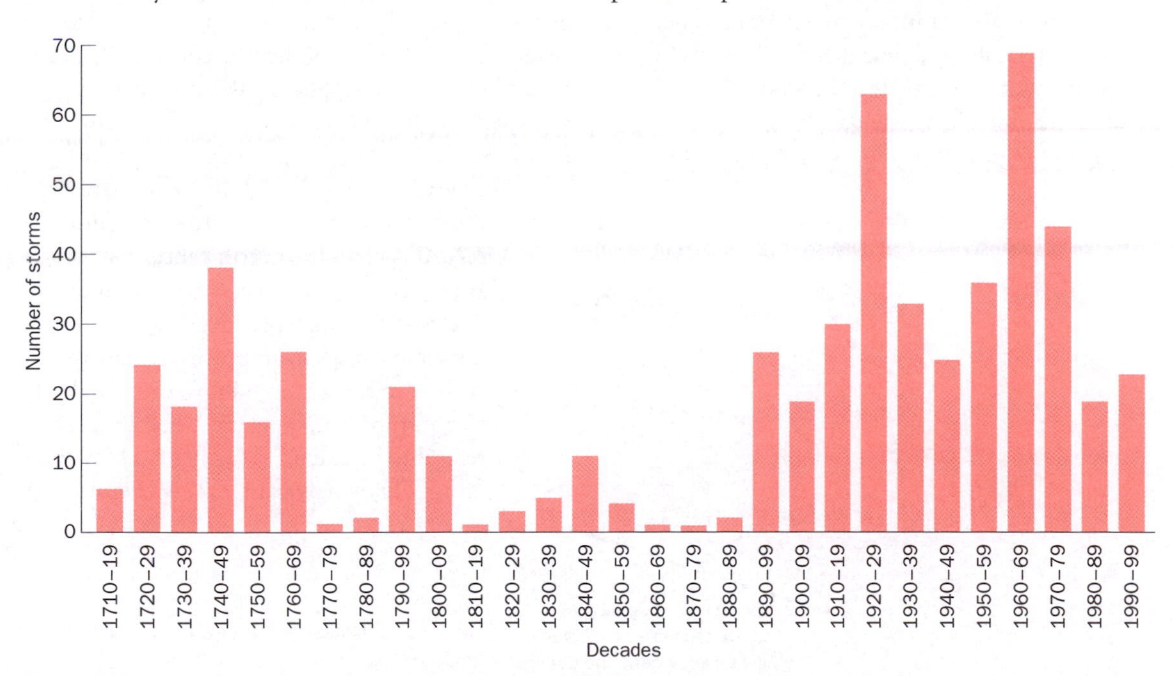

Figure 6.5 Storm frequency at Dublin, 1715–2000, as inferred from newspaper sources

of storm systems during times before the establishment of the modern observational network.

6.3.3 Natural proxy sources

Although climatic conditions may be inferred from measurements of natural phenomena, without an ability to date the material concerned it is impossible to construct a chronology or calibrate a record. Some phenomena lend themselves to dating, however. Among these are the growth cycles of plants such as trees and sediments and ice layers laid down in an undisturbed sequence over lengthy time periods.

(a) Tree rings

An annual growth cycle is produced by certain trees which manifests itself as a ring visible in a cross-section through the trunk. The outermost ring under the bark is the most recent addition. The ring is distinctive because it marks the boundary between the small, dense cells of slow growth during the autumn and winter and the larger, lighter coloured, less dense cells of rapid spring and summer growth. The wood added between the outside of one dark ring and the outside of the next thus represents the growth which occurred between two spring periods. The number of such rings can be counted inwards from the bark to provide an approximate age for the tree. The distance between rings is a measure of how successful the growing season was. If local limiting factors existed, the ring may be small. Such factors might include insect attack, changes in root competition, management practices, shade, etc. But, frequently, similar patterns of ring thickness are seen over wide areas, suggesting that a macroscale factor, such as climate (cold, soil waterlogging, drought) is responsible. The relationship of the outermost rings to present climate can be established by correlation with nearby instrumental records. As with the principles of phenological climatic reconstruction, the transfer function established from the present holds true for the past.

The best climatic information is obtained from trees growing near to their altitudinal or moisture supply limits where the best indicators of climate-induced stress will be apparent. Usually three cores are obtained and the rings compared, either in terms of their thickness or by using x-rays to estimate their cell density. Allowance must also be made for the life cycle of the tree concerned. Trees tend to grow more vigorously when young, and statistical techniques are used to eliminate this factor. Some trees reach a great age and provide a very long ring record. In North America the bristlecone pine can provide a climate record back to the fourth century BC. However, it is not necessary to confine the analysis to living trees. Overlapping of distinctive rings from previously cut wood, such as rafters of old houses, or even dead tree stumps from boglands, enables the record to be extended further back by the process known as cross-dating (Figure 6.6).

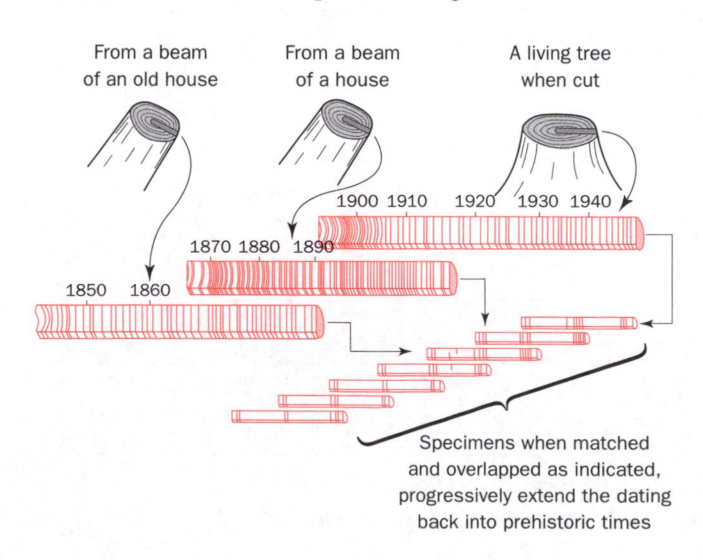

Figure 6.6 Cross-dating tree rings from various sources
Source: Fritts (1976)

European oak and Californian bristlecone pine chronologies now extend back 8,000–10,000 years.

(b) Pollen grains

The concept of tolerance of a plant to climate also explains how pollen may be employed to reconstruct past vegetation assemblages, and by inference past climates. Pollen is particularly suited to long-term climate reconstructions firstly because of its ubiquity. A single plant may produce several million pollen grains. A rye plant may for example produce 23 million grains. Thus pollen will spread widely, falling slowly to the surface. If undisturbed, a sequential build-up from year to year forms layers with the oldest at the base and the most recent at the surface. Second, pollen grains resist decomposition because of their tough outer skin. This may render a grain intact for thousands of years if it comes to rest in an oxygen-free environment, such as a lake bed or peat bog. There is therefore the capability of identifying pollen from distant time periods. Third, each plant species produces distinctive pollen grains. This makes it possible to count the proportion of a particular layer represented by a particular species. Subject to some caveats re productivity differences, the proportion of the landscape occupied by a particular species may thus be estimated. The vegetation mix, or assemblage, is indicative of climatic conditions prevailing at the time the pollen 'rain' fell. For example, tropical grassland or coniferous forest assemblages would translate to the particular climatic conditions necessary to sustain them. Frequently, dating of the pollen assemblage zones is facilitated by the presence of organic deposits such as antlers, bones, etc. These enable the climate variations to be dated, usually with a rather coarse resolution of decades or centuries. Not unrelated to the principles of palynology is the analysis of beetle remains. In this case the skeleton of the beetle is identified and its climatic requirements known from present-day analysis. The presence of different species enables fine-tuning of the climatic conditions to be achieved (Figure 6.7).

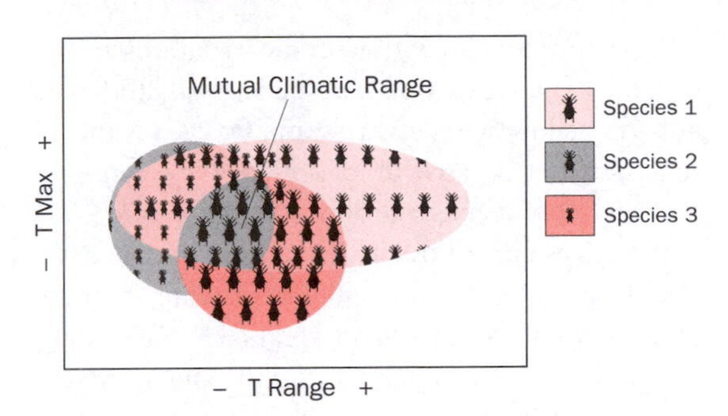

Figure 6.7 Overlapping tolerance ranges for beetles for climate reconstruction

(c) Oxygen in ice and sediments

One of the most powerful means of reconstruction of palaeotemperatures is provided by analysis of the isotopic composition of layered sediments from oceans and lakes and from ice accumulations. Isotopes occur as a result of slight variations in the numbers of neutrons in the atomic nucleus of a particular element, with most elements naturally occurring as a mixture of such isotopes. In the case of oxygen, for example, the most common form of the element is composed of 8 neutrons and 8 protons in the nucleus. This gives an atomic weight of 16, an atomic structure denoted as ^{16}O. Any sample of a compound containing oxygen atoms would have approximately 99.759 per cent of them exhibiting this configuration. A small percentage of oxygen atoms in the sample, about 0.037 per cent, would have one extra neutron in the nucleus, giving the isotope of oxygen known as ^{17}O, and the remainder, 0.2039 per cent, would have two extra neutrons, ^{18}O. Chemically, the isotopes have similar properties. The key differences arise from the slightly heavier nuclear masses of ^{17}O and ^{18}O. It is these slight weight differences which provide the capability for reconstructing the global temperature for much of the past 2–2.5 million years.

Several elements with isotopic variations are capable of being used in isotope analysis studies aimed at climatic reconstruction. Oxygen is the most popular because of its presence in water and therefore its passage along so many different pathways as part of the hydrological cycle. The quantities of ^{17}O are too small to provide accurate measurement tools and so the ratio of $^{18}O/^{16}O$ is the basis of isotope analysis studies. The normal oxygen isotopic make-up of water is an $^{18}O/^{16}O$ ratio of 0.2039/99.759. When this ratio changes it means that the heavier ^{18}O isotopes are either over-represented or under-represented.

When water evaporates, such as from the oceans or large lakes, the heavier ^{18}O-based water molecules find it more difficult to enter the vapour phase than their lighter ^{16}O-based counterparts. The remaining water thus becomes enriched in the 'heavier' molecules and relatively depleted in the 'lighter' ones. Condensation is also selective, the heavier molecules return first and the residual water vapour in the atmosphere is depleted in the heavier forms. These will eventually make their way back to the oceans to restore the balance. If, however, the precipitation gets locked up as snow and ice on the land, the oceans become depleted of ^{16}O-based molecules and the ice on land enriched in them. As a glaciation develops, therefore, the $^{18}O/^{16}O$ ratio in the oceans rises. Conversely, the ratio falls in the ice content of the glaciers. Thus, the more extensive global ice masses are, the more these contrasts expand. The $^{18}O/^{16}O$ ratio is therefore capable of being used as a global geological thermometer.

Key ideas

1. The instrumental record provides a relatively short perspective on recent climate change, particularly outside of Europe and North America.

2. A variety of documentary sources can be employed to reconstruct past climates prior to the instrumental era. These include: harvest dates, crop prices, phenological data and ships' logs.

3. Natural proxies for climate also exist whereby past climates can be inferred from tree ring thicknesses, pollen deposits or beetle remains. The oxygen isotopic composition from ice cores or shells in oceanic sediments shows strong correlations with global temperature and has enabled the temperature changes of the past 2 million years to be reconstructed.

6.4 The climate record

6.4.1 Ocean and ice cores and the Pleistocene record

Approximately 2 million years ago to 10,000 years ago, extensive areas of the northern hemisphere were covered by ice, conventionally spanning the period known as the Pleistocene epoch (section 8.2). If the post-glacial period is included, i.e. the last 10,000 years, the interval from 2 million years to the present is called the Quaternary period. During the ice age conditions which prevailed during the last 800,000 years, and certainly during the last 400,000 years (see Figure 8.5), the earth's climate has been characterised by lengthy periods of cooling typically lasting about 90,000–100,000 years punctuated by shorter interglacial intervals of about 10,000 years. The last glacial episode began about 110,000 years ago and lasted until about 10,000 BP. During glacial episodes, up to 30 per cent of the land areas of the globe are covered by ice expanding out of North America and Europe. The Laurentide ice sheet centred over Baffin Bay and the Scandinavian ice sheet centred over the Gulf of Bothnia extended well into the offshore oceans during several glacials, reaching the continental shelf in the eastern and western Atlantic Ocean. The transition from cold to warmer conditions appears to occur very rapidly on occasion while even the warm interglacials exhibit occasional bursts of rapid and intense cooling.

Much of the climate history of the Pleistocene has been unravelled using *ocean core* investigations. Plankton, both plant (phytoplankton) and animal (zooplankton) and other tiny life forms, exist in huge numbers in the ocean. As the animal plankton complete their life cycle, some of their remains settle out on the ocean floor, accumulating slowly at rates of 1–2 mm per century. Unlike the land, the ocean floor may be undisturbed for millions of years, producing layered sequences of marine organism remains. Some zooplankton forms (particularly the foraminifera group of species) are indicative of warmer ocean temperatures (temperate), being present only when sea temperatures are above a certain value, while others are found in colder Arctic and pack-ice environments. One type of zooplankton (species of Pachyderma) actually changes its form according to water temperatures, coiling its shell to the left in cold waters and to the right in warmer waters (see Figure 8.8). Thus by mapping the distribution of the *abundance* of such indicator species, water temperatures can be derived over large areas.

In addition to measuring the abundance of various zooplankton species, their oxygen content can also be analysed to produce climate information from the past. The shells of many tiny marine organisms are made of calcium carbonate. To manufacture these, the organisms extract oxygen from the waters near the surface in which they live. The isotopic composition of the shells thus enables the isotopic composition of the ocean in which they lived to be inferred, and with it global temperature. Several deep ocean cores have now been extracted and the global temperature trends of the past 2 million years can be ascertained (Figure 6.8 and see also Figure 8.5). The cycle of glacial–interglacial is clear, with an abrupt transition from glacial to interglacial and a rather more gradual cooling down from interglacial to glacial. However, both modes of climate are clearly subject to abrupt interruptions on occasion.

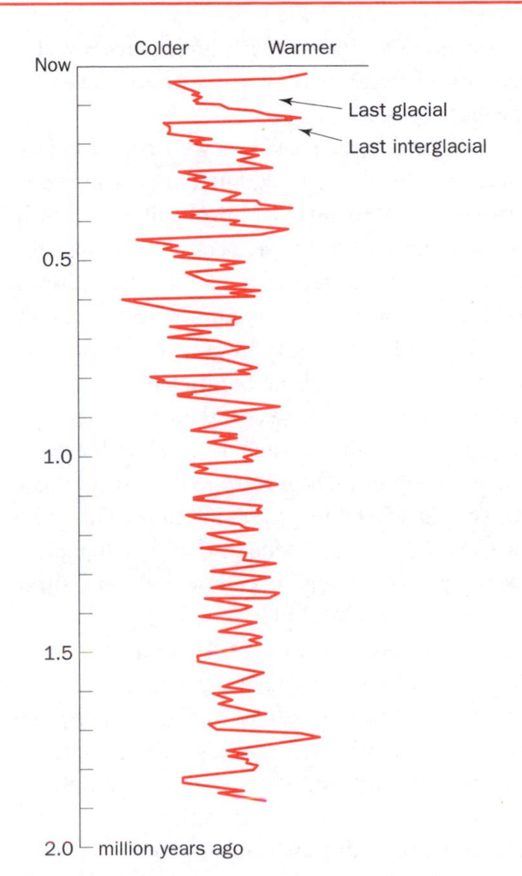

Figure 6.8 Global temperature trends of the past 2 million years as inferred from ocean core isotope analysis
Source: Van Andel (1994)

Greater resolution is possible if *ice core* data are used. In the polar regions the annual accumulation of snowfall becomes compacted in deep layers which can be counted back from the surface (Plate 6.1). The isotopic composition of each annual layer of ice records the isotopic composition of the snowfall for that year. As explained previously, where a layer is depleted in ^{18}O isotopes the temperature conditions in which the snow occurred must have been cold and vice versa. Close to the ice dome is the best for drilling, since summer melting is least and shearing of the ice layers beneath the surface is minimised. Both in Antarctica and Greenland very long cores have now been recovered. The Greenland Ice Sheet Project (GISP) core in the interior of Greenland finally struck bedrock after over 3 km of ice core was recovered, providing a record of the past 200,000 years of climate at this location. In the Antarctic (Vostok), the ice cores are even deeper, revealing climate history over 400,000 years (see Figure 8.5). Part of both these cores reveal dramatic changes, particularly towards the end of a glacial event.

By 14,000 BP the ice sheets were beginning to retreat and by about 13,000 BP temperatures in north-western Europe were probably just slightly less than those of today (see Figure 8.15). Birch and pine trees had recolonised Britain and Ireland and further south in Europe more heat-demanding species were beginning to migrate north from their glacial refugia around the Mediterranean. This climatic amelioration was dramatically interrupted by a sudden cooling known as a Heinrich event which restored much of Europe to cold conditions from about 12,700 to 11,500 BP. In the GISP2 ice core, the $^{18}O/^{16}O$ ratio tells us that this dramatic cooling produced a fall in temperature in central Greenland of about 20 °C.

As indicated in section 8.6.1, the warm conditions of around 14,500 BP would have melted the Laurentide ice sheet and released icebergs into the North Atlantic. The melting icebergs released cold, low-salinity, and therefore light sea water across the North Atlantic, an action which is believed to have diminished the vigour of the Gulf Stream, forcing it to shut down much further south. The formation of deep water (heavy salty downward-moving water) to drive the oceanic conveyor belt thus took place at lower latitudes, and the North Atlantic and north-west Europe froze. Figure 8.15 shows that unlike western Europe, cooling over central Greenland from about 14,500 BP took place in a series of stages before the final plunge to cold temperatures between 12,800 and 11,500 BP. During this cooling and certainly during the coldest period, ice sheets expanded over the North Atlantic. It is argued that with continued cooling and ice sheet expansion (such as during the cold Heinrich event of 12,700–11,500 BP) the growing ice sheets eventually broke up as they spread out over the ocean shelves, helping to release more icebergs into the North Atlantic. Upon melting over the North Atlantic, these icebergs release further fresh water into the surrounding ocean, helping to sustain the prevention of deep water formation and the penetration of the Gulf Stream in the North Atlantic. Analysis of the ocean cores in the North Atlantic for this time (12,700–11,500 BP) reveals a sudden change in sedimentation. Instead of the shells of marine organisms, angular inorganic debris is encountered (Plate 6.2). These are now believed to be ice-rafted debris dumped from massive iceberg surges during the coldest phase of the cooling episode.

During the Heinrich event of 12,700–11,500 BP, often called the Younger Dryas, glaciers formed again on the uplands of Scotland and Ireland and ice sheets advanced out of the Highlands of Scotland to the margins of the Central Lowlands. Tree growth was destroyed and tundra conditions returned to many parts of western Europe (see Box 8.4). It is important to emphasise that it is possible that a previous warming period generated the initial iceberg surge that ultimately resulted in cold stadial conditions lasting over a millennium. It is considered unlikely that global warming could achieve a similar result in the near future, but the sensitivity of the ocean conveyor belt to freshwater forcing is not yet fully understood. Concerns remain that the shutting down or at least the decline of the North Atlantic Drift could occur as the polar ice cap disintegrates.

Rapid warming both within, and especially after, a cold stage is equally evident, and may also relate to circulation flips in the thermohaline circulation of the oceans. In the case of the ending of the Younger Dryas event, a dramatic change in temperature is evident in the ice core data (see Figure 8.16). A temperature shift of about 10 °C in about 10 years is seen in central Greenland, together with an approximate doubling in the rate of snow accumulation. Once again the capability for the climate system to exhibit abrupt changes is evident.

The rapid climate changes of the Pleistocene considerably affected the distribution and evolution of plants and animals. The processes of natural selection favoured those capable of adapting to the changes. Animals which were mobile, omnivorous and with greater brain power succeeded. Foremost among these were early humans whose growing ecological dominance and growth in numbers contrast with the extinction of many other species. Riding the wave of glacial/interglacial oscillations, early humans moved polewards from the low latitudes to exploit the new habitats laid bare by the retreating ice and tundra. Although technological developments in fire, clothing, shelter and tools helped their success considerably, the chronological coincidence of major climate changes and major cultural changes in the period of glacial/interglacial oscillations is striking.

6.4.2 Post-glacial climates

The period from around 10,000 years ago to the present is called the Holocene. Compared to the rapid climate shifts of the glacial period, the Holocene has witnessed a period of relative climate stability. Such has been the overall climate constancy of the last 10,000 years that the period has been linked to the fostering of rapid human development and the rise of agriculture and civilisation.

Despite the general climate uniformity of the Holocene, however, moderate changes in temperature and precipitation can be discerned. Warming occurred rapidly during the Holocene after the cold snap of the Younger Dryas, and vegetation zones shifted polewards as new lands became available for colonisation. Sea levels rose rapidly as ice melted, submerging many coastal plains. This would have occurred in sporadic storm surge events. The opportunities of catching fish and evaporating sea water to produce salt to preserve their catch may have resulted in rather a greater percentage of the human population living near the coast than today and their vulnerability to inundation may have been considerable. Many ancient legends of a great flood may reflect this. Certainly many areas such as the broad plain that became the North Sea were inhabited, and a 1 m sea-level rise per century may have translated into a loss of many lives. By 7,000 BP sea level had risen to cut off the islands of Britain, Ireland and Denmark from continental Europe. This event also virtually brought to an end the immigration of plant and animal species from Europe. Those that had made the journey defined what is referred to as the 'native species'. That many species of plant and animal did not is evidenced by the floristic and faunistic impoverishment of the islands today by comparison with mainland Europe. For instance, Ireland has only 890 native flowering plant species, the UK has 1,550 while France has as many as 4,500.

The warmest times of the present interglacial were reached about 5,000 BP when temperatures generally were 1–2 °C above those typical of the last half of the twentieth century. The peak warmth was reached in some parts of the tropics somewhat earlier, and cave paintings from the Sahara depicting hippopotami hunts and cattle herding show the extent to which the moister equatorial climates had migrated polewards, to areas that are today hyperarid. Elephants, giraffes and rhinoceroses roamed in Egypt and the discharge of the Nile was considerably greater than that of today. After about 5,500 BP the drying out of North Africa seems to have occurred and only places such as Egypt with its reliable flooding of the Nile each year could prosper. Elsewhere in the world similar climatic stresses were apparent and concentration of people and animals around oases and river valleys became pronounced. Many of the great riverine civilisations of Asia and Africa date from this time. The rivers of Mesopotamia, the Nile, the Hwang-Ho and Indus offered the capability of feeding substantial populations while areas beyond only offered failing pastures and hunting. It has even been suggested that it was the refugees from these exterior areas that provided the essential slave labour necessary to maintain the irrigation systems on which the riverine civilisations depended.

The mid-Holocene maximum also saw people and plants spreading polewards. Agriculture becomes established in Europe and China, reaching up the hillsides to heights above what would be considered viable today. Many of the great megalithic monuments of Britain and Ireland can be dated from this time. Forests extended to the Atlantic edges of north-west Europe and even reached some valleys in Iceland that are today ice-covered. By 4,000 BP this was in retreat and neolithic farmers were abandoning their higher cultivation sites. By the first millennium BC, a change to cooler, wetter conditions had become established as the storm tracks moved south.

6.4.3 Historical climate change

Among the earliest sources of climate information for Europe are the writings which date from Roman times. These indicate that for much of the last millennium BC, during which imperial Rome grew and spread, climates in the eastern Mediterranean seem to have been colder and moister than today. These conditions would undoubtedly have helped the harvests of the Greeks and also those of the other early rivals to Rome in North Africa such as the Carthaginians and Phoenicians. The early Roman writer Livy records snowy winters when the Tiber froze and noted how beech trees were common around the city. By the first century AD Pliny was commenting on how such trees were confined to the cooler mountains. This warming seems to have manifested itself further east as increased drought and the drying up of pastures in central Asia. It was a time of plagues and mass migration of people westwards, ultimately to undermine Rome itself. After about AD 400 a slow return to cooler conditions seemed to occur in much of Europe, marked by storminess and coastal floods in the west.

(a) The Medieval Warm Period

Towards the end of the first millennium AD a warming set in which was particularly marked in northern Europe. Known as the Medieval Warm Period, or the Little Optimum, this lasted for about three centuries from AD 900 to 1200. Until quite recently this was thought to be a period of global warmth. However, this idea has now been revised as information from outside Europe has become available, and while the Medieval Warm Period is associated with positive temperature anomalies in Europe, it is more strongly associated with drought elsewhere. It is, however, clear that temperatures in the higher latitudes of the northern hemisphere were about 1 °C warmer than the mid twentieth century, and that probably this period represents the warmest few centuries there since the Post-Glacial Climatic Optimum around 6,000 years ago. The Medieval Warm Period came at a time when documentary sources were becoming more generally available to aid climate reconstruction. These sources indicate that the episode was marked by latitudinal and altitudinal expansion of settlement and cultivation. In Europe, for example, vines were grown much further north than today. The Domesday Book of William the Conqueror documents 38 vineyards in Britain in 1066, some as far north as York. Fossil forests in Canada 100 km north of the present tree line, reports of swarms of locusts in Austria, and tillage at heights over 350 m on Dartmoor convey a picture of general warmth in the middle/high latitudes of the northern hemisphere. But it was in the marginal lands of northern Europe that the impacts of these benign climates appear to have manifested themselves most (Box 6.1).

(b) The Little Ice Age

Numerous studies suggest that a sharp downturn in climate after about 1550 occurred extensively in the northern hemisphere. Even before this, storms and coastal floods in some parts of Europe may be taken as the first symptoms of what has now become known as the Little Ice Age. Death tolls of over 100,000 were reported in several storms affecting the low-lying lands around the North Sea, and sand movements overwhelmed a number of coastal settlements. The cooling trend persisted until the middle of the nineteenth century by which time its intensity could be recorded by the growing instrumental network. Mean global temperature dropped by 0.5 °C, with European winter temperatures about 1.3 °C colder between 1500 and 1800 than the early twentieth century. At least in the northern hemisphere, the Little Ice Age was probably the coldest three-century period of the entire post-glacial. In the more northerly latitudes, suffering was most pronounced. In Iceland and

BOX 6.1

THINKING FURTHER

The Vikings: why adaptation to climate change is essential

Most models of global climate predict a change in temperature over the next few decades as a result of the enhanced greenhouse effect of about 0.2 °C per decade. Such warming is expected to be most pronounced at the high latitudes as melting ice transforms the reflectivity of the surface into a darker more absorbing surface. This ice-albedo feedback effect is one of a number of feedbacks which have restricted the utility of global circulation models over the past decade and led to some uncertainty as to the rate of warming which can be anticipated during forthcoming decades. It is thus interesting to look backwards to a time when the high latitudes were warmer than today. Of course, a true analogue of the future climate is not possible, but a glimpse of who might be the winners and losers as global warming proceeds is obtained. Such a warm period occurred in the interval AD 900–1200, sometimes known as the Little Optimum or Medieval Warm Period.

(a) Good times of the Medieval Warm Period

Harvests were undoubtedly easier. This is important since harvests in those days were a matter of life or death. Only if food security could be obtained could attention turn to other things. Certainly it was much easier to subsist in these northern lands, where the harvests are mainly limited by the short cool summers, than at present. Wheat could be grown in the north of Norway and oats in Iceland. Barley, oats and rye are recorded in Greenland. In much of Norway, population growth was considerable, and the number of farms increased substantially, especially in the northern regions and in the more sheltered south around Oslo. However, on the coastal fringes of the west the limitations of the physical environment could not be overcome easily, even by a more benign climate, and the options were more restricted. It was from these parts that most of the Viking expeditions to the Atlantic islands south and west originated. By the first half of the ninth century, Viking kingdoms had been set up on the Hebrides, the Northern Isles and northern Scotland and from these bases raids on other parts of western Europe were mounted.

For the inhabitants of these northern lands the weather was crucially important. When did the ice melt on their fjords in spring? When did they need to be back before the winter freeze-up? Such records were often meticulously kept in the sagas. From Iceland comes the Landnam saga, while from Greenland the Greenlander saga is an invaluable source of information. Since climatic inferences can be drawn from such sources a comparison with present-day instrumental records enables reconstruction of the rudiments of climate for this period.

(b) Viking expansion

The era of Scandinavian expansion commences at the end of the eighth century, probably coinciding with advances in sailing techniques which seem to have occurred around that time. No compasses were available to these early mariners and yet the exploration of the northern oceans commenced, undoubtedly helped by calmer conditions than both before and after the period. By the mid ninth century, fleets of Vikings were sailing up the Seine, sailing into the Mediterranean, exploring the rivers of northern Russia (the word Rus means oarsmen) and sacking cities such as Pisa, Lisbon and Paris. To the west, the first recorded Viking exploration to Iceland was in 860 by a farmer, Floki Vilgerdason, who found that Irish monks had got there before him. These peoples were termed the *papas* – clearly a corruption of the Irish for 'father'. This was a common discovery in Atlantic Europe at the time as Irish monks sought to find contemplative solitude. Interestingly, the tradition of documentation which characterises both Iceland and Ireland is not present to the same extent in Norway.

However, in 860 the warming of the Medieval Warm Period was not yet under way and Vilgerdason's settlement failed. Floki returned to tell of an island where the fjords were 'choked with ice'. By the end of the next century, however, the warming was well under way and Iceland was an altogether more hospitable place. By 1095 over 77,000 settlers were established. It was from Iceland that in 982 Erik the Red was banished for killing two men and set sail further to the west where vague stories of another land had been told. By 982 the warming was well advanced and Erik settled on the southern tip of the large island which he called Greenland. It is one of the quirks of history that the island which today is so inhospitable is called Greenland, while the island on which a long-established viable community exists is given the harsher name of Iceland. Was this the result of the stages during the Medieval Warm Period in which they were settled? Iceland was at the beginning, before the warming had become pronounced, Greenland when it was close to its peak. Or perhaps Erik was trying to lure other settlers to his new fiefdom; or perhaps the sunlight glinting off the icefields appeared green. In any event two thriving settlements were established, numbering over 6,000 people by 1200. The spectacle of first contact between the Norse settlers and the Inuit hunters must have provided an interesting contrast in cultures – one a sedentary agriculturalist – a prisoner of their climate – and the other a nomadic hunter – rolling with the vagaries of climate in search of mobile food sources. By 1125 the Greenland Norse were so well established that they sent back home for a bishop, traded a live polar bear for him and built a small cathedral. It was but a short distance for ships blown off course to end up in northern Newfoundland or the North American coast (Vinland).

(c) Hard times and the Little Ice Age

But as the Medieval Warm Period wore on, a downturn in climate was becoming apparent. In the northern latitudes the sea ice was beginning to return, making trade increasingly difficult. It was becoming too cool to save the hay in many years and the all-important harvest was becoming problematical. By 1342 the main sailing route from Scandinavia to Iceland, close to the Arctic Circle, had to be abandoned in favour of a more southerly route. The Danish parliament even debated whether to evacuate Iceland. In Greenland, the encroaching permafrost can be inferred from the increasingly shallow burial depths in the graveyards. By the thirteenth century the colony was isolated, reachable only in very good summers, and facing extinction. On the same year that Columbus made landfall in North America, Pope Alexander VI expressed his concern for this northern outpost of Christendom:

> The church of Garda is situated at the ends of the earth in Greenland, and the people dwelling there are accustomed to living on dried fish and milk for lack of bread, wine and oil . . . shipping to that country is very infrequent because of the extensive freezing of the waters – no ship having put into shore, it is believed, for eighty years.

By 1540 the last inhabitants had died off. Their bodies were found in more recent excavations of the graveyard buried in permafrost dressed in out-of-date European clothing. The average height of adult males in the graveyard had declined by 13 cm since the pioneers arrived.

(d) A lesson not learned

Was the demise inevitable? Perhaps not. Had the Greenland Norse abandoned their cattle and switched to the sea they might have survived. Had they adopted the flexible lifestyle of the Inuit they would almost certainly still be there. Even persevering with cattle and dressing in wool rather than furs showed their stubborn unwillingness to adapt to changed climatic conditions. Conservatism doomed these peoples. The lesson as we enter the third millennium is obvious. We must be prepared to adopt radical ideas to fit with the changing threats and opportunities of a greenhouse-warmed world.

Greenland the cooling had come earlier than elsewhere. The possible evacuation of Iceland was debated by the Danish parliament, and by 1540 the Greenland colony had perished (see Box 6.1). Even quite late on in the Little Ice Age the population of Iceland was demonstrably vulnerable – from 1753 to 1759 a 25 per cent drop in the population of the island occurred.

In Scotland, during the Medieval Warm Period, the agricultural margin had moved high up into the glens. Now it retreated. Runs of consecutive harvest failures in the 1690s and 1780s led to famine years. In the 'ill years of King William's reign' from 1693 to 1700, the oats harvest failed for seven out of eight years in the uplands. Between one- and two-thirds of the population died. Such impacts suggest a greater reduction in temperature was being experienced in these northern lands, probably in the region of 5 °C. It seems likely that the polar waters had spread southwards close to the Shetlands, and on a number of occasions between 1690 and 1728 Eskimos appeared in their kayaks in the vicinity of the Orkneys. On one occasion during this interval an Eskimo was sighted in the mouth of the River Don near Aberdeen.

Though spared the worst ravages of the Little Ice Age, areas further south were also severely affected. In England in the 1690s, the growing season was reduced by 5 weeks and snow lay on the ground for up to 30 days of winter. In 1657/58 parts of southern England were snow covered for 102 days. Winter 1683/84 was probably the coldest, with the ground being frozen to a depth more than 1 m and slabs of sea ice 5 km in extent appearing in the English Channel. The Thames froze on at least 20 occasions during the Little Ice Age, on many occasions to a depth that permitted wheeled traffic to cross it. These were the times when elaborate frost fairs were held on the river. During 1683/84 the ice was 28 cm thick and the river was laid out in streets with stalls and venues for horse and coach races, puppet plays, bull baiting and a variety of other activities. In the rural parts, though, people had little to celebrate. Deaths exceeded births for 70 years from 1660 to 1730, with little substantial change thereafter until 1780. Life expectancy fell by 10 years and the age of marriage for women, 23 in 1830, was 30 around the turn of the seventeenth century.

Further south in Europe, the Mediterranean Sea around Marseilles froze in 1595 and 1638. Viticulture was abandoned in the north due to frost and in 1708/9 all the orange trees were lost in Provence. Both the Alpine and Scandinavian glaciers advanced downslope, destroying forests and cultivated land. In the Alps, snow covered the ground for three times the length of time it does in today's winters. When the snow still lay on the ground in March and April, fodder supplies were often used up and livestock had to be slaughtered. Wheat, potatoes and milk were often in short supply and famines were recorded in 1769–71 and again in 1816–17. Similar glacial advances occurred in the Rockies where the area of glaciers doubled and they advanced downslope by 400 m. Though conditions were probably as extreme in parts of North America as in Europe, there are not as many data available from either documentary or instrumental records. Samuel Champlain, the founder of Quebec, reported thick ice on the edges of Lake Superior in June 1608, perhaps the harshest of some notably severe winters around this time. In 1816, which has come to be known as 'the year without a summer', unusually cold continental Arctic air masses dominated the eastern part of Canada and the USA throughout the summer, bringing heavy snow in June and frost in July and August which devastated crops. The extremely cold conditions of 1816 have been related to volcanic activity, specifically to the eruption of Tombora (Indonesia) in the previous year of 1815 (see section 6.5.3).

Even during the depths of the Little Ice Age, hot, dry summers and mild winters did occur. The frequency with which these milder intervals punctuated the generally colder conditions was, however, significantly less than in more recent times. Marked regional variations were also in evidence and whereas in the past the Little Ice Age was thought to be a global phenomenon, the best evidence currently available suggests it was principally a northern hemisphere event.

6.4.4 Contemporary trends

The availability of extensive instrumental records makes the establishment of trends in climate since the Little Ice Age much more confident. The most authoritative collation of global records has been carried out by the Intergovernmental Panel on Climate Change (IPCC), the latest report (2001) of which draws a number of key conclusions regarding recent climate trends as follows:

- Globally temperatures have increased by approximately 0.6 °C since the end of the nineteenth century (see Figure 1.1). This represents the greatest change of any century of the past millennium at least.
- Warming has been concentrated into two periods, the first from 1910 to 1945 and the second from 1976 to the present. In both these periods the rate of warming in the northern hemisphere was twice that of the southern hemisphere. Slight global cooling occurred for a time during the middle of the last century from about 1946 to 1975.
- A rate of warming of about 0.2 °C/decade is currently being observed, with the bulk of the warming occurring in nocturnal minimum as opposed to daytime maximum temperatures. Most warming is also occurring in the higher latitudes compared to the low latitudes.
- The 1990s was probably the warmest decade of the past millennium; 1998 is likely to have been the warmest year since instrumental records began, with 2002 and 2003 the next warmest.
- Associated with these temperature changes, many glaciers have been retreating. In some locations such as Norway and New Zealand the reverse has been observed as additional snowfall more than compensates for melting. Typically, the ice season in the northern hemisphere rivers and lakes is about 19 days shorter than in the mid nineteenth century. Arctic sea ice has been thinning by an average of 4 cm/year, though the extent of ice in the Antarctic has not changed significantly.
- Precipitation over land areas of the world has increased by about 2 per cent over the past century. This is to be expected since warmer air can hold more water vapour than colder air.
- Northern Europe and northern North America have experienced the greatest increases in precipitation, with areas in the southern hemisphere such as Argentina and Australia also experiencing increases.
- In the tropics and subtropics trends are less clear. Runs of drought years have characterised some locations such as Sahelian Africa, and changing long-term trends in Indian monsoonal rainfall are also apparent. However, no convincing linear trend is apparent in most parts of the tropics with respect to rainfall changes.

Key ideas

1. The Pleistocene epoch has been characterised by lengthy glacial episodes lasting about 100,000 years and shorter interglacial intervals of 10,000 years. Abrupt changes from glacial to interglacial modes occur with significant fluctuations also apparent within each mode.

2. Iceberg surges in the North Atlantic are capable of disrupting the oceanic heat conveyor. The 'Younger Dryas' cooling period is believed to have resulted from such a 'Heinrich event'.

3. Though not deterministic, climate change may have played a significant role in the rise to ecological dominance of the human species.

4. Major climate fluctuations in recent centuries have included the Medieval Warm Period (AD 900–1200) and the Little Ice Age (AD 1450–1850). Marginal environments such as the high latitudes show more marked responses to climate fluctuations and the experiences of inhabitants of these areas emphasise the importance of adaptation to changing climates.

5. Rapid global warming of about 0.2 °C/decade is currently under way, with a clustering of the warmest years in the instrumental record apparent in the period since 1990.

6.5 Causes of climate change

6.5.1 Climate change in the climate system

The atmosphere is a chaotic system. Despite the availability of immensely powerful supercomputers, forecasting weather more than about 10 days ahead is subject to significant errors, and accurate forecasts of weather for future periods much longer than this are likely to remain elusive for a long time to come. This is because the behaviour of a chaotic system is so sensitive to the initial conditions from which a forecast starts that future predictions have multiple possibilities for even a slight change in these initial conditions. This is why good quality observational data are so important for meteorologists and why the maintenance of as dense a network of observing stations as possible aids forecast quality. Climate is less chaotic and responds to changes in both external and internal forcing. The nature of that response is still not fully understood. In particular, the extent to which the climate system could resist forcing for a time and then 'flip' to a new equilibrium unexpectedly is a cause for concern (see Figure 6.2).

The processes leading to a climate change can be examined on different timescales and also in terms of whether they are external or internal to the climate system. It has been shown in section 3.4 that changes in the net radiation or available energy at the surface of the earth plays an important role in climate change. Basically, energy conditions at the surface can be altered by two sets of factors. First, the input of solar radiation, and how it is seasonally and geographically distributed, may be considered part of the 'external' factor input or boundary conditions. Second, the way in which the earth-bound component parts of the climate system (atmospheric composition, land/ocean/ice surface, biosphere, human activity) function to store, distribute and release this energy input may be considered as 'internal' forcing mechanisms producing climate change. Complex feedback loops may characterise these relationships, for example between atmosphere and ocean, which make understanding and predicting climate change an as yet imperfectly understood exercise (see section 7.3).

6.5.2 External causes of climate change

(a) Direct solar radiation input

The most important external influence on the climate system is the delivery of solar radiation. Flickers in the solar furnace or changes in the earth's orbit around the sun could be expected to produce profound changes in climate. For long the sun was believed to send an unchanging flow of radiation to the sunlit hemisphere which delivered 1,370 W/m^2 on a level horizontal surface at the top of the atmosphere. The figure of 1,370 W/m^2 was referred to as the solar constant. It is important to note that this figure is reduced to 342 W/m^2 when the input is averaged over the entire globe (see Figure 3.2). Satellite measurements have recently suggested that the solar constant does vary more than previously thought. During the period 1986–91 the solar constant increased by 0.1 per cent, while from 1992 to 1996 it decreased again. Many of the changes occurred in the UV part of the spectrum and some recent research suggests that increased solar radiation in the UV wavelengths might have an influence on the Hadley cells (see Chapter 4), possibly producing a poleward migration of the mid-latitude storm tracks.

The best-known causes of changes in solar output arise from disturbances on the solar surface known as sunspots. These complex magnetic phenomena are more fully discussed in section 3.6 where it is explained that the number of sunspots typically follows a cyclical trend of 11 years, embedded in longer cycles of 22 years, 80–90 years and other longer cycles around 180 years. Up to 200 sunspots may be detectable at the maximum stage of the solar cycle, with numbers falling almost to zero during the minimum. Though they block the flow of heat from the sun's interior to its surface, the negative influence of sunspots on overall solar output is overcompensated for by the simultaneous occurrence of adjacent whiter, hotter areas known as faculae. When the sunspot number is high, solar output is higher and vice versa.

Attempts to relate sunspots with climatic trends have up till now proven inconclusive. Droughts in the Great Plains of North America are well correlated with the 22-year cycle, with other suggested links including El Niño and sea surface temperatures. The most striking match has come during the Little Ice Age. From 1650 to 1730 the sunspot numbers declined almost to zero. This period, known as the Maunder Minimum, corresponded to some of the most severe winters of the Little Ice Age. However, the phasing of the Little Ice Age and the global extent of it are not universally agreed, and considerable uncertainty regarding the role of sunspots as climate change determinants continues. Thus, while a number of correlations with climatic phenomena appear convincing (see section 3.6), correlation alone does not imply causation. Indeed, the changes to the solar constant which have been measured over the past century due to the 11-year sunspot cycle amount to about 0.1 per cent, much too low to drive the temperature changes seen over that period.

(b) Indirect solar inputs: orbital variations

Fluctuations in solar income to the earth also occur due to cyclic variations in the earth's orbit around the sun. These orbital variations arise as a result of the gravitational attraction of the sun, moon and other planets, particularly the giant planets of Jupiter and Saturn. As a result, the earth's orbit displays three key cyclic variations which affect the seasonal radiation income at certain latitudes.

(i) Eccentricity or orbital stretch

The first of the orbital features constituting what today is known as the astronomical theory of climate change, involves the eccentricity of the orbit. The earth's orbit around the sun is not perfectly circular, but rather an ellipse. Accordingly the earth is nearer the sun at certain times of the year and further away at others. The point of closest approach, the perihelion, currently occurs on the 2–3 January, the northern hemisphere winter, when 1,400 kW/m^2 of solar energy arrives at the top of the atmosphere. The point of furthest distance from the sun, the aphelion, occurs on 5–6 July, the southern hemisphere winter, when 1,311 kW/m^2 of solar energy arrives at the top of the atmosphere. This means that at present the earth as a whole gets 7 per cent more solar radiation in January than July. At the stage when the earth is at its most elliptical orbit this figure increases to 25 per cent. The orbit changes from its most elliptical to almost a perfect circle, back to its most elliptical (i.e. one full cycle) over a period of about 96,000 years. It should be emphasised that for any place on the earth the annual receipt of radiation is unaffected by this orbital feature. If a place receives a bit less in the winter season it is recompensed during the summer season. Only the seasonal receipt is altered by the eccentricity changes.

(ii) Obliquity or axial tilt

The second cycle involves changes in the tilt of the earth's axis. Obliquity is a measure of the angle between the earth's axis of rotation and the plane in which the orbit takes place. Presently this is 23.5° which is why the sun is directly overhead the tropic of Cancer (23.5° N) at noon on 21 June and directly overhead the tropic of Capricorn (23.5° S) at noon on 21 December. A full cycle of the axial tilt varies from a minimum of 21.8° to a maximum of 24.4° and back over a period of 41,000 years. The importance of the axial tilt is that it controls the seasonality of the earth's climates. Without any axial tilt there would be no seasons. On the other hand, if the axial tilt were 54°, everywhere on the earth, from the North Pole to the South Pole, would get equal amounts of solar radiation in the course of a year. At present the axial tilt is decreasing by about 0.00013° per annum. When the tilt is greater, the more the high latitudes 'lean' towards the sun during their summer and 'lean' away during their winter. Consequently, seasonal contrasts are higher, especially in polar and high latitudes, when tilt is close to its maximum and as axial tilt decreases seasonal contrasts diminish. Once again the total annual radiation receipt for the earth as a whole is not affected by these orbital characteristics, merely the seasonal latitudinal distribution.

(iii) Precession or earth wobble

The third and final orbital characteristic arises because the earth wobbles or gyrates on its axis somewhat like a spinning top or spinning gyroscope. The axis of the North Pole currently points to the North Star, Polaris; but the wobble reorientates this to other stars over a long period. At a distant time in the future the North Pole will point to the star Vega before returning again to point to Polaris. The full period cycle of this orbital characteristic, known as the precession of the equinoxes, is 22,000–23,000 years with a weaker periodicity signal at 18,800 years. The name 'precession of the equinoxes' is given to this characteristic because, as a consequence of the 'wobble', the elliptical orbit itself rotates slowly around. The positions of the perihelion and aphelion thus also move, changing the positions on the orbit of the times of year when the earth is closest and furthest away from the sun. For example, today the aphelion is in July; 10,500 years ago (i.e. one half-cycle) it was in January. The solstices and equinoxes thus move around the orbit, again altering the receipt of radiation seasonally and latitudinally. These three orbital characteristics, depicted in Figure 6.9, interact with each other to provide long-term cyclical changes in radiation receipt at particular latitudes and seasons on the earth which are now known to be extremely instrumental in influencing long-term changes in climate.

(c) Discovery

In 1842 a French mathematician, Adhémar, sought to explain the alternation of glacial and interglacial climates using the precession of the equinoxes. He reasoned that when the winter solstice occurred at the time the earth was furthest from the sun in its elliptical orbit (i.e. at the aphelion), winters would be longer and summers shorter in the hemisphere concerned. A glacial episode would result in that hemisphere. In the opposite hemisphere shorter winters and longer summers would promote interglacial conditions. A cycle of 23,000 years for glaciations would exist, with glaciations alternating between the two hemispheres. Since, at present, the aphelion or furthest distance occurs in the southern hemisphere winter, a glaciated continent at the South Pole seemed to back up Adhémar's theory. What Adhémar did not appreciate, however, was that the shape of the earth's orbit also changed over time from an ellipse to a near circle. When the orbit was close to circular the earth would at all times be the same distance from the sun and so his theory on precession would have been irrelevant.

(i) James Croll

It was a Scottish mathematician, James Croll, who took up this aspect over 20 years later. Croll was a carpenter, a hotel owner, an insurance salesman and held a variety of other jobs until he became a janitor in the Andersonian Institute in Glasgow, now the University of Strathclyde. There he came across Adhémar's work in the library and he began to deliberate on the effect of orbital eccentricity. Croll calculated the changes in orbital eccentricity from a circular to an elliptical orbit over the past 3 million years. He reasoned that only when the orbit was in a highly elliptical phase would

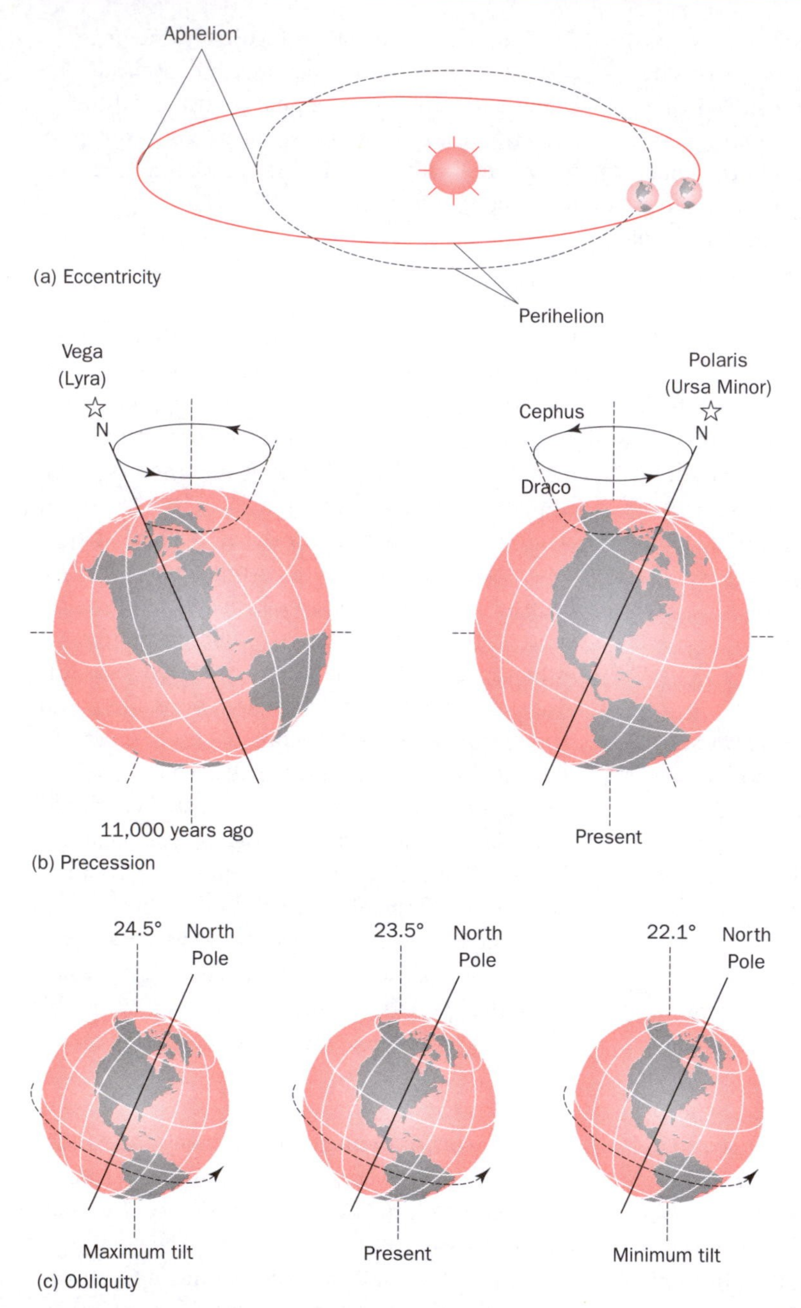

Figure 6.9 The astronomical theory of climate change: aspects of the earth's orbit

precession of the equinoxes have any effect in inducing glaciation. When in circular mode there would be no differences in winter radiation inputs between the two hemispheres since both would be equidistant from the sun. A glaciation, once triggered, would be enhanced by the positive ice-albedo feedback effect. Furthermore, Croll added the third key aspect of the earth's orbital behaviour: the axial tilt cycle. The more the tilt, he suggested, the more the summer heat could melt the accumulation of ice and snow at high latitudes. The less the tilt the less the summer heat reaching the poles, and the more the previous winter's snow and ice accumulation could survive to the next cold season.

According to Croll's ideas, conditions favourable for glaciation would occur when both the precession, tilt and eccentricity conditions were favourable, approximately every 80,000 years. He thus suggested the peak of the last ice age was about 80,000 years ago. This, even by the 1890s, was being openly questioned and the astronomical theory of ice ages was consigned to the category of an interesting but not very useful hypothesis and several more decades were to pass before Croll's ideas were resurrected.

(ii) Milutin Milankovitch

Milutin Milankovitch was a Serbian mathematician who became interested in the influence of orbital changes on climate around the time of the outbreak of the First World War in 1914. He calculated the solar radiation changes which would occur at each stage of the three cycles for a number of different latitudes. What became clear from this was that nearer the poles it was the

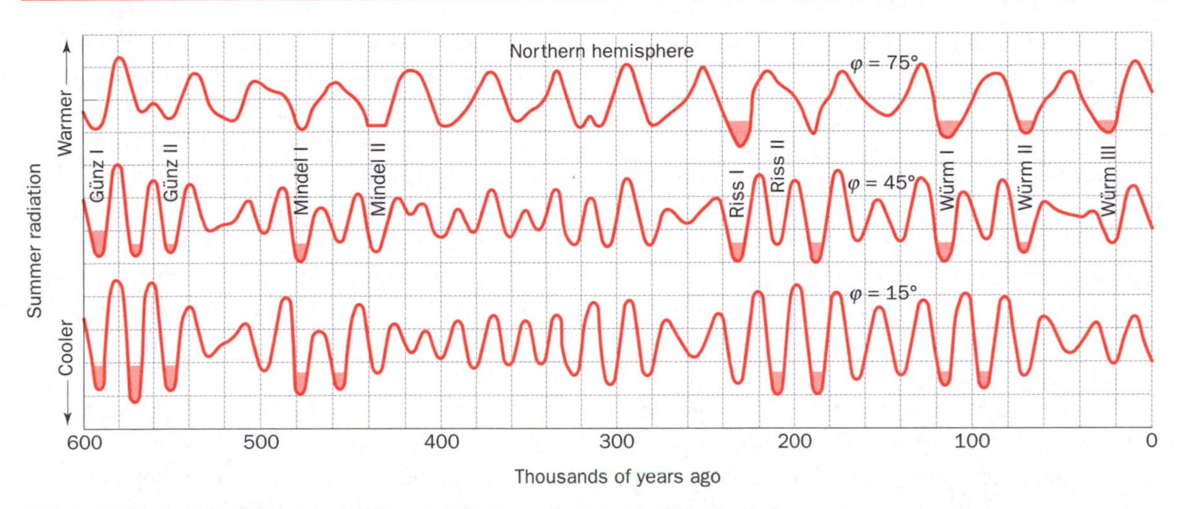

Figure 6.10 Radiation receipt at different latitudes calculated by Milankovitch

41,000-year axial tilt cycle that was most important, while nearer the equator it was the 23,000-year precession cycle that appeared dominant (Figure 6.10). Summer was confirmed as the crucial season when reductions in radiation income would have the greatest effect.

Milankovitch showed his work to a German climatologist, Vladimir Koppen, who matched up the suggested dates of recent glaciations with evidence of glacial events in the valleys just north of the Alps. At that time this evidence suggested four glacial events had occurred, termed the Würm, Riss, Mindel and Günz after the rivers in which their deposits had been found. The answer seemed to have been found to the mystery of the Pleistocene glaciations. Milankovitch's calculations resulted in a set of predictions for climate change over the past 650,000 years. The key dates he suggested for the last three glaciations were 25,000, 72,000 and 115,000 years BP. As new techniques became available to date sediments from these distant times, the astronomical theory would either be corroborated or discredited. The latter occurred and even before his death in 1958 Milankovitch's theory was widely rejected.

Deposits from the most recent cold phase were amenable to dating by the newly developed technique of radiocarbon dating by the 1950s. This placed the last major glacial advance (glacial maximum) at 18,000 BP, some 7,000 years later than Milankovitch had predicted. For the next two decades his ideas were largely ignored. It was with the advent of deep ocean drilling in the 1970s that new evidence came to light concerning long-term changes in global climate. The oxygen isotope analysis of these cores found four embedded cyclic components with periods of 100,000, 43,000, 23,000 and 19,000 years. The 100,000-year eccentricity cycle seems to be the dominant one. Every 100,000 years during the past 800,000 years a major expansion of ice has occurred. Superimposed on this rhythm are smaller advances which seem to occur at intervals of 23,000 and 41,000 years, corresponding to the precession and axial tilt cycles respectively. The eccentricity, obliquity and precession cycles seemed to be vindicated. Variations in the earth's orbit were deemed to be the 'pacemakers of the ice ages'.

(d) Continuing issues

Yet a number of unanswered questions remained concerning Milankovitch's theory. There are questions concerning the strength and timing of the 100,000-year eccentricity cycle and many climate changes during the ice age are of much shorter duration than those imposed by orbital forcing. Many such questions and others are addressed in section 8.3, but two general ones demand attention here.

(i) Why synchronous glaciations?

The astronomical theory should induce glaciation in only one hemisphere at a time. The other hemisphere should be a beneficiary of the 'opposite' radiation regime prevailing at the particular time concerned. Yet the evidence seems to suggest that glaciations are roughly synchronous in both polar regions.

The answer may lie in the different geographies of the two poles. In the northern hemispheres cool summers are indeed crucial to glacial advances. The winter snowfall lies on the land areas surrounding the Arctic Ocean and fails to melt before another cold winter augments ice volume further. In the Antarctic cool summers are not so crucial. Winter snowfall occurs mainly on a water surface in the Southern Ocean. Cold winters are needed to freeze more ice from the seas surrounding the landmass of Antarctica. On these ice bodies snowfall can accumulate and interior ice can advance. Cold northern summers and cold southern winters occur in phase in the Croll–Milankovitch cycles and hence glacial episodes occur roughly simultaneously in both hemispheres. Because the north polar area is water surrounded by land, and the south polar area is land surrounded by water they behave differently to warming and cooling influences. There is incidentally a lesson here in how global warming may affect both areas. Current models suggest that ice will melt in the north polar areas but accumulate over Antarctica as warming proceeds.

(ii) Rare events

The second puzzle is why glaciations are only a feature of the earth's climate in relatively recent times. Certainly there were glacial events much earlier in earth's history, but the current cycle seems to date only from the last 2 million years or so, the geological epoch known as the Pleistocene. The search for a trigger factor(s) to initiate the present series has led to suggestions that the Alpine orogeny, the relatively rapid uplift of major fold mountains such as the Alps, Rockies, Andes and Himalayas, may have been responsible. The elevation of large areas of land into higher colder parts of the atmosphere may have disrupted the upper westerlies (see Chapter 4) sufficiently to allow the astronomical influences to take control. But this is still conjectural and a range of other influences such as long-term changes in atmospheric CO_2 concentration and changes in ocean circulation brought about by continental drift may also be implicated (see Chapter 8). All this suggests that astronomical or orbital forcing, while providing an important framework for directing long-term climate change, does not act alone. Rather their influence is modified by other influences and by complex feedback linkages which need to be considered in any explanation of long-term climate changes.

6.5.3 Internal natural causes of climate change

In section 3.5, it was shown that a range of internal factors could also alter radiation conditions at the surface and thus promote climate change. These include both natural and human changes to atmospheric composition, land surface character (with a case study on the urban heat island in section 3.6) and alterations to the biosphere. Examples of the climate effect of changes in atmospheric composition by natural and human agencies will now be discussed in the final sections of this chapter.

(a) Volcanic activity

When the volcano Krakatoa erupted on 26 August 1883, much of the island disappeared overnight as huge amounts of debris were ejected into the atmosphere. Estimates of the volume ranged from 6 to 18 km^3, and the material reached a height of 27 km. Some 160 km away the capital city of Indonesia, Djakarta, was in darkness for 5 hours and even as far away as Montpellier in France, a decline in the strength of solar radiation was measured which persisted for about 4 years. This was not the first time a link between climate and distant volcanic events had been noted in France. Almost exactly a century before Krakatoa the first ambassador to France of the newly established United States of America had suggested that the harsh winter of 1783–84, following a foggy summer in which the Parisian sunshine was unusually faint, was linked to a major eruption of Laki, in Iceland. Literally, a smoking gun seemed to exist linking climatic cooling to major volcanic eruptions.

(b) Role of sulphate particles

When a volcano erupts it sends both particulate matter (dust, ash, condensed water droplets) and gases (water vapour, CO_2, sulphur dioxide) high into the atmosphere. Some gases and the heavier particles only reaching levels below 10 km have a short residence time in the atmosphere and are quickly removed. Scavenging by raindrops rinses them out and air motions may also contribute to their fallout. Gases which are quite light, especially sulphur dioxide, can however linger for much longer if they are injected to heights beyond the troposphere, into the lower stratosphere. Recall that unlike the troposphere, the stratosphere becomes warmer with height. This means it is quite stable and material injected into it is not subject to rapid dispersion. The sulphur dioxide emissions from a large explosive event react with water vapour in the presence of sunlight to form tiny reflective sulphate droplets and particles which can remain suspended in the atmosphere for lengthy periods. Aerosol particles with diameters of less than about 10 μm can remain suspended for approximately a year, while those less than 2 μm in diameter may remain airborne for up to a decade. The aerosols tend to concentrate in a band about 25 km high and gradually spread out to form a stratospheric dust veil which envelopes first the hemisphere in which the eruption occurred and, within about 4 months, the entire globe. Spectacular sunsets, red skies and blue moons can often be observed in the wake of a major eruption as the suspended particles scatter the light of the sun. Sulphate aerosols are so efficient at scattering and reflecting solar radiation because of their small size (similar to the wavelength of solar radiation). They can scatter incoming radiation up to 10 times more effectively than would otherwise be the case. As a result, solar energy availability at the surface is reduced. The main effect of volcanic activity is thus to cool the troposphere (though increased absorption in the stratosphere may result in warming there).

(c) How much cooling?

The magnitude of cooling depends on a number of factors. The most important include the latitude and hemispherical location of the volcano, the size and explosive power of the eruption, the sulphur content of the emissions, the timing of the emission in terms of other global changes under way and finally the clustering or frequency with which eruptions are occurring over a period.

(i) Location

Eruptions occurring at high latitudes have less significance for global climate than their counterparts nearer the equator. This is because the global wind circulation predominantly moves air polewards, so that high-latitude eruptions may not be as effective in forming a hemispheric veil. The global wind circulation is also slow to move the products of an eruption across from one hemisphere to another. Thus the northern hemisphere reacts more quickly to an eruption in the northern hemisphere than to distant ones in the southern hemisphere. On the other hand, the southern hemisphere shows little response to eruptions in the northern hemisphere. It also exhibits a slower response to eruptions in the southern hemisphere, possibly because it is so dominated by oceanic influences on climate which dampen temperature changes of all kinds.

(ii) Explosive power

A non-explosive eruption that produces predominantly a lava flow has little climatic significance. The more violent the eruption, the more likely that material will end up in the stratosphere. For example, Mount St Helens (USA, 1980) exploded sideways. It was also located at a relatively high latitude. It therefore had a hardly noticeable effect on world climate. Tambora (Indonesia, 1815), close to the equator, produced 1,000 times as much sulphuric acid aerosol as Mount St Helens and was followed by 'the year without a summer' in which exceptionally late frosts destroyed crops in North America and French vineyards had their latest wine harvest since at least the fourteenth century.

(iii) Sulphur content

The sulphur content of the eruption is crucial and there is a good correlation between the amount of sulphur gases emitted and the resultant degree of global cooling. Tambora released over 200 Mt of sulphur gases and cooled the planet by 0.7 °C. Mt Pinatuba (Philippines, 1991) was only one-tenth as prolific, but still produced global cooling of 0.3–0.5 °C lasting 2–4 years. While this amount seems relatively small and short-lived, it is approximately equal to the total warming effect of human-induced climate change over the past century.

(iv) Timing of event

The climatic impact of a volcanic eruption can also accelerate ongoing changes as well as masking them. If cooling is already occurring, such as the approach of a glacial event due to astronomical forcing, an eruption may trigger more rapid and long-lasting changes. The largest volcanic eruption of the past million years was probably Mt Toba in Indonesia which erupted about 73,000 years ago as the world was rapidly cooling towards the last glaciation. This episode would have reduced daytime sunlight to that of a moonlight night, and the cooling produced would have been expected to enable winter snowfall in higher middle latitudes to survive summer melting. The ice-albedo feedback effect associated with this increased snow cover may well have accelerated the onset of the glacial event.

(v) Duration

On geological timescales, massive continental basaltic lava flows may have been capable of producing such large amounts of sulphur gases that their lack of high-level injection may have been compensated for by long duration emissions. The continental flood basaltic flows of the north-western United States (Oregon, Washington) 15–17.5 million years ago is estimated to have emitted 1.2×10^{13} kg of sulphur into the atmosphere over a period of several thousand years. Some 65 million years ago a much larger basalt flood eruption took place to form the Deccan Plateau of India and lasted sporadically for about 6 million years. The climate cooling and other effects of this massive eruption have been linked to the demise of the dinosaurs. The basalt flow eruptions in Siberia around 250 million years ago were even larger than those in India and have been associated with wiping out 95 per cent of all life on earth at the time. With these prolonged basaltic eruptions, it would appear that even though the emissions were from low-lying areas, the removal of sulphates from the atmosphere was cancelled by ongoing emissions over a long time period.

It has often been suggested that the climate (cold, low levels of sunlight) and environmental (acid rain, plant and animal death) impact of the basaltic flow eruptions, though considerable, was probably enacted at a local and regional rather than at a global scale. However, certain scientists now implicate later, long-term and cumulative CO_2 emissions from the basaltic eruptions as perhaps the greatest threat to life on earth. So rather than prolonged sulphate cooling, long-term CO_2 warming may have been the lethal factor in the demise of many forms of life at certain periods in earth's history.

(vi) Clustering

The acidity of annual layers in the polar ice cores provides a means of analysing past eruptions and a good correlation exists between periods of low volcanic activity and above-average temperatures. The Medieval Warm Period seems to coincide with a lull, while the Little Ice Age was marked by higher levels of volcanic activity. The early decades of the twentieth century and the middle and late 1990s were also periods of rapid warming accompanied by relatively few volcanic eruptions. However, the local and regional effects of volcanic activity can be complicated and while on average hemispheric temperatures fall by 0.2–0.5 °C for 1–3 years after a major eruption, this does not apply everywhere. During the winters after the 12 largest volcanic eruptions from 1893–1992 warming by as much as 4.3 °C occurred over Eurasia and North America, while cooling of 1–2 °C occurred over the Middle East, North Africa and Greenland. Winter warming seems to occur as a result of the increased latitudinal temperature gradient between the stratosphere in the low latitudes (warmed by the aerosol-absorbing solar energy) and the polar stratosphere. As was discussed in Chapter 4, a thermal gradient between the equator and pole creates a pressure gradient which drives a strong westerly airflow aloft. These stronger westerlies inhibit blocking and direct mild maritime air masses deeper into the continents of Eurasia and North America.

6.5.4 Internal human-induced causes of climate change

(a) Global warming

In Chapter 2 it was reported that 7.9 Gt of carbon was being added annually to the atmosphere in the form of CO_2 and that the build-up of this gas from a pre-industrial level of 280 ppm to concentrations of approximately 370 ppm has now occurred. It is very likely that such a concentration has not been exceeded for the past 15 million years. The significance of increased CO_2 concentrations is that, unlike oxygen and nitrogen which account for about 99 per cent of the atmosphere, CO_2 (together with water vapour and other trace gases) effectively absorbs outgoing long-wave radiation. This is trapped and reradiated back downwards, resulting in the surface temperature for the earth becoming warmer than would otherwise be the case. Even without human-related increases in CO_2 concentrations, surface temperatures are about 33 °C higher than they would otherwise be (i.e. 15 °C as opposed to −18 °C). This is known as the natural greenhouse effect, section 3.2.1). It should also be noted that the more the build-up of heat in the lower levels of the atmosphere occurs, the less there is available to heat the higher levels. Thus, tropospheric warming is accompanied by stratospheric cooling.

(i) Carbon dioxide

A Swedish chemist, Svante Arrhenius, suggested in 1896 that if the concentration of CO_2 in the atmosphere were to double, global average temperature would increase by about 5 °C. Remarkably, this figure approximates quite closely to what current models, based on much more sophisticated techniques, also suggest. The work of Arrhenius was considered somewhat hypothetical at the time and few people believed the changes in atmospheric composition on which they were predicated would ever occur. However, measurements of CO_2 which commenced in the 1950s at the South Pole and in the Hawaiian Islands at Mauna Loa confirmed that rapid increases in concentration were taking place as a result of human activities. The rapidity of change was further highlighted by analysis of the CO_2 concentration detected in trapped air bubbles in ice cores. These confirmed a growth rate of about 1.5 ppm/year with signs of a recent acceleration in this trend since the late 1990s. As explained in Chapter 2, processes involving the ocean and terrestrial biosphere remove about 40 per cent of the additional CO_2 loading on the atmosphere due to human activities. CO_2 dissolves in water and atmospheric CO_2 is incorporated into the oceans especially as breaking waves trap air pockets beneath the surface waters for a brief period. Mixing to lower levels takes much longer – of the order of centuries or more. In addition, as the normal processes of biological activity take place near the surface, organisms not consumed by predators, or decomposed back to inorganic form, gradually settle out of the surface layers to build up on the seabed where they may take no further part in the carbon cycle for perhaps millions of years. In their place, close to the surface, more CO_2 can be drawn down from the atmosphere to restore the equilibrium between oceanic and atmospheric concentrations. In a similar vein, the terrestrial biosphere also responds to increased CO_2 by increasing photosynthesis in some families of plants. Like the oceans, a negative feedback effect is apparent, with increased atmospheric levels stimulating increased terrestrial uptake. However, both of these sinks become less effective as atmospheric CO_2 concentrations increase to the levels projected for mid century, particularly if the accompanying climate changes and vegetation changes are also incorporated into the models used.

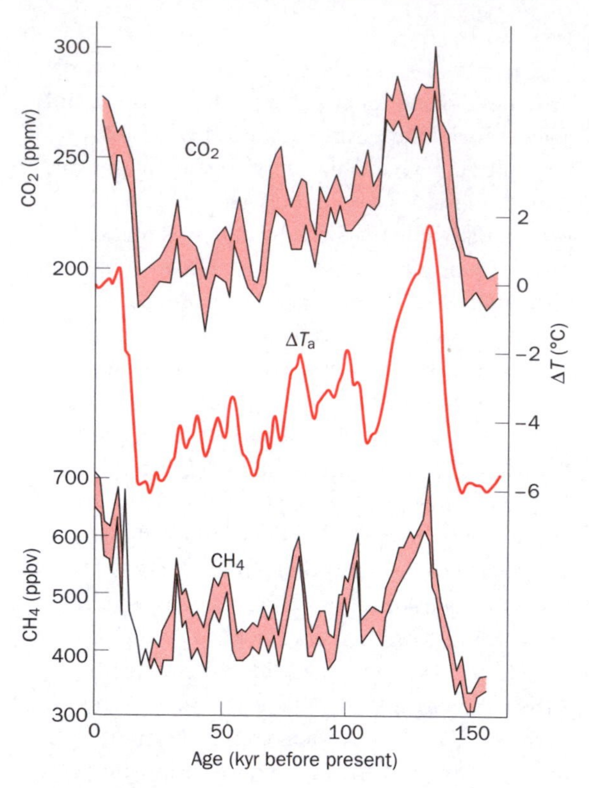

Figure 6.11 Temperature, methane and CO_2 concentrations during the last 150,000 years or one glacial cycle. Evidence from the Vostok core

Source: National Ice Core Laboratory, University of New Hampshire: http://www.exploratorium.edu/climate/cryosphere/data2.html

(ii) Methane

Methane is the second most significant enhanced greenhouse gas and has already more than doubled in concentration since pre-industrial times. About 60 per cent of the annual emissions of about 600 Gt can be linked to various human-related activities, with much of the remainder due to natural emissions from marshes and other wetlands. Although its concentration today of 1.8 ppm appears low in comparison to CO_2, on a molecule for molecule basis it is almost an order of magnitude more powerful at absorbing long-wave radiation. Examining the methane content in ice core bubbles provides dramatic evidence of how greenhouse gas concentrations are higher during interglacials when biospheric processes are in full swing (Figure 6.11). Even during previous interglacials, however, concentrations seldom approached the pre-industrial level of 0.7 ppm, less than half today's value. In addition to industrial releases of methane from coal mining operations and oil and natural gas exploration and transport, sources of methane are also strongly related to agriculture, particularly paddy rice growing and cattle rearing (Table 6.2). These are not always easy to quantify, though novel approaches are presently being developed (Plate 6.3). The combination of both industrial and agricultural sources exemplifies the political problem of seeking to control greenhouse gas emissions simultaneously in the developed and developing worlds.

(iii) Other greenhouse gases

A number of other gases are generally minor contributors to the enhanced greenhouse effect. Nitrous oxide (N_2O) has a concentration of about 314 parts per billion (ppb) in the troposphere and

Table 6.2 The global methane budget (Gt/year)

Sources	
Natural	
Wetlands	115
Termites	20
Oceanic emissions	10
Other	15
Anthropogenic	
Coal mining, natural gas, petroleum industry	100
Rice paddies	60
Ruminants	85
Animal wastes	25
Sewage treatment	25
Landfills	40
Biomass burning	40
Sinks	
Atmospheric removal	530
Removal by soils	30

Source: After Houghton (1997)

has shown only a modest increase of about 13 per cent since pre-industrial times. Emissions from soils are the main source and increases in this may be related to growing intensification of agriculture, particularly increased fertiliser use. Hydrofluorocarbons (HFCs) have become widely used as substitutes for the chlorofluorocarbons whose production has been almost eliminated as a result of the Montreal Protocol. Though concentrations are very low, measured in parts per trillion (ppt), rapid increases are currently occurring. Some HFCs have long atmospheric residence times. A similar situation exists with perfluorocarbons (PFCs) and sulphur hexafluoride, almost exclusively anthropogenic in origin. Increases in the range 1–3 per cent per annum are presently occurring for some species in these families and while concentrations are still tiny, atmospheric residence times are over 1,000 years and a contribution to the greenhouse effect will still be occurring from today's emissions well into the fourth millennium AD.

The growing importance of another greenhouse gas, tropospheric ozone, has also been acknowledged. Tropospheric ozone concentrations have increased by about 35 per cent since pre-industrial times. Over remote areas of the tropical Pacific and Indian oceans, levels are typically around 10 ppb, while downwind of major sources of precursor pollutants, concentrations an order of magnitude higher are often observed. Background levels in many parts of Europe now approach or exceed critical threshold levels above which yield reductions in some crops can be expected and summertime episodes of high concentrations frequently breach human health-related limit values.

Ozone forms as a secondary pollutant under the action of sunlight from a range of fossil fuel-related emissions such as nitrogen dioxide, carbon monoxide and reactive organic gases known as volatile organic compounds (VOCs). It also gets mixed down from the stratosphere on occasion, though about 50 per cent of tropospheric ozone in the northern hemisphere is anthropogenic in origin. Because of the importance of local emissions, and the short lifetime of a matter of days, ozone trends are geographically quite specific to particular regions, and the global trend is indistinct at present. However, the enhanced greenhouse effect component provided by tropospheric ozone is expected to increase substantially during the next few decades as precursor emissions increase, particularly in the rapidly developing economies of Asia.

(iv) Cooling agencies

Greenhouse gas global warming is moderated to a certain extent by a number of other human-related processes that act to cool the planet. Such processes are shown in Figure 6.12 and range from stratospheric ozone destruction to enhanced human emissions of sulphate aerosols in the troposphere.

Ozone is a greenhouse gas, and therefore its depletion in the stratosphere will tend to result in global cooling. As described in Chapters 2 and 3, chlorofluorocarbon compounds emitted from human activities at the surface will continue to degrade the ozone shield located in the stratosphere for some decades to come. As these

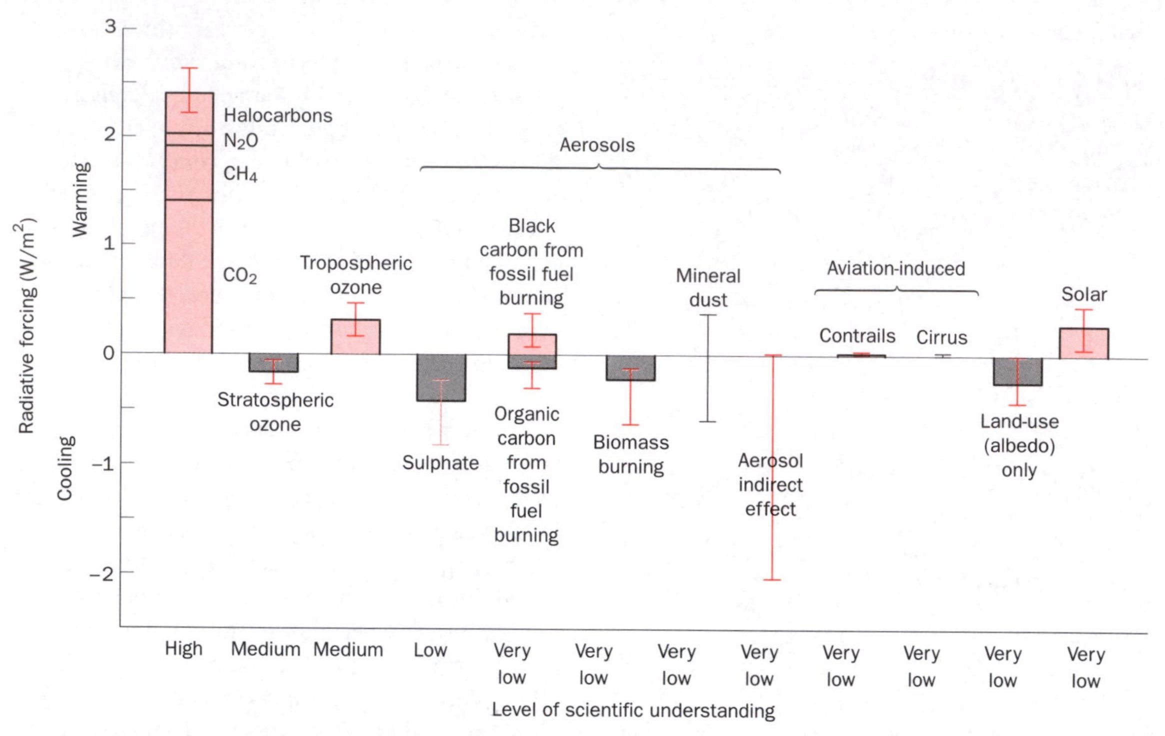

Figure 6.12 Radiative forcing of climate change in recent centuries
Source: Houghton *et al.* (2001, p. 8)

harmful emissions reduce, they are being replaced by other halocarbon compounds which are themselves greenhouse gases, especially hydrochlorofluorocarbons (HCFCs) and hydrofluorocarbons (HFCs). The overall temperature response is dependent on whether the negative radiative effect of ozone depletion is outweighed by the positive radiation effect of the CFCs, HCFCs and HFCs. Current thinking is that at high latitudes, where ozone depletion is greatest (see Box 2.1), the net effect is for cooling, while at low latitudes where ozone depletion is least, the opposite is the case. The so-called 'ozone hole' observable in the polar zones during springtime is thus something which does not contribute to global warming. In its absence (or as it is rectified by reduced CFC emissions) an accentuated warming trend could be expected in these regions.

Human increases in the concentration of a range of tropospheric aerosols (very small particles and droplets) have been implicated in global cooling, although their potential radiation forcing is not considered to be large enough to reverse the warming trend. The effect of aerosols is not completely understood but they are known to act directly by reflecting, scattering and absorbing radiation, and indirectly by helping to form clouds which also modify the pathways and exchanges of energy in much the same way. The cooling effect (mainly from reflecting solar radiation back to space) of sulphate aerosols from fossil fuel and biomass burning has been acknowledged by many scientists. Sulphate aerosols may have helped the earth to cool or to stop warming from 1945 when industrial activity increased, to around 1970, when their effect was swamped by the rising concentrations of CO_2 (see Figure 1.1). The potential future cooling effect of sulphates is incorporated into climate change scenarios at both global and regional levels. Recent modelling studies by the Hadley Centre have shown that Africa may warm proportionally more in the twenty-first century than the other continents (except Antarctica) because it has by far the lowest emissions of industrial sulphate (Box 6.2).

(b) Future greenhouse gases and uncertainties

Clearly the radiative effect of greenhouse gases and their residence times in the atmosphere vary enormously. If policies to control emissions are to be put in place, it is desirable to know which gases to target to achieve the best results over a given period. Global warming potential (GWP) provides an index of the relative radiative effect of a particular greenhouse gas integrated over a chosen time period, such as 20 or 100 years. The benchmark is usually taken to be CO_2 and other gases are expressed on a kilogram for kilogram basis. A sample selection of GWPs is shown in Table 6.3. The implications are that removing a tonne of methane from the atmosphere today would contribute over 60 times as much benefit to reducing global warming over the next 20 years as removing the same amount of CO_2.

Understanding the scientific processes involving greenhouse gases is a first step towards anticipating the future course of climate. However, this alone is insufficient to provide confident future climate scenarios. A source of major uncertainty lies in projecting how much atmospheric loading of greenhouse gases will change in the years ahead. Considerations such as population growth, economic activity, energy use, types of energy sources and land use changes are extremely difficult to forecast accurately. Uncertainties in the effectiveness of mitigation measures, such as the Kyoto Protocol, also exist. Yet trends in these categories will determine how fast greenhouse gas concentrations will change, and how much radiative forcing will occur. A number of attempts to provide future emission scenarios have been made, each based on sometimes radically different sets of assumptions. The current best practice involves 40 future scenarios determined according to a variety of assumptions concerning population growth, economic development and technological progress. Four 'storyline families' were constructed to place in broad context the emission changes being proposed (see Box 7.5). Six specific scenarios derived from these families are widely used as inputs to global climate models. These epitomise the difficulties involved in projecting future CO_2 concentrations, with their estimates for 2100 ranging from 541 to 970 ppm. Social science as much as science poses problems for charting the likely course of the earth's climate over the next century.

A grasp of the multiple causes of climate change and a confident projection of how future forcing of the climatic system is likely to occur enables a modelling exercise to be undertaken to establish future climate scenarios. These are the tools which bring scientific understanding into the arena of policy formulation and enable choices to be made at all levels of environmental management. A sample future climate scenario for Ireland is shown in Plate 6.4. This indicates that current mean January temperatures are projected to increase by 1.5 °C mid century by which time the extreme south and south-west coasts may have a mean January temperature of 8.0–8.5 °C. The scenarios suggest July temperatures will increase by 2.5 °C by 2055. While increases in precipitation are predicted for the winter, marked decreases of over 25 per cent during the summer are projected. The ability to provide such scenarios is increasingly important. They obviously have vital applications in many activities where long-term planning is required, such as in water supply and flood control, the building of major civil engineering structures such as dams, bridges and drainage systems, as well as in medium-term concerns such as agriculture, water quality and ecosystem management planning. Future climate modelling is increasingly a necessary component of informed forward planning and the principles of modelling methodology are examined in the following chapter.

BOX 6.2

CASE STUDY

Climate and war – when science and politics clash

Aerosol loading of the atmosphere can also come about other than through volcanic eruptions. Dust blown up from surface sandstorms in the Sahara occasionally falls quite far north in Europe, having been transported by stable southerly airflows. Dramatic plumes of such dust can occasionally be seen on satellite images (Plate 11.2). Smoke from warfare has also been postulated as a potential agent of climate change as the smoke particles could block incoming radiation.

Apart from long-term volcanic activity, major dust injection into the atmosphere as a result of an asteroid impact had for long been suggested as instrumental in the extinction of the dinosaurs some 65 million years ago. Indeed on 30 June 1908 a meteorite some 50 m in diameter came to earth in Tunguska in Siberia. Some 2,000 km^2 of forests were destroyed by an explosion roughly equivalent to a 15 megaton nuclear bomb. According to this theory, with a sufficiently large asteroid collision in the late Cretaceous Period, enough sunlight could have been obscured by the dust loading to damage sufficiently the plant life on which the dinosaurs depended for food, leading to their starvation. This viewpoint implied that any species, including the human species, could be rendered extinct as a result of catastrophic climate change induced by a sudden major dust loading in the atmosphere.

A nuclear exchange was suggested in the 1980s as having the potential to generate a massive smoke pall which would form a stratospheric veil similar to a major volcanic eruption, though on a vastly increased scale. For several weeks after the war, it was suggested, light-absorbing soot particles would block the sun, producing darkness over large parts of the northern hemisphere. If the nuclear exchange had occurred in winter, large tracts of the northern hemisphere would experience subzero temperatures. A drop in temperature of 20–40 °C would result in widespread crop failures and famine. Advocates of the nuclear winter theory suggested that the 'heat equator' would move polewards to where the burning cities lay, reordering the Hadley cells and changing the planetary wind circulation drastically.

Gradually the nuclear winter theory has unravelled. It became clear that the modelling exercise on which it was based was simplistic and flawed. The model of the earth used for climate simulation was featureless and did not incorporate any oceans or continents. Neither did it take into account day and night, clouds or winds, and also tended towards worst-case scenarios for assumptions in several key areas where hard information was lacking. When later more sophisticated techniques were employed it became clear that a 100 megaton nuclear exchange was only one-millionth the blast of an asteroid impact, and that a 'nuclear fall' rather than a nuclear winter was the most likely result.

Before the science caught up with events, however, the politicians had seized on the doomsday predictions. More money was spent on public relations firms, videos and advertising than on the science. The public became sensitised to issues on nuclear non-proliferation. Arms control treaties were signed. Ironically, what many regarded as a flawed theory gained political and popular acceptance and the end product was beneficial for all humankind. The ethical issues the nuclear winter episode highlights are important to consider in terms of when scientists should disseminate their work and how objectivity can sometimes be sacrificed unwittingly in the rush to achieve a desirable political objective.

Greenhouse gas	Atmospheric lifetime (years)	GWP 20 years	GWP 100 years
CO_2	c.100*	1	1
Methane	12	62	23
Nitrous oxide	114	275	296
CFC11	45	6,300	4,600
HFC–23	260	9,400	12,000
SF_6	3,200	15,100	22,200

Table 6.3 Direct global warming potential of some greenhouse gases relative to CO_2 on a weight-for-weight basis

*CO_2 is not destroyed chemically in the atmosphere, but redistributed via the carbon cycle. Turnover time varies enormously depending on process involved. 100 years is therefore only a 'guide' value.

Source: After Houghton *et al.* (2001)

Key ideas

1. Cyclic variations in the earth's orbit around the sun affect seasonal radiation income at certain latitudes. The eccentricity, obliquity and precession cycles have been strongly linked to the glacial–interglacial climate changes of the Quaternary Period.

2. Explosive volcanic eruptions, close to the equator, and with a high sulphur content, may produce global cooling up to about 0.5 °C persisting for a few years.

3. Rapidly increasing concentrations of greenhouse gases as a result of human activities are implicated in current warming trends. Comparisons of the relative contributions of each gas over a specific time period can be made using the concept of global warming potential, though major uncertainties in estimating future emissions of each persist.

Further reading

Drake, F. (2000) *Global Warming: The science of climate change*. Arnold, London, 273 pp.

Houghton, J. (1997) *Global Warming: The complete briefing*, 2nd edn. Cambridge University Press, Cambridge.

Houghton, J. T., Ding, Y., Griggs, D. J., Noguer, M., van der Linden, P. J. and Xiaosu, D. (eds) (2001) *Climate Change 2001: The scientific basis.* Contribution of Working Group I to the Third Assessment Report of the Intergovernmental Panel on Climate Change (IPCC). Cambridge University Press, UK, 944 pp.

Lamb, H. H. (1995) *Climate, History and the Modern World*, 2nd edn. Methuen, London, 433 pp.

O'Hare, G. (2000) Reviewing the uncertainties in climate change science, *Area*, **32**(4): 357–69.

Useful websites

Basic information on climate change:
http://www.cru.uea.ac.uk/cru/info/

Tree rings and climate change:
http://vathena.arc.nasa.gov/curric/land/global/treestel.html

Ice cores and climate reconstruction:
http://calspace.ucsd.edu/virtualmuseum/climatechange2/03_2shtml

Vikings and the Little Ice Age:
http://www2.sunysuffolk.edu/mandias/lia/vikings_during_mwp.html

Volcanic eruptions and climate change:
http://www.dar.csiro.au/publications/greenhouse_2000e.htm

Astronomical theory of climate change:
http://joides.rsmas.miami.edu/files/AandO/Zahn_ODPLegacy.pdf

Greenhouse gases and climate change:
http://www.ieagreen.org.uk/

CHAPTER 7

Modelling the climate system

7.1 Projections of future climate

Models are used to make *projections* of atmospheric concentrations of greenhouse gases and aerosols and hence of future climate. The modelled outputs are scenarios, or possibilities, of what the state of the atmosphere or climate might be like in the future: they are not precise or reliable enough to be termed 'predictions' or 'forecasts'. Most modelled outputs of future climate have been made at the global level, but increasingly scenarios at the regional (India, UK) and site-specific level (river basin, city) are being made. A summary of the latest (2001) global modelled projections from the Intergovernmental Panel on Climate Change (IPCC) Third Assessment Report (TAR) include the following.

7.1.1 Greenhouse gases

- Emissions of CO_2 due to fossil fuel burning will dominate the trends in atmospheric CO_2 concentration during the twenty-first century.
- CO_2 concentrations are likely to increase during the twenty-first century from 365 ppm (today) to 540 ppm (low estimate), or to 970 ppm (high estimate) (Figure 7.1(a)). The CO_2 concentration in pre-industrial times (AD 1750) was 280 ppm.
- As a result the global mean radiative forcing due to all greenhouse gases will continue to increase through the twenty-first century with the CO_2 contribution projected to increase from around half today to about three-quarters.
- The projected increase in aerosols forcing, both direct and indirect, is projected to be smaller in magnitude than that of CO_2.
- Modelled scenarios of the concentrations of non-CO_2 greenhouse gases by the year 2100 also vary, with methane (CH_4) changing by –190 ppb to +1,970 ppb over current levels (1,760 ppb); and nitrogen oxide (N_2O) by +38 ppb to +144 ppb above present concentration of 316 ppb.
- There is also a wide range in the estimated concentrations of tropospheric ozone (O_3) which according to some top estimates will rival CH_4 as a greenhouse gas in the future; hydrofluorocarbons (HFCs, e.g. CHF_3, CF_3CH_3); perfluorocarbons (PFCs, e.g. CF_4, C_2F_6) and sulphur hexafluoride (SF_6) relative to year 2000 (Figure 7.1(a)).
- Stabilisation of CO_2 at 450, 650 or 1,000 ppm would require global anthropogenic release of CO_2 to drop below 1990 levels within a few decades, about a century or about two centuries, respectively, and continue to decrease thereafter.

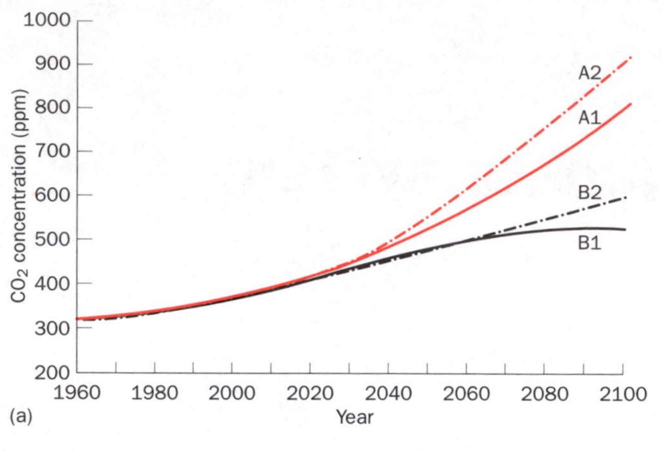

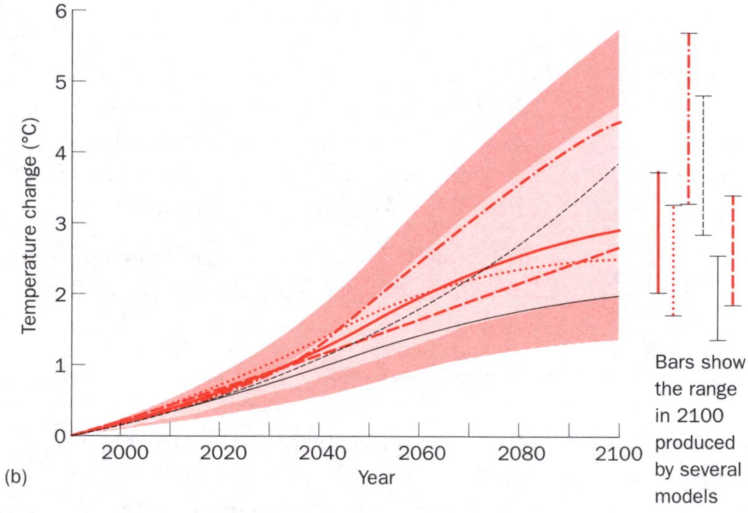

Figure 7.1 (a) Four possible scenarios of the future global concentration of CO_2 (ppm) for the twenty-first century (see Box 7.5 for explanation) and (b) full range of modelled projections of the rise in global mean annual surface temperature (1.4–5.8 °C) for the twenty-first century
Source: IPCC (2001)

7.1.2 Temperature

- Globally averaged temperature is projected to increase (using all scenarios) by 1.4–5.8 °C from 1990 to 2100 (Figure 7.1(b)).
- This temperature range is greater than that predicted in 1996 due to lower projected levels of sulphate (a cooling agent) in the atmosphere.
- The projected rate of warming is much larger than the observed rise in temperature during the last century and is very likely to be without precedent during the last 10,000 years based on palaeoclimatic data (Chapter 6).
- Over the next few decades, global warming is likely to lie in the range 0.1–0.2 °C per decade but this rate is projected to increase in some climate scenarios in the latter part of the century (Figure 7.1(b)).
- Nearly all land areas will warm more rapidly than the global average, particularly those at northern high latitudes in the cold season. Warming over northern North America and central Asia will exceed global mean warming by more than 40 per cent. Warming is projected to be less than the global mean in South and South-East Asia in summer and in southern South America in winter.
- Most warming will be due to increases in the night minima rather than increases in the daily maxima. As a result diurnal temperature ranges in many locations are likely to decrease.

7.1.3 Other parameters

- Many models project more El Niño-like behaviour in the eastern Pacific with warming more here than in the western Pacific. This will engender higher rainfall in the eastern Pacific Basin compared to the western Pacific Basin.
- Global average water vapour content and precipitation are projected to increase during the twenty-first century. By the middle of this

century, it is likely that precipitation will have increased over northern mid to high latitudes (following observed trends) and Antarctica in winter. In contrast, at low latitudes there will be both regional increases (e.g. northern India) and decreases (Sahel) over land areas. Larger year-to-year (interannual) variations in precipitation are very likely over most areas (e.g. the Asian monsoon) where an increase in mean precipitation is projected.

- It is possible that extreme weather events (drought, great heat, intense rainfall, flooding) are likely or very likely to increase in a globally warmed world in the twenty-first century. Again this may reflect increases in observed weather extremes during the last few decades of the twentieth century.
- Most models show a weakening of the global thermohaline circulation which will serve to cool parts of the North Atlantic. Nevertheless, a complete shutdown of the North Atlantic conveyor is not predicted so that there is still a net warming over Europe due to increased greenhouse gases. Beyond the year 2100, the global thermohaline circulation could completely alter or collapse.
- Northern hemisphere snow cover, glaciers, ice caps and sea-ice extent are all projected to decrease further during the twenty-first century.
- Ironically, Antarctica might gain mass because of enhanced snowfall in a warmer environment (e.g. from −50 to −30 °C).
- Future global mean sea levels are difficult to predict (they are a function of increased water expansion under global heating and water additions from snow and glacier melt) but are projected to rise by about 0.1–0.9 m between 1990 and 2100.

The above summary projections vary from one major climate model to another, and the collective outputs show a very wide range. The wide range is a result of two factors. First, the scenarios are based on widely varying estimates of future greenhouse gas and aerosol emissions (Figure 7.1(a)). Second, there is much uncertainty concerning the detailed workings of the climate system. The role of various feedbacks in the climate system, for example cloud, water vapour, ice, soil carbon, are difficult to quantify precisely. One result of this is that the climate sensitivity of different models varies. Sensitivity can be defined as the ease with which certain inputs of greenhouse gases, etc. can alter the climate. Another way of defining sensitivity of a climate model is the rate at which climate will stabilise after a doubling of CO_2. The ability of climate models to project future climate is thus still subject to much uncertainty, but the reliability of outputs has improved greatly over time with increasing computer power and more sophisticated understanding and mathematical modelling of the workings of the climate system (Table 7.1). In order to fully appreciate the validity (both strengths and weaknesses) of the above-modelled scenarios, and the ability of recent models to produce projections of climate at the regional and local level, it is useful to have a detailed understanding of the development of techniques and methods used in the modelling of future climate change.

Key ideas

1. The most recent projections of future greenhouse gas emissions and climate change are published in the Third Assessment Report of the IPCC (2001).
2. Projections of future CO_2 range from a rise from about 365 ppm today to a low estimate of 540 ppm to a high estimate of 970 ppm by the year 2100.
3. Present mean global annual surface temperatures (about 15.5 °C) are expected to rise by a low estimate of 1.4 °C to a high estimate of 5.8 °C from the year 1990 to 2100.
4. A list of other future projections is available in relation to other greenhouse gases, e.g. CH_4, N_2O, tropospheric O_3, CFCs and other aspects of climate change, i.e. precipitation, extreme events, El Niño, monsoons, thermohaline circulation, snow and ice and sea level.

Table 7.1 A chronology of climate modelling developments

Decade	Significant milestone
1890s	Arrhenius links global warming to atmospheric CO_2
1920s	Method of numerical weather forecasting first described
1950s	First numerical weather prediction schemes Supercomputer (IBM 701) first used specifically for weather forecasting
1960s	Atmospheric general circulation models (AGCMs) Ocean general circulation models (OGCMs) EBMs first described
1970s	Multi-layer oceans added to GCMs Supercomputer (CDC 6600) breaks the MFLOP* barrier Greenhouse modelling with GCMs
1980s	Palaeo-data sets first employed for model validation Satellites provide global observations First statistical downscaling schemes (SDS) GCMs become pre-eminent First regional climate models (RCMs)
1990s	Supercomputer (Cray C90) breaks the GIGAFLOP** barrier Transient climate model simulations Models of atmospheric chemistry and aerosols Coupled ocean/atmosphere general circulation models (O/AGCMs) Atmospheric Model Intercomparison Project (AMIP) Development of sea-ice and land-surface schemes Coupled O/AGCMs without flux correction Project to Intercompare Regional Climate Simulations (PIRCS)
2000s	Supercomputer (Cray SV1-32) breaks the TERAFLOP*** barrier Coupled carbon cycle included Multi-model ensemble simulations

Computer power is measured in floating point operations per second (FLOPS): * (10^6); ** (10^9); *** (10^{12}).

7.2 Introduction to modelling

Models are simplifications of reality, through which the real world is expressed at smaller scales by replica structures, or in abstract terms by mathematical equations. Reducing the scale or complexity of an environmental system helps us get to grips with processes that are beyond our limited physical or conceptual grasp. Models provide a framework within which to test hypotheses about a system, and a means of assessing the relative importance of component processes. Models also help us to forecast the future behaviour of a system as a consequence of actual or hypothesised disturbances to the present status of the system. However, the level of skill of forecasts is ultimately constrained by our (partial) understanding of system functioning.

Climate models apply simplified representations of the physical laws governing mass and energy exchanges in the ocean–atmosphere system (Chapters 2, 3 and 5), enabling us to better understand and predict the consequences of greenhouse gas emissions. Model sophistication has developed rapidly over the last few decades as a result of advances in computing power and the growing availability of observational data for model verification (Table 7.1). Climate models have also become integral to policymaking at international levels and, as such, are subject to considerable scrutiny by the global change community. Decisions about climate change mitigation and adaptation are complicated by the fact that no model can fully explain all aspects of the climate, across all space and timescales. Furthermore, it is acknowledged that there are significant uncertainties surrounding the future emissions that underpin climate model experiments.

The following sections provide an introduction to different types of climate model used for global to site-specific projections of climate change. The main timescale of interest will be climate dynamics at decadal timescales, and beyond. This develops discussions in previous chapters on natural and human-induced changes to the climate system. The final section examines the principal uncertainties attached to projections of future climate change arising from climate model experiments.

7.2.1 Energy balance models (EBMs)

(a) Zero-dimensional models

The simplest possible climate model would consider only the incoming and outgoing energy flows to the globe, and compute the resultant globally average annual temperature at the surface. In this zero-dimensional energy balance model (EBM), the rate of change of temperature depends on just two factors. One is the *net radiation balance* which, as shown in Chapter 3, is the difference between the incoming absorbed solar radiation and the net outgoing long-wave radiant energies to space, and second is the *heat capacity* of the earth. As we have seen previously, incoming energy (or absorbed energy) is a function of the solar flux, and the planetary (ground and atmosphere) albedo. The energy emitted is estimated using the Stefan–Boltzmann law (see Box 3.3) which indicates that emissions from perfect radiators (so-called black bodies), at a given wavelength, are proportional to the temperature of the emitting body. The law can be refined to take into account the infrared transmissivity (a function of the degree of absorption) of the atmosphere, and the degree to which the earth is an imperfect radiator. The third element, the heat capacity of the earth, is determined mainly by the very large heat storage of the ocean, involving features such as the average depth of the uppermost (or mixed) layer, the density of water, and specific heat capacity of water at constant pressure.

The above components imply an equilibrium global temperature of 287 K (14 °C) – a good estimate of the observed surface temperature today (about 15 °C). The model can also be manipulated to investigate the effects of changes in the system. For example, a new equilibrium temperature could be calculated for a change in the solar flux. Alternatively, the behaviour of the system can be modified by allowing factors (e.g. the planetary albedo) to change in relation to mean global temperature. For example, as the temperature rises (perhaps due to external solar forcing) the area of the planet covered by snow and ice might decrease. This would reduce the global albedo, allowing more energy absorption at the surface, leading to further warming and ice melt – a positive feedback (see Box 7.2). Conversely, a reduction in the solar flux might lead to a positive feedback working in the opposite direction, i.e. an increase in snow and ice cover, a higher planetary albedo, less absorption and further cooling.

(b) One-dimensional EBMs

An extension to the simple model would be to incorporate horizontal energy transfers through heat exchanges across the latitudes. In this one-dimensional model the atmosphere is averaged vertically in the east–west direction, while multiple processes of heat transport by the atmosphere and ocean are represented by north–south diffusion. This energy flow depends on the latitudinal temperature gradient, and is usually expressed as proportional to the deviation of the zonal temperature from the global mean temperature. Each latitudinal zone could have a different net radiation or energy balance and albedo that vary with season. By relating albedo to temperature it is possible to represent high-latitude temperature feedbacks involving changes of sea-ice extent.

Despite their simplicity, one-dimensional EBMs have provided valuable insights into the role of horizontal heat transport feedbacks as well as the fragility of the equilibrium climate (Box 7.1). These models are also particularly useful for investigating the radiative effects of changes in the atmosphere's composition (e.g. greater reflection of incoming solar radiation and hence reductions in energy transfer through the atmosphere associated with volcanic eruptions or emissions of sulphate aerosols over industrialised regions). Furthermore, by varying variables such as the solar flux (an external factor) or the temperature–albedo

BOX 7.1

THINKING FURTHER

Daisy World

Energy balance models provide a simple means of exploring different equilibrium states of the global climate system. Daisy World is a radiation balance model that shows how the biosphere interacts with the atmosphere. The model has been used to explore the Gaia hypothesis (see section 2.1.2) that abiotic (or non-living) and biotic (living) components of the environment are strongly coupled. Furthermore, that biota actively regulate local environments to realise preferred conditions is a basic premise. The imaginary planet of Daisy World is occupied by black and white species of daisy, grazed in a non-selective way by a single species of herbivore. Local temperatures are greater above black daisies than white daisies because of the lower albedo (i.e. less reflected solar energy). Local temperature differences also mean that black daisies have a faster growth rate at lower levels of solar radiation than white daisies (Figure 7.2(a)). Note that clouds and gaseous emissions from phytoplankton are thought to act like the white daisies in Daisy World by creating feedback loops that keep the earth's temperatures habitable.

With constant death rates, the rate of expansion of area covered by the two daisy populations depends on the availability of fertile land, the local temperature, and indirectly on solar radiation. By systematically increasing levels of incoming solar radiation it is possible to show that optimum temperatures for daisy growth are maintained for longer in Daisy World than on a planet without daisies (Figure 7.2(b)). However, beyond a critical temperature (approximately 30 °C), even the high albedo of surfaces occupied by white daisies is unable to offset increasing solar radiation, and both populations crash. Thereafter, the planet becomes lifeless and temperatures spiral out of control – a grim warning for the stewards of biodiversity on planet earth!

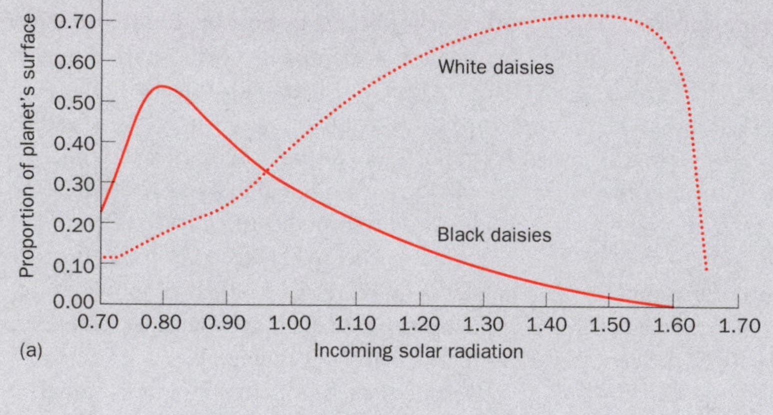

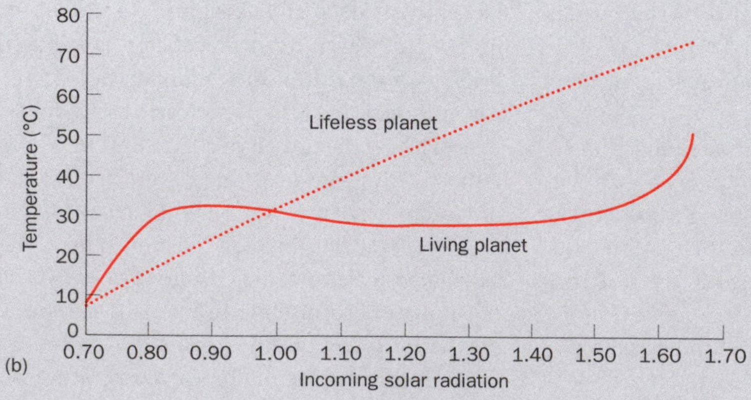

Figure 7.2 Daisy World simulations of (a) black and white daisy coverage, and (b) mean surface temperatures on a living and on a lifeless planet under conditions of increasing incoming solar radiation. Units given are in terms of ratios of the present energy flux

BOX 7.2

THINKING FURTHER

Climate model feedback mechanisms

Climate feedbacks outlined in section 7.1, play an important role in converting changes in radiative forcing into a change in climate. The primary mechanism is through alterations to radiative damping. Non-linear behaviour in model systems means that climate projections are sensitive to initial conditions and/or parameterisation of important processes such as cloud radiative feedbacks. Atmospheric feedbacks also control overall climate sensitivity of the model to greenhouse gas forcing. The most important *rapid* feedback mechanisms included in GCMs are:

- *Atmospheric water vapour* is a greenhouse gas that increases in the boundary layer as the climate becomes warmer. This is because a warmer atmosphere can hold more moisture. Increased concentrations of atmospheric water vapour will, in turn, trap more heat, resulting in a positive feedback. Inclusion of water vapour feedback roughly doubles the warming expected for an atmosphere with fixed water vapour.
- *Atmospheric temperature profiles (or lapse rates)* govern the relationship between surface temperature and long-wave infrared (thermal) emission to space. A decrease in lapse rate with warming would cause the upper atmosphere to warm faster than the lower troposphere, increasing infrared emissions from the latter, meaning less surface warming is required to maintain radiative balance.
- *Water vapour distributions* in the upper and lower troposphere determine the climate sensitivity to increased water vapour concentrations. This is because water vapour is more effective as a greenhouse gas in the upper atmosphere where temperatures are lower. Modelled increases in water vapour in the free troposphere above the boundary layer are the most important cause of greenhouse gas warming.
- *Clouds* introduce both positive and negative feedbacks on climate and are the greatest source of uncertainty in climate model projections (section 7.1). High clouds exert a net warming by absorbing outgoing infrared thermal radiation from the surface and re-emitting more thermal radiation to the atmosphere than to space. Low clouds exert a net cooling by reflecting incoming solar radiation. This effect is greatest when the underlying surface albedo is low. Cloud radiative feedbacks also depend on cloud thickness and cloud brightness. Generally speaking the thicker the cloud the more radiation it will absorb, warming the planet; while the brighter the cloud (i.e. with higher numbers of smaller aerosol particles and therefore more tiny cloud droplets) the more sunlight it will reflect back to space, cooling the planet.

The main *slow* feedbacks in climate models are:

- *Areal extent of snow and ice* determines the surface reflectivity. Less sea ice and seasonal snow cover associated with climate warming, reduces the surface albedo, favouring greater warming (assuming cloud cover remains the same, see above). Some models represent changes in snow albedo arising from ageing or temperature dependence. The effects of sub-grid scale variability in sea-ice cover and thickness are being incorporated but land-ice dynamics and thermodynamics are currently ignored.
- *Vegetation* distribution and vegetation architecture also affect the surface reflectivity,

▶

and hence exert positive or negative feedbacks on climate depending on the difference in albedo between the invading and retreating biomes. The latest generation of models couple the terrestrial energy, water and carbon cycles yielding demonstrable improvements in land–atmosphere fluxes. Increased CO_2 concentrations could directly affect plant physiology, leading to lowered transpiration across the tropics, reduced cloud cover and hence increased regional warming.

- *Carbon cycling* between the terrestrial biosphere and oceans is sensitive to climate warming. Enhanced chemical weathering on land or dissolution of carbonate sediments in the ocean, along with ecosystem processes, alter sources and sinks of CO_2 and CH_4 and hence atmospheric radiative forcing. Higher temperatures favour the release of carbon from soil stores (see section 2.3).
- *Ocean currents* redistribute heat and moisture and hence affect regional patterns of climate change and cloud feedbacks. Changes in regional temperatures and precipitation, in turn, affect the distribution of sea ice and freshwater influx to oceans. Most models suggest a slowing of the Atlantic thermohaline circulation over the present century.

relationship (an internal factor) it is possible to explore the inherent stability of climate. However, the first generation of EBMs were found to be stable only for a narrow range of perturbations from present-day conditions. For example, slight reductions in the present solar flux could result in an ice-covered earth. Associated albedo changes could then effect unrealistic energy balances with global temperatures falling to minus infinity!

The inherent instability of EBMs arises from two factors. First, the energy balance equations used in EBMs are non-linear and must be 'tuned' to mimic present-day climate conditions. This means that models behave realistically under 'normal' conditions, but are less likely to perform well outside known boundaries. The non-linearity of the equations can also cause slight changes in the energy balance to spiral out of control. Second, EBMs do not always acknowledge the different heat capacities of the various components of the climate system. For instance, as previously discussed, the thermal or heat capacity of water is much larger than air, so the ocean responds at a slower rate to changes in energy than the atmosphere. This implies that the thermal inertia (resistance to change in temperature) of the ocean–atmosphere system is far greater than that of the atmosphere alone. Deep ocean mixing and heat transport mean that the thermal response time of the global ocean is of the order of decades. Therefore, any EBM that does not represent the thermal behaviour of the ocean will tend to be oversensitive to short-term variations in the energy balance.

(c) Upwelling-diffusion EBMs

Some of the limitations of one-dimensional EBMs have been addressed by more sophisticated models consisting of three system components: a global atmosphere, with surface ocean layer, underlain by a deep ocean (Figure 7.3). In upwelling-diffusion energy balance models (UD EBMs) the ocean is treated as a one-dimensional column with a well-mixed shallow layer representing the global horizontal average, and a smaller region of deep water formation. Below a critical temperature, polar water sinks to the bottom of the ocean, and upwells elsewhere. At the same time there is downward diffusion of heat from the mixed layer. In addition, there is a transfer of latent and sensible heat, along with long-wave infrared radiation from the overlying atmosphere to the surface ocean layer. The sinking and compensating

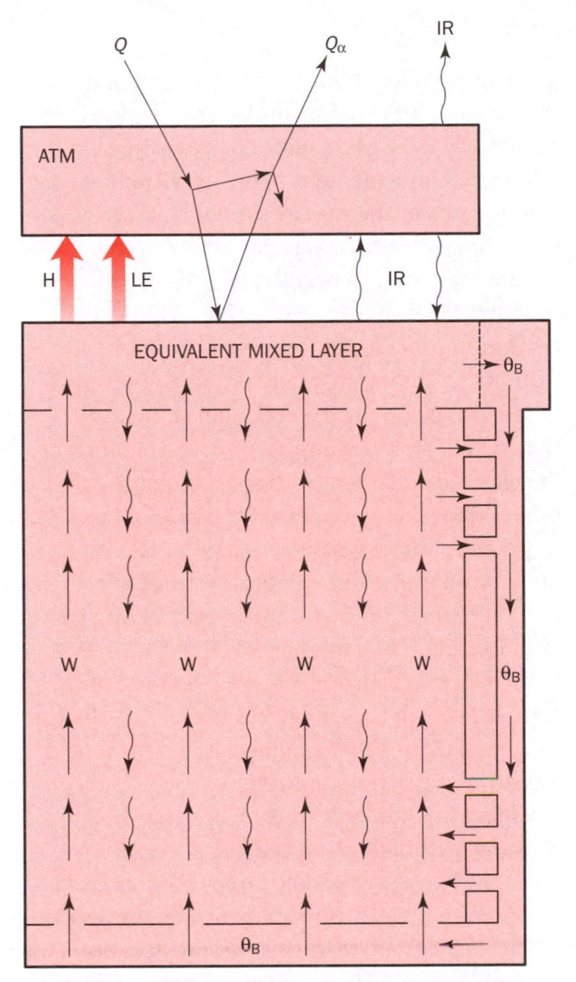

Figure 7.3 Illustration of a one-dimensional upwelling-diffusion model consisting of boxes for the global atmosphere and mixed ocean layer and underlying deep ocean. Bottom water forms at temperature θ_B, sinks to the bottom of the ocean and upwells W. The temperature profile in the deep ocean is governed by the balance between the upwelling of cold water and downward diffusion of heat from the mixed layer. Latent (LE) and sensible (H) heat, and infrared radiation (IR) are transferred between the atmosphere and the surface. Q is incoming solar radiation and Q_α the proportion that is reflected back to space
Source: Harvey (2000)

upwelling flows are crude representations of global-scale convective overturning in the ocean, and reflect simplified views of surface and deep water currents of the global thermohaline circulation. Nevertheless, this mechanism introduces thermal inertia to the model's ocean–atmosphere system.

(d) Applications of EBMs

Although EBMs are less physically realistic than more complex models (see below), they have several practical advantages. Since most physical processes are represented in highly idealised ways, EBMs can be run quickly on microcomputers (as opposed to the supercomputers required by the global and regional climate models discussed below). This enables the exploration of a far wider range of climate scenarios and model configurations. For example, UD EBMs have recently been used to emulate the mean global temperature response of more sophisticated climate models to changes in emissions of greenhouse gases and aerosols, carbon cycling and ocean mixing. Ninety per cent of the model configurations show a warming of the global mean surface temperature within the range 1.7–4.9 °C by 2100, i.e. very close to the outputs from more sophisticated models (section 7.1). Other experiments have shown that mean global surface temperature rises since the nineteenth century are best explained by combining solar, volcanic and anthropogenic forcing rather than by any individual factor (see Figure 7.14).

Energy balance models are also used to investigate specific elements in the ocean–atmosphere system (e.g. the role of carbon cycling or sulphate aerosol forcing). Unlike the more complex coupled ocean–atmosphere general circulation models (see section 7.2.2), EBMs do not generate internal climate variability. In other words, EBMs produce climate change signals that are not obscured by the 'noise' of high-frequency climate variability (i.e. day-to-day or year-to-year variations). For these reasons, simple EBMs have figured prominently in the assessment reports of the IPCC. Even so, one-dimensional EBMs do not provide any indication of the spatial or geographical pattern of the climate as it responds to future radiative forcing. This requires more complex general circulation models.

7.2.2 General circulation models (GCMs)

General circulation models (GCMs) represent the three-dimensional climate system using four primary equations describing the movement of energy (*Newton's first law of thermodynamics*) and momentum (*Newton's second law of motion*), along with the conservation of mass (*continuity equation*) and water vapour (*ideal gas law*). Each equation is solved at discrete points on the earth's surface, at a fixed time interval (typically 10–30 minutes), and for several layers in the atmosphere defined by a regular grid. For example, the UK Meteorological Office Hadley Centre third-generation ocean–atmosphere GCM (HadCM3) has an atmospheric model with a horizontal resolution of 2.5 × 3.75° and 19 vertical levels, and ocean model with a horizontal resolution of 1.25 × 1.25° and 20 vertical levels (Figure 7.4).

(a) Model processes

The computational burden of solving equations at thousands of grid points means that the horizontal resolution of GCMs is quite coarse – typically ~300 km. Unfortunately, many important components of the climate system have scales much finer than this (e.g. convective clouds, coastal breezes) and must be parameterised (i.e. simplified or averaged). Although higher-resolution process models are available (see section 7.2.3) they are at present prohibitively expensive to run for the entire globe. Instead, parameters provide simplified representations of the effect of small-scale processes on large-scale responses and are necessary features of all GCMs. In practice, many parameters used in GCMs must be obtained through observations of the current climate, so their future applicability is questionable. The representation of clouds is particularly challenging, not least because of their complex role in the energy balance and feedbacks (see Box 7.2) arising from increased atmospheric moisture under climate change.

Figure 7.5 shows the major processes that are typically included within a single horizontal grid box of a GCM. As well as simplifying key processes through parameterisation, climate models also average atmospheric conditions over the entire grid box. For example, precipitation is assumed to occur at a uniform rate everywhere within the cell, leading to an overestimation

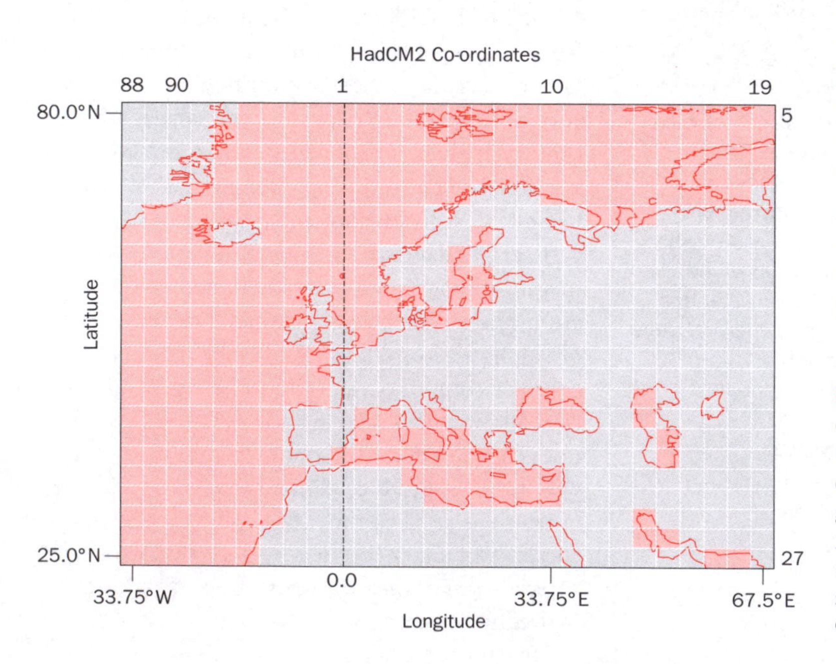

Figure 7.4 Grid representation of Europe employed in both the HadCM2 and HadCM3 global climate models. The 2.5° × 3.75° grid resolution equates to about 240 × 278 km in the vicinity of Ireland

Source: Climate Impacts LINK project on behalf of the Hadley Centre

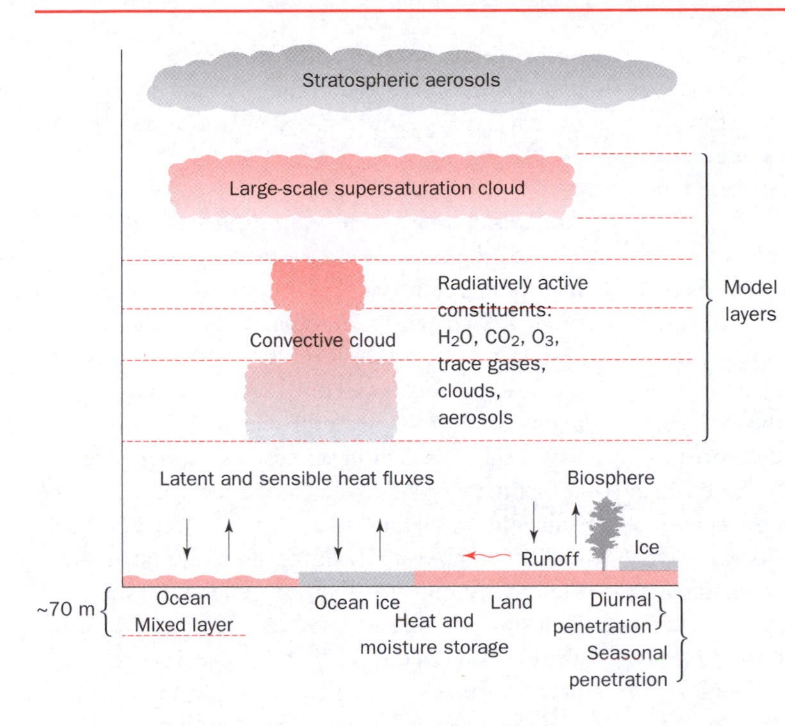

Figure 7.5 Key processes represented at the grid-box scale of a general circulation model
Source: Hansen *et al.* (1983, pp. 609–62)

of rainfall frequencies and underestimation of intensities compared with reality – the so-called 'drizzle' effect. As Figure 7.5 shows, GCMs compute (i) energy transfers through the atmosphere (involving water vapour and cloud interactions), (ii) the direct and indirect effects of aerosols (on radiation and precipitation), (iii) changes in snow cover and sea ice, (iv) the storage of heat in soils and oceans, (v) surface fluxes of heat and moisture, and finally, (vi) the large-scale transport of heat and water by the atmosphere and ocean. Some GCMs incorporate land-surface schemes including the freezing and melting of soil moisture, and the regulation of evaporation by plant stomata (due to variations in temperature, vapour pressure and CO_2 concentration). More sophisticated models include carbon cycling and atmospheric chemistry for trace gases (such as CH_4, N_2O, CFC_{11}, CFC_{12} and $HCFC_{22}$). Preliminary results from the climate model HadCM3L suggest that the inclusion of interactive sulphur and carbon cycling lead to accelerated global warming once terrestrial carbon is released into the atmosphere.

(b) Equilibrium and transient experiments

Until the last decade most climate modelling investigations of global warming were based on so-called equilibrium experiments (see Table 7.1). In these simulations, the GCM is given an observed climatology (e.g. radiation balance, moisture, pressure condition) and atmospheric composition, then run until it reaches an equilibrium state that compares favourably with the present-day average climate. Next, the amount of CO_2 in the model's atmosphere is instantaneously doubled (or even quadrupled) and the model is run until a new equilibrium is reached. The difference between the first and second equilibrium states represents the model's analysis of how climate would change, given a doubling (or quadrupling) of atmospheric CO_2. The main advantage of these types of experiment is that they are relatively cheap because they do not require sophisticated ocean models and take fewer computations. They are also more straightforward to compare because the climate-forcing scenario (i.e. doubled CO_2) is always the same. However, equilibrium experiments do not give any indication of the transient or time-dependent climate response to future greenhouse gas concentrations. In other words, the experiment will not show how the climate might evolve between the present and doubled CO_2 scenarios, or how long it would take to do so. Equilibrium studies also overlook the possibility that different regions will respond at different rates to radiative forcing.

To model the time-dependent (or transient) response of the atmosphere to time-varying atmospheric compositions of greenhouse gases requires more realistic representations of oceanic effects. Simple 'slab' or 'mixed-layer' ocean models – the type commonly used for equilibrium experiments – incorporate only shallow oceans (usually 50–100 m deep) with thermal inertia that are much less than the real ocean. Conversely, fully coupled ocean–atmosphere models incorporate three-dimensional ocean models that simulate the effects of oceanic thermal inertia on the overlying atmosphere. In coupled models, the deep ocean stores heat while ocean currents redistribute energy and salinity, delaying changes in the atmospheric response to climate forcing. Complex ocean models are based largely on the same equations as the atmosphere but require higher resolutions than atmospheric models to resolve important motions. Conversely, ocean models need to consider far longer timescales than atmospheric models, and are much more sensitive to the effects of terrain (in this case the ocean bottom and margins).

(c) Linking model components

(i) Ocean and atmosphere coupling

Coupling ocean and atmosphere models is not a trivial problem. First, the ocean and climate models must exchange information on energy, mass and momentum – ideally in a synchronous way – between different spatial and temporal scales. Second, because the response time of the ocean is much longer than the atmosphere, it is necessary to 'tweak' the ocean model physics to catch up with the atmosphere at the start of an experiment. Third, mismatches in energy and mass fluxes between the atmosphere and ocean models can cause the ocean climate to drift from the observed ocean behaviour, requiring flux adjustments to compensate for systematic errors. Finally, establishing the initial conditions of the ocean is complicated by the fact that greenhouse gas emissions since the beginning of the Industrial Revolution have already affected the slow response of the real ocean. In other words, ocean temperatures observed in the 2000s incorporate the effects of greenhouse gas emissions since the 1860s.

(ii) The coupled ocean–atmosphere model

State-of-the-art GCM experiments now incorporate fully coupled ocean–atmosphere models, high-resolution ocean models that maintain poleward heat transport (thus negating the need for flux correction) reproduce key ocean–atmosphere features such as El Niño, have negligible drift in sea surface temperatures compared with observed climatology, and produce stable thermohaline circulations. However, model realism and sensitivity of the global climate to radiative forcing are highly dependent on the inclusion of climate feedbacks (see Box 7.2). As already examined in section 7.1, these are physical processes that either enhance (positive feedback) or dampen (negative feedback) the climate response to an initial perturbation. Recognising the importance of feedbacks, the modelling community is incorporating improved representations of the cryosphere and biosphere in climate models. For example, earth system models of intermediate complexity (EMIC) bridge the gap between EBMs and GCMs and are used to explore important climate feedbacks arising from biogeochemical cycles over several millennia. However, other important land-surface and atmospheric feedbacks operate at spatial scales finer than the grid size of GCMs, and must be addressed using higher-resolution models.

7.2.3 Regional climate models (RCMs)

Regional climate models (RCMs) simulate climate features dynamically at resolutions of 20–50 km given time-varying atmospheric conditions modelled by a GCM bounding a specified region or domain (Figure 7.6). Large-scale atmospheric fields simulated by the GCM (e.g. surface pressure, wind, temperature and vapour), at multiple vertical and horizontal levels, are fed

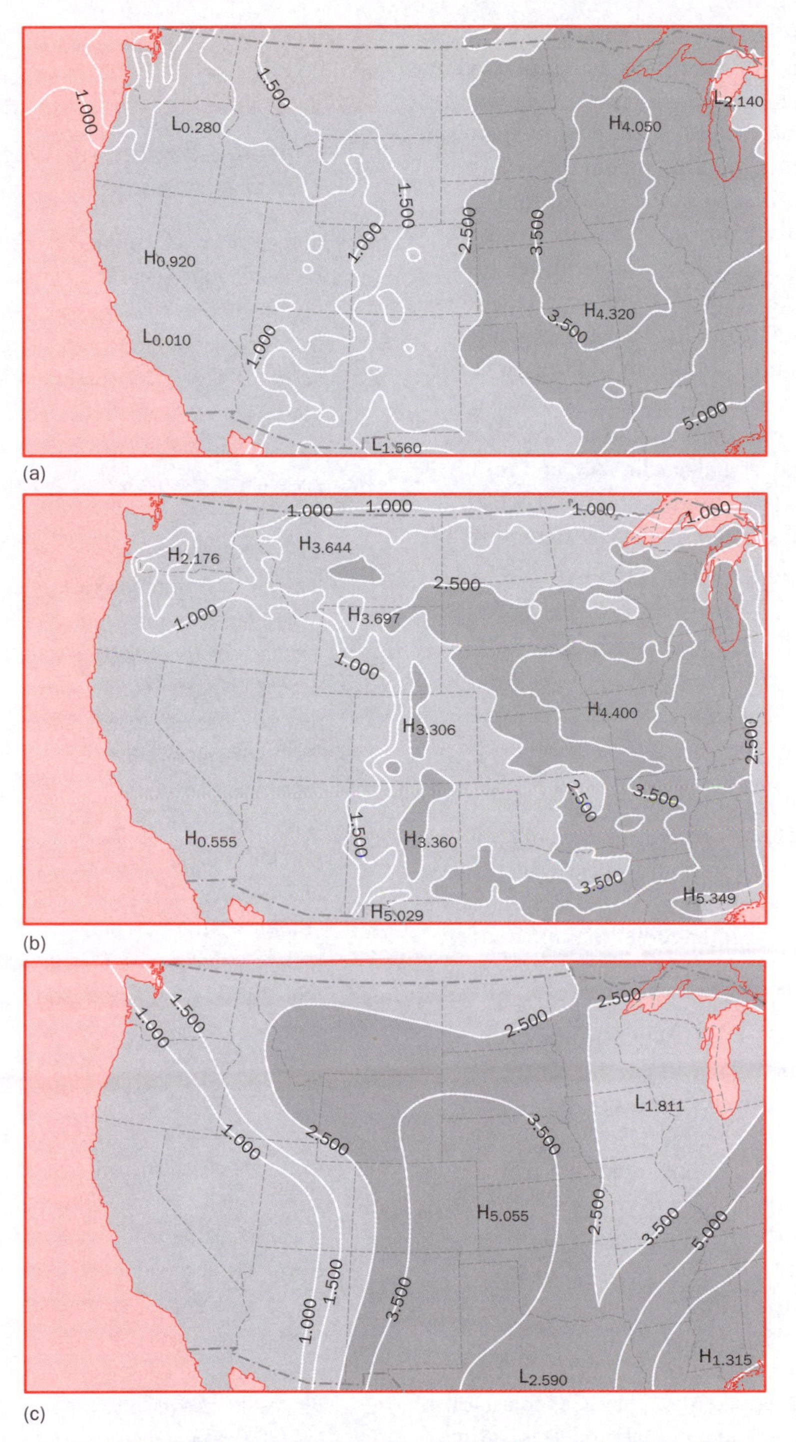

Figure 7.6 Average summer precipitation (mm/day) over the Central Plains of the USA: (a) observations; (b) RCM; (c) GCM. Note that with better representations of the land-surface topography the RCM better captures the observed pattern of summer rainfall compared with the coarser resolution GCM

Source: Adapted from Giorgi and Mearns (1999, pp. 6335–52)

into the boundary of the RCM through a lateral buffer zone. This information is then processed by the higher-resolution RCM such that the internal model physics and dynamics can generate patterns of climate variables that differ from those of the 'host' GCM. The *nesting* of the RCM within the GCM is typically one way. In other words, the behaviour of the RCM is not permitted to influence the large-scale atmospheric circulation of the GCM. To date, RCMs have been used for a wide variety of applications in addition to the projection of future climate change in selected regions of the world, including numerical weather prediction, studies of palaeoclimates, and the effects of land-surface modification(s). For example, the UK Climate Impacts Programme (UKCIP02) scenarios were produced by the regional climate model HadRM3 using four alternative emission scenarios (see Box 7.5 and section 9.4.3).

(a) Model processes

The main advantage of RCMs is their ability to model regional climate responses to changes in land-surface vegetation or atmospheric chemistry in physically consistent ways (Box 7.3). Because of higher spatial resolution and hence improved representation of surface elevations, RCMs also resolve important atmospheric processes (such as orographic rainfall) better than the host GCM. This is particularly important for the representation of extreme events such as intense precipitation and tidal surges. However, RCMs are computationally demanding, requiring as much processor time as the GCM to compute equivalent scenarios, and are not easily transferred to new regions. The results from RCMs are also sensitive to the choice of the initial conditions (especially soil moisture and soil temperature) used at the start of experiments. In most RCMs the soil 'spin-up' period (or time taken for the GCM to establish stable behaviour in the regional model) is of the order of a few seasons to a year. During this period the RCM may produce spurious results. This is

BOX 7.3

THINKING FURTHER

Soil–vegetation–atmosphere–transfer schemes

In recognition of the importance of land-surface attributes to regional and global climates, current models now explicitly include soil–vegetation–atmosphere–transfer (SVAT) schemes. Figure 7.7 shows the main processes involved in an interactive land surface and atmosphere scheme. SVATs are generally one-dimensional models describing vertical (but not lateral) moisture, nutrient, momentum and energy fluxes at the soil–vegetation–atmosphere interface. The main processes represented in SVAT schemes include: soil moisture transfers (involving throughfall, canopy drip, surface evaporation and runoff, capillary and gravitational drainage, and transpiration); transpiration by the canopy (depending on stomatal resistance, soil water supply, atmospheric humidity); aerodynamic resistance or surface drag (related to canopy architecture); canopy albedo (which determines the fraction of incident solar radiation absorbed at the surface); and runoff (currently represented crudely in most schemes as the excess of precipitation over evapotranspiration).

Incorporating SVAT models into climate models presupposes global (or regional) data sets describing soil and vegetation types, and land-surface attributes at the spatial scale of the host climate model. Nonetheless, interactive SVAT schemes, that couple exchanges of carbon between the biosphere and atmosphere, are being included in the latest generation of climate models in recognition of the critical importance of carbon cycling to the future climate.

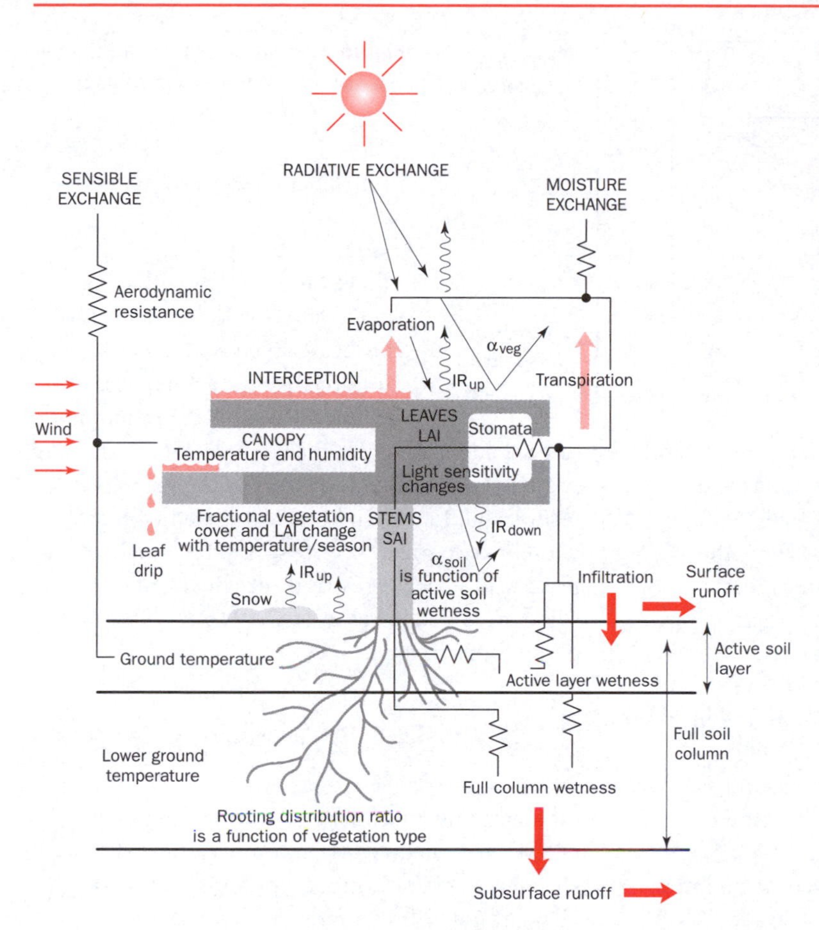

Figure 7.7 Schematic of SVAT model processes. The major features incorporated into a scheme for interaction between the land surface and the atmosphere suitable for inclusion in an RCM. LAI is the leaf area index, SAI is stem area index and IR stands for infrared radiation

Source: From Wilson *et al.* (1987)

because soil water content and temperature affect the climatology of the model by regulating surface sensible and latent heat fluxes. In comparison, the time taken for atmospheric spin-up depends on the RCM domain size, the season and vigour of atmospheric circulation, but is normally only a few days.

The regional climate modelling community has debated the relative merits of implementing the full physics of the GCM within the RCM, as opposed to developing RCM physics that differ from the host GCM. The use of same physics schemes in the nested RCM and driving GCMs maximises the compatibility of the models, but some of the GCM processes (e.g. convective clouds) may not be valid at the higher resolution of the RCM. The main advantage of using different physical schemes at the GCM and RCM scales is that each set of processes has been developed for the respective model resolutions. However, in these cases it can be difficult to interpret any differences between the RCM and GCM results, because they may be due to differences in resolution and/or model physics.

(b) Boundary forcing

The quality of regional climate simulations depends not only on the validity of the RCM physics but, more critically, on errors in the GCM boundary information. This is a classic case of 'garbage in, garbage out'. For example, gross errors in an RCM's precipitation climatology may arise if the GCM misplaces the mid-latitude jet and associated storm tracks. However, with the advent of supercomputers and continued improvements in the horizontal scale of GCMs, the quality of the simulated large-scale circulations has substantially improved compared with earlier experiments. Even so, regional climate modellers advocate the initial testing of RCMs using reanalysis data (these are observed climate variables that have been assimilated by a climate model and are often termed 'perfect boundary conditions'). If multiple GCM simulations are available, the experiment showing the best performance across the region of interest then offers the best chance of minimising the effects of gross boundary errors.

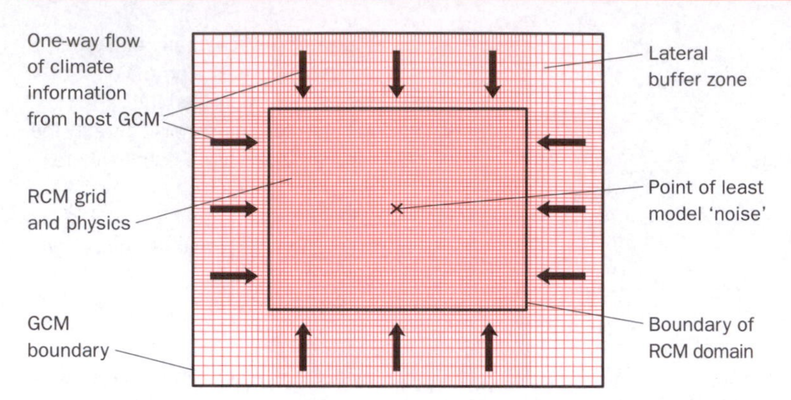

Figure 7.8 Nesting an RCM within a GCM using a variable resolution grid

The choice of model domain and grid spacing are two of the most important decisions when undertaking RCM experiments. Ideally, the domain should be large enough to allow the free development of mesoscale atmospheric circulations, and the grid spacing fine enough to capture detailed topographic and coastal features such as sea breezes. In practice, domain size and grid spacing are constrained by computational factors, with simulation times increasing exponentially with increasing vertical and horizontal resolution. The actual location of the domain should capture the most significant circulations that affect climate over the region of interest (e.g. low-level jets or storm tracks). Finally, the area of greatest interest should be as remote as possible from the lateral buffer zone (Figure 7.8). This is because model noise is greatest at the lateral boundaries where the finer resolution grid of the RCM abruptly meets the coarser grid of the GCM. A number of techniques exist for coupling the two discordant scales including variable grids, interpolation and spectral methods. A further consideration of the lateral forcing is the interval of updates from the large-scale GCM fields. For example, in summer, updates are needed at least every 6 hours to approximate the differential heating associated with the diurnal cycle.

Finally, it was noted that the present generation of RCMs are one-way nesting approaches because the regional model does not feed back into the global model. In other words, local-scale atmospheric responses to land-surface changes or higher-resolution topography are not allowed to have any influence on the large-scale circulations of the GCM. Even so, feedbacks from relatively small regions of enhanced resolution are thought to be of limited significance in most situations, and probably not justified in terms of the increased technical difficulty and cost associated with fully coupling two mismatching scales. Regional climate feedbacks associated with future distributions of sulphate aerosol forcing, however, are the subject of ongoing research.

7.2.4 Statistical downscaling (SDS) models

Previous sections have described techniques for estimating future climate conditions at global (EBM), regional (GCM) and subregional (RCM) spatial scales. These models provide 'broad-brush' views of how climate variables, such as areal average rainfall or regional temperatures, might change in the future. However, for many practical applications, far greater spatial (and temporal) detail is often required. For example, assessments of climate change impacts on river flows require daily precipitation scenarios at catchment or even station scales. As has been shown, RCMs embedded within GCMs represent the physical dynamics of the atmosphere at horizontal grid spacings of 20–50 km, but are computationally demanding, expensive to run and currently impractical for obtaining long climate simulations or suites of scenarios. Furthermore, even at 50 km resolution the HadRM3 model represents the whole of south-east England with less than 20 grid boxes, and Greater London with a single grid box. In the process, the RCM averages important land-surface variations and topographic detail.

As a consequence of these limitations, climate modellers and impact assessors have developed a range of statistical techniques for deriving more detailed climate scenarios directly from coarse resolution GCM output. These are known collectively as statistical downscaling methods. Several alternative approaches have emerged (see below), including scaling methods, regression-based techniques, weather pattern classification and weather generators. In most cases, the basic principle is to use large-scale climate variables from the GCM (e.g. mean sea-level pressure) to estimate small-scale meteorological variables (e.g. station precipitation or wind speeds). As in the case of RCMs, scenarios downscaled via these methods depend heavily on the validity of the host GCM. It should also be noted that statistical downscaling often requires extensive observational data sets for model training and significant amounts of pre-processing for the GCM output.

(a) Scaling methods

The most straightforward procedure (sometimes called 'unintelligent' downscaling) involves three steps. First, a baseline climatology is established for the site or region of interest. Depending on the application this might be a representative long-term average such as 1961–90, or an actual meteorological record such as daily maximum temperatures. Second, changes in the equivalent temperature variable for the GCM grid box closest to the target site are calculated. For example, a difference of 2.5 °C might occur by subtracting the mean GCM temperatures for 1961–90 from the computed mean of 2061–90. Third, the temperature change suggested by the GCM (in this case, +2.5 °C) is then simply added to each day in the baseline climatology.

Although the resultant scenario incorporates the detail of the station records as well as the areal average climate change of the specified GCM grid box, there are problems with this method of scaling. The downscaled and the baseline scenarios only differ in terms of their respective means, maxima and minima; all other properties of the data, such as the range and temporal sequencing, remain unchanged. The procedure also assumes that the spatial pattern of the present climate remains unchanged in the future. Finally, the method does not apply to precipitation records in a straightforward way because the addition (or multiplication) of observed precipitation by GCM precipitation changes can affect the number of rain days, the size of extreme events, and even result in negative precipitation amounts!

(b) Regression-based methods

Regression-based downscaling provides a convenient method of relating (large-scale) cause and (small-scale) effect. For example, Figure 7.9 shows the relationship between observed pressure elevations, i.e. at the 500 hPa geopotential height, h (a measure of the thickness of the atmosphere), averaged over eastern England and observed maximum daily temperatures in summer, T, at Nottingham, England. Historically, maximum temperatures at Nottingham increase non-linearly as the thickness of the atmosphere increases. The inset regression equation specifies the average relationship between observed h and T, and could be used to downscale future maximum

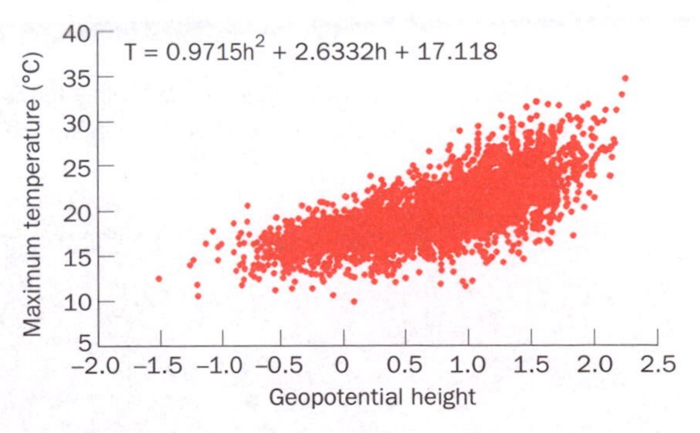

Figure 7.9 Observed relationship between summer maximum daily temperatures at Nottingham, England, and geopotential heights over eastern England, 1961–90

temperatures at the same site given future values of h projected by a GCM. The regression method applies equally well to other surface variables such as wind speed, sunshine amounts and humidity, but may require different predictor variables (e.g. large-scale wind speed and direction). It is also assumed that the regression relationships developed for the present climate remain equally valid for the future climate despite expected changes in radiative forcing.

(c) Weather pattern methods

The weather pattern methods use surface and upper-atmosphere pressure patterns to classify the daily circulation into predefined weather types (such as the Lamb weather types described in section 9.2.4). The observed climate record (e.g. station daily rainfall totals) is then split according to the associated weather type, and an average for all days in each weather type computed. For example, Table 7.2 shows the chance of rainfall occurrence, average daily rainfall amount and temperature at Durham, UK under different Lamb weather types. On average, the anticyclonic weather type has the smallest chance of rainfall (just 10 per cent), and the cyclonic type the greatest (62 per cent). The cyclonic type also has the highest rainfall amounts, on average 7.1 mm per rain day.

This implies that a future climate scenario in which there are more cyclonic days (and fewer anticyclonic days) would result in greater rainfall totals at Durham. Similarly, more westerly days and fewer easterly days would tend to increase the mean daily temperature.

This downscaling method assumes that the historical weather type properties remain valid under future climate conditions. The method also presupposes that pressure patterns produced by the GCM can be translated into weather patterns consistent with the observed pressure data. It is also assumed that weather patterns capture the main processes governing local weather, and neglects other factors such as changes in atmospheric humidity. However, a major advantage of circulation-based downscaling is that it can be applied to a wide variety of environmental indicators, including air quality, ecological and hydrological indices.

(d) Weather generators

Because circulation methods use sequences of weather patterns (whether observed or from GCM experiments) they can produce more realistic sequences of downscaled variables such as persistent wet or dry spells. Weather generators use statistics like those in Table 7.2 but do not require daily weather pattern sequences to drive

Table 7.2 Average chance of rainfall, average daily rainfall amounts and average daily temperatures at Durham, UK associated with the seven main Lamb weather types, 1881–1990. The percentages of days in each weather type are also given

Weather type	Frequency (% days)	Chance of rain (% days)	Rainfall (mm/day)	Temperature (°C)
Anticyclonic	18	10	3.9	9.3
Cyclonic	13	62	7.1	9.9
Easterly	4	40	5.2	7.5
Southerly	4	41	5.0	9.9
Westerly	19	33	4.1	10.1
North-westerly	4	23	3.8	9.3
Northerly	5	40	4.7	7.3

the downscaling. Instead, weather generators employ pseudo-random number series to sample the rainfall or temperatures from each class by chance. Using the statistics shown in Table 7.2, the random number generator would determine first the weather pattern of the present day, second the chance of rainfall, and third, if rainfall occurs, the amount. In more sophisticated models, the weather pattern and rainfall of the previous day can partly determine the outcome of the current day. In the long run, the weather generator will reproduce the same mean statistics as the circulation method but from a different day-to-day sequence of events (Box 7.4).

The generator can also produce daily weather for future climate scenarios once GCM results have been used to estimate the new frequency of each weather pattern. However, the future chance of rainfall and average rain day amounts are assumed to remain unchanged from the training period. A key advantage of the method is that it can produce very large sets of daily weather which are useful for investigating extreme events such as long wet spells or heatwaves. The main disadvantage is that basic weather generators do not always reproduce slowly varying properties of climate behaviour, such as decade-to-decade changes in rainfall intensity over the British Isles associated with changes in the strength of the North Atlantic Oscillation.

BOX 7.4

THINKING FURTHER

Modelling random processes

GCMs and RCMs are examples of process–response models in which the behaviour of the climate system is fully deterministic. In other words, a given input or combination of parameters *always* produces the same model output. However, many climate processes display apparently random or chaotic behaviour that can be mimicked by simple statistical models. For example, maximum daily air temperatures may be represented using the following equation:

$$T_t = rT_{t-1} + aH + R$$

where T_t is the mean temperature on day t, r is the correlation between successive days' temperature, T_{t-1} is the temperature on the previous day, a is a model parameter, H is an index of atmospheric circulation and R is a random component that accounts for unexplained meteorological factors. Given an initial temperature, the constants a and r, observed circulation H, and a pseudo-random number on each day R, it is possible to statistically model the daily temperature series at a station (Figure 7.10). Because of the random element, R, each simulation of the model differs slightly from the next despite the fact that they were all given identical starting conditions. However, each is an equally valid representation of the *distribution* of temperatures at the site (i.e. the mean and variance of the three runs shown in Figure 7.10 are statistically indistinguishable from each other for long simulations). Random processes are an important ingredient of weather generators, and are particularly useful in situations where the underlying physical mechanisms are poorly understood, or where there are insufficient data to fully describe the process. Daily rainfall at single or multiple sites is often modelled in this way because rainfall occurrence and intensity are the result of the large-scale synoptic situation (which is generally well understood) and 'chance' factors (which are not well defined due to the complex interplay of local topography and meteorology).

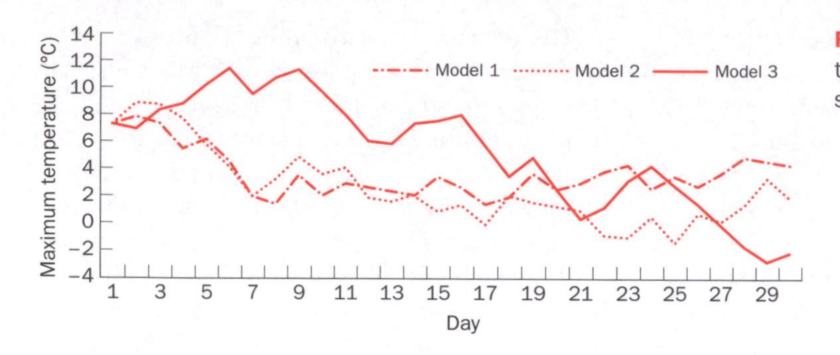

Figure 7.10 Maximum daily temperatures generated by a simple statistical model

Key ideas

1. Models are simplified representations of the physical stores and fluxes governing the spatial and/or temporal behaviour of an environmental system.

2. Energy balance models (EBMs) are useful tools for investigating the influence of changes in atmospheric composition, net incoming and outgoing radiation, ocean heat transport, and albedo feedbacks on global mean temperatures and sea-level rise.

3. EBMs have been used extensively by the IPCC to explore the effects of different greenhouse gas emission scenarios.

4. General circulation models (GCMs) represent the three-dimensional climate system using equations governing the flow of energy, momentum, mass and water vapour solved at regular grid points on the earth's surface and at multiple heights in the atmosphere.

5. Due to their coarse spatial resolution (~300 km) important processes such as cloud formation must be simplified (or parameterised) in GCMs.

6. State-of-the-art GCMs use feedbacks between atmospheric chemistry, radiant energy balances, snow cover and sea ice, carbon cycling, land-surface hydrology and ocean mixing to resolve global patterns of climate variables such as temperature and rainfall.

7. Regional climate models (RCMs) are 'nested' within coarser resolution models and compute climate variables for patches of the earth's surface given information about pressure, wind, temperature and vapour fluxes across the boundary of the patch.

8. Due to their relatively high spatial resolution (~20 km), RCMs provide better representations of orographic rainfall and low-level jets than the host climate model or GCM.

9. RCMs are as computationally demanding as GCMs to run, but are useful for investigating local climate responses to greenhouse gas forcing and/or land-surface changes (such as transitions from natural to irrigated vegetation cover).

10. Statistical downscaling (SDS) methods employ observed relationships between large-scale atmospheric features and local meteorology as the basis for estimating future changes in climate.

11. Like RCMs, scenarios derived from SDS are largely dependent upon the realism of the drivers supplied by the host GCM, and are typically forced in one direction (i.e. any local climate changes projected by the SDS do not feed back into the GCM).

12. SDS methods are computationally undemanding, but may need large amounts of data for calibration and assume that the scaling relationships remain valid under a changing climate.

7.3 Estimating future climate change

Evidence in support of the claim that human activities cause global climate change comes from two main sources. First, there are long-term observational records of changing atmospheric composition (principally CO_2 concentrations), surface temperature (see Figure 1.1) and precipitation, sea-ice extent (section 8.1), biological indicators, and so on. As we have already seen in section 6.3, few records are truly global in their coverage, or of adequate duration or quality (homogeneity) to establish whether or not a trend really exists. Even with sufficient data or robust statistical techniques for trend detection, some sceptics argue that it is still a quantum leap to attribute cause (i.e. rising concentrations of greenhouse gases) and effect (i.e. global warming). This is because the global climate is inherently variable, and it must be shown that recent behaviour lies outside of natural variability.

The second line of evidence in support of human-induced global warming comes from climate model data. It is a widely held belief that models provide the better opportunity of attributing climate change to increasing greenhouse gases. In 1995, the IPCC Second Assessment Report (SAR) made the controversial scientific statement that 'the balance of evidence suggests that there is a discernible human influence on global climate' largely on the basis of climate model results. This statement was strengthened and qualified still further by the 2001 IPCC Third Assessment Report (TAR). But, as we shall see, model predictions of future climate change are far from certain, and are subject to many caveats. Understanding the origin of these uncertainties, and their relative significance, should be uppermost when evaluating the results of climate model experiments. The following sections will examine the main elements in the 'cascade of uncertainty' that pervades climate modelling.

7.3.1 Defining baseline climatology

Any legitimate assessment of climate model realism presupposes the existence of accurate regional or global baseline climatologies. In addition, the climatology should be gridded and ideally of the same resolution as the climate model in question. For example, the CRU CL 1.0 Global Climate Data Set (based at the University of East Anglia) consists of a mean monthly climatology for the period 1961–90, for 11 surface variables at 0.5° latitude by 0.5° longitude resolution. The data set is for global land areas (excluding Antarctica) and constructed from 19,800 station records for precipitation, and 3,615 for wind speed. This climatology has since been augmented by the CRU CL 2.0 and CRU TS 2.0 data sets which provide respectively finer resolution (10 minute latitude/longitude) and longer series (1901–2000) albeit for slightly fewer land surface variables.

Unfortunately, there are no universally accepted climatologies because each national agency has its preferred observational network, interpolation method and period of record. The uncertainties are greatest for variables such as precipitation and wind speed because they are highly variable in both space and time. Furthermore, reliable long-term records exist only for the earth's land area since the beginning of the twentieth century. Even during this period, stations may have moved or observers adopt different recording practices. In addition, observational networks are densest in the coastal regions of North America, Europe and parts of Australia and Africa, with less coverage in continental interiors. As a result there are differences between climatologies, so even the baseline against which a model's climatology is to be compared is open to debate.

7.3.2 Future emission scenarios

Assuming that model and observed climatologies compare favourably for the present climate, it is then necessary to construct plausible scenarios for factors that will affect the future climate.

BOX 7.5

CASE STUDY

IPCC Special Report on Emission Scenarios

Model projections of future global mean temperature and sea-level changes depend on future estimates of greenhouse gas and sulphate aerosol emissions. In March 2000, the IPCC approved a new set of greenhouse gas plus sulphate aerosol emission scenarios to update and replace the IS92 (1992) scenarios used in the 1996 IPCC SAR. The new scenarios, presented in the IPCC Special Report on Emission Scenarios (SRES), have much lower emissions of sulphur dioxide than the IS92 scenarios. Unfortunately, the SRES scenarios were too late for the modelling community to use in GCM experiments prior to the 2001 IPCC Third Assessment Report. Instead, four preliminary SRES 'marker' scenarios A1, A2, B1 and B2 were released (see Figure 7.1(a)). These scenarios cover a range of future demographic, economic and technological 'storylines':

- The A1 storyline describes a future world of very rapid economic growth, a global population that peaks in the mid twenty-first century and thereafter declines, and the rapid introduction of new and efficient technologies. The scenario also envisages increased cultural and social interaction, with a convergence of regional per capita income.
- The A2 storyline describes a very heterogeneous world, characterised by self-reliance and preservation of local identities. Population continues to grow but economic growth and technological change are slower than other storylines.
- The B1 storyline describes the same population dynamics as A1, but envisages a transition towards service and information economies, with lower material consumption and widespread introduction of clean and efficient technologies.
- The B2 storyline describes a world with lower population growth than A2, accompanied by intermediate levels of economic development, with less rapid and more diverse technological change than in B1 and A1.

It is important to recognise that none of the scenarios is more likely than another; they are all plausible descriptions of socio-economic trends that could affect future emissions of greenhouse gases.

In particular, future anthropogenic emissions of greenhouse gases and aerosols must be specified. For example, Figure 7.1(a) illustrates and Box 7.5 outlines four world storylines used by the 2001 IPCC TAR to describe ways in which global social, economic and technological structures might evolve in the twenty-first century. Each storyline implies quite different greenhouse gas emission trajectories and hence climate change scenarios. Even for a specified emission scenario, it is uncertain how the emission path translates into atmospheric concentrations of the radiatively active greenhouse gases. Furthermore, natural sources such as volcanic eruptions may intermittently augment the atmospheric burden. The uncertainties are compounded by our poor understanding of the global carbon cycle, atmospheric aerosols and trace gas chemistry – processes that are just beginning to be included in climate models.

7.3.3 Model uncertainty

(a) Climate sensitivity

Even if future emissions and greenhouse gas concentrations were known, there would still be considerable uncertainty in the climate response. This is because the direct and indirect effects of aerosols on radiative forcing are poorly understood. For the same patterns of forcing, different climate models generate very different global and regional climate responses, as denoted by their respective climate sensitivities. This is the equilibrium change in global mean surface temperature associated with a doubling of atmospheric equivalent CO_2 concentration. The 2001 IPCC TAR suggests that climate sensitivity lies between 1.5 and 4.5 °C, depending on the differing rates of heat uptake in each model. The effective climate sensitivity is probably a better measure because it recognises that model sensitivity can vary through time. Models with high effective climate sensitivity tend to have a large net heat flux into the ocean that delays the climate response, introducing time-dependent cloud feedbacks. Even for similar climate sensitivities, different climate models yield very different regional climate change patterns, especially for precipitation (Plate 7.1).

(b) Climate variability

Further uncertainties are introduced into climate model scenarios due to unforced, internally generated, climatic variability. Even in the absence of external radiative forcing the heat content of the climate system remains constant, but it is continuously redistributed between the oceans and the atmosphere, and between different states such as water vapour to liquid, or water to ice. Such redistributions occur on timescales from seconds to millennia and are referred to as 'noise'. Each climate model simulation, therefore, comprises the response to radiative forcing (the signal) along with an unpredictable component (the noise). The signal-to-noise ratio indicates the extent to which the human-induced and natural signatures can be discriminated. The ratio is relatively high for surface temperature, but low for precipitation.

(c) Multiple runs

One way of extracting more robust signals from climate model experiments is to undertake multiple simulations with the same forcing but slightly different initial conditions. Each model experiment (or ensemble member) will produce the same underlying trend but the precise pathway differs due to internal climate variability (Figure 7.11). The ensemble mean smooths out much of this variability but is computationally expensive to obtain. Nonetheless, large ensemble experiments are now being conducted by the UK Met Office's Hadley Centre with a view to developing probabilistic predictions of climate change based on a *single* climate model.

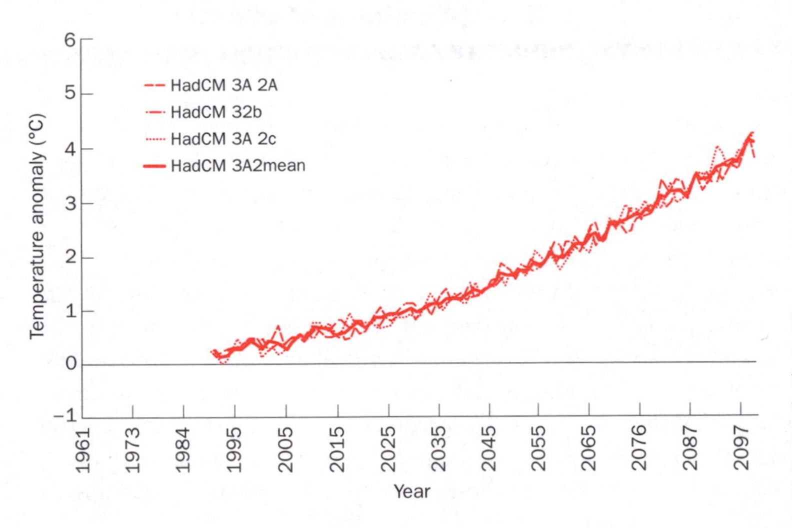

Figure 7.11 Comparison of ensemble members and their respective means for the global average surface temperatures of the HadCM3A2 and HadCM3B2 experiments

Source: Climate Impacts LINK project on behalf of the Hadley Centre

7.3.4 Feedbacks and abrupt change

The IPCC cautions that no emission scenario is more likely than another. However, for formal risk analyses, it is often the extreme outcomes that are of greatest interest, and sometimes 'surprise' events may arise from non-linear feedbacks or abrupt changes in the ocean–atmosphere system due to increased greenhouse gas concentrations (see Box 7.2). For example, most coupled climate models suggest that the thermohaline circulation (THC) weakens as greenhouse gas concentrations increase. Although the level of weakening varies between models, it seems to depend on the rate of CO_2 increases. Since the THC is responsible for poleward heat transport in the Atlantic, a weakened THC could produce a cooling of north-western Europe and the north-eastern part of North America. Even stronger reductions in the THC might result from significant changes in freshwater inputs (and hence the salinity of the North Atlantic) due to meltwaters from the Greenland ice cap and mountain glaciers (sections 4.4 and 8.6). In any event, the probability attached to such non-linear responses cannot be determined using conventional modelling techniques.

7.3.5 Overview

Despite all the uncertainties described above, climate model output still provides valuable information for designing mitigation and adaptation strategies in the face of global climate change. But it is important to separate uncertainties in climate scenarios that may be reduced by further research (such as those due to incomplete understanding of physical processes), from those due to the unknowable future of the world's socio-economic state. It is not, therefore, unreasonable to expect revisions in climate change projections as scientific understanding and modelling techniques evolve. For example, the higher global mean temperature changes published in the IPCC TAR (2001) compared with those in the SAR (1996) are largely due to reduced estimates of future sulphur dioxide emissions (and hence the sulphate aerosol cooling effect declines). Data from climate model experiments are only the starting point for all impact studies and, as we have already seen, all regionalisation techniques (e.g. regional climate modelling, statistical downscaling) introduce their own uncertainties to the regional climate scenario. Further uncertainty is attached to the choice and application of impact models. The final section examines some of the most common uses of climate model results.

Key ideas

1. Models provide convincing evidence of human-induced climate change, but the exact rate and pattern of future warming are uncertain due to unknowable human behaviour (expressed in terms of greenhouse gas emissions), natural forcing (due to variations in solar output and volcanic eruptions), and large-scale climate system feedbacks.
2. Climate model experiments involving families (ensembles) of simulations with slightly different initial conditions for ocean temperatures help to assess the magnitude of natural climate variability relative to human-induced climate trends.

7.4 Applications of climate model results

The above sections discussed a hierarchy of climate models – ranging from zero-dimensional energy balance models through coupled ocean–atmosphere models – used to project future climate change from specified greenhouse gas emission scenarios. General circulation models are also the basis for all regional climate simulations derived using dynamical or statistical downscaling. However, it was shown that many uncertainties are attached to predictions of future climate change at global and, especially, regional scales. So, while there is little doubt that the climate is changing, much debate surrounds the extent to which observed climate trends and land-surface impacts are being forced by anthropogenic factors and/or natural variability. For example, long-term changes in flooding and drought hazards associated with regional climate changes have been the subject of considerable analysis (Box 7.6). To date, most evidence for climate change impacts has been inferred from secondary models *driven* by climate model projections.

7.4.1 Climate model evaluation

Rigorous evaluation of climate model performance should precede all attempts to elucidate impacts in climate-sensitive systems. To this end, the climate modelling community has undertaken several large-scale evaluation studies to compare simulations of the present-day climate with current observational data. For example, the Coupled Model Intercomparison Project (CMIP)

BOX 7.6

THINKING FURTHER

Detecting climate change in river flow records

The causes of changes in historic river flows and consequences of climate change for water resources generally have been the subject of much speculation. Despite the impression often given by the media, it is not possible to conclusively attribute individual flood events to global climate change. It is not even possible to state that climate change is having a discernible influence on *average* river flows across the globe. But it is possible to argue that observed increases in river flow around the North Atlantic seaboard, and reductions in the Sahel, are *consistent* with future climate change scenarios projected by many GCMs. Even so, there are still many reasons why it is difficult to unambiguously link changes in river flow to changes in climate:

- Worldwide, there are very few continuous river flow records of one century or more, with the majority having less than 30 years of data. There are data-rich regions (such as North America and Europe) and data-poor regions (such as Africa and central Asia), so global trends may be spatially biased.
- It is hard to distinguish between the effects of climate change and other factors due to human activities such as land-use change, river regulation or groundwater abstraction. Perceived increases in flood and drought frequency/severity may simply reflect society's increased exposure or vulnerability to hydrological hazards (such as more people living in floodplains).
- What is considered *normal* river flow depends very much on the period of record used to construct the baseline (Figure 7.12). River flows can also be highly variable from year to year, meaning that it is difficult to distinguish between natural variability and the effects of climate change. Furthermore, large-scale climate mechanisms such as El Niño (see section 5.5) lead to flood-rich periods separated by flood-poor years, making trend detection problematic.
- The hydrological response to climate forcing depends on river catchment geology, land use, vegetation, soil properties and snowpack behaviour (see section 10.4). This implies that climate impact assessment is best undertaken on a river-by-river basis.

For the above reasons, most evidence for the impact of climate change on river flows originates from river catchment models driven by climate model results.

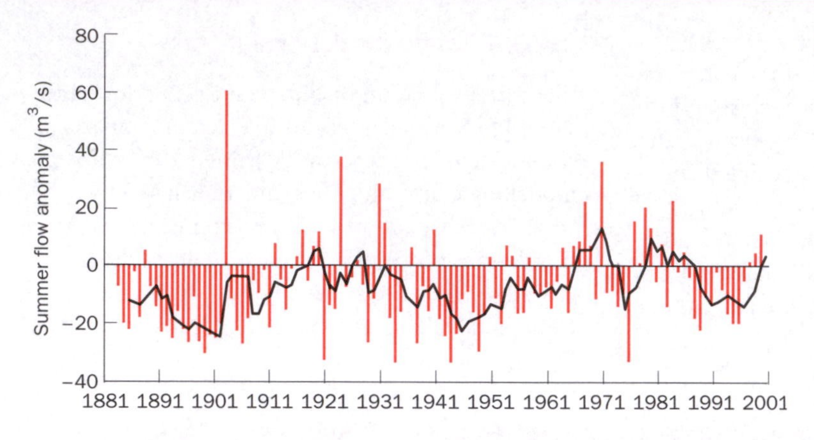

Figure 7.12 Summer mean flows in the River Thames, UK, 1881–2001

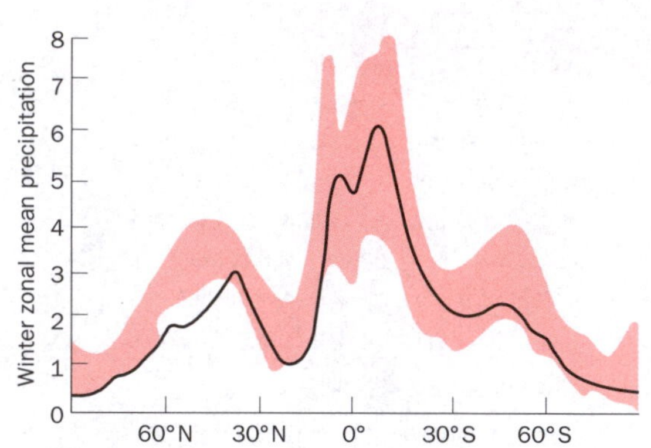

Figure 7.13 December–January–February zonal mean precipitation(mm/day) simulated by CMIP1 model control runs (shaded area) compared with observations (solid line)

Source: Adapted from Lambert and Boer (2001)

recently evaluated 20 such models worldwide and concluded that the current generation of models provide credible simulations of the present annual mean climate and seasonal cycle at continental scales for most variables relevant to climate change. The most problematic variables continue to be clouds and humidity, and to a lesser extent, precipitation (see Figure 7.13). As noted in section 7.3, the task of model evaluation continues to be hindered by the paucity of observations away from occupied land areas. Conversely, satellite data provide global coverage but are only available since the 1970s. Furthermore, satellite observations of the atmosphere's thermal structure have been the subject of much controversy, with both mainstream scientists and greenhouse sceptics (a small group of scientists who believe that global warming is a myth or at least is caused by factors other than anthropogenic gases) citing the data in support of their opposing arguments!

Other evaluation methods place less reliance upon individual climate variables. For example, a recent technique for identifying potential fingerprints of human-induced climate change involves comparing expected spatial patterns of warming in the lower and mid troposphere, and cooling in the stratosphere, with climate model output. These experiments indicate that simulations involving both CO_2 *and* sulphate aerosols yield patterns of temperature change that are closer to expectation than CO_2 alone. If both aerosol effects and solar forcing are incorporated, model predictions of global warming are in even closer agreement with observations.

Thus, it can be seen in Figure 7.14 that if only one factor, i.e. greenhouse gases, were involved in global warming, the observed mean global surface temperature rise of the last century would closely follow the ever-increasing concentration of greenhouse gases in the atmosphere. The fact that it does not indicates that other factors are involved. When the effects of aerosols are considered (cooling influence) the modelled temperature trend falls well below the observed values of temperature change, at least from the 1920s. The addition of aerosols, however, allows us to associate them with a period of planetary cooling after the 1940s to about 1965. During this period it is thought that large emissions of sulphate aerosol from growing industrial activity after the Second

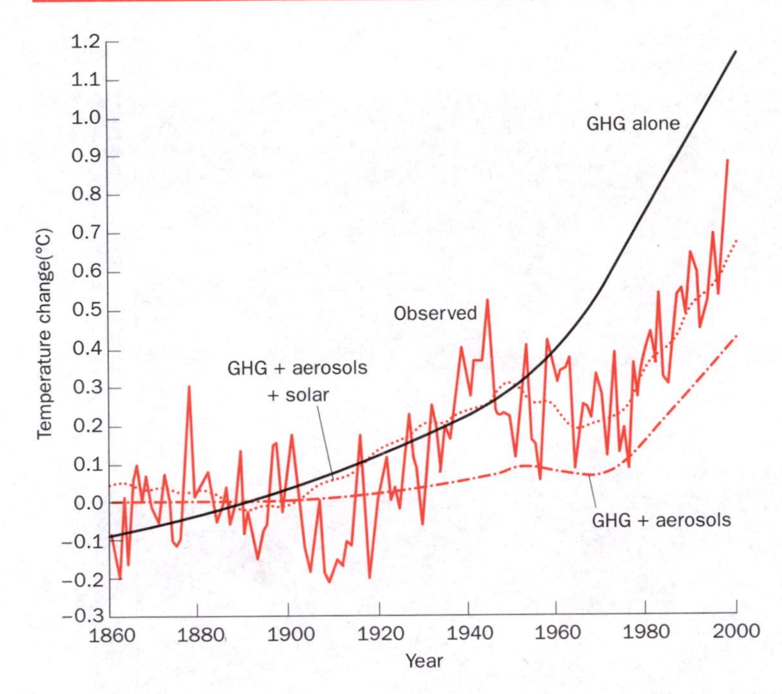

Figure 7.14 Comparison of observed (thin black line) and model-predicted global temperature changes (°C) from 1860 to 1999, with greenhouse gases (GHGs) alone, GHGs plus aerosols, and GHGs plus aerosols and solar forcing (dotted lines)
Source: Wigley (1999)

World War served to suppress global temperatures. When the impact of solar activity is added to the other two factors, the match between the modelled factor trends and the observed warming is pretty close. The inclusion of the effect of variations in solar output is especially clear during the period between about 1910 and 1940 when the first rapid increase in global temperatures took place (the second occurred after 1970). In the future the incorporation of volcanic aerosols (cooling) may improve some aspects of simulated twentieth-century climate variability still further.

The IPCC TAR (2001) concluded that, although coupled models have improved significantly since the SAR (1996), no single model should be considered best overall. This is because some models perform better in one region and less well in another, and some models perform well for a particular variable but relatively poorly for others. Ideally, a range of models should, therefore, be considered when undertaking climate impact assessments even though a consensus of opinion about the climate system response to radiative forcing is beginning to emerge.

7.4.2 The IPCC climate change scenarios

As indicated in section 7.1, the global mean surface temperature is projected to increase by 1.4–5.8 °C over the period 1990–2100 (this compares with 1.0–3.5 °C reported in SAR). As global temperatures increase, the extent of snow and ice in the northern hemisphere decreases, and atmospheric water vapour increases. Land surfaces warm faster than the ocean, with the cooling effect of tropospheric aerosols moderating both local and global warming (Figure 7.15). Increasing surface air temperatures are accompanied by more frequent extreme high temperatures and less frequent extreme low temperatures (see for instance Figure 12.1). In many areas, there is a decrease in the diurnal temperature range, due to nocturnal lows increasing more than daytime highs. Finally, daily variability of surface air temperatures decreases in winter, but variability increases in summer over northern hemisphere land areas.

For precipitation, there is a general increase in intensity but the signal-to-noise ratio is much weaker than for temperature. Mid-continental areas experience a general drying during summer with increased likelihood of drought due to a combination of increased temperatures and decreased precipitation. Enhanced interannual variability of northern summer monsoon precipitation (see Figure 11.16) is expected, and globally, heavy rainfall events contribute

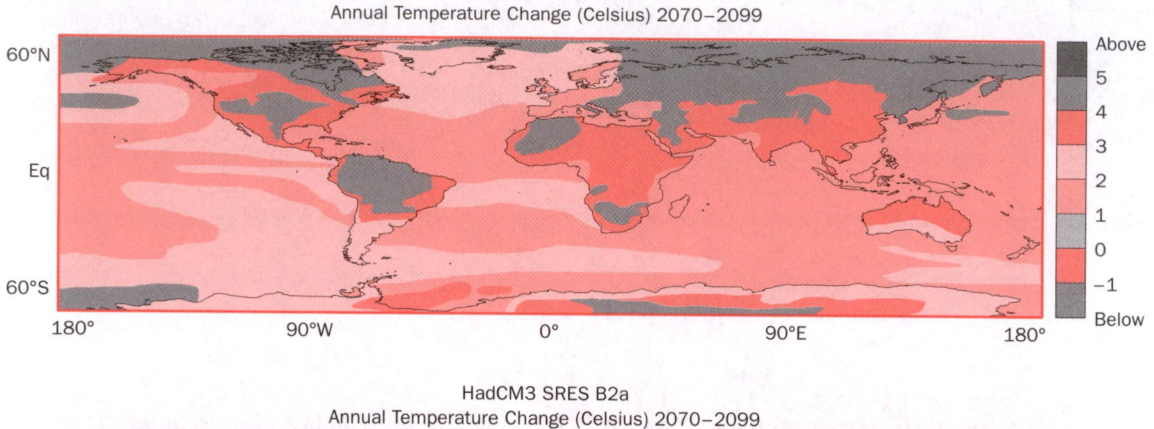

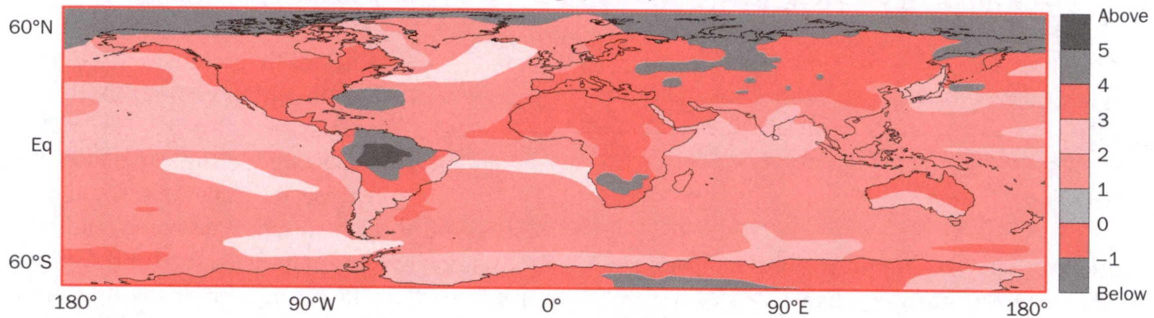

Figure 7.15 Annual mean temperature changes projected by HadCM3 by the 2080s under the A2 and B2 SRES emission scenarios

Source: Climate Impacts LINK project on behalf of the Hadley Centre

proportionately more to seasonal precipitation totals. Projected changes in the frequency and intensity of both mid-latitude storms (possibly greater frequency but unchanged intensity), and tropical cyclones (possibly unchanged frequency but greater intensity) still show differences among models.

Other commonalities include a weakening of the THC which contributes to lower rates of warming in the North Atlantic regions. If the radiative forcing is large enough and applied long enough the THC even collapses entirely in some experiments. Many models indicate an El Niño-like response in the tropical Pacific, with greater warming in the eastern Pacific accompanied by an eastward shift of precipitation. Some models show increases in ENSO interannual variability associated with large radiative forcing (but it is acknowledged that the simulation of ENSO-like phenomena in GCMs is still in its infancy). Overall, changes in future vegetation and cloud cover remain some of the largest sources of uncertainty in model projections of future climate change. For instance, palaeoclimate studies indicate that vegetation feedbacks strongly influence climate responses to radiative forcing.

7.4.3 Regional climate change scenarios

Given the large uncertainties associated with future greenhouse gas emissions and climate model sensitivity, it can be helpful to standardise model results. The pattern-scaling method takes regional patterns of temperature (or precipitation) change provided by one or more GCMs and scales them in proportion to the global mean warming. The technique relies on the tendency for space–time patterns of climate change in GCMs (driven by increasing radiative forcing, with or without aerosols) to be well represented by a *fixed* spatial pattern. In other words, for a given GCM, the regional pattern of climate change is dominated by large-scale global feedbacks rather than by local forcing, so the *pattern* is largely insensitive to the radiative forcing.

Under this assumption, tools such as MAGICC/SCENGEN can be used to quickly explore a wide range of emission scenarios, to compute time variations in global mean warming, and finally to scale regional patterns of temperature change in selected GCM(s) to the global mean change of an EBM. Although the technique can be used to combine patterns of climate change from several GCMs, precipitation scaling is more problematic. In this case the spatial pattern of change *does* depend on local forcing, meaning that less of the pattern variability is explained by the global mean temperature change.

7.4.4 Scenarios for impact assessment

For the present discussion, the final level in the 'cascade of uncertainty' involves generating a regional climate change scenario for impact modelling (e.g. water resources, agricultural, biodiversity). Once again, it must be recognised that the climate change impact will depend on the technique chosen to construct the scenario at the space–time scales of the system of interest. Different techniques yield very different scenarios even when driven by the same host GCM. Imperfect impact models introduce additional sources of error and uncertainty through further simplifications and/or parameterisation of environmental processes such as rainfall–runoff or snowpack accumulation and melting. To date, the most commonly used methods for scenario generation and impact modelling include, as explained below, climate change analogues, sensitivity analyses, direct use of GCM output, and indirect downscaling of GCM output. None of these techniques allows the surface response to feedback into the GCM experiment – all are therefore undertaken 'off-line'. At present, fully integrated impact modelling can only be undertaken within the limited framework of a SVAT model (see Box 7.3).

(a) Climate change analogues

These are constructed by identifying climate records that may typify the future climate of a given region. In this case, the climate model(s) provides qualitative information that can guide analogue selection (e.g. wetter, drier, hotter, more extreme). The analogue is then obtained from either past climate data (temporal analogue) or from another region (spatial analogue). A major advantage of the analogue approach is that the climate scenario (and associated impacts) may be described in far greater temporal and spatial detail than might otherwise be possible. For example, the hot/dry summers of 1976 and 1995, and the mild/wet winters of 1990/91 and 1994/95 provide useful temporal analogues for future climate changes in the UK. Although the summer (winter) rainfall deficit (surplus) for individual analogue years may be more extreme than the projected climatological average, the *likelihood* of such extreme events is expected to increase. Because the UK drought of 1995 actually occurred, data are available for a wide range of impacts, providing detailed insights into the possible consequences such events.

(b) Sensitivity analyses

Sensitivity analyses begin with the premise that model projections of regional climate change are so uncertain that it is better to apply plausible changes to observed data. In this way, a flood simulation model might be used to investigate the effect of –20, –10, 0, +10 and +20 per cent changes to historic precipitation intensities. At the same time, or in a separate set of experiments, tidal surge heights might be varied by +10, +20, +30 or +40 per cent. In this way, the impact of critical variables (whether in isolation or in combination) may be explored. The main advantages of the sensitivity approach are ease of application, transferability and comparability. The main disadvantages are that the specified climate changes are largely arbitrary (but ideally consistent with GCM projections), and the magnitude of the impact response will depend on the record chosen to define 'normal' conditions.

(c) Direct GCM output

Perhaps the most common method of scenario construction involves the use of GCM climate variable changes in conjunction with a baseline climatology. The simplest procedure is to interpolate gridded climate model output to the point of interest. In which case, the main consideration is the choice of smoothing technique (e.g. inverse distance, kriging, splines) and availability of baseline data. Finer spatial detail may be resolved by assimilating topographic information (elevation, aspect, slope), or temporal detail through weather generation techniques (see section 7.2.4). Alternatively, the coarse resolution climate changes of the GCM may be applied directly to the baseline climatology *without* interpolation. In this case, climate change signals are assumed to apply uniformly over the GCM grid box, and across all elevations. The main appeals of interpolation are simplicity and consistency with the host GCM. However, all the limitations noted under 'unintelligent' downscaling apply (section 7.2.4).

(d) Indirect GCM output

Finally, dynamical or statistical downscaling are increasingly used for climate impact assessment. As noted previously, downscaling using RCMs is computationally demanding and time-consuming. On the other hand, statistical downscaling requires significant data pre-processing and identification of robust predictor–predictand relationships. Whereas RCMs produce climate change scenarios of 20–50 km resolution for specified domains, statistical methods can deliver point-resolution scenarios. Although the latter may seem appealing for point-process impact modelling (e.g. soil erosion, flood estimation, crop modelling, small islands) all such downscaling methods are subject to serious caveats. In particular, both approaches depend entirely upon the realism of the host GCM. However, as cautioned in the introductory comments of this chapter (section 7.1), the most serious danger associated with the use of any climate model output is to treat a climate change *projection* as if it were a *prediction*.

Key ideas

1. Climate change detection in environmental systems is often confounded by data limitations, by year-to-year natural variability, by artificial (human) influences and by local variations in the pattern of responses.

2. Climate change assessments reflect the 'cascade' of uncertainty, beginning with the estimation of future emissions, modelling the climate response at global and then regional levels, and ending with the choice of impacts model.

Further reading

Crowley, T. J. (2000) Causes of climate change over the past 1000 years. *Science*, **289**: 270–7.

IPCC (2001) *Climate Change 2001: The scientific basis. Contribution of Working Group I to the Third Assessment Report of the Intergovernmental Panel on Climate Change* [Houghton, J. T., Ding, Y., Griggs, D. J., Noguer, M., van der Linden, P. J. and Dai, X., Maskell, K. and Johnson, C. A. (eds)]. Cambridge University Press, Cambridge, 944 pp.

Lambert, S. J. and Boer, G. J. (2001) CMIP1 evaluation and intercomparison of coupled climate models. *Climate Dynamics*, **17**: 83–106.

Leung, L. R., Mearns, L. O., Giorgi, F. and Wilby, R. L. (2003) Regional climate research: needs and opportunities. *Bulletin of the American Meteorological Society*, **84**: 89–95.

McGuffie K. and Henderson-Sellers A. (2001) Forty years of numerical climate modelling. *International Journal of Climatology*, **21**: 1067–109.

New, M., Hulme, M. and Jones, P. (1999) Representing twentieth-century space-time climate variability. Part I: Development of a 1961–90 mean monthly terrestrial climatology. *Journal of Climate*, **12**: 829–56.

Wigley, T. M. L. and Raper, S. C. B. (2001) Interpretation of high projections for global-mean warming. *Science*, **293**: 451–4.

Wilby, R. L. and Wigley, T. M. L. (1997) Downscaling general circulation model output: a review of methods and limitations. *Progress in Physical Geography*, **21**: 530–48.

Useful websites

The scientific basis and worldwide impacts of climate change:
http://www.unep.ch/ipcc/

Up-to-date scenarios of climate change:
http://ipcc-ddc.cru.uea.ac.uk/

History, causes and impacts of climate change:
http://www.cru.uea.ac.uk/cru/cru.htm

Reports and data on changes in mean sea level:
http://www.pol.ac.uk/psmsl/

Climate and hydrological change:
http://www.ghcc.msfc.nasa.gov/

Climate models used by UK Meteorological Office:
http://www.met-office.gov.uk/research/hadleycentre/index.html

National Center (USA) for Atmospheric Research:
http://www.ncar.ucar.edu/ncar/

CHAPTER 8

Weather, climate and climate change in the high latitudes Polar regions

8.1 The polar climate

The northern coastal areas of Eurasia and North America together with Greenland as well as Antarctica have a polar climate. The main features of this climate region are year-round low temperatures and relatively low precipitation amounts. The single most important factor affecting the climate of the polar regions is their high latitudinal location. Located polewards of the Arctic and Antarctic circles (66.5° N and S respectively) the Arctic and Antarctic regions suffer a fundamental deficiency in the input of sunlight or solar radiation. This deficiency is clearly expressed in a total absence of sunlight for 5–6 months near the poles during the winter season and a low-elevation sun with weak inputs of solar radiation during the summer. We have already seen (section 3.3.2 and Figure 3.4) that because of such weak solar receipts both polar regions exhibit a permanent net radiation deficiency. This means that incoming amounts of (absorbed) solar radiation fall far short of the continuing net loss of heat to space. Despite the import of large amounts of sensible and latent heat by winds and ocean currents to make good this energy deficit, temperatures in the polar regions remain cold.

8.1.1 Processes and controls

(a) Why so cold?

Various factors conspire to make the polar regions both cold (see Figure 3.7) and dry. First, the high latitudinal location and low angle sun of the polar regions (see Figure 3.6(a)) means that annual inputs of solar radiation are low (see Figure 3.5). Second, physical models of the general circulation of the atmosphere show us that the high-latitude polar regions are dominated by surface anticyclones, i.e. by almost year-round high-pressure circulations. High-pressure conditions encourage descending airflow with surface air divergence. Downward-moving air over the poles creates warming near the surface and encourages evaporation (the warming is relative, the polar regions are still cold). These conditions limit the mechanisms necessary for precipitation generation which need surface convergence and convective updraught. Third, the clear skies produced by the descending warming air of the semi-permanent highs encourage massive heat loss or outgoing terrestrial radiation from the surface to space. Fourth, surface air chilling is also encouraged because of the high albedo or reflectivity of the polar surface (see Table 3.1) where large percentage amounts of incoming solar radiation are reflected from a permanent or semi-permanent ice and snow cover. As a high percentage of sunlight is reflected, little of the sun's heat is absorbed to warm the surface and the air above. Fifth, the polar snow and ice surface also act as a good radiator or emitter of outgoing thermal radiation. A sixth factor explaining the cold of the polar areas are long winter nights that facilitate a large amount of heat loss from the surface. In fact, the polar regions lose so much thermal radiation to space and receive so little heat from incoming solar radiation that there is a permanent net radiation deficit not only for the polar earth–atmosphere system but also at the polar surface itself (see Figure 3.4).

(b) Temperature inversions

Another feature that helps to explain and reinforce the characteristics of the polar climate is the temperature inversion. Year-round chilling of air at the polar surface together with adiabatic warming of air at mid levels from air descent serves to build strong temperature inversions over the polar surface. These temperature inversions with cold dense air at the surface and warmer less dense air above stabilise local air masses. They help explain the low rainfall amounts of the polar climate by effectively preventing the conditions necessary for precipitation, i.e. air instability and convection. Most precipitation in the polar regions comes not during the winter when the temperature inversions are strong but during the short 'summer' period when they are weakened, and when they can be penetrated or displaced by poleward-moving, rain-bearing, mid-latitude low-pressure systems (depressions).

(c) Air masses

It may be recalled from section 4.3.1 that an air mass is a large mobile body of air with broadly uniform weather conditions, i.e. with respect to pressure, temperature, moisture, humidity and wind. The location and defining features of the major air masses can be found in Figure 4.12 and Table 4.1 respectively. The interior polar regions, north of about 80°, are dominated by semi-permanent high-pressure cells incorporating very dry (continental) and cold Arctic air masses (cA). The high-pressure cells of the lower-latitude coastal polar regions are dominated by cold polar air masses (P). The polar air can be further subdivided according to moisture content. The more maritime regions, for example across the North Atlantic, are influenced by maritime polar air (mP) while the more landlocked polar regions (e.g. northern Canada, northern Siberia) are affected by continental polar air (cP). One of the most significant climate boundaries in the polar region is where milder air from lower latitudes meets the higher latitude colder air. Because the milder air does not mix easily with the colder air, a frontal zone (line of separation) forms at the junction between the two air masses. The most important frontal zones in the polar region include the *polar front*, separating milder air from the cold polar air and the *Arctic/Antarctic front* separating mild or cold air from very cold Arctic air. The polar front over the northern North Atlantic is instrumental in delivering a good proportion of the region's precipitation.

8.1.2 The upper westerlies (the Circumpolar Vortex)

(a) Links with surface weather

As outlined in section 4.1.4, the upper westerlies form a quasi-continuous belt of high-altitude, fast-moving westerly winds moving around both poles, between latitudes 40 and 70°. Such winds are given other names including the Circumpolar Vortex. Because of earth rotation, the upper westerlies of each polar region gyrate (as in a vortex) in a series of great waves (Rossby waves) with their peaks towards the poles and their troughs towards the equator. The main generator for the upper westerlies relies on a pressure difference between the surface and the upper air. The cold high-pressure cell over the polar surface produces shrinkage in the air column above it, and leads to the development of a deep low-pressure system at higher levels in the polar atmosphere. The thermally created, high-elevation, low-pressure cell which sits over the polar surface high is the main driver of the upper westerlies.

The polar surface high-pressure cell can intensify and weaken over time and can thus affect the strength of the upper westerlies. When the surface high strengthens so too does the upper low-pressure cell which accelerates the movement of the upper westerlies. The opposite applies when the surface high loses strength, thus weakening the upper low and the associated upper winds. There is some evidence to show that cause–effect relationships between pressure conditions at the surface and at higher elevations can be reversed. Recent stratospheric cooling over the poles, possibly linked to greenhouse gas global warming, has been linked to a strengthening in the upper low and upper winds respectively. It has been suggested that such upper air forcing can work its way down to the surface by intensifying in turn the surface high-pressure cell.

(b) Short- and longer-term climate change

The upper westerlies over both poles vary in intensity on a seasonal basis. For example, they strengthen in the winter compared to the summer as the polar air chills and the surface high pressure intensifies at the poles. The upper westerlies also show annual, decadal and longer-term variations between weaker (negative) and stronger (positive) phases. Recent evidence has shown that these longer-term phases are governed by hemispheric-wide surface pressure gradients between the poles and the mid latitudes. Switches in the intensity of surface atmospheric pressure gradients between weak and strong phases across the northern hemisphere are referred to as the Arctic Oscillation (AO) while those over the South Pole are collectively called the Antarctic Oscillation (AAO).

8.1.3 The Arctic and Antarctic oscillations

(a) Negative and positive phases

The varying semi-continuous winds of the northern and southern Circumpolar Vortex have wide-ranging effects not only on both polar regions, but also on the climate of both hemispheres. When the AO strengthens as it has done over the Arctic since the mid 1980s (Figure 8.1) triggered by colder air and lower pressure in the upper air, the upper waves intensify, become straighter and more continuous, and shift polewards. The acceleration of wind flow in the upper westerlies

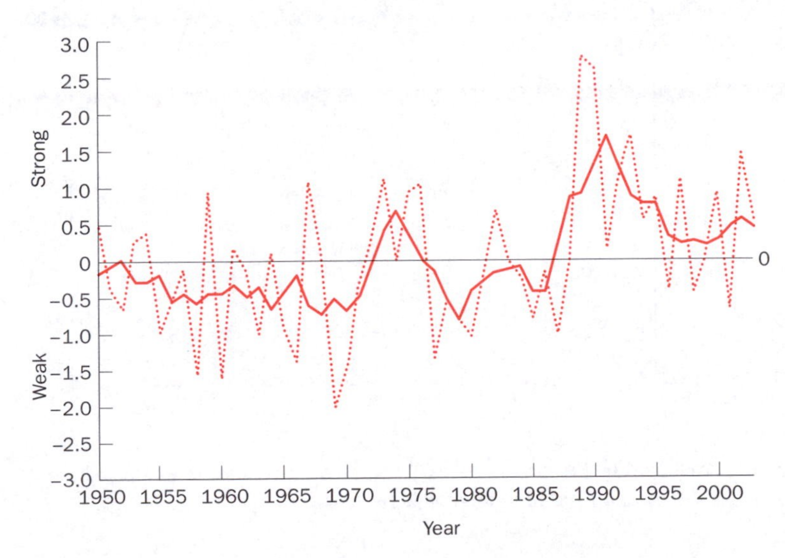

Figure 8.1 The mean winter (Jan.–Feb.–Mar.) AO index, compared to the long-term average, 1950–2003. Dotted line is the annual trend; the solid line indicates the 5-year running mean; the line 0–0 is the long-term average
Source: NOAA website

prevents the penetration of warmer air into the polar regions and the movement of cold polar air out of the vortex. As a result, frigid winter air does not plunge as far south over North America and Eurasia, meaning warmer winters over much of the United States east of the Rocky Mountains and over Scandinavia and Siberia. A positive AO and an intensification of the upper westerlies also mean that central polar regions including Greenland, northern Canada and the western Arctic Ocean have become colder.

A polar shift in the upper westerlies (jet streams) during a positive phase in the AO also helps to steer ocean storms further north and has brought warmer and wetter weather in recent years to the maritime fringes of the Arctic region including northern Alaska and Scandinavia. This has meant that in recent years more heat has been advected from southern latitudes over the North Atlantic and Arctic oceans. One result of this is that regional sea temperatures have increased. Between 1979 and 1997 there have been signs that warming during the winter and spring seasons over the eastern Arctic which is most strongly influenced by northward-moving ocean currents has been as much as 1 and 2 °C/decade. AO forcing of higher North Atlantic and Arctic sea temperatures, together with more widespread global warming in the high latitudes, have possibly been responsible for significant declines in Arctic sea-ice extent and thickness in recent years.

When the AO loses strength with relatively warm temperatures and high pressure in the upper air over the pole, weaker, more wave-like, more southern and more discontinuous upper-level westerly winds (jet streams) develop. During negative phases in the AO (Figure 8.2), as during the winters of 1964/65–1969/70 and much of the decade between 1976/77 and 1985/86, weaker upper westerlies allow more air mass exchange

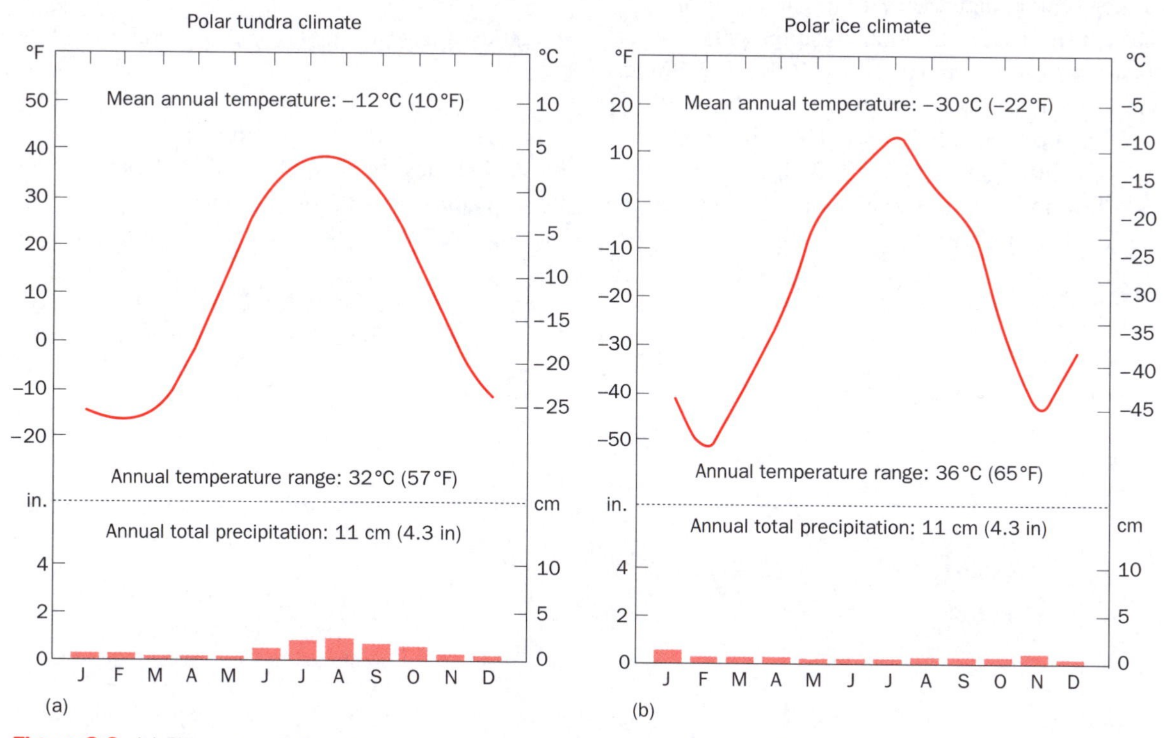

Figure 8.2 (a) The polar tundra climate as represented by Barrow, latitude 71° N, located on the north Alaskan coast; and (b) the polar ice climate at Eismitte, 71° N and located in interior Greenland

Source: Ahrens (1993, pp. 368–9)

between the polar and neighbouring subpolar and middle latitude climates. As a result, more cold air penetrates Europe and North America from the North Pole making these regions colder. At the same time, more southerly warmer weather systems (depressions and anticylones) can penetrate the northern polar and high-latitude areas bringing increasing temperatures to them.

Table 8.1 reinforces the above discussion and shows a long-term control exerted by the AO on temperatures in the North Atlantic and western Greenland (interior polar region) over the last 150 years or so. During those winters (Dec.–Feb.) between 1865 and 1990 when mean temperatures in western Greenland (as represented by Jakobshavn) were colder by at least 4 °C than those in the North Atlantic (as represented by Oslo, Norway), the AO was often strongly positive. In contrast, during those winters from 1869 and 1996, when Jakobshavn was at least 4 °C warmer than Oslo, a relatively weak AO with negative indices often prevailed. Some of the coldest winters in

Table 8.1 Correspondence between the strength of the winter AO index (Dec.–Feb.) and those years when Greenland (Jakobshavn) was at least 4 °C colder or 4 °C warmer than northern Europe (Oslo)

Years with Greenland temperature less than northern Europe winter (Dec.–Mar.) temperature	AO index (Dec.–Mar.) in strong or positive phase	Years with Greenland temperature above northern Europe winter (Dec.–Mar.) temperature	AO index (Dec.–Mar.) in weak or negative phase
1865–66	0.54	1869–70	–3.01
1873–74	2.32	1870–71	–1.01
1883–84	1.49	1874–75	–1.35
1886–87	0.45	1876–77	0.05
1889–90	1.78	1878–79	–2.22
1893–94	3.89	1880–81	–3.80
1895–96	1.12	1892–93	–1.07
1897–98	1.02	1894–95	–3.97
1898–99	0.03	1925–26	0.11
1905–06	2.06	1927–28	0.63
1906–07	2.06	1928–29	–1.03
1908–09	0.00	1939–40	–2.86
1909–10	2.10	1940–41	–2.31
1910–11	0.29	1941–42	–0.55
1913–14	2.69	1946–47	–2.71
1914–15	1.48	1947–48	1.34
1920–21	1.63	1958–59	–0.37
1924–25	2.39	1962–63	–3.60
1936–37	0.72	1968–69	–4.89
1948–49	1.87	1974–75	1.63
1970–71	–0.96	1976–77	–2.14
1971–72	0.34	1978–79	–2.25
1972–73	2.52	1979–80	0.56
1982–83	3.42	1981–82	0.80
1983–84	1.60	1984–85	–0.63
1988–89	5.08	1985–86	0.50
1989–90	3.96		

Source: A. Dawson, University of Coventry, UK

Table 8.2 Gale days and mean gale-day frequency at stations in north and south Iceland. Gale days measured for those winters (Dec.–Feb.) between 1860 and 1990 when Greenland was 4 °C colder than the Iceland station (suggesting a positive AO) and 4 °C warmer than the Iceland station (suggesting a negative AO)

	Greenland colder (4 °C) than Iceland, i.e. with strong AO index		Greenland warmer (4 °C) than Iceland, i.e. with weak AO index	
	Grimsey (N Iceland)	Stykkisholmur (SE Iceland)	Grimsey (N Iceland)	Stykkisholmur (SE Iceland)
Number of winters	25	32	24	29
Gale days	817	1,299	356	601
Mean gale-day frequency per winter	32.68	40.59	14.83	20.72

Source: A. Dawson, University of Coventry, UK

northern Europe have been experienced during periods with a negative AO, when cold pulses of air from the Arctic can penetrate far south over the high latitudes. The winters of 1946/47 and 1962/63 with strongly negative AO indices were two of the coldest winters experienced by the UK in the last 100 years. Table 8.2 also shows that during positive AO winters with a northern shift in the polar front, more frequent and stronger mid-latitude depressions can penetrate from the high latitudes into the northern polar regions. Over the period 1840–1990, when winter temperatures in Jakobshavn were lower than those in Oslo (a measure of a strong AO index) almost 1,300 gale days were recorded in southern Iceland and over 800 in the northern part of the country. During those years when winter temperatures in Jakobshavn were higher than those in Oslo (a measure of a weak AO index) the number of gale days falls dramatically, with southern Iceland recording just over 600 gales and north Iceland just over 350 gale days.

It is important to emphasise at this stage that the large hemispheric-wide Arctic and Antarctic oscillations incorporate and influence other smaller regional-scale atmospheric oscillations in both hemispheres. The most important of these is the North Atlantic Oscillation (NAO) that operates, as its name suggests, over the northern North Atlantic. Its strength depends on the pressure gradient between the semi-permanent Azores High situated off the coast of Portugal and the semi-permanent Icelandic Low situated over Iceland. Like the AO, the NAO has strengthened during the last 20 years or so, with effects on the maritime polar regions similar to those shown for the AO. The NAO also has real significance with regard to the weather and climate of the mid latitudes of the northern hemisphere. This is discussed in Chapters 9 and 10.

8.1.4 Climate types

(a) Polar tundra climate

Although cold is endemic and widespread over the polar regions (see Figure 3.7), two main types of polar climate are usually identified, i.e. the polar tundra climate and the polar ice cap climate. In the polar areas furthest from the North and South Pole, i.e. the northern coastal areas of North America and Eurasia, the coastal margins of Greenland and the north Antarctic Peninsula, a polar tundra climate is found. Dominated by polar air masses (mP and cP), the distinctive feature of the polar tundra climate is that mean temperatures of the warmest month are above freezing but below about 10 °C. Winters are cold to very cold with very short days. Summers are cloudy and cool but have a thermal advantage in long day-lengths.

Nevertheless, in the warmer coastal tundra climates, the chance of frost occurring during the summer is never absent.

The weather and most of the rain is mainly controlled by depressions which are weakly developed in summer: in winter much of the polar tundra climate is north of the main depression tracks and most rain is carried into the region by occluded fronts. Summer is, for most areas, the season of maximum precipitation. At Barrow, Alaska (see Figure 8.2(a)), mean annual temperatures are −12 °C. The annual temperature range is large, with mean winter temperatures as low as −28 °C, and summer temperatures around 4 °C. Vardo, in northern Norway (70° N, 31° E), is influenced by the relative warmth of the North Atlantic Drift (NAD) and has higher monthly mean temperatures of −6 °C in January and 9 °C in July.

Another distinguishing feature of the polar tundra climate is that plant growth can take place during the brief summer season where long days are characteristic. Moreover, annual rainfall is generally low in the polar tundra climate with amounts less than about 20 cm (but rising to 50 cm in maritime areas like northern coastal Norway). This rainfall is usually adequate for plant growth, however, because of the very low rates of evaporation that occur in such cold zones. Tundra vegetation including mosses, lichens, acid grasslands, short woody vegetation and scattered low trees are found and give their name to this climate type. Much of the soil and ground in polar tundra climate regions is permanently frozen (permafrost) to about 50 m below the surface. But summer heat, with 3 months often around 10 °C, is often sufficient to thaw out the top metre or so of soil, leading to wet and swampy landscapes. Tundra landscapes and the ecosystems they contain are fragile and easily and permanently damaged under human impact. When used by heavy transport vehicles, for instance, the terrain often becomes unstable. The friction from the tyres can melt the permafrost, leading to extensive muddy and uneven areas. Global warming is also melting and de-stabilising the permafrost.

(b) Polar ice cap climate

The north polar region beyond 80° latitude including the interior regions of Greenland and all of Antarctica (with the exception of the Antarctic Peninsula) have an even harsher climate, i.e. an ice cap climate where the land/sea is either composed of ice sheets or permanently frozen ground. Dominated by Arctic air all year round, temperatures in the polar ice cap climates rarely rise above freezing level, even during the summer (see Figure 8.2(b)). Precipitation remains very low, with often less than about 10 cm per annum. In winter, conditions fluctuate between long periods of very cold and dry high-pressure weather and shorter spells of dull bleak snowy weather as decaying depressions with occluded fronts move into the polar region. The accumulated depth of ice in some places, however, can measure thousands of metres. This fact has been used to great effect as these great ice layers have been used in climate reconstruction (see section 6.4.1).

8.1.5 Changes in precipitation and sea-ice extent

Because there are relatively few meteorological stations or long-term climate bases in the high latitudes, changes in precipitation (snowfall) over the last 100 years or so are not particularly reliable. Nevertheless, current evidence and modelling studies suggest that while snowfalls may have decreased slightly in the Arctic as a result of global warming and AO forcing, the total amount of precipitation may have increased by 5–10 per cent with slightly higher amounts of rainfall. No such change is detectable, however, in the Antarctic region. But snowfalls may have increased here during the last few decades. This latter trend could be the result of the Antarctic being so cold (−20 to −30 °C) so that slight increases in temperature from global heating could increase the moisture-holding capacities of the Antarctic air and thus provide a greater potential for higher snowfall amounts.

Changes in sea-ice extent are perhaps a more reliable index of climate change in the polar regions. Satellite data from the 1970s combined with other longer (100-year) regional data sets show that spring and summer ice extent in the northern hemisphere (Arctic) has declined by about 10–15 per cent since the 1950s. The thickness of the Arctic Sea ice cover in summer has also been reduced by 40 per cent over the same time period. The most dramatic single example of ice break-up is the recent (Sept. 2003) splitting in two of the largest ice shelf in the Arctic, the 30 m thick Ward Hunt ice shelf located on the north coast of Ellesmere Island at the top of the Arctic archipelago. Declines in Arctic sea-ice extent and reductions in sea-ice thickness are consistent with the recent pattern of high-latitude temperature change, that includes a warming over most of the subarctic land areas as a result of global heating and stronger AO forcing. It is possible, however, that the recent, marked reductions in Arctic sea-ice thickness (taken from submarine measurements) are part of a multi-decadal fluctuation of thinner and thicker ice layers rather than reflecting a long-term global warming trend.

In contrast, there are no available records to suggest any long-term (100-year) change in Antarctic land and sea-ice extent/volume. This is despite the fact that for the Antarctic as a whole, temperatures may have risen by about 0.5 °C over the last half-century, and regionally by as much as 2 °C in the last 40 years over the Antarctic Peninsula. Recent evidence shows that Antarctic sea-ice extent fell somewhat in the mid 1970s, but rose again slightly after 1979. Some observations have concluded that there has been a slight increase in ice cover over Antarctica in the last 50 years as a consequence of small increases in snowfall. This is despite the fact that there have been some spectacular signs of Antarctic sea-ice shrinkage involving the retreat and collapse of massive sea-ice shelves and the calving of large new icebergs in the mid 1990s (Box 8.1).

BOX 8.1

CASE STUDY

Retreat of ice shelves on the Antarctic Peninsula

Global warming over the Antarctic continent as a whole appears to have been perhaps 0.5 °C in the last 50 years. Because the Antarctic remains so cold (average temp –20 °C) this warming has not significantly affected the overall volume or cover of the Antarctic ice sheet. Indeed, recent warming over the continent may have slightly increased the snowfall rate and the cover of permanent ice. On the other hand, there has been significant regional warming in the Antarctic. The Antarctic Peninsula (Figure 8.3) may have warmed by as much as 2 °C since the 1960s. This regional warming whose cause has yet to be fully accounted for, has led to the spectacular break-up and retreat of a number of massive sea-ice shelves in the region.

Dramatic collapses

Ice shelves fringe most of the Antarctic continent where there are bays or islands to contain them. The thermal limit of sea-ice development corresponds closely to a mean January (summer) air temperature of 0 °C or to an annual mean air temperature of about –5 °C. As a result of Antarctic Peninsula warming of up to 2 °C in recent years, this temperature limit has migrated south so that a number of ice shelves have melted and retreated. Figure 8.3 shows changes in the areal extent of eight sea-ice shelves on the Antarctic Peninsula over the last 50 years. It can be seen that five northerly ones in a zone where most warming has taken place, i.e. Muller, Wordie, Prince Gustav, Larsen Inlet and Larsen-A ice shelves, have all suffered progressive retreat via ▶

▶ a series of regular annual melting (calving) events over many years without substantial advance. The progressive retreat of these ice shelves eventually ended in spectacular final stage collapse in 1995, leaving in each case only a small residual shelf. After the collapse, James Ross Island, situated off the northern end of the Antarctic Peninsula, is now circumnavigable by ship for the first time since it was discovered in the early nineteenth century.

Continued peninsular warming since the mid 1990s has put the stability of some of the more southern ice shelves (Wilkins, Bach, Larsen-B and Larsen-C) also in jeopardy. Plate 8.1 shows the spectacular break-up of the northern section of the Larsen-B ice shelf since 1995. A consequence of the disintegration has been a plume of thousands of icebergs adrift in the Weddell Sea. What is significant is the speed with which the shelf has disintegrated and continues to disintegrate. A total of 3,250 km^2 of shelf area (equivalent to 500 billion tonnes of ice sheet) broke up in a 35-day period beginning between 31 January and 5 March 2002. Over the last 5 years the shelf has lost a total of 5,700 km^2, and is now about 40 per cent the size of its previous minimum stable extent. This case study has shown that ice shelf extent may be a sensitive indicator of regional climate change. Given continued warming and the view that climatic forcing may not need to be very strong to produce ice shelf retreat, Larsen-B will probably continue to retreat and eventually Larsen-C may well behave in a similar way.

Future stability of the West Antarctic ice sheet (WAIS)

We cannot be certain about the relationship between global warming and ice shelf melting in the Antarctic because our records of ice sheet dynamics and temperature change are not long enough to establish statistically significant correlations. It remains unclear, for instance, whether such ice shelf retreat is part of a natural cycle of decay and growth or is due to climate forcing by global warming. On millennial timescales these retreats may not be unique or even unusual, and ice fronts may have oscillated several times during the last 10,000 years.

We are also uncertain about the stability of the main West Antarctic ice sheet (WAIS) (the East Antarctic ice sheet is more stable) in the face of greenhouse gas global temperature rise. The IPCC maintains that despite global warming, there has been no overall change in sea-ice extent over the Antarctic in the last 50 years, and that such warming in the short and mid term may actually have increased snowfall and ice cover over the region. Nevertheless, a number of modelling studies suggest that the WAIS will eventually succumb to global heating over the coming centuries. A number of scenarios suggest that with progressive increases in global temperature, the Antarctic ice shelves will continue to melt at their base, and ice streams will become more active. The extensive Ross ice shelf in western Antarctica could disintegrate from anything between 50 and 200 years. The duration time for the total melting of the WAIS ranges from a low figure of 250–400 years to a moderate one of 500–700 years. The future meltwater contribution of the Antarctic ice sheet to sea-level rise is considerable, averaging between 60 and 120 cm per century. These melt speeds would result in total rates of sea-level rise beyond the year 2100 that are double or triple the median values projected by the IPCC for the twenty-first century. The total disintegration of the WAIS would ultimately result in a massive increase in sea levels of between 4 and 6 m. Such rates of change would pose a serious challenge to the ability of societies to adapt to rising sea levels and would devastate many extensive coastal ecosystems. The most important of these include many of the world's largest cities, low-lying coral islands, coastal marshes and natural wetlands, food-producing deltas in the tropics and mangrove zones.

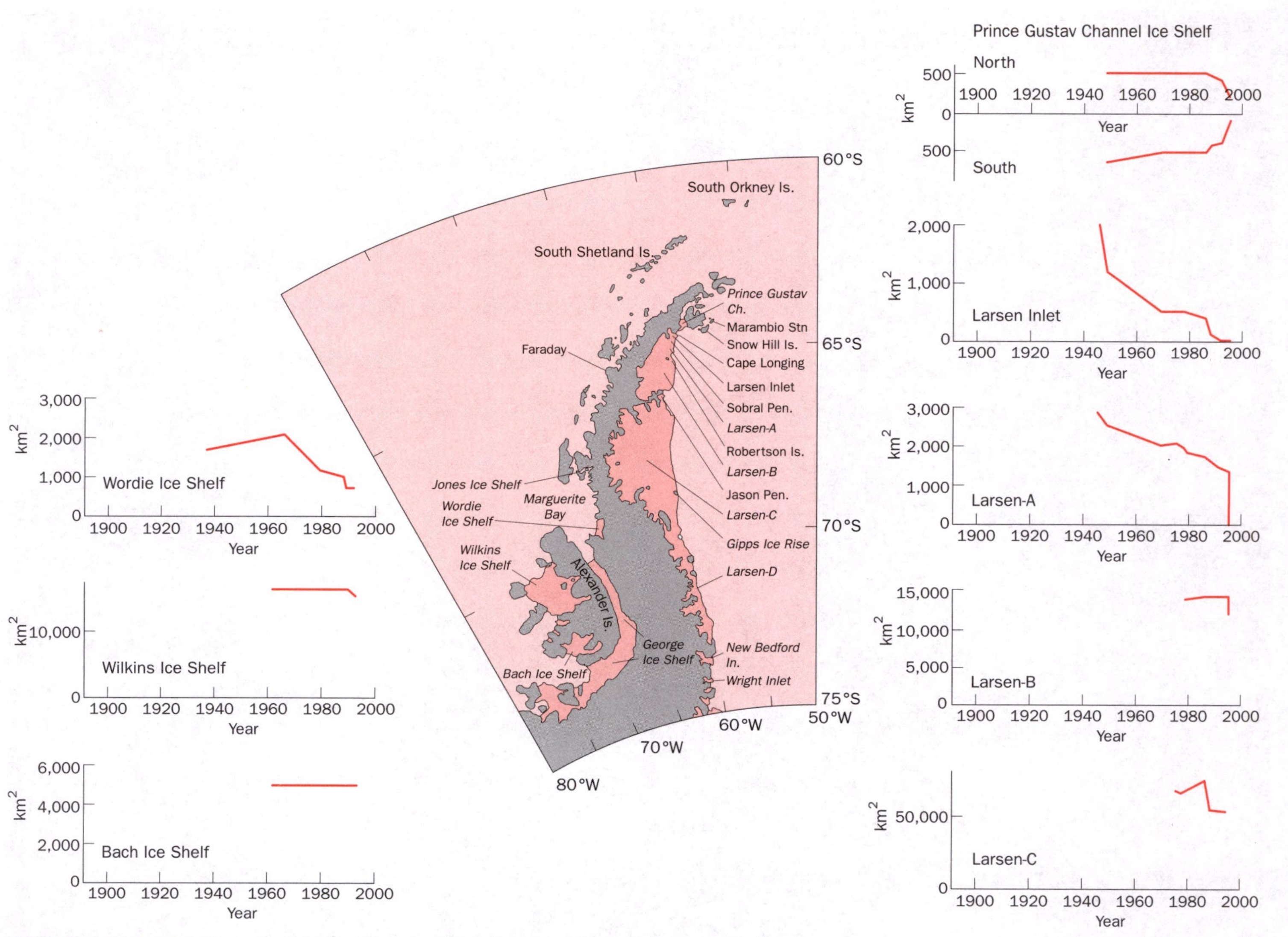

Figure 8.3 Changes in the extent (km^2) of ice shelves in the Antarctic Peninsula, 1900–95

Source: Vaughan, D.G. and Doake, C.S.M. British Antarctic Survey (BAS), NERC, Madingley Road, Cambridge, CB3 OET

8.1.6 Cold polar air and stratospheric ozone

The natural production and loss as well as the human accelerated depletion of polar stratospheric ozone have been described in Box 2.1. This section re-examines the effects of human activity (global warming) on the stratospheric ozone layer and compares the polar regions in terms of their ability to make and break ozone.

(a) Global warming favours ozone loss!

Evidence exists to show that other factors additional to direct chemical annihilation by CFC and related gases may be reinforcing the current trend of stratospheric ozone decline, so that its reversal and recovery may be difficult to achieve. It has been shown that the release of greenhouse gases in the atmosphere (see section 6.5.4) which cause global warming in the troposphere, results in a progressive cooling of the stratosphere. As previously shown, such high-level atmospheric cooling provides ideal conditions for chemical ozone annihilation. Moreover, as concentrations of ozone are reduced in the stratosphere, less solar energy can be absorbed in this atmospheric layer, so that the stratosphere will continue to cool. Further stratospheric cooling promoted by a more intense polar vortex generated by a positive AO is currently predicted to prolong Antarctic and Arctic ozone depletion. It needs to be emphasised again that stratospheric ozone loss, contrary to popular opinion, is *not* a cause of global warming. The additional amount of UV light striking the earth's surface as a result of ozone decline is pretty insignificant in terms of global heating (but not in terms of being harmful to living systems). Recent evidence suggests that rather than causing global warming, greater receipts of UV radiation at the surface probably induces net global cooling.

(b) Antarctic versus Arctic

Ozone depletion over the southern pole (the Antarctic) is more prominent than over the North Pole (the Arctic) although serious losses of Arctic ozone have begun to occur since the late 1990s. Ozone destruction is greater over the Antarctic as a result of the greater strength of the southern Circumpolar Vortex (upper westerlies). The southern Circumpolar Vortex is stronger than the northern one because of the greater expanse of continuous ocean around the South Pole that offers little friction to the upper winds. The northern oceans, on the other hand, are interrupted by major mountainous landmasses such as the north–south trending Rockies that can greatly reduce the speed of the upper circumpolar westerlies by frictional drag.

The greater strength of the southern vortex lowers respective ozone levels in two distinct ways. First, most of the ozone found over the poles is actually formed in the tropics and is brought to the polar regions by stratospheric winds. However, the greater strength of the southern vortex provides a greater barrier to this import and lessens overall accumulations. As previously suggested, a strengthening of the Antarctic Oscillation (AAO) in recent years has increased the velocity and continuous nature of the southern vortex and has helped to isolate cold Antarctic stratosphere air from the warmer air of the middle latitudes. During the long dark Antarctic winter, temperatures inside the southern vortex can drop to −85 °C. This frigid air is critical to the formation of *polar stratospheric clouds* (PSCs). The ice particles that compose these clouds act as sites where the CFC-driven chemical and photochemical reactions that accelerate ozone loss can take place. Chemical reactions are activated at these sites at the beginning of the Antarctic spring (e.g. September and October) when temperatures are at their lowest (after a long winter) and when there is sufficient sunlight (UV light) to begin the photochemical degeneration process. Ozone levels normally recover during the Antarctic summer

(Dec.–Feb.) when CFC concentrations are less and there is more UV light to regenerate ozone levels.

Key ideas

1. Two types of polar climate can be identified: polar tundra climate where the average temperature of the warmest month is above freezing but below 10 °C, and polar ice cap climate where mean temperatures for every month are below freezing.

2. The north and south polar regions remain cold and dry because of weak year-round heating by the sun and domination by quasi-permanent surface high-pressure systems.

3. Surface high-pressure circulation over both poles is linked to changes in the intensity of upper air movement in the form of the upper westerlies (the Circumpolar Vortex).

4. When the upper westerlies weaken, more mid-latitude depressions can move into the polar region bringing wetter and less cold conditions.

5. When the upper westerlies intensify as they have done in the last 20 years or so the poles become colder and drier with fewer mid-latitude depressions moving into the region.

6. The Arctic Oscillation (AO) and the Antarctic Oscillation (AAO) are variable hemispheric-wide pressure gradients between the poles and the lower latitudes.

7. There is a close positive link between the strengthening and weakening of the AO and the AAO and the strength of the upper westerlies over each respective pole.

8. The South Pole as a whole has shown little recent evidence of global warming and associated reductions in ice extent.

9. In contrast, evidence shows that reductions in ice cover and thickness associated with possible global warming over the North Pole have taken place since the 1950s.

10. The cold polar regions provide ideal conditions for the destruction of stratospheric ozone.

11. Human-induced stratospheric ozone destruction is greater over the South Pole (since the 1970s) than over the North Pole (since the 1990s) because of the former's colder temperatures.

8.2 Longer-term climate change: ice ages

8.2.1 Describing the ice age

Current evidence of global warming and modelling predictions of future climate change (section 7.1), suggest that in a world warmed by greenhouse gases, most of the heating will occur in the high latitudes. If the high latitudes continue to warm they will have great significance for the world's environment. Global warming in polar and subpolar regions is already melting the ice caps, thinning large areas of sea ice, and permafrost is melting and shrinking in the subpolar regions. A runaway melting of the Antarctic ice shelves could raise sea levels by as much as 4–6 m in 100–200 years, flooding most of the great cities and deltas of the world with drastic consequences. Supporting evidence that the polar regions may be vulnerable to significant, even runaway, climate change in the near future can be found from rapid climate changes in the not too distant past.

From about 2.5 million years ago till about 10,000 years ago, the earth has been in an ice age. The most remarkable aspect of this ice age (called the Quaternary Ice Age) is that it has been witness to some of the most marked global/regional climate changes in the geological record. Temperature changes of up to a maximum of 16 °C between alternating hot and cold periods within the ice age event have been recorded. Such warm and cold phases have been associated with cyclical advances and recessions in the polar ice caps. Over the past 800,000 years, for instance, there have been six to seven regular main cold glacial advances and warm interglacial (between glacial) retreats (Figure 8.4).

(a) Glacial and interglacial cycles

Figure 8.5 shows temperature fluctuations over Antarctica during the last 400,000 years of earth history. The temperature data, obtained from ice core evidence in Antarctica, indicates the occurrence of four main glacial cycles of earth cooling (ice advance) and earth warming (ice retreat). It can be seen that each of the last four glacial cycles (cooling and warming period together) lasted about 100,000 years. It seems to

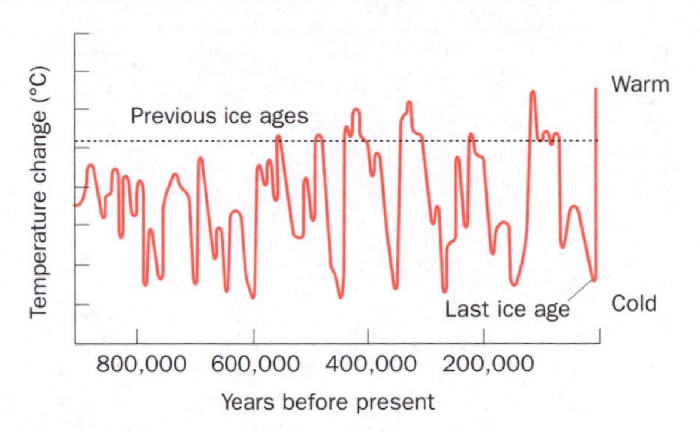

Figure 8.4 The advance and retreat of ice sheets in the northern hemisphere over the last 800,000 years. The peaks in temperature represent interglacial periods, the troughs glacial maximum conditions

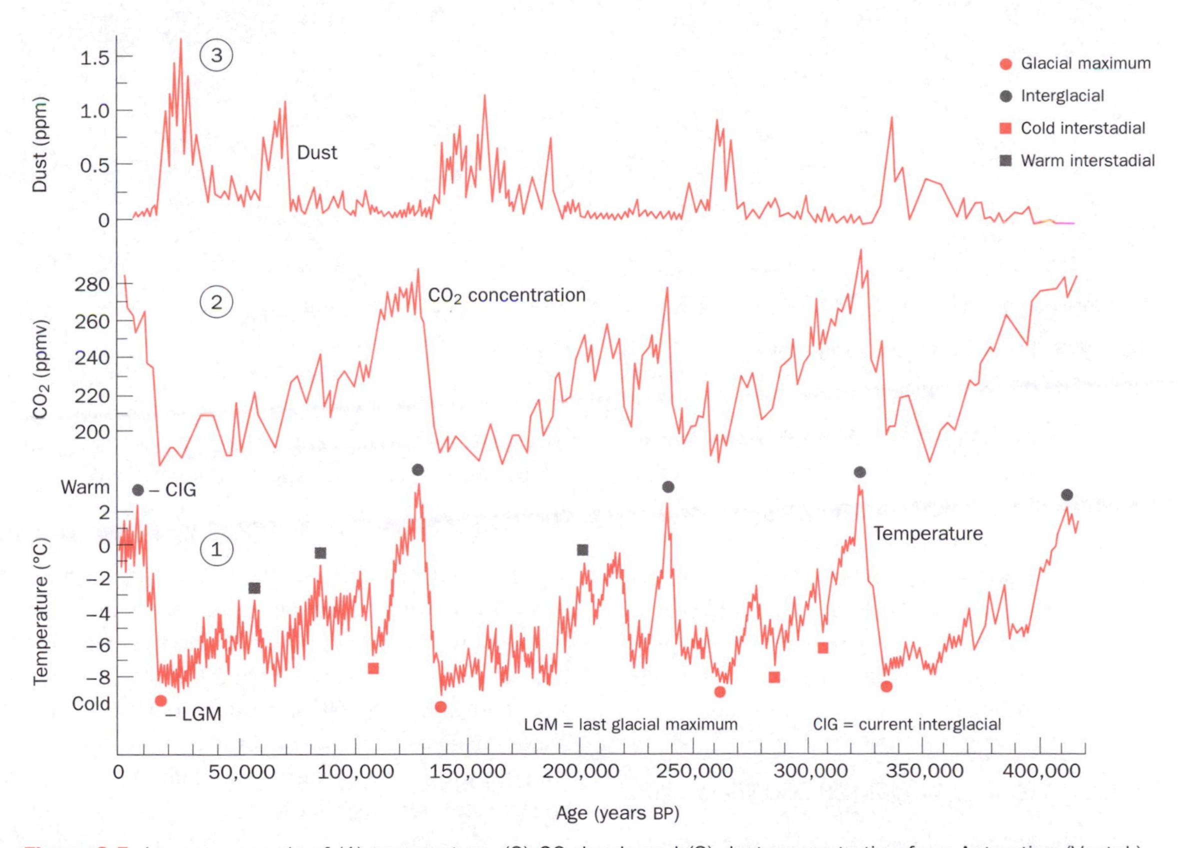

Figure 8.5 Ice core records of (1) temperature, (2) CO_2 levels and (3) dust concentration from Antarctica (Vostok) over the last 400,000 years (i.e. four glacial–interglacial cycles). Interglacial phases denoted by black circles, full glacial conditions by red circles, examples of interstadials by black squares and stadials by red squares

Source: Adapted from Petit *et al.* (1999, pp. 429–36)

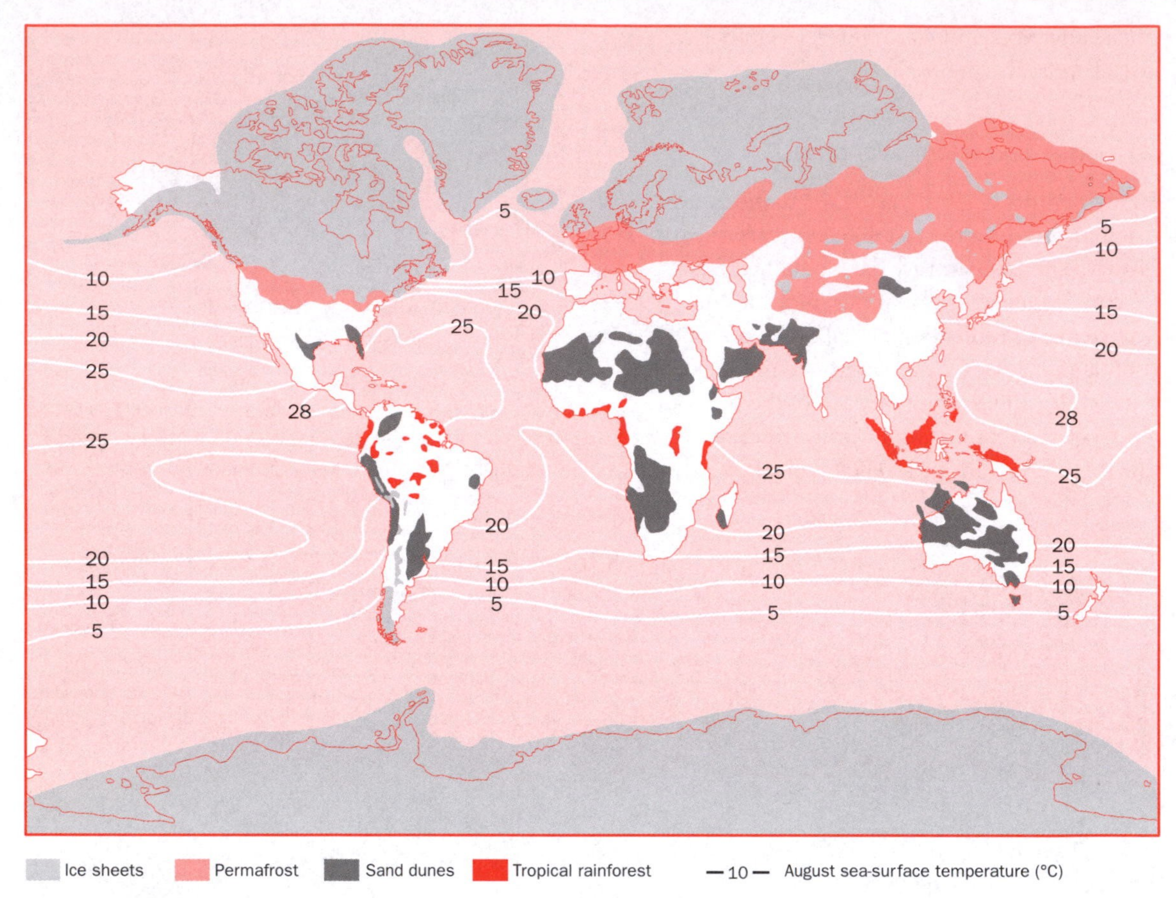

Figure 8.6 The world at full glacial maximum (FGM) 18,000 years ago
Source: Wilson *et al.* (2000, p. 18)

take about 90,000 years for the earth to cool about 8–10 °C from a warm interglacial to a state of full glacial conditions (both shown in Figure 8.5). By contrast, the warming transition from full glacial maximum conditions to an interglacial is a much shorter event lasting only several thousand years. Thus each glacial cycle is asymmetrical in shape with a long period of cooling followed by a much shorter warming event.

(b) Last glacial maximum (18,000 years ago)

During the last glacial maximum (furthest glacial advance) about 18,000 years ago (Figure 8.6), the average surface temperature of the planet fell by about 8 °C (glacial minimum temperature) compared with today. This drop in global temperature induced ice sheets to spread out apparently simultaneously from both polar regions. The ice advance was greatest, however, over the continental landmasses in the northern hemisphere, where great ice sheets eventually covered much of North America and northern Eurasia (compare Figures 8.6 and 8.7). Such glacial advance brought, and was associated with, major changes in the global climate in terms of temperatures, wind patterns, ocean circulation, and moisture/aridity conditions.

As well as larger tracts of permanently frozen ground beyond the northern glacial margins, the global climate was also much drier than today with relatively large areas of desert in the lower

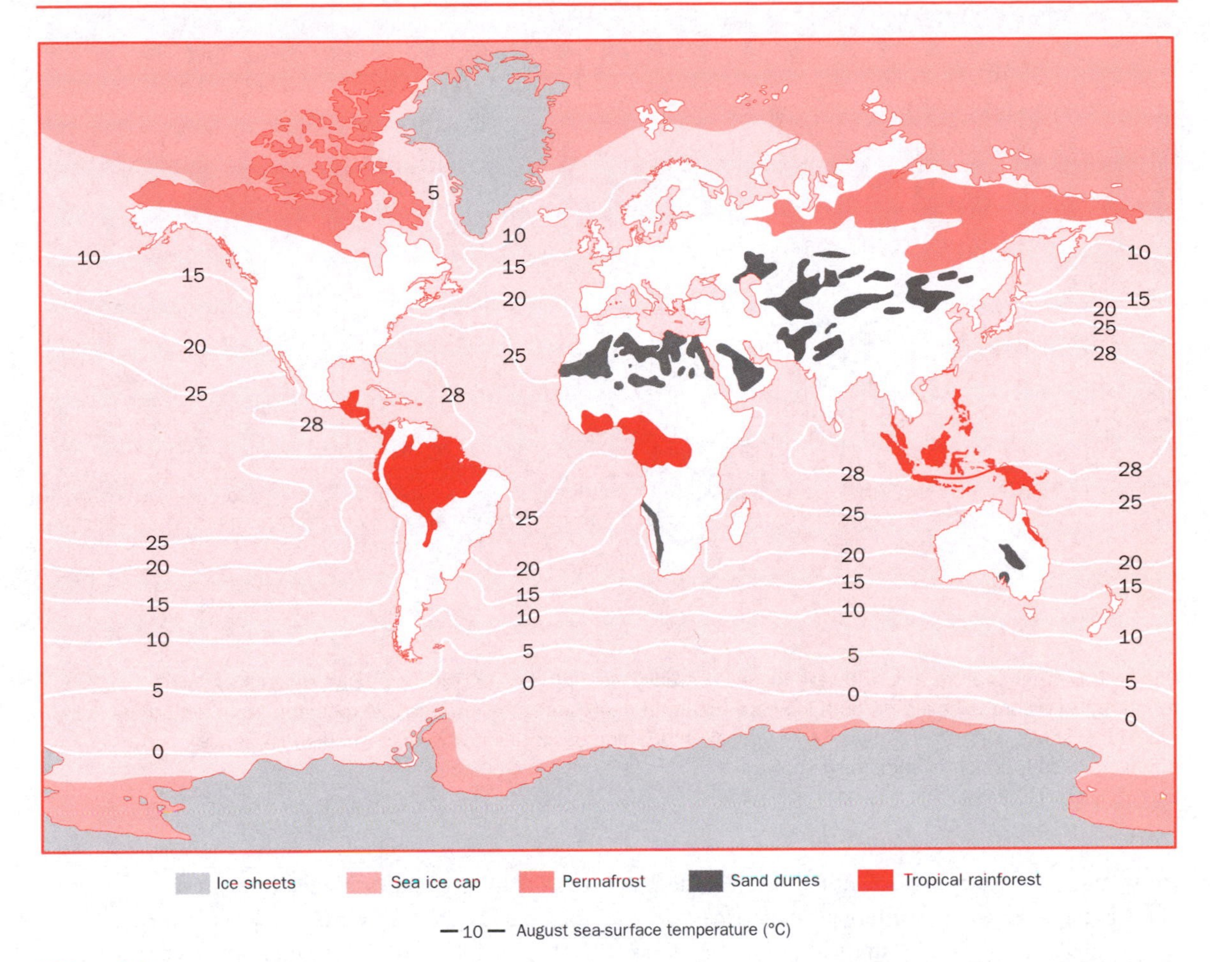

Figure 8.7 The world today
Source: Wilson *et al.* (2000, p. 19)

latitudes. We can surmise this from dust concentrations obtained from ice cores and marine sediments. Dry or desert areas are rich in dust that can be easily removed by wind currents to the high atmosphere and eventually to the polar ice sheets. Figure 8.5 shows that during the coldest parts of the glacial cycles (full glacial conditions) the concentrations of dust in the Antarctic ice core reached their highest levels. The presence of a drier climate during glacial conditions can also be suggested since a massive amount of ocean/surface water would be locked up as ice in the expanded glaciers and would reduce water vapour supplies in the air. Lower rates of evaporation from colder waters especially those in the northern oceans as well as from a reduced ocean cover (sea levels fell by an average of about 100 m compared with today) would also contribute to planetary aridity.

(c) Shorter-term events: stadials and interstadials

The temperature fluctuations over the last four glacial cycles (see Figure 8.5) also reveal a more detailed pattern of smaller-scale, shorter-lived warming and cooling events. Periods which

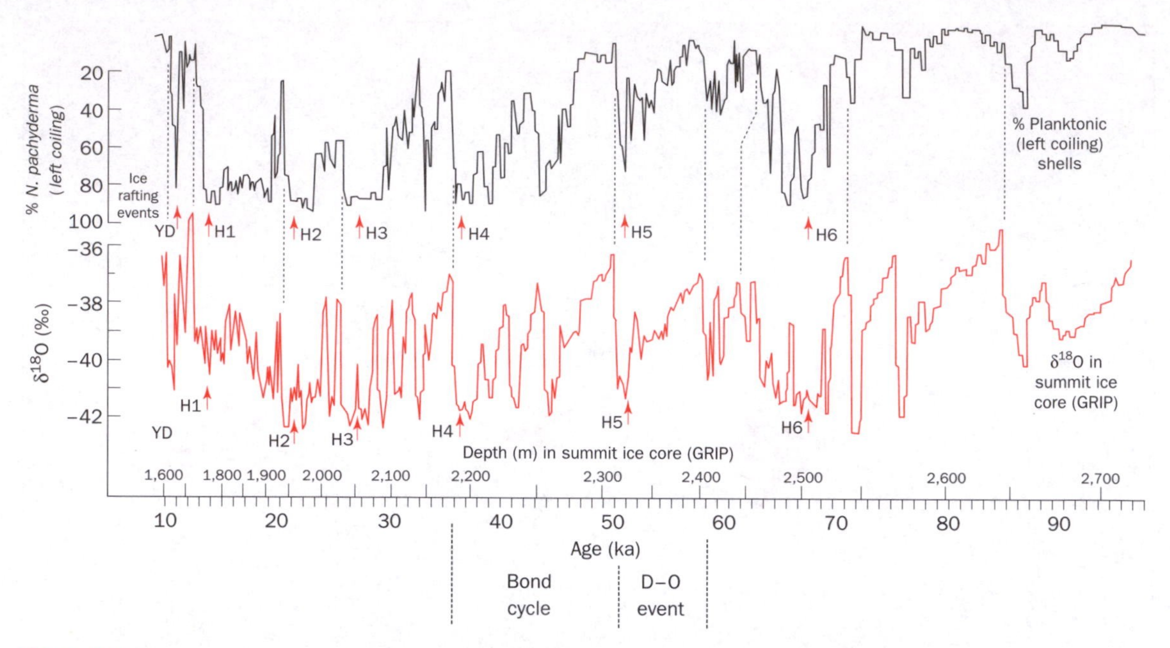

Figure 8.8 Temperature record during the last ice advance between 90,000 and 10,000 years ago. The temperature record is based on both ice core (^{18}O) and deep sea sediment analysis (percentage of left-coiling planktonic shells compared to right-coiling ones). The timing of six Heinrich events (marked H) as well as the Younger Dryas Heinrich event is also shown

Source: Bond *et al.* (1993, pp. 143–7)

show moderate warming (i.e. not as large as the warming associated with interglacials) are called interstadials (i.e. examples marked as black squares in Figure 8.5). Cold events that are not as severe as the full glacial coolings are termed stadials (i.e. examples marked as red squares in Figure 8.5).

As many as 24 interstadial–stadial events from the Greenland ice core (each warm interstadial event is linked to or ends in a colder stadial event) have occurred over the last full glacial cycle (115,000–10,000 years ago). The detailed temperature shifts in the last 21 interstadial–stadial events between 90,000 and 10,000 years ago are shown in Figures 8.8 and also 8.16. The temperature patterns shown in Figure 8.8 are based on ice core (^{18}O) data from Greenland (Greenland Ice Core Project, GRIP) backed up with ocean sediment data (the percentage of left-coiling plantonic *Pachyderma* shells) from the North Atlantic. It can be seen in Figure 8.8 that the temperature changes associated with the interstadial–stadial events that are often called Dansgaard–Oeschger (or D–O) events are very rapid. This rapid and continuous switching from warm interstadial to cold stadial conditions is a pervasive feature of the temperature record and gives the glacial cycles their well-known saw-toothed shape. More detailed analysis from Figure 8.8 reveals that the 'jump' in warmth at the start of the interstadial is very rapid indeed, taking place over a few decades or less, but is followed by a more gradual temperature decline involving a stepwise series of smaller cooling events. These more gradual cooling stages are often then followed by a very rapid (also only taking a few decades) and often large cooling event sometimes showing a temperature drop of a further 3 °C which returned conditions to the colder glacial state. These very cold stadial periods are called Heinrich events (Box 8.2). See also Plates 8.2 and 8.3.

BOX 8.2

THINKING FURTHER

Heinrich events and ice rafting

Figure 8.8 shows the six most recent cold stadial Heinrich events (H1–H6) to have been identified in the ice core record between 70,000 and 14,000 years ago. A more recent post-glacial Heinrich event can also be seen in the record beginning about 12,700 years BP and ending about 11,500 years BP. This Heinrich event is called the Younger Dryas and is identified by the symbol YD. Heinrich events appear and disappear very quickly in the ice core records. By analogy with the well-studied Younger Dryas period (see Box 8.4), they may be regarded as possibly beginning and ending with sudden climatic 'jumps' taking just a few decades (see YD and H5). In terms of their periodicity, recent detailed studies have shown that Heinrich events occurred in the North Atlantic about every 7,000–10,000 years on average between 70,000 and 10,500 years ago.

Because of their extremely cold climates, Heinrich events occurred at times of low sea surface temperatures (SSTs), low sea salinity concentrations and low ocean plant activity (low plankton productivity). Because of their extremely cold conditions, they are also associated with periods when very large numbers of icebergs were released from continental and sea shelf ice sheets from northern Canada, Greenland, Iceland, Scandinavia and the British Isles into the North Atlantic. Some of the Canadian icebergs travelled 3,000 km across the Atlantic almost reaching Ireland – an enormous distance testifying to the extreme cold of the North Atlantic seas and the very large amounts of drifting ice. Some debate exists concerning the conditions that released the icebergs in the first place. One theory is related to a build-up of geothermal warming under the growing ice sheets which eventually causes ice sheet failure and iceberg release. A second more popular theory considers climate change as the main trigger. According to this theory, the extremely cold conditions of the Heinrich event cause ice sheet expansion. As the ice sheets expand over the ocean shelves they eventually break up, releasing icebergs into the North Atlantic. It is interesting to note, however, that current iceberg release from the polar ice caps may be due to global warming (see Box 8.1), so iceberg armadas may be the result of both warm and cold periods.

As the icebergs moved southwards from the northern ice sheets across the relatively warmer North Atlantic they melted. On melting these icebergs left traces of their presence by depositing large amounts of ice-rafted debris (IRD), especially limestone fragments, in the fine-grained marine sediments of the North Atlantic. Much of this IRD is distinctive, consisting of limestones very similar to those exposed over large areas of eastern Canada today (see Plates 8.2 and 8.3). As shown in Plate 8.2, the IRD can be as thick as 40 cm close to the continental source regions, but thins as it is deposited further and further from the continental mainland. Some of the ocean bed deposits (Heinrich layers) also contain material that could only come from separate ice sheets in Greenland, Iceland, the British Isles and Scandinavia. Fragments of basaltic glass can be matched with lavas in Iceland and chalk deposits can be connected to chalk formations in the British Isles and the North Sea. The launch of the iceberg armadas was fairly frequent. High-resolution studies of ocean cores reveal that iceberg surges occurred at intervals of 1,000–3,000 years between 10,000 and 38,000 years ago and with lesser frequency of 7,000–10,000 years between 70,000 and 38,000 years BP.

The duration of each interstadial–stadial event though generally relatively short-lived varies in detail. Some of the D–O events may have been just a few decades, while others vary in length from a few centuries to several thousand years. The warm interstadials which last 500–2,000 years exhibit a temperature variation of 7 °C but are typically 5 °C cooler than at present. Finally, longer cooling cycles lasting from 5,000 to 20,000 years called Bond cycles have also been identified in the ice age record. As shown in Figure 8.8, each of these cycles has a period equal to the time between successive cold Heinrich events. They are made up of (i) a series of progressively cooler D–O events (giving a saw-toothed shaped cooling trend) and (ii) the final rapid cooling of the Heinrich event. It can also be seen that as each Bond cycle ends in a Heinrich event it is then followed by a very high amplitude warming phase.

Key ideas

1. From 2.5 million years ago till about 10,000 years ago (a period known as the Pleistocene) the earth has been in an ice age.

2. During the last 800,000 years there has been a fairly regular sequence of six to seven cold glacial advances and warm interglacial retreats.

3. It takes about 90,000 years for the earth to cool 8–10 °C from a warm interglacial period to a cold glacial period but only several thousand years to warm back to an interglacial.

4. During the last full glacial maximum (FGM) about 18,000 years ago, large ice sheets spread out from the poles over both hemispheres, but especially over the northern hemisphere where there is a greater landmass extent.

5. During a major glacial advance, tropical forest ecosystems are greatly reduced, areas of permafrost and desert increase substantially and sea levels can fall by as much as 100 m.

6. When the earth cools gradually over 90,000 years towards an FGM the climate record actually shows a wide variation in temperature with distinctly warmer periods (interstadials) alternating with distinctly colder periods (stadials).

7. When a warm interstadial phase is followed directly by a cold stadial phase, the temperature sequence is called a Dansgaard–Oeschger (D–O) cycle.

8. Very cold stadial phases associated with ice rafting and sediment deposition in the North Atlantic are termed Heinrich events.

9. A sequence of progressively colder (D–O) events is called a Bond cycle.

8.3 Explanation of ice ages

8.3.1 External and internal factors

Theories about the origins and causes of the Quaternary Ice Age and its multiple glaciations and deglaciations have been outlined in section 6.5.1. Like most theories of climate change, two broad categories of factors can be identified in the explanation of the ice age. First, there are a group of external forcing factors such as changes in the amount/distribution of sunlight falling on the earth. Second, there are a class of internal factors, i.e. ocean currents, CO_2 levels, ice extent, that are part of the natural climate system. Although the various climate change factors are analysed separately here, it should be borne in mind that they often act together with reinforcing effects (positive feedback) or dampening effects (negative feedback) on each other.

External variations in the amount of sunlight received by the earth can result directly from changes in the sun's radiation output (see the solar constant, section 6.5.2) or more indirectly from changes in the orbital characteristics of the earth in relation to the sun. Although direct changes in the sun's radiation emission have been implicated in former climate change and may be involved in current global warming (section 3.6.1), they have rarely been seen as a significant contributor to the creation of ice-house climates. In contrast, changes in amount of solar radiation falling on the earth as a result of changes in the earth's orbit around the sun have been a popular explanation of glacial and interglacial cycles.

8.3.2 External factors and Milankovitch theory

As outlined in section 6.5.2, the Milankovitch theory suggests that glacial–interglacial cycles are linked to important variations in the earth's orbit around the sun. The orbital changes in question, i.e. changes in orbital stretch (eccentricity), axial tilt and earth 'wobble' (precision), produce regular or cyclical changes in the effective distance of the earth from the sun, and hence the quantity or distribution of solar radiation received by the earth–atmosphere system. Although the distortions make only a small difference to the total amount of sunlight reaching the earth, they have a large effect on the amount of sunlight arriving at different parts of the globe, and at different times of the year. For instance, the mathematician Milankovitch believed that orbital variations producing changes in the intensity of the seasons in the northern hemisphere were the main control in the waxing and waning of northern high-latitude ice sheets. In particular, he believed that summer temperatures in the high latitudes (between 60 and 80° N), largely controlled by precessional changes, hold the key to the onset of glaciations. If, as predicted by his theory, a long cyclical series (perhaps several thousand years) of summers were cold enough (occurring every 19,000–23,000 years) winter snows would not completely melt, and so permanent snowfields would grow into glaciers. If a long cyclical series of summers were warm enough, previous winter snow and ice accumulations would begin to melt, resulting in the thinning and retreat of glaciers.

(a) The parameters combined

Milankovitch also hypothesised that the earth–atmosphere system heats up and cools down in response not only to precession-led changes in summer temperatures but as a result of the combined cyclical effects of the three orbital parameters of orbital stretch, axial tilt and earth wobble. A simple model can be presented (Figure 8.9) showing how the three parameters with their different cycle periods might act together. The model clearly shows that at certain times, the three factors (orbital stretch or *eccentricity* with 100,000-year cycle; *axial tilt* with 41,000-year cycle and *earth wobble* with a 23,000-year cycle) act positively together to increase the amount of insolation received by the earth to produce a warm or interglacial phase. At other times, the three factors combine to reduce the amount of received insolation so that a cold or glacial phase occurs. At other times the three factors may act and counteract to produce a local warming or an interstadial phase or local cooling resulting in a stadial event.

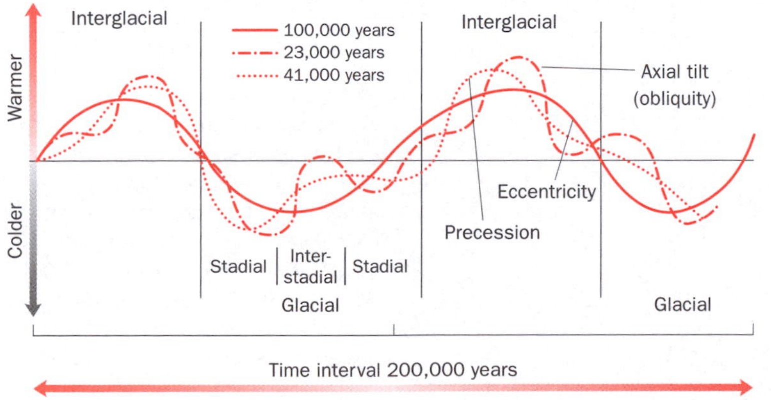

Figure 8.9 Possible combined effect of the Milankovitch orbital cycles on ice age temperatures
Source: Whyte (1995)

8.3.3 Validating the Milankovitch theory

(a) Strengths of the theory

It has been estimated that a good deal of the variance (about 60 per cent) in the record of global temperature over the last 1 million years (i.e. the period embracing ice ages) occurs close to frequencies identified in the Milankovitch theory. This suggests that the climate system is not strongly chaotic but responds in a predictable way to seasonal and regional variations in incoming solar radiation brought about by orbital forcing. We are left to conclude that the Milankovitch theory provides a valid explanation of long-term climate changes. The principal areas of association between orbital forcing and climate change during an ice age include:

1. Over the past 800,000 years, the global ice volume has peaked every 100,000 years or so (see Figure 8.4). This coincides with the eccentricity variation of 100,000 years.
2. As shown in Figures 8.5 and 8.10, ice age temperatures as well as possibly other associated parameters such as global dust concentration, sodium levels and greenhouse gas levels correspond closely to the Milankovitch periodicity of 100,000 years.
3. Smaller increases and decreases in temperature and ice volume as well as variations in other associated factors have occurred at 23,000 and 41,000 years that correspond to the precession and tilt cycles (Figures 8.5 and 8.10).
4. Current evidence shows that the Milankovitch cycles correspond to the timing of the glacial maximum.
5. The eccentricity variation (100,000 years), the only orbital change to alter the amount of radiation received, appears to drive the glacial cycles, with the other orbital variations causing smaller changes in ice volume.

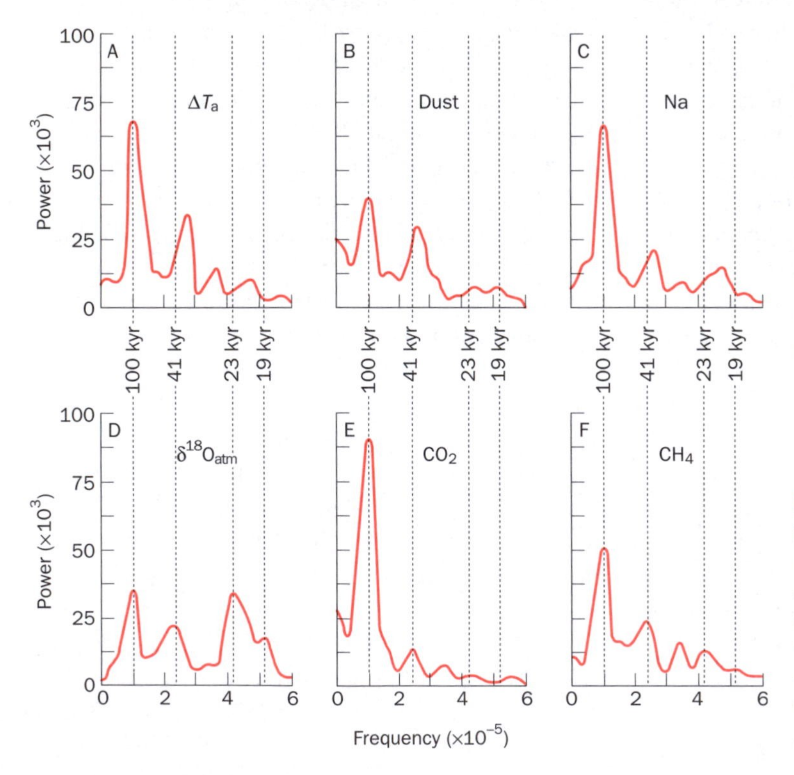

Figure 8.10 Correspondence during the ice age of spectral peaks of isotopic temperature of the atmosphere, dust content, sodium concentration and CO_2 levels with Milankovitch periodicities of 100,000, 41,000 and 23,000 years. Data from the Vostok ice core, Antarctica
Source: Petit *et al.* (1999, pp. 429–36)

(b) Problems with Milankovitch theory

Although the Milankovitch theory has been aptly described as the 'pacemaker' or 'pulse beat' of ice age climate, it nevertheless fails to explain a number of aspects in the timing and scale of a number of glacial episodes. These include:

1. The 100,000-year change in eccentricity is not entirely regular, oscillating between 96,000 and 125,000 years and even 400,000 years, whereas the climate record shows fairly regular 100,000-year cycles during the last 800,000 years.

2. The 100,000-year cycle is quantitatively the weakest of the three orbital changes in terms of its ability to alter the amount/distribution of solar radiation received, but it seems to have the greatest impact on earth's climate.

3. The astronomical cycles rise and fall smoothly (see Figure 8.9) while the ice record has a clear saw-tooth or asymmetrical shape to its growth and decay (see Figure 8.5). Ice growth typically takes 80,000–90,000 years, perhaps triggered by precessional cooling of northern hemisphere summers that fail to melt the previous winter's snow/ice accumulation. Ice growth is then quickly followed by ice decay that takes only a few thousand years in a period of precessional strengthening of northern summers.

4. The theory does not account for a change in the timing of major phases of ice advance and retreat. These seem to vary from a dominant periodicity of about 41,000 years before 800,000 years ago to the current one of 100,000 years after 800,000 years ago. Ice sheets also became larger from about 800,000 years ago.

5. The actual amount of cooling and warming during glacial episodes is too great to be caused by orbital forcing factors alone. The climate changes are often very rapid, not at all on timescales similar to astronomical cycles. The ice core data from Greenland and Antarctica indicate extremely rapid changes in temperature on occasion, by as much as 10 °C in 10 years.

6. Major cooling between interglacials can only be fully explained by other sustaining factors including ice albedo effects and CO_2 withdrawal from the atmosphere by ocean phytoplankton.

7. Mid-term and smaller increases (interstadial) and decreases (stadial) in temperature and ice volume may well correspond to the precession and tilt cycles (see Figure 8.9), but they also correspond to other agencies including variations in atmospheric CO_2 (see Figure 8.5) and methyl sulphate (see Figure 8.14) influenced by ocean plankton.

8. Many short-term, i.e. rapid, climate changes of about 5–7 °C occurring over 50, 100 and 1,000 years in the record of the northern hemisphere glaciers are not explained by the theory which is a good indicator for longer-term change (10,000–100,000 years).

9. If orbital forcing alone was the key to ice ages, continuous ice advance and retreat would be a dominant event in the geological record. Ice ages are, however, relatively rare events in earth history. Orbital theory does not explain why the last 2.5 million years of earth history provided such good conditions for the onset of ice ages, i.e. for the regular series of glacial advances and retreat.

10. According to Milankovitch theory, seasonal and regional variations in solar radiation receipt should kick-start glaciation in the northern hemisphere to be followed by the slow spread of glaciation in the southern hemisphere. Modern evidence shows that this is not the case, with southern glaciations expanding and contracting simultaneously with those in the north. Some recent work has shown that the southern hemisphere may hold the key to glaciation, with evidence of southern glaciations expanding before those in the north.

Key ideas

1. Milankovitch theory involving earth orbital variation around the sun is often used to explain ice ages.

2. Three types of earth orbital variation identified by Milankovitch include (i) eccentricity or orbital stretch, (ii) axial tilt or obliquity and (iii) the precession of the equinoxes or earth 'wobble'.

3. The most critical Milankovitch trigger for ice accumulation and the onset of an ice age relates to precession where a long series of cool northern hemisphere summers fails to sufficiently melt each winter's snowfall.

4. Milankovitch theory is a useful theory explaining about 60 per cent of temperature variations during an ice age.

5. Milankovitch theory does not explain the existence or the detailed temperature variations of an ice age and has been subject to criticism for these reasons.

8.4 Reassessing theory: Milankovitch is not enough!

8.4.1 Tertiary cooling: a vital prerequisite

While it is accepted that orbital variations in solar radiation receipt are responsible for global cooling as the earth enters a glacial episode, and for significant fluctuations in ice extent during ice ages, the theory cannot explain the existence of ice ages. It seems that there needs to be in place certain environmental prerequisites before an ice age can begin. These include a favourable disposition in the location of the continents and the right amount of earth cooling before the world can respond to orbital forcing and trigger an ice age. There have been relatively few cool intervals of earth history lasting several tens of millions of years during which glaciers repeatedly expanded and contracted. There are records of major glaciations 650–700 million years ago (Proterozoic), 450 million years ago (Ordovician) and 250–300 million years ago (Carboniferous–Permian). During the much longer intervening periods of earth history when the earth was warm (much warmer than today) orbital forcing would have had little effect on the world's climate. During the last 15 million years (Cenozoic) conditions have been cool enough for glaciations especially during the last 2.6 million years (Pleistocene). Let us now examine those factors thought to be responsible for such dramatic cooling.

8.4.2 Plate tectonics and ocean circulation

Plate tectonics and continental drift continually alter the distribution of the various continents, and the directions of the main warm and cold ocean currents. Whether the world, and in particular its extremities (the poles), become too hot for ice ages or cold enough for them, depends on the disposition of the planet's oceans and continents. Periods of earth history with open seas over one or both poles are not conducive to ice age climates: the marine polar climates so developed with warm ocean currents able to penetrate them remain too warm. On the other hand, when land covers the poles, the continental climates so generated can develop the extremes of temperature necessary for glaciation. During the last 10–20 million years, the world has been ripe for the development of repeated glaciations. This is because during the current (unique) geological era the world has not one but both of its poles covered in land: there is a continental landmass over the South Pole, and a fragmented land mass with an enclosed rim of islands effectively preventing the penetration of warm water, surrounding the North Pole. Thus, the cold continental climates which landmasses bring to polar regions are a prerequisite for global cooling and the development of glaciations.

The disposition of ocean currents and the closing of oceanic gateways have also been implicated in current debates of the earth as an ice-house climate. The polar climate with its out-blowing winds and cold conditions has relatively low amounts of precipitation. An adequate supply of snowfall to glaciated regions is thus crucial for the development of ice caps. It has been suggested that the closing of the gap between North and South America (the isthmus of Panama), about 4.6 to 2.5 million years ago, may have helped to initiate northern hemisphere glaciation by increasing snowfall over the northern Arctic region. The process could work in the following way. Closure of the Panama gateway would have prevented warm water escaping westwards from the Atlantic into the Pacific and increased the temperature of the surface waters of the Caribbean. As a result, more evaporation would have taken place over the Caribbean, eventually increasing its salinity. Increased salinity in the Caribbean could in turn have increased the strength of the Gulf Stream with more deep water formation in the North Atlantic (section 4.4.1). This whole process would have resulted in more moist air reaching colder higher latitudes and higher precipitation (snowfall) over the growing ice caps of the Arctic.

8.4.3 Tectonic uplift (mountain building)

Uplift of large areas of the earth's crust, either by vertical elevation of large crustal blocks (epeirogenic uplift) or by horizontal compression and folding (orogenic uplift) can alter climate in at least three ways by:

1. raising regions above the regional snowline/glaciation limit allowing glaciation to be initiated;
2. modifying global and regional atmospheric circulation patterns; and
3. increasing the rates of chemical (and physical) weathering in the elevated areas encouraging the sequestration of CO_2 from the atmosphere.

(a) Cold elevations

The Quaternary glaciations that began about 2.6 million years ago were initiated at the end of a 15–20 million year or so period of earth cooling. It is likely that such earth cooling resulted from the direct effects of mountain building on global temperatures. Areas in the North Atlantic subject to Quaternary glaciation including Labrador, Greenland, Britain and Scandinavia were uplifted by as much as 2 km during the Tertiary. When these areas were elevated they were raised above the regional snowline. The result was that snow accumulating on them would not have completely melted during the summer months, leading to glaciation. In addition, because these upland areas lie in the key latitude belt (60–80° N) for Milankovitch forcing they would also have accumulated snow and ice during periods with cool summers.

(b) Wind currents

Mountain building episodes have also been implicated in accelerating the onset of the ice age by altering global wind circulation patterns. Late Tertiary mountain building (15–20 million years ago) in the Himalayas and Tibet, and in the North American Rockies, had a powerful climate effect on much of the northern hemisphere (Figure 8.11). Modelling studies have shown that the mountain barrier effect of these regions (height average at 5,000 m) would make northern latitudes colder, and formerly moist areas drier. Increasing cold conditions were generated over North America by the north–south trending Rocky Mountains which effectively (i) reduced warm air streams being advected over central America from the Pacific Ocean and (ii) helped channel increased cold air flows from the Arctic south over the central plains of the USA.

The raising of the Tibetan plateau is also associated with the development of much stronger and wetter monsoons over South Asia and with increasing aridity over the

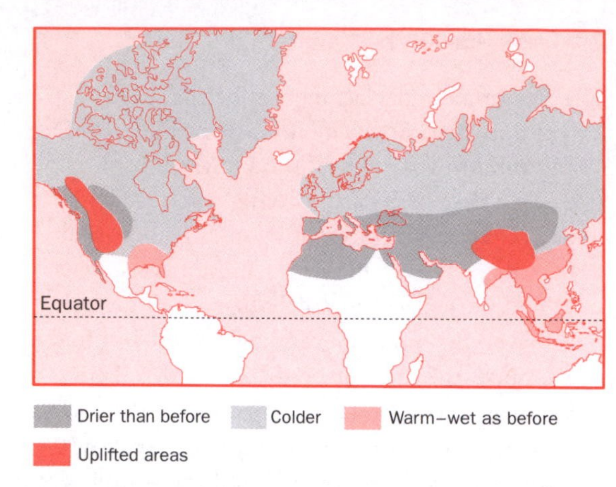

Figure 8.11 Computer simulation of climate changes induced by Tertiary uplift of the Tibetan plateau and western North America
Source: Wilson *et al.* (2000, p. 146)

Mediterranean, the Near and Middle East and over central China. Dust levels in the atmosphere in northern high latitudes increased about 3.6 million years ago (indicated from sediment cores from the north Pacific) possibly as a result of this enhanced Asian aridity. Such higher concentrations of dust in the air could have reflected more incoming solar radiation, induced cooler global conditions and helped in the initiation of the ice age.

8.4.4 Long-term changes in greenhouse gases

Mountain building by orogenic or epeirogenic uplift was not just a physical phenomenon. Chemical forces set up by mountain building particularly by the raising of the Tibetan plateau may have greatly altered the chemical composition of the atmosphere, especially in relation to the concentration of CO_2 and other greenhouse gases. Mechanisms that regulate the concentration of CO_2 in the atmosphere influence the greenhouse effect and thus the surface temperature of the planet.

(a) Geochemical processes and CO_2

Over long periods of earth history, the concentration of CO_2 in the atmosphere is determined by the action of various geochemical processes. These processes include:

1. the physical and chemical weathering of rocks that are rich in silicate minerals (e.g. clays, granite, basalt);
2. outgassing of CO_2 and other gases from volcanic activity;
3. biological productivity involving the burial of organic plant and animal remains in sediments.

Chemical weathering processes help to reduce CO_2 levels in the atmosphere. Rain removes CO_2 from the air as dilute carbonic acid (H_2CO_3). At the surface, this acid combines with silicate minerals (clays) to convert the CO_2 to bicarbonate (HCO_3). These bicarbonates are then dissolved in water and carried to the oceans where they are incorporated into the shells of living creatures. Eventually on their death these sea creatures will be deposited on the ocean floor and laid down as sedimentary rocks.

Carbon dioxide is returned to the atmosphere by tectonic activity. Great temperatures and pressures deep in the earth's moving crust compress, buckle and melt the sedimentary layers, releasing CO_2. The CO_2 is returned (outgassed) back to the atmosphere at mid-ocean ridges (in association with ocean floor spreading) and volcanoes (at the continental plate edges). Normally, during warmer wetter geological periods more chemical weathering occurs, with more CO_2 being lost from the atmosphere to the sedimentary rocks. This loss of CO_2 has a regulatory (negative feedback) cooling effect on the earth–atmosphere system. During cooler geological periods, less chemical weathering produces a net return of CO_2 to the air, thus raising global temperatures.

The dramatic decline in global temperature from the late Tertiary (20 million years ago) to 2.5 million years ago that resulted in the Quaternary glaciations may also be attributed to the action of the geochemical model. Certain factors for instance may have made the Tibetan plateau and its river systems one of the most powerful of the earth's weather machines, and by helping to remove CO_2 from the atmosphere, one of its most efficient cooling machines. The large territorial extent of the Tibetan plateau and its mean elevation today of some 4 km suggest that its uplift was an exceptional event in earth history. High rates of chemical weathering with the removal of carbonate by rivers to the sea have long been known to take place in warm humid climates, for example on Tibet's warm and wet southern margins. Recent work has revealed that relatively fast rates (faster than expected) of chemical weathering also operate over Tibet's extensive higher and colder areas. It can be assumed therefore that during the uplift of Tibet, chemical weathering would have removed huge amounts of CO_2/carbonate sediments and carried them to the Indian Ocean by the Ganges and Indus river systems. Subsequent rapid sedimentation rates in the Indian Ocean ensured the long-term burial and sequestration of carbon-rich silicate rocks and thus the removal of CO_2 from the atmosphere.

Key ideas

1. Several factors operating over long geological time periods are thought to be necessary for an ice age.
2. These include (i) continental drift resulting in 'landmass' effects and limited warm ocean circulation across both polar regions, (ii) tectonic activity causing more extensive highland zones with colder climates, and (iii) enhanced geochemical weathering resulting in low global atmospheric CO_2 concentrations.

8.5 Reinforcing the Milankovitch signal

Another problem with Milankovitch theory is that orbital changes in the receipt of radiation are not sufficient by themselves to explain either the scale or the details of the cooling and warming involved. It is now thought that other factors internal to the climate system act to augment and sustain the orbital effects through positive and negative feedback mechanisms into large-scale climate change. These factors include ice albedo effects and biological and hydrological factors linked to CO_2/methyl sulphide changes during the Quaternary Ice Age.

8.5.1 Ice albedo effects

One of the most popular theories is the ice albedo mechanism (Figure 8.12). This hypothesis suggests a positive feedback mechanism between growing ice sheets, increased solar reflection, climate cooling and more ice build-up. When there is relatively little sunlight falling on ice sheets during the summer – for instance during a Milankovitch cycle of cool northern hemisphere summers – they do not have a chance to melt, and would keep growing year after year from fresh snow accumulations. A big growing ice sheet cools the surrounding region both by directly chilling the air that blows over it, and by reflecting more of the sun's warmth back to space. The cooler temperatures created would in turn encourage more ice accumulation and greater solar radiation reflection and loss. More solar radiation reflection would continue to cool the planet and increase ice accumulation. After being initiated by a period of cool northern hemisphere summers (at I_1 in Figure 8.12(a)) the expanding ice sheets would eventually – in 80,000–90,000 years or so – grow into the massive ice sheets typical of a full-scale ice age. The positive reinforcing albedo effect of the growing ice sheet would not be interrupted by the occurrence of subsequent episodes of warmer Milankovitch summers.

The ice albedo theory also contains a positive feedback mechanism which sounds the death knell of the large glacial continental ice sheet. Ice sheets will only continue to grow if precipitation exceeds ice/snow melting. During periods of maximum glaciation (at I_2 in Figure 8.12(a)), ice sheets (especially in the northern hemisphere) occupied

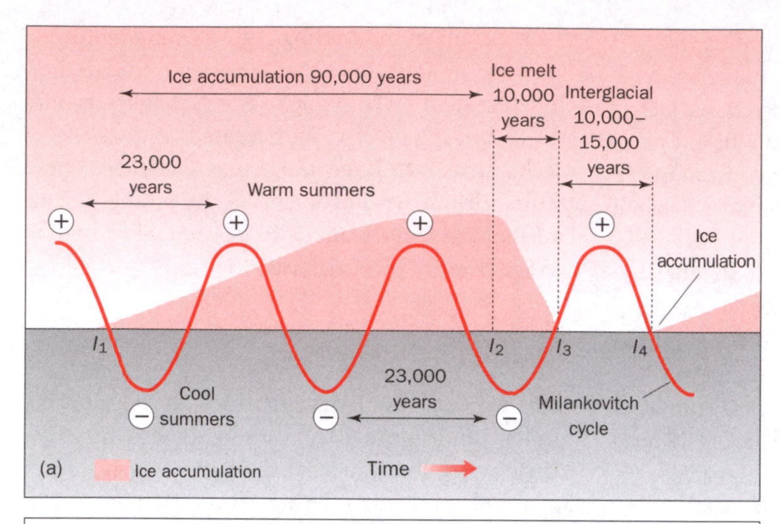

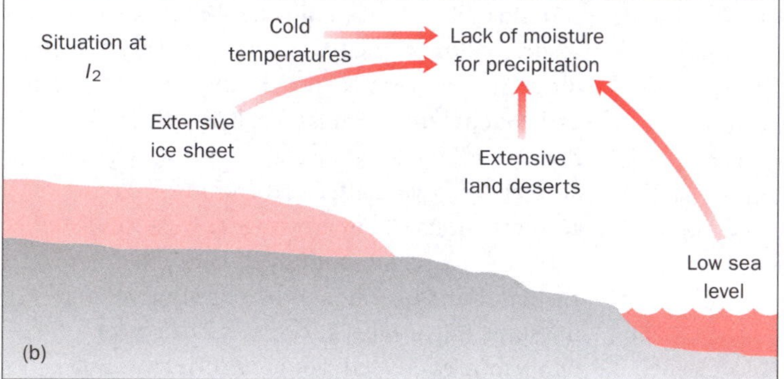

Figure 8.12 (a) Schematic diagram showing the relationship between ice accumulation and loss and the possible effects of ice albedo and Milankovitch (orbital) forcing; (b) conditions which would initiate ice sheet decline during an ice age

their greatest extent, sea levels were at their lowest level and the global climate was at its most arid. These changes all served to minimise the supply of moisture and precipitation to the ice sheet (Figure 8.12(b)) and effectively prevented it from growing further. After ice sheet stabilisation, all it would take to begin the process of ice thinning would be a favourable (Milankovitch related) series of warm northern hemisphere summers. With warmer summers more ice would begin to melt (I_2 to I_3) than would be replaced by snowfall. This would allow the thinning and contracting ice sheet to reflect less of the sun's heat back to space. Moreover, greater amounts of solar energy would be absorbed by the increasing expanse of ice-free surfaces around the melting ice sheet and ambient temperatures would increase. As temperatures rose more ice would melt, further accelerating the absorption of sunlight and the raising of temperatures. This positive feedback cycle of shrinking ice sheets, higher temperatures and further ice shrinking during the end of a glacial phase would continue until the system moved into an interglacial (I_3 to I_4). During the warm interglacial the retreating (and now small polar located) ice sheet would finally stabilise as a result of precipitation supplies and ice accumulation being once more equal to the prevailing rates of ice melt. As shown from the simplified model in Figure 8.12, ice sheet melting and thinning is a much more rapid process than ice sheet growth, taking only several thousand years to reach full interglacial, i.e. non-ice sheet advance conditions.

8.5.2 CO_2 and glacial–interglacial cycles

Analysis of air trapped in the Vostok polar ice cores (section 6.4.1) has revealed a fluctuation in atmospheric CO_2 (and other greenhouse gases such as methane) during glacial–interglacial cycles. Figure 8.5 shows a close correspondence between atmospheric concentrations of CO_2 and changes in temperature throughout the last 400,000 years. Low concentrations of CO_2 of 180–200 ppm occur during glacial episodes and high levels of 280–300 ppm during interglacials. The corresponding figures for methane are 300–350 ppb during glacials and 650–700 ppb during interglacials. The present ambient concentration of CO_2 is about 370 ppm as a result of fossil fuel burning and land-use change. This concentration is more than 25 per cent higher than concentrations during interglacials and throughout pre-industrial times.

The statistical association between levels of atmospheric CO_2 and ambient temperatures during the last four main glacial–interglacial cycles is so close (closer than the Milankovitch signals) that mechanisms explaining the links between the two systems have developed. As we shall see, however, much uncertainty still exists concerning the CO_2/temperature fluctuation cycles. We are still not sure about precise cause and effect mechanisms. For instance, which event comes first: do ambient temperatures change first to be followed and reinforced by altered CO_2 concentrations, or do CO_2 levels alter first, helping to drive the glacial–intergacial temperatures?.

The terrestrial biosphere stores more carbon during interglacial periods than during glacial periods. Therefore, the terrestrial biosphere is unlikely to be the source or cause of differences in atmospheric CO_2 between glacial and interglacial periods. The cause must lie therefore in the ocean. Two main hypotheses are used to explain glacial–interglacial variations in atmospheric CO_2. These are (i) dissolution and outgassing of CO_2 in the ocean, (ii) nutrient enrichment of marine ecosystems in high latitudes.

(a) Dissolution and outgassing

Carbon dioxide gas is more soluble in cold water than in warm water. This process supports the evidence which shows that atmospheric CO_2 levels were lower in cold glacial than in warm interglacial periods. However, CO_2 is also less soluble in salty water than in fresh water. It is thought that salty water was more prevalent during glacial than interglacial periods. Thus, each of these effects may cancel each other out. It has been proposed that extended sea-ice layers during glacial times place a 'cap' on the ocean which prevents the outgassing of upwelled CO_2-rich waters, especially around the Antarctic. This mechanism could explain the parallel increases of Antarctic temperature and CO_2 during deglaciation.

(b) Nutrient enrichment of marine ecosystems

Billions of tiny plants called phytoplankton living in the surface sunlit layers of the ocean utilise carbon during their growth by a process known as photosynthesis. The carbon that is fixed into the organic tissues of these plants is derived from the CO_2 in the atmosphere or dissolved in the oceans. One theory maintains that variations in the growth rate of phytoplankton in the upper layers of the world's oceans can drastically alter atmospheric concentrations of CO_2 during the glacial–interglacial sequences. This theory is termed the 'biological pump' (Box 8.3).

BOX 8.3

THINKING FURTHER

The biological pump

This theory suggests that microscopic plant (phytoplankton) growth in the oceans can drastically affect the concentration of CO_2 in the atmosphere and thus the temperature of the planet. When phytoplankton activity is high, CO_2 is progressively removed from the atmosphere; when such activity is low, CO_2 is returned to the atmosphere. The main limiting factor controlling the productivity of ocean plankton today, as in glacial times, is not the concentration of CO_2 in the water or even ocean temperatures, but the supply of available mineral nutrients in the ocean waters such as Ca, NO_3, PO_4, Fe and Mg. Nutrient availability, and therefore plankton productivity, is generally greatest in Arctic and cold ocean areas rather than in warm oceans because of ocean layer instability and the processes of seasonal overturning and upwelling.

In temperate and high latitudes, water rich in nutrients is transferred from deep water to levels near the surface during winter. This is because the cold winter surface water (chilled during the winter) becomes heavier and sinks below the warmer less dense water in the layers below which eventually rise to the surface.

Such seasonal overturning of water levels ensure a rich supply of nutrients to the sunlit surface. As sunlight increases in the following spring an explosive growth of the phytoplankton population occurs known as the 'spring bloom'.

There are other areas of the ocean where cold nutrient-rich water is available at the surface for sunlight to stimulate phytoplankton growth. These are the cold upwelling zones associated with cold ocean currents. One such area of plankton-rich waters lies off the coast of Peru and northern Chile. Here dynamic processes including strong offshore easterly winds, rather than temperature density contrasts, push warm surface ocean water to the west and replace it with a supply of deeper nutrient-rich cold water from below. In contrast, open warm tropical oceans are deficient in nutrients and show low rates of plankton productivity. Compared to the cold oceans, tropical waters are 'biological deserts'. This is because the top photic layers of such oceans remain nutrient deficient and therefore plankton free. This is a result of year-long ocean surface heating by the tropical sun which keeps the warmer lighter top waters more or less permanently at the surface. With little overturning or upwelling of nutrients from deeper layers to the upper layers, supplies of surface nutrients are quickly used up by the phytoplankton.

Further evidence for the varying action of the biological pump between glacials (high activity) and interglacials (low activity) comes from another substance locked in the air bubbles of ice sheets. This is the gas methyl sulphuric acid that originates from decaying ocean plankton. Its concentration is a sign of the rate of phytoplankton activity: high levels indicate a lot of growth and decay while low concentrations point to the opposite effect. As the global temperature rose at the end of the last glacial maximum about 18,000 years BP and as CO_2 in the atmosphere rose, the methyl sulphuric acid concentration decreased.

(c) The biological pump

As the phytoplankton grow and reproduce, they are consumed by herbivorous animal plankton (zooplankton) and these in turn are eaten by carnivorous zooplankton and so on up the food chain. Plant and animal debris from these living systems sinks in the ocean (Figure 8.13). Some decomposes and is released as mineral nutrients back to the surface waters. A small fraction (perhaps 1 per cent) falls to the ocean floor where it can be buried for hundreds, thousands and even for millions of years. The accumulative net effect of the biological pump is therefore to remove carbon from the atmosphere to the lower depths of the ocean and eventually into deep ocean sediments. The loss of CO_2 can be a continuous process. As the amount of carbon in the surface waters is reduced more CO_2 from the atmosphere can be drawn down into the water to restore the previous balance. What is hypothesised therefore is that when phytoplankton activity is high, CO_2 is removed from the atmosphere and concentrations fall; when phytoplankton activity is low, CO_2 is returned to the atmosphere and its concentration rises.

(d) Phytoplankton and methyl sulphide

Increases in glacial ocean productivity may be associated with the initiation of another feedback mechanism, in addition to the productivity–CO_2 connection. When phytoplankton activity is high, higher amounts of non-sea salt (NSS) sulphates are produced by marine plankton. Ice-core records show that high levels of NSS sulphate are deposited within snow and ice when temperatures are cooler, and lower NSS sulphate amounts when warmer temperatures exist (Figure 8.14). NSS sulphates form dimethyl sulphide aerosols in the atmosphere. These sulphate aerosols are the primary source of cloud condensation nuclei (CCN) in the marine atmosphere. Great ocean productivity would therefore increase cloud cover, increasing planetary albedo, and initiating further global cooling. Nevertheless, the magnitude

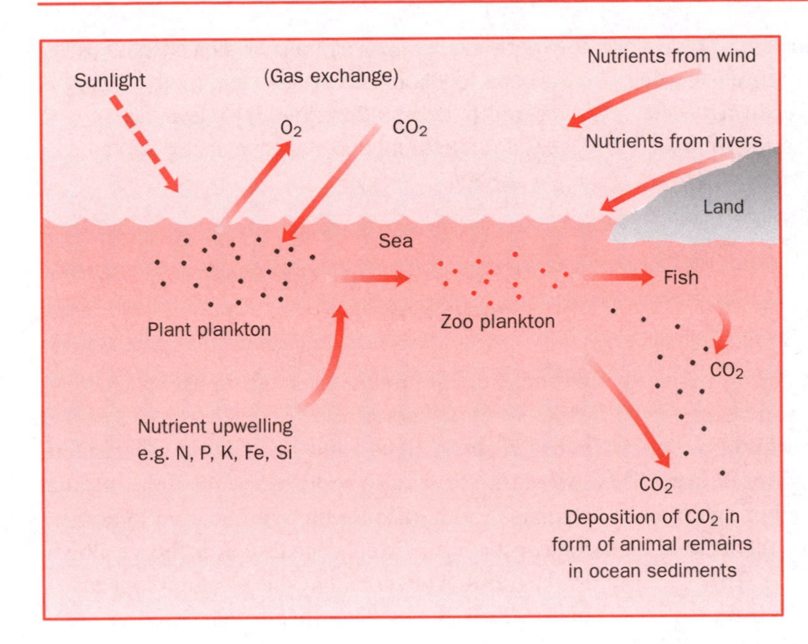

Figure 8.13 Schematic model showing the removal of CO_2 from the atmosphere under enhanced rates of ocean biological productivity (the biological pump)

(e) Factors in high glacial productivity

A number of hypotheses have been suggested to support the idea that phytoplankton activity could have been accelerated during cold glacial phases. Most of these theories relate to increasing the supply of nutrients (a limiting factor in phytoplankton growth) in the surface layers of the ocean where the phytoplankton are most concentrated.

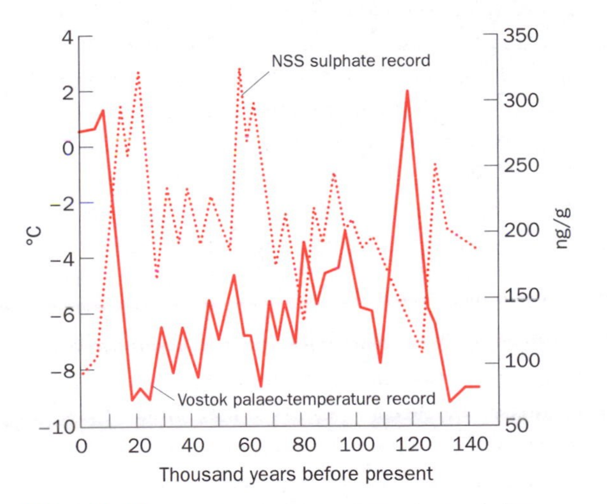

Figure 8.14 The relationship between Antarctic ice core temperature and non-sea salt (NSS) sulphate
Source: Global Climate Change Student Guide, ARIC, Dept of Environmental and Geographical Sciences, Manchester Metropolitan University

of this feedback is uncertain since increased cloud cover may enhance greenhouse warming of the lower atmosphere. Thus the reality of this mechanism must remain for the time being in some doubt.

(i) Thermal overturning

During the colder periods of the Ice Age (glacials and stadials), slow but progressive cooling of the surface layers of the ocean could have led to ocean layer instability over wide areas. The enhanced nutrient supply associated with such regions of ocean overturning would have stimulated biological productivity. Enhanced phytoplankton production in turn would have been responsible for maintaining CO_2 at the lower level of concentration, and therefore in sustaining lower glacial temperatures. In contrast, warm and stable ocean surface waters associated with the higher temperatures of interglacial climates would be by comparison deficient in nutrients and plankton production. With such environmental conditions, CO_2 would be released from the oceans to the atmosphere where under higher concentration they would sustain the warmer temperatures.

(ii) Wind–ocean overturning

During cold glacial phases, temperature and therefore pressure gradients between the equator and the poles (see Figures 8.6 and 8.7) would have been greater than in warmer interglacial periods, resulting in a more invigorated global wind circulation system. Greater wind strengths, including stronger trade winds, could have overturned the ocean surface waters more easily, leading to more effective nutrient upwelling and phytoplankton activity during the cold glacial periods.

(iii) The iron-enrichment hypothesis

Experiments have suggested that ocean phytoplankton growth may be limited because of deficiencies in the supply of certain key nutrients like iron, phosphorus and silicates. Some areas of the oceans, especially the oceans around Antarctica, have very low concentrations of iron. If these areas had more iron dissolved in the water during the ice ages than they do now there could be more biological activity in the water. This could withdraw large amounts of CO_2 from the atmosphere and cool the planet. With more arid climatic conditions and windier conditions prevailing during glacial episodes, more desert dust rich in iron could have been blown from the land to the oceans. Such iron enrichment could have fertilised the southern oceans, dropped ambient CO_2 levels and helped cool the planet.

(iv) The ocean-shelf phosphorus theory

During glacial episodes, sea levels fell at least 100 m, exposing a much greater area of continental shelf to subaerial erosion. Rivers may have been less active and carried smaller mean and peak discharges during colder and drier glacial phases. Nevertheless, wide areas of freshly exposed continental shelf would have been susceptible to wind and river erosion and provided a rich nutrient source including phosphorus for the ocean phytoplankton.

Key ideas

1. It is thought that a number of other factors working over medium-term scales would be necessary to reinforce and sustain the development of an ice age triggered by orbital forcing.
2. These include cooling mechanisms such as the ice albedo feedback mechanism and the biological pump which lowers atmospheric CO_2 levels as a result of nutrient-enhanced phytoplankton production in the oceans.

8.6 Shorter-term fluctuations and the ocean conveyor

Orbital climate forcing mechanisms associated with Milankovitch theory operate over timescales of between about 10,000 and 100,000 years. The same sort of timescales also apply to several of the internal feedback mechanisms which can be used to boost the effects of the orbital signal such as the ice albedo effect and to a lesser extent the ocean biological pump. These factors can therefore be used in combination to explain the 100,000-year glacial–interglacial cycles and the subcycles of 43,000 and 23,000 years. As shown in Figure 8.8, ice core and marine sediment records from the North Atlantic region show that climate during the last glacial period oscillated rapidly between cold stadial (D–O and Heinrich events) and warm interstadial states that lasted several thousand years to several decades. For these shorter-term temperature fluctuations, the role of internal factors becomes more significant than any external solar forcing. Various mechanisms involving changes in ocean circulation, together with modifications in atmospheric concentrations of greenhouse gases or haze particles, and ice albedo forcing of land and sea-ice extent, have been invoked to explain these sudden regional and global climate transitions.

8.6.1 Global thermohaline circulation

Unlike the continents, the world's oceans are physically continuous. They are linked by a global system of surface and deep water currents (Figures 4.18 and 4.19 and Box 4.5). Because these currents shift heat and salt around the planet they comprise what is termed the global or ocean thermohaline conveyor belt (section 4.4.1). The world's surface currents (e.g. Gulf Stream, Kiro Siwu, Humboldt) are mainly driven by wind and are linked to the deep ocean currents (e.g. along the bottom of the Atlantic, around Antarctica) by ocean density contrasts brought about by variations in sea water saltiness and temperature. For instance, the northward-pushing Gulf Stream is linked to a deeper ocean current flowing southward along the floor of the Atlantic at a point in the North Atlantic where heavy salty surface Gulf Stream water descends to meet the deep ocean current.

The circulation of the North Atlantic (as part of this global conveyor belt) is thought to play a major role in either triggering or amplifying the rapid climate changes recorded in the Greenland glacial record (see Figure 8.8). The workings of the North Atlantic conveyor belt and its association with climate change, will now be re-examined in the light of the previously described detailed changes in temperature during the glacial period.

(a) Gulf Stream action: current mechanisms

The Gulf Stream initially and then the North Atlantic Drift (NAD) carry warm and relatively salty water from the Caribbean north-eastwards to the seas between Greenland, Iceland and Norway (see Box 9.3). This dense salty water becomes even more dense as it cools by evaporation in its northward journey so that by the time it reaches the northern North Atlantic region it is both salty and cold, and therefore dense enough to sink down into the deep ocean. The 'pull' exerted by this descending deep water helps to maintain the strength of the Gulf Stream, ensuring a current of warm surface water to the North Atlantic. Westerly winds blowing over this warm water send mild air masses across the European continent. These mild air streams help to keep Europe much warmer, especially in winter (by as much as 10 °C), than the region would be given its high latitudinal position. Estimates suggest that as much as 50 per cent of the region's winter energy budget (the remaining 50 per cent comes directly from the sun) is derived from the action of the warm Gulf Stream.

(b) Melting ice and Gulf Stream shutdown

During warm interglacial phases, Gulf Stream action and deep water formation in the extreme North Atlantic, i.e. in the Greenland–Iceland and Norwegian Seas, are strong. As a result, continental and sea ice sheets continue to melt and retreat, releasing massive amounts of fresh water into the northern oceans (including the Arctic). Freshwater incursions from melting ice sheets over North America or from armadas of icebergs across the North Atlantic eventually dilute the waters of the North Atlantic making them less salty. Less salty and therefore less dense water conditions reduce deep water formation and Gulf Stream dominance, and initiate cooling in the North Atlantic region. Cooling proceeds because with the North Atlantic conveyor shut down (or being less vigorous) less heat is advected by ocean currents and westerly winds to the North Atlantic from the Caribbean. As cooling proceeds ice sheets develop and grow across the North Atlantic. With continued cooling and ice sheet expansion (such as during a cold Heinrich event) the growing ice sheets eventually break up as they spread out over the ocean shelves helping to release more icebergs into the North Atlantic. Upon melting over the North Atlantic, these icebergs release further fresh water into the surrounding ocean helping to sustain the prevention of deep water formation and the penetration of the Gulf Stream in the North Atlantic.

Model simulations have shown that the whole process of conveyor belt shutdown, as outlined above, could occur very rapidly, in the space of a few decades, or even over several years. The result would be a very sudden climate change to colder conditions with a decline, for instance, in the northern hemisphere of mean annual temperatures of 5–8 °C. If this is the case then this mechanism of conveyor belt shutdown could be used to explain the many rapid cooling events that have happened in the region during the last glacial event (100,000 years) and possibly during previous glacial epochs (last 400,000 years).

(c) Other feedbacks

The colder climate would also be drier because less moisture would reach Europe from a cold sea rather than from a warm Gulf Stream. The increasingly extensive ice sheets of the North Atlantic (Greenland, Scandinavia) would further cool the region, giving a massive high-pressure zone that would be additionally effective in diverting warm Gulf Stream air and moisture away from the mid latitudes of Europe. With the biological pump possibly more active during cold phases, rapidly falling concentrations of atmospheric greenhouse gases would reinforce and perhaps accelerate the cooling trend further. Evidence for this can be found in measurements of air trapped in the Greenland ice cores (GRIP) that have exposed marked fluctuations in CO_2 and methane during more than 20 shorter-term cold stadial/warm interstadial sequences. For instance, levels of CO_2 have been found to be up to 20 per cent less and methane up to 30 per cent less during the Younger Dryas cold period (12,700–11,500 years ago) than during the preceding and subsequent interstadials (Box 8.4 and Figure 8.15). A final feedback mechanism accelerating the

BOX 8.4

CASE STUDY

The Younger Dryas Heinrich event

(a) Warmth before the cold

The Younger Dryas cold period lasted between about 12,700 and 11,500 years ago (Figure 8.15). It is so called because of the rapid spread of an arctic/alpine flower, *Dryas octopetala*, over the UK and much of Europe when tundra conditions returned. As shown in Figure 8.15, for 6,000 years before the start of this event (from 18,000 years ago) the earth had been warming up from the end of the last ice age. This warming and associated deglaciation was helped from about 13,000 BP when the astronomical cycle of solar radiation increased receipts of solar radiation substantially in the northern hemisphere. The Bølling interstadial, a particularly warm event lasting from 14,500 to 13,000 years ago (Figure 8.15), shows that part of the early interglacial warming had been very fast. In the warmer interglacial conditions before the Younger Dryas, trees had begun to colonise the UK from Europe. Coniferous forest was widespread in Scotland and much of England was covered in deciduous woodland. The Scottish Highlands may have lost all their glacial ice and therefore been completely deglaciated.

(b) The cold plunge

During the Younger Dryas period, the polar front which had been pushed as far north as Iceland during the previous Bølling interstadial, shifted south again towards Iberia. As a consequence, temperatures fell by as much as 7 °C in a mere 50 years and subarctic/tundra conditions returned to the UK and much of northern Europe. Winter (January) temperatures could have been as low as –17 to –20 °C for much of Britain, while summer (mean July) temperatures may have been no higher than 8–10 °C. Such low temperatures destroyed much of the forest and woodland vegetation that had re-established itself over the UK. With the return of colder conditions and higher snowfalls, a major ice cap up to 600 m thick built up over the central Scottish Highlands while many small glaciers occupied valleys in the Southern Uplands, the Lake District, Snowdonia and even the Brecon Beacons. Broad regions of the earth including South America and the Antarctic as well as the North Atlantic (Figure 8.15) experienced almost synchronous climate changes (rapid cooling and warming) over a period of about 30–50 years, showing that the Younger Dryas and perhaps other Heinrich episodes were global events. Cold conditions ▶

in the higher latitudes of the northern hemisphere were matched by a weakening monsoon over Africa and India (see Chapter 11). This resulted in a contemporary phase of aridity in East Africa, accompanied by abrupt falls in lake levels. Climate instability in regions as far away as the Andes is indicated by shifts between cold phase grassland and warmer phase forest at high altitudes. However, there is increasing evidence that cooling and warming trends associated with the Younger Dryas period in the Antarctic known as the Antarctic Cold Reversal *preceded* the Younger Dryas cooling event in the North Atlantic by at least 1,000 years (Figures 8.15 and 8.16). In addition, movement into and out of the Antarctic Cold Reversal was a much smoother event, revealing smaller and less abrupt temperature changes. It seems increasingly likely that the Southern Ocean provides the main regulator of the major oscillations in climate over the past 2.5 million years.

(c) Abrupt return to warmth

The northward shift of the polar front and the return to warm climate conditions over the North Atlantic at the end of the Younger Dryas (11,500 years ago) was also very abrupt. The Greenland ice core and North Atlantic land/sea sediment records show that central Greenland temperatures rose by as much as 7 °C or more in a few decades. Positive feedback may have contributed to the rapidity of these climate changes, with growing swamps and biotic activity contributing considerable methane and CO_2 emissions and thus increasing the natural greenhouse effect, as climate warmed. The event was also associated with decreased storminess shown by a dramatic fall in dust in the ice core and an increase in precipitation by about 50 per cent. Pollen records from peat bogs in northern Europe and the UK show that a forest cover – first coniferous and then deciduous – once again returned to the UK as grassland and tundra vegetation receded.

(d) What caused the Younger Dryas?

The Younger Dryas was probably a consequence of a sudden influx of fresh water from the decaying ice sheet in North America into the North Atlantic. For most of the period of retreat of the Laurentide ice sheet, meltwater would have found its way to the Gulf of Mexico via what was to become the Mississippi–Missouri river system. But with retreat northwards to the present US–Canada border, the meltwater found a new means of exiting – the Gulf of St Lawrence. This sudden influx of light, cold, fresh water must have spread out above the heavier more saline Atlantic water and effectively shut off the Atlantic conveyor belt. The sinking of Atlantic water to power the oceanic conveyor belt must have occurred much further south, plunging Europe back into glacial conditions. Cold surface waters would have chilled and stabilised the air above and inhibited the formation of depressions. The glacial anticyclone that developed would have formed a block preventing the ingress of maritime air into Europe.

The overturning time for the ocean is approximately 1,000 years (which was also roughly how long the Younger Dryas cooling lasted) and feedback to the atmosphere may thus take place several hundred years after a major oceanic change has occurred. This makes for difficulties in incorporating the ocean into climate models based on predicting the response of the atmosphere to new conditions such as increased greenhouse gas loading. However, it is the speed of climate change when it does occur, and the unexpected turns which can be taken to produce the Younger Dryas event, which is a salutary lesson of how much is unknown regarding the operation of the climate system in its broadest context. The lesson to be learned is that climate surprises have occurred in the recent past and are equally likely to be features of the immediate future as 'greenhouse earth' evolves.

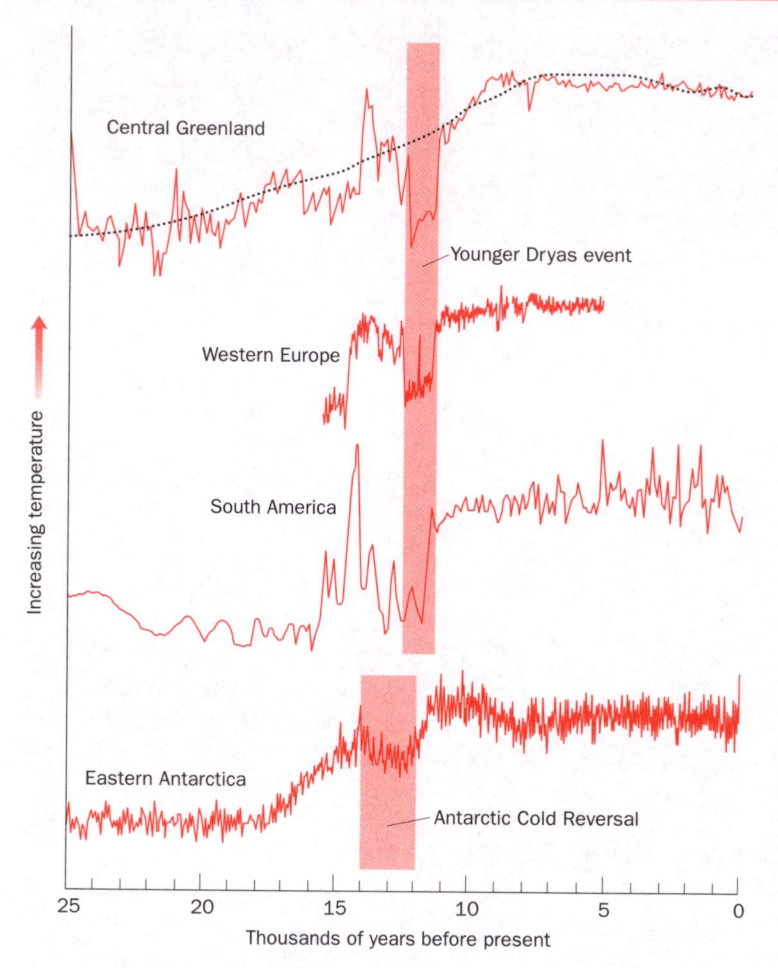

Figure 8.15 The Younger Dryas cold event from about 12,700 to 11,500 years ago. Rapid warming before the Younger Dryas (the Bølling interstadial) and rapid changes of temperature at the onset and close of the Younger Dryas are shown. Note the global imprint of the Younger Dryas and the associated but earlier Antarctic Cold Reversal

Source: Houghton *et al.* (2001)

cooling phase could come from the higher concentrations of dust in the atmosphere during colder periods (see Figure 8.5). High concentrations of atmospheric haze during cold events would reflect incoming solar radiation and produce greater cooling.

(d) Restarting the Gulf Stream

Model simulations have also shown that a sudden switch could also occur in the opposite direction helping to start up and strengthen the Gulf Stream and deep water formation in a matter of years or decades. The cooling caused by conveyor belt shutdown and ice albedo feedbacks reduces the melt rate and increases the amount of sea-ice formation. More salty water is left behind in the top surface layers of the ocean when growing ice sheets extract more and more fresh water from the oceans. The increase in salty water eventually re-establishes North Atlantic deep water (NADW) formation causing Gulf Stream return and rapid warming in the northern latitudes. As the ice sheets begin to melt, positive feedbacks again work to reinforce the process. Ice albedo effects (i.e. less ice, more solar absorption at the surface, more surface heat retention) serve to warm the climate. A possible decline in biological activity in the warmer northern oceans would help to increase the concentration of atmospheric CO_2 and warm the region further. Increasing heat and moisture of the northern regions would also help to liberate methane, another greenhouse warming gas, from growing swamps and other biological source regions. With less contrast in temperature between the equator and poles, the warmer climate would have less dust particles in the atmosphere possibly because of less storminess and dryness: with less haze, more solar radiation would be absorbed by the ground surface and temperatures would increase in the region. Eventually, the combined temperature increase from these agencies again leads to an ice melt and freshwater input into the North Atlantic sufficient

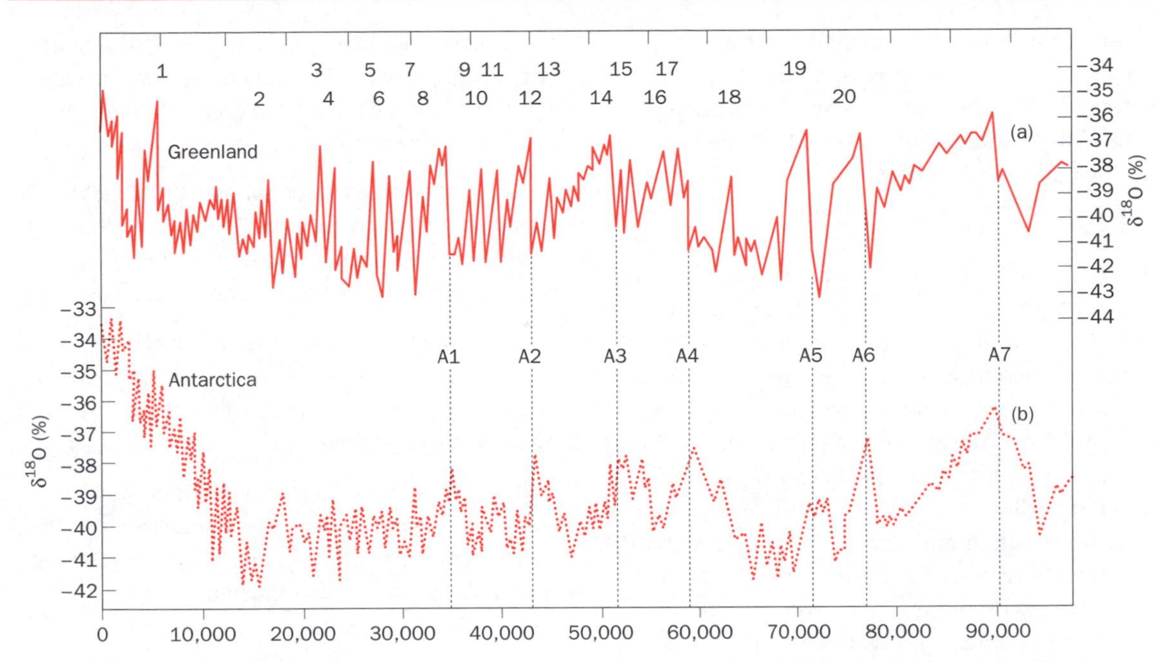

Figure 8.16 The timing of warming and cooling periods (interstadials and stadials) in the North and South poles during the last glacial period. Evidence based on ice core data from Greenland and Antarctica
Source: Blunier and Brook (2001, p. 110)

to begin to shut down NADW formation and Gulf Stream action.

It is worth noting that the links between the rapid fluctuation in warm/cold temperatures in the Greenland ice core and the on/off switching mechanism of the North Atlantic conveyor belt have not been fully worked out. Evidence has shown, for instance, that the North Atlantic conveyor belt was not always shut down during the Younger Dryas cold stadial (Heinrich) event.

(e) Links between Arctic and Antarctic regions

Modern evidence demonstrates that despite their overall synchronous glaciations (section 6.5.2), large but fairly regular differences in the timing of millennium-scale climate variability between West Antarctica and Greenland was a distinctive feature of the last glacial record. Wide and rapid fluctuations in temperature during an ice age seem to be more a feature of the north than the south polar region. Ice-core records from Antarctica (Byrd) reveal a smaller number of interstadial events than appear in the Greenland (GRIP) ice cores. Figure 8.16 shows that from 90,000 to 10,000 years ago, 21 interstadials have been identified for the north polar region (see also Figure 8.8), while only 7 have been recorded in Antarctica. Warming into an interstadial in the Antarctic is not only much slower but appears to take place before similar events in the Arctic. The Antarctic warmings 1–7 precede interstadials (D–O events) 8, 12, 14, 16/17, 19, 20 and 21 by as much as 1,500–3,000 years. In general, during the gradual warmings of the Antarctic interstadials, Arctic temperatures were cold or cooling.

The semi-regular and out-of-phase temperature relationships between the north and south polar regions discount the role of the atmosphere in causing the differences, and suggest a bipolar see-saw in thermohaline oceanic exchange between the hemispheres. Modelling studies have shown that a strong North Atlantic conveyor (associated with a rapid warming event in the Greenland ice core) causes a significant cross-equatorial transfer of heat from the South Atlantic to the North Atlantic. This export of heat from the South Atlantic may eventually trigger an Antarctic cooling phase. Since warming in West Antarctica seems to *precede* that in Greenland by several thousand years, another bipolar mechanism suggests that the trigger for climate change lies in the southern hemisphere and particularly in the south polar ocean conveyor belt. Stronger than usual upwelling of warm water in West Antarctica over several thousand years could lead to a gradual Antarctic warming. Such warm upwelling in the southern seas could weaken warm advection by the North Atlantic conveyor belt and eventually lead to the onset and maintenance of cold stadial conditions in the north polar region.

Key ideas

1. Short-term fluctuations of climate during the last Ice Age such as rapid switching from warmer interstadial to much colder stadial periods over 50–100 years to several thousand years are most likely related to oscillations in the ocean circulation.

2. A good example of rapid oscillation in the ocean circulation relates to changes in the North Atlantic conveyor belt (changes in ocean surface temperature, salinity and density) and the shutting down and starting up of the Gulf Stream.

3. One theory suggests that the warm Gulf Stream could shut down at the end of an interstadial event, in response to prolonged ice melt and an abundance of less dense fresh water in the North Atlantic, leading to the onset of a cold stadial (Dansgaard–Oeschger or Heinrich) event.

4. By contrast, at the end of a cold stadial episode, the Gulf Stream could start up again, as a consequence of prolonged ice accumulation and an increasing supply of salty and denser surface waters in the North Atlantic, eventually leading to the onset of a warmer interstadial climate.

5. Small but significant out-of-phase changes in temperature and ice cover between the north and south polar regions may also be linked to a bipolar see-saw in the thermohaline ocean circulation.

6. Rapid warming and an interstadial climate associated with a strong North Atlantic conveyor may trigger ocean cooling in the South Atlantic and an Antarctic cooling phase.

7. Conversely, strong upwelling of warm water in the South Atlantic could lead to a weakening of the North Atlantic conveyor belt and the onset of colder stadial conditions in the high latitudes of the northern hemisphere.

Further reading

Anderson, D. (2000) Abrupt climate change. *Geography Review*, March: 2–6.

Broecker, W. and Denton, G. H. (1990) What drives glacial cycles? *Scientific American*, **262**: 43–50.

Drake, F. (2000) *Global Warming*. Arnold, London, 273 pp. Ch. 3, pp. 68–101.

Ruddimann, W. (2001) *Earth's Climate: Past and future*. Freeman Press, New York, 465 pp.

Williamson, P. and Gribben, J. (1991) How plankton change the climate. *New Scientist*, 16, March: 48–52.

Wilson, R., Drury, S. and Chapman, J. (2000) *The Great Ice Age*. Routledge, London, 267 pp.

Useful websites

WIKIPEDIA:
http://www.wikipedia.org/wiki/Ice_age

NOAA:
http://www.ngdc.noaa.gov/paleo/slides/slideset/index19.htm

National Snow and Ice Data Centre, University of Colorado:
http://nsidc.org/iceshelves/larsenb2002/

British Antarctic Survey:
http://www.antarctica.ac.uk/met/bas_publ.html

Oak Ridge National Laboratory:
http://www.esd.ornl.gov/projects/qen/nercEUROPE.html

CRIEPI News:
http://criepi.denken.or.jp/en/e_publication/pdf/den357.pdf/den357e.pdf

NCAR:
http://tao.atmos.washington.edu/wallace/ncar_notes/

Encyclopaedia of the Atmospheric Environment:
http://www.doc.mmu.ac.uk/aric/eae/Climate_Change/Older/Ice_Ages.html

CHAPTER 9

Weather, climate and climate change in the mid-latitude oceanic margins North-west Europe

9.1 Climate of north-west Europe

The task of analysing the weather and climate of the oceanic margins of north-west Europe is complicated by virtue of the region's situation at a crossroads of contrasting atmospheric and oceanic influences. To the west are the moderating effects of the North Atlantic; to the east the continentality of the vast Eurasian landmass. From the north come incursions of cold, relatively dry polar and Arctic air masses; from the south, warm and moist air masses of tropical origin.

Given this diverse template, the climatology of north-west Europe may be explained in terms of several factors including the frequency and seasonality of mobile weather features such as air masses, depressions, anticyclones and fronts. Regional variations across this domain reflect local features such as mesoscale synoptic systems, distance from the western seaboard and topography. The following sections examine the most important physical controls and resulting features of the region's climatology.

Key ideas

1. The climatology of north-west Europe reflects the complex interplay of contrasting air masses, depressions, anticyclones and fronts, as well as oceanic and topographic influences.

9.2 Processes and controls

9.2.1 Air masses

It may be recalled from section 4.3.1 that an air mass can be defined as a large mobile body of air whose physical properties, such as temperature, lapse rate, moisture content, and all-round weather features are relatively homogeneous over thousands if not millions of square kilometres. These great air masses originate in major high-pressure cells in tropical and polar regions (see Figure 4.12) and gradually adopt the properties of the underlying surfaces of their source region (see Table 4.1). Despite their general homogeneity over large horizontal distances, air masses are gradually changed by the earth's surface when they move out of their source regions to invade other areas (see Figure 4.12). The extent of air mass modification depends on the original character of the air mass, the trajectory of the air with respect to its source region, the degree of heat and moisture exchanges with the underlying surfaces, and the extent and depth of mechanical mixing. As shown in Box 4.4, the climate of the UK is influenced by the composition and frequency of four major air masses, i.e. maritime tropical (mT), maritime polar (mP), continental polar (cP) and continental tropical (cT), together with two hybrid air masses, i.e. returning maritime polar (mPw) and maritime Arctic (mA). The climate of north-west Europe is also determined by the character and frequency of invasion by these air masses. This includes the Mediterranean, with its dominance of continental tropical (cT) air in summer and mild Atlantic air (mT and mPw) in winter, as well as the northern temperate maritime regions of western Scandinavia with its all-year dominance of maritime polar (mP) and maritime tropical (mT) air.

Although air mass analysis can assist in our understanding of the climate of regions (see Box 4.4) and can make contributions to weather forecasting, especially in terms of temperature change, the technique is of limited value in describing and predicting precipitation variation. This is especially the case in the mid latitudes where the dominant precipitation-controlling features are the polar front (i.e. where certain air masses come together) and the depression.

9.2.2 Frontal systems and depressions

When two fluids with different densities come together they do not readily mix, for example oil and water. Similarly two air masses with different temperatures, and consequently different densities, also resist mixing, and a transition zone between them, known as a front, becomes established. The term 'front' was first used by the 'Bergen school' in the 1920s to denote the surface of separation between two air masses of markedly different temperature and humidity. Since this time, it has been recognised that much of the weather of the mid latitudes is explained in terms of the formation and movement of such frontal boundaries. The passage of such fronts is accompanied by measurable weather changes including variations in pressure, wind speed and direction, cloud type and precipitation. The most important interface for north European weather is between the polar maritime (mP) and tropical maritime (mT) air masses which collide in the North Atlantic along the polar front. Figure 9.1 shows the central role of the polar front in relation to the origin and development of a typical mid-latitude depression. When the polar frontal zone becomes wavelike (stage B) due to the friction of the opposing air mass flows (mT and mP), disturbances at the apex initiate the formation of frontal depressions.

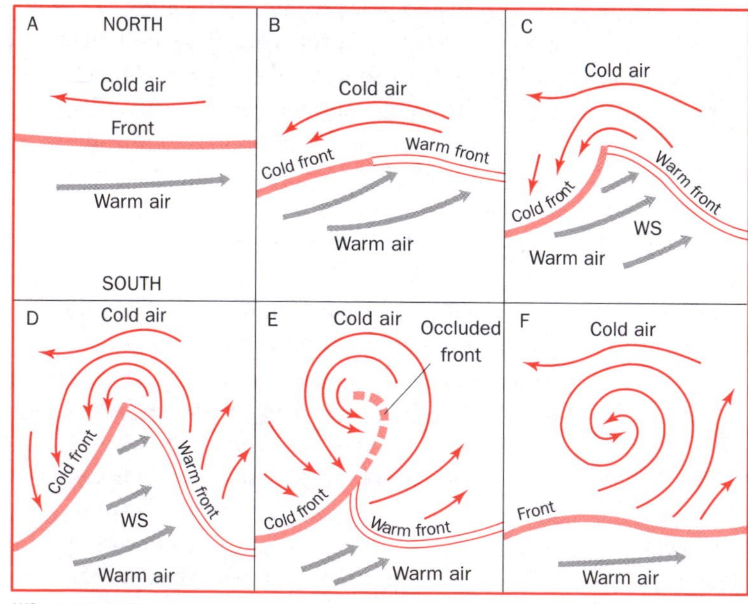

(a) Frontogenesis (frontal development)

A depression (or cyclone) is a synoptic feature with anticlockwise (in the northern hemisphere) air circulation around a low-pressure centre (Plate 9.1). Near the surface, the air movement is said to be convergent because the air spirals towards the centre, resulting in

Figure 9.1 Six typical stages in the development of a mid-latitude depression. The depression is moving generally from west to east along the polar front, reaching maturity at stage D and occluding at stage E
Source: Barry and Chorley (1998)

uplift and cloud formation. At higher levels in the atmosphere the air movement is divergent, moving away from the centre and then descending outwards. The depression typically has a lifespan of 4–7 days and a diameter of 1,500–3,000 km (a property of daily weather charts referred to as a 'synoptic-scale' feature).

According to the classic Bergen model of depression development, the polar front becomes divided into warm and cold sectors (Figure 9.1, stage A) with a distinct low-pressure region forming at the crest of the wave (stage B). A warm front occupies the leading edge of the wave that is moving generally from west to east or south-west to north-east. The warm front is followed by the warm sector (ws), i.e. the warm air enclosed by the wave on its southern side. Finally, the cold front leads the trailing cold air mass at the rear of the developing wave (stage B/C). As the depression deepens the cold front, which moves faster than the warm front, begins to catch the warm front (stage C to E), and the area of the warm sector begins to decrease (stage D to E). An occluded front, or occlusion, forms where the warm front is overtaken by the cold front, with the warm air being completely lifted off the ground (stage E). The length of occluded front gradually increases away from the centre of the depression and, as it does so, the waveform at the surface is eliminated (stage F). Thereafter, the depression dissipates and the polar front may be re-established, but further south than before. Depressions seldom form in isolation, but rather in families of three or more, each younger than its predecessor and forming progressively further south on the trailing cold front of the older one.

Figure 9.2 shows in more detail the horizontal and vertical structure of the mature depression (Figure 9.1, stage D). In the example shown, the low-pressure centre with anticlockwise airflows around it and centred at the apex of the warm sector can be clearly seen. The wedge of warm sector air is denoted by its generally high temperatures (between 15 and 20 °C) compared to the colder air which falls in temperature with distance from the warm sector air (from about 15 °C to –5 °C). At the warm front, the warm lighter air is moving forwards and *upwards* away from the surface over the heavier colder air with which it comes into contact (see cross-section C–D). It can also be seen that the cold air at the rear of the depression is undercutting

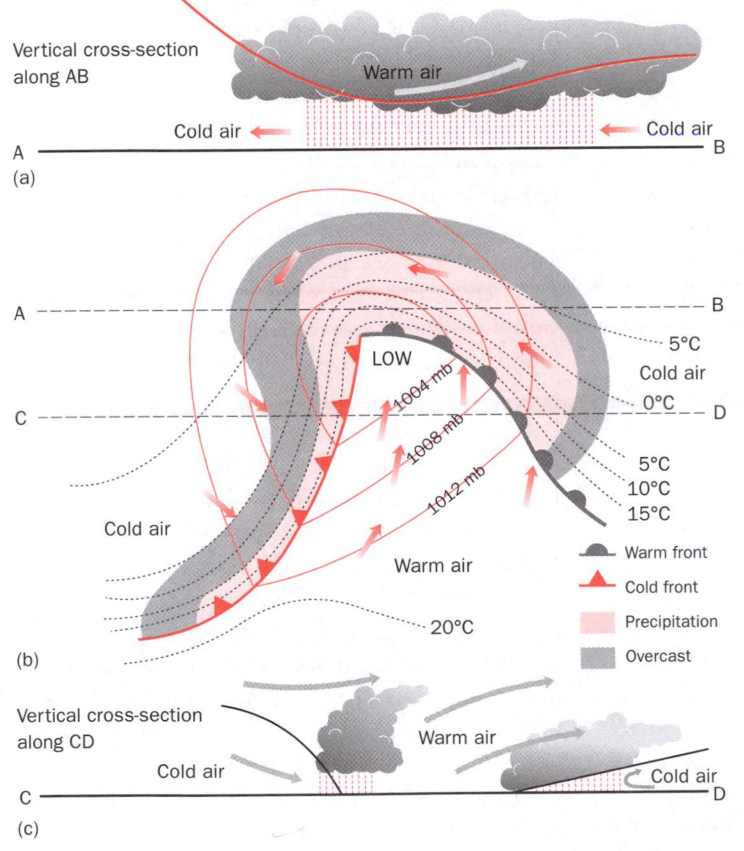

Figure 9.2 Vertical and plan structure of a mature mid-latitude depression showing weather features
Source: O'Hare and Sweeney (1986)

the warm air and pushing it forwards and upwards at the cold front. As the air rises at the warm and cold fronts, cooling and condensation take place and clouds begin to form, resulting in zones of rainfall in advance of the warm and cold fronts. During the process of occlusion referred to above (Figure 9.1, stage E), it can be seen that the wedge of warm air is undercut at the surface by the advancing cold air and lifted completely from the ground (Figure 9.2, cross-section A–B). Here clouds also form in the rising and cooling air so that the occluded front is also associated with rainfall in its vicinity. In order to take our understanding of weather conditions associated with the mid-latitude depression further, we need to explore the mechanisms and processes which encourage air instability (rising air, release of latent heat) within the developing depression.

(b) Frontal weather

The passage of a depression brings a distinct sequence of weather conditions that depends upon the vertical motion and stability of air masses in the warm and cold sectors. As shown in section 4.2.2, airflow within a developing depression is closely connected to waves and jet streams in the upper westerlies. Strong airflow and divergence on the rising limb of an upper westerly wave draws surface air upwards, helping to encourage the initiation and development of a low-pressure centre or depression at the surface. If the air within a depression is rising overall (but especially at the frontal zones) due to the action of an upper westerly jet, an anabatic depression results with very active ana-fronts (Figure 9.3(a)). Here the air is rising and cooling, leading to condensation, cloud formation and the release of latent heat. As described in section 2.3.1, latent heat release into the ambient atmosphere increases air buoyancy, further encouraging air uplift, deeper cloud development and longer and more intense rainfall. At the warm front, however, which has a very gentle gradient (less than 1°), air rises relatively slowly, favouring the development of multi-layered shallow cloud such as cirrus (Ci), followed by cirrostratus (Cs), altostratus (As) and then nimbostratus (Ns). It is worth noting here that frontal zones themselves are 2–3 km wide and are more than just single lines of air mass separation. As shown in Figure 9.3, they are depicted as having a dual surface and are composed of dry air drawn from the upper atmosphere. It can also be seen in Figure 9.3 that air rising up and over the frontal zones is able to penetrate through them so that clouds form not only along each front but through them as well. The arrival of the altocumulus during the development of mid-latitude depressions is often accompanied by drizzle, the nimbostratus by continuous but often light rainfall. It can be seen in Figure 9.2(b) that as the warm front passes overhead the wind veers from a south-easterly to a more southerly direction, temperatures rise, pressures stabilise, and rainfall becomes intermittent or halts in the following warm sector air (mT).

The gradient of the cold front is steeper (approximately 2°) than the warm front. This causes more rapid vertical air motion and deeper air instability at the cold front. Strong instability and the release of latent heat maintain instability to high levels in the atmosphere which result in deeper and more pronounced cloud formation. The accompanying precipitation tends to be relatively short-lived but more intense compared with the warm front. With rapid upward air motion (Figure 9.3(a)), the cold ana-front is often marked by nimbostratus (Ns) or cumulonimbus (Cn), brief heavy rainstorms, and thunder.

Conversely, as shown in Figure 9.3(b), when the upper westerly jets are weak with air within the depression descending relative to each frontal surface we have a kata-front situation. In this case, only stratocumulus and light rainfall occur across the system. As the cold front passes (Figure 9.2(b)), we are left with the trailing cold air (mP) of the depression. Here, the sky may clear abruptly, the wind veers sharply from the south-west to north-west, temperatures fall and pressure begins to rise. Over the Atlantic, the trailing cold air is often represented by many scattered cumulus cloud

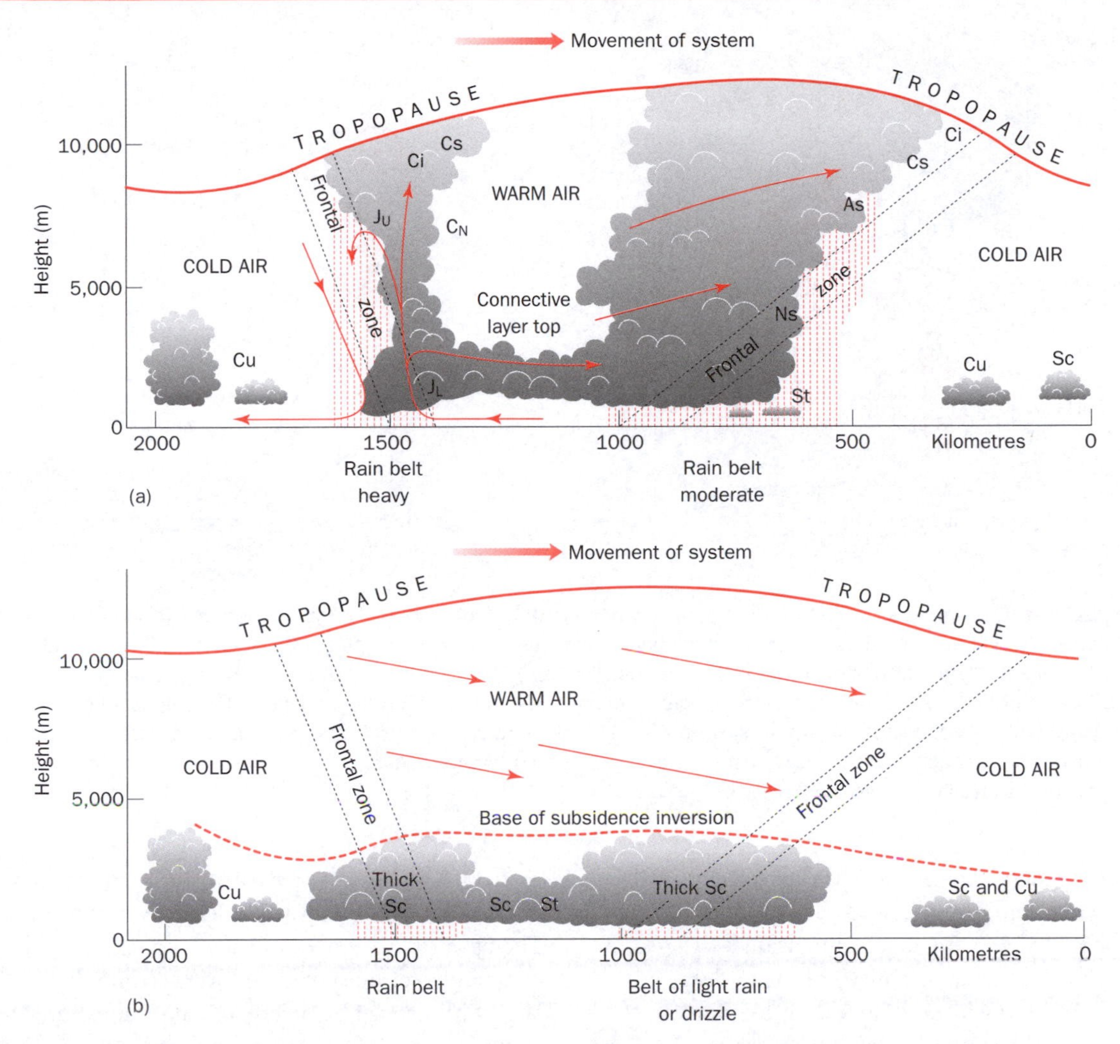

Figure 9.3 Cross-section of a depression (a) with active jet streams (upper jet = J_U, lower jet = J_L) and warm ana-fronts where the air is rising relative to each frontal surface and (b) with cold kata-fronts with air descending in the system relative to the frontal zones

forms. These occur as a result of surface heating of the cold southward-moving (mP) air mass which becomes unstable as it moves across the warm North Atlantic ocean.

(c) The 'conveyor belt' model

The exact sequence of events shown in the classic warm- and cold-front models (Figure 9.3) is highly idealised. First, forecasting rain belts at fronts is complicated by the fact that most fronts display ana- and kata-behaviour throughout their length and at various levels in the troposphere. Second, radar observations suggest that frontal precipitation is controlled not only by upper-level westerly jet streams (section 4.2.2) but also by lower and middle altitude level jet flows.

One important lower jet stream to affect the weather of the mid-latitude depression is found about 1 km above the surface (900 mb), and flows northwards into the warm sector of the depression parallel to, and ahead of, the cold front, at

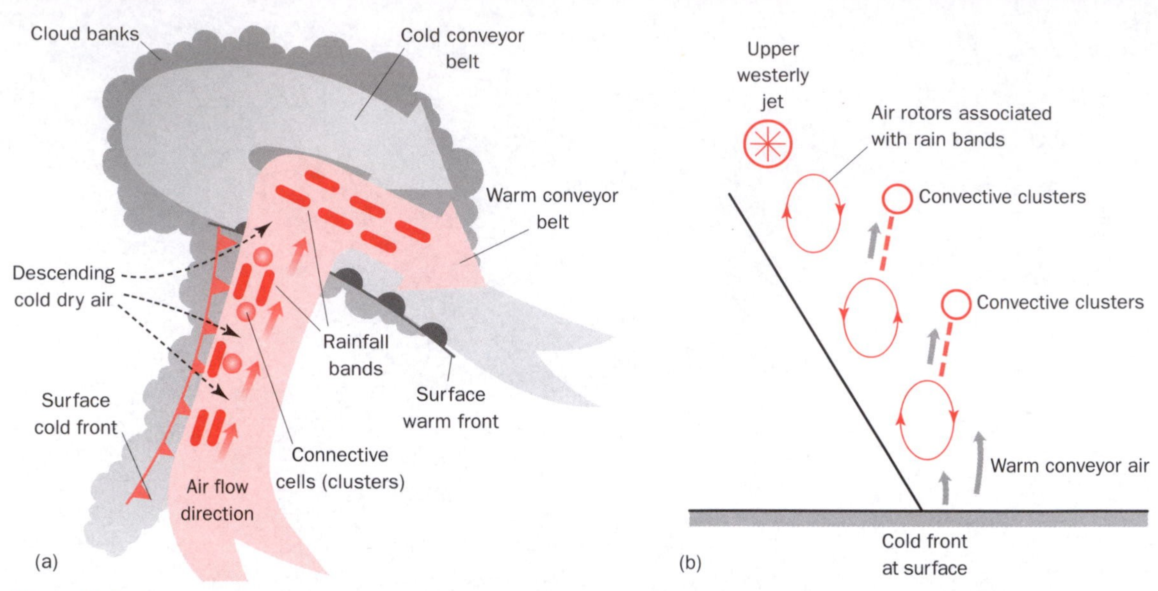

Figure 9.4 (a) A mature depression illustrating low-altitude (the warm conveyor) and mid-altitude (cold conveyor) jet streams. Precipitation features including rainfall bands and clusters parallel to, and in advance of, the warm and cold fronts respectively are also shown. Note that the upper westerly jet that crosses the system at high altitude above the two lower-level jets and influences convergence and air uplift at the surface (see Figure 4.11) is not shown. (b) Individual but connected air rotors rising up over the cold front producing rain bands. Clusters of more intense rain occur where additional convective cells break upwards from the rotor cells
Source: (a) Adapted from Semple (2003)

horizontal speeds of ~30 m/s (Figures 9.3(a) and 9.4). The term 'conveyor belt' is used to describe this northward-moving three-dimensional large-scale motion of air that is typically 1–3 km deep, 200–300 km wide and thousands of kilometres long. It is important to note from section 4.4.1 that the term 'conveyor belt' is also used to describe the transfer of water in the oceans as in the ocean conveyor belt. The atmospheric conveyor belt or jet flow is further defined as a 'warm' conveyor belt since it brings warm (and moist) air from the south into the warm sector of the depression. 'Cold' conveyor belts at low to middle levels in the atmosphere (700 mb) can also be identified in the developing mid-latitude depression (see Figure 9.4(a)). A typical one originates ahead of the warm front in cold subsiding air to the north-east of a developing cyclone.

The warm moist air of the low-level conveyor rises at about 7 m/s from around 900 mb (1 km elevation) within a narrow strip ahead of the cold front and moves polewards over the warm front. The upward movement of warm moist conveyor air at and over both cold and warm fronts is largely influenced by divergence aloft in the upper westerly jets which are positioned above each front just below the tropopause boundary (Figures 9.4(b) and 4.11). As the air rises, cooling and condensation take place so that deep cloud banks form in the vicinity and just ahead of the main frontal zones. A common feature of these rain zones is that they are separated into rain bands and individual rain clusters. The observed rain bands (50–100 km wide) which form along both the cold and warm fronts (Figure 9.4(a)) may be the result of air rolling up over the fronts as separate but connected rotary cells of air with each cell producing cloud and precipitation (Figure 9.4(b)). These rain bands are connected with the well-known *intermittent* nature of rain at the frontal zones. Atmospheric instability within the system can be further enhanced as descending cold dry air from the southward-moving cold conveyor in the middle troposphere can overrun the lower warm conveyor belt. This action triggers increasingly steep lapse rates and broad-scale potential instability especially at the cold front, where cellular clusters of small-scale convective cells can develop from the rising rotating air currents (Figure 9.4(b)).

Finally, the seeder–feeder cloud mechanism may also increase the precipitation yield of lower clouds by seeding them with ice particles falling from upper cloud layers. Droplets falling from the higher seeder cloud grow rapidly within the lower feeder clouds. They grow quickly because of a constant washout of ice and rain droplets from the feeder clouds to the surface below. This mechanism is most effective in near-surface, moist, fast-moving airflows and may be initiated by orographic uplift. Hence, the 'conveyor belt' model is an important advance on the classic frontal theory because it highlights the significance of the three-dimensional structure of frontal systems.

(d) Frontal decay

Occlusion signifies the final stage of a front's existence (Figure 9.2(a)). Frontal decay occurs when there cease to be differences in the temperature and humidity properties of adjacent air masses. This can arise through the system assimilating air of the same character, or through the simultaneous stagnation of neighbouring air masses over a similar surface. Alternatively, the air masses may follow paths that are in series or parallel, in which case their properties will tend to equalise.

9.2.3 Anticyclones

Anticyclones are regions of high pressure and surface divergence, in which the barometric pressure progressively declines outwards from the core of the system. The anticyclone typically appears on a weather chart as a set of widely spaced, concentric isobars. Surface winds are light and variable, with air spiralling clockwise (in the northern hemisphere) from the centre. The system is characterised by descending air, and generally clear skies with local cumulus cloud formation in summer whenever there is strong surface heating (Plate 9.2). In winter, clear skies and long nights allow a considerable loss of long-wave radiation to space, cooling the surface. Surface cooling in turn chills the air in contact with it, so that the surface air can be colder than that immediately above. Under such *ground-based* temperature inversions, frost and radiation fogs are common. Large-scale air descent under anticyclonic subsidence may also cause warm air to accumulate above the cooling air at the surface, thus reinforcing the development of surface inversions. Such subsidence inversion restricts vertical mixing, providing favourable conditions for widespread air pollution episodes (see Box 9.1).

(a) Travelling and blocking anticyclones

The temporary anticyclones of the mid latitudes should not be confused with the quasi-permanent features associated with the descending limbs of Hadley cells (Chapter 4). Nor should they be grouped among the persistent high-pressure winter anticyclones of central Asia or Arctic Canada. Instead, travelling anticyclones are mobile, synoptic-scale features that are of comparable dimensions to depressions. These form between members of a depression family, and provide a clear weather respite between successive depression passages. The last member of the family is commonly followed by a cold air mass of polar origin that sets up a region of high pressure.

Deep warm anticyclones, interspersed by stationary deep occluding cold depressions, may also form in the mid latitudes as a consequence of the Rossby wave index cycle (section 4.1.4). Under a high zonal index there is a strong westerly airflow at middle latitudes with minimal north–south air mass exchange. However, under conditions of low zonal index the jet develops waves (Rossby waves), with ridges and troughs becoming more pronounced

BOX 9.1

CASE STUDY

Atmospheric circulation and air quality in London

The UK government's National Air Quality Standards (NAQS) set defined levels of air quality that avoid significant risks to health for eight pollutants (benzene, 1,3-butadiene, carbon monoxide, lead, nitrogen dioxide, sulphur dioxide, particulates (PM_{10}) and ozone). Air quality is monitored by the National Automated Monitoring Networks and Non-automatic Networks at over 1,500 sites across the UK, with data available as far back as 1972 for some sites. Even so, the reliable detection of air quality trends is complicated by a combination of short data sets, changes in instrumentation, representativeness of monitoring sites and the strong control exerted by weather patterns on pollution episodes.

According to Defra (Department of Environment, Food and Rural Affairs) there has been a decline in the number of days in the UK when air pollution was classified as moderate or high at urban sites from an average of 59 days per site in 1993, to 21 days in 2001. However, air quality remains at unacceptable levels in many parts of London. For example, particulate concentrations in Marylebone exceeded NAQS on over 30 days in 1998 and 1999. Furthermore, a recent study has found links between air pollution and social deprivation in London, with higher concentrations of nitrogen dioxide and particulates found in areas exhibiting higher social deprivation.

Traffic emissions of nitrogen dioxide and particulates have replaced sulphur dioxide from coal-burning as the most significant air quality problems currently facing London. The highest concentrations of nitrogen dioxide (and its related pollutant nitrogen oxide) are found in central London and along busy road corridors. Background concentrations of particulates are highest in the east of the Thames region due to *secondary* pollution (secondary pollution forms from the chemical reactions of primary or original pollutants). The high levels of secondary pollutants east of the Thames region are composed of sulphate and nitrate particles formed from chemical reactions in the atmosphere, that originate from mainland Europe. Concentrations of sulphur dioxide are also highest in the east Thames corridor due to emissions from a number of power stations and a refinery.

Tropospheric or low-level ozone is a secondary pollutant formed from the reaction of sunlight on precursor gases such as nitrogen dioxide and volatile organic compounds (VOCs), enhanced concentrations of which are produced by transport and industrial processes. Tropospheric ozone is of concern because unlike ozone in the high stratosphere, it is a highly toxic ground-level gas that causes damage to human health, crops and vegetation. Complex processes of tropospheric ozone production, transport and decay, lead to marked differences in pollutant concentration at rural and urban sites in the UK, and among different weather patterns. Peak ozone concentrations (>60 ppb) in central London typically occur under stable warm summer anticyclonic weather which favours *in situ* photochemical processes (sunlight + nitrogen dioxide leads to ozone production) and long-range low-altitude transport under inversion conditions from continental sources (Figure 9.5). Surrounding rural sites also experience high concentrations under anticyclonic weather because of the advection of tropospheric ozone from urban sources. Rural areas additionally have high concentrations of ozone because of a lack of precursor pollutants such as nitrogen oxide which is known to destroy ozone.

Westerly airflows and cyclonic depressions can also elevate ozone concentrations at rural sites through turbulent down-mixing. Conversely, the lowest average ozone concentrations in central London have historically been associated with northerly and north-westerly airflows with their deep mixing layers and hence ability to disperse the surface pollutant. However, all weather patterns display strong seasonalities in monthly mean ozone concentrations, peaking in late spring/early summer and at a minimum in winter.

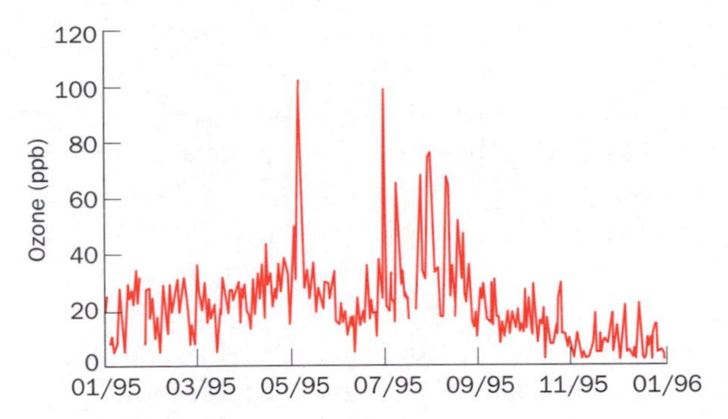

Figure 9.5 Maximum hourly concentrations of ozone at Russell Square Gardens in London during 1995. The UK air quality objective is 50 ppb, not to be exceeded more than 10 times per year as the daily maximum of the running 8-hour mean

until there is breakdown and cellular fragmentation of the upper westerlies (see Figure 4.8). The jet stream may then become divided into two branches, one passing to the north of the warm high-pressure cell over Norway and the other moving south into the Mediterranean. This feature is referred to as a blocking anticyclone because it prevents the normal west–east passage of depressions in the westerly flow. They can persist for several weeks, and may drift slowly eastward.

A major area of blocking often occurs over Scandinavia in spring. Under these circumstances depressions are steered northwards towards the Norwegian Sea or into southern Europe. When this pattern occurs in winter, easterly airflows along the southern perimeter of the anticyclone (the general airflow is clockwise around the system) can bring extreme cold and heavy snowfall to north Europe as in January–February 1947. With summer blocking, precipitation totals are below average over most of Europe. For example, the 1975–76 drought in Britain was the result of a persistent blocking anticyclone over north-west Europe, bringing warm, dry air from the south. Conversely, 1816, 'the year without a summer', experienced below-average temperatures and widespread crop failures in the UK and much delayed wine harvests over France. This was because of a strengthening and extension of the polar anticyclone that forced depressions to follow a more southerly track across central England and Wales. It is interesting to note that the cold year of 1816 is often associated with strong volcanic activity from Mt Tambora in Indonesia which erupted the previous year in 1815 (see section 6.5.3).

(b) Anticyclonic weather

The character of anticyclonic weather is strongly influenced by the location of the blocking. The British Isles are seldom covered by a region of high pressure; anticyclones are most commonly centred over the south-east in January, to the north-east in May, and to the south-west in July. The unusually persistent high pressure over northern Europe during 1975–76 brought warm easterly winds (i.e. summer cP air) and drought to south-east England. However, Scotland and Ireland were more influenced by southerly and westerly airflows from the high pressure, and received near normal rainfall over the same period. If high pressure is centred over the north-west, as in the last week of June 1995, then fine weather is more likely to prevail over Scotland and Ireland, and south-east England experiences cool and cloudy north-easterly winds. During the last week of June 1995, varying cloud amounts led to large temperature gradients across the country, when Tynemouth reached only 11.1 °C while ~300 km away a station near Ben Nevis logged 28.6 °C.

Anticyclonic weather is also dependent upon the season. In summer, high sunshine amounts promote strong surface warming and the evaporation of low cloud within several hours of sunrise. The presence of an inversion can further inhibit cumulus cloud formation even on fine days. Sea breezes will also fail to develop in the presence of a deep temperature inversion. Otherwise, anticyclones generally provide the necessary precursors to abundant sunshine and weak regional airflows that are necessary for localised air circulation. In winter, anticyclonically induced easterly air (cP) flows across the North Sea may bring moisture and low cloud over the UK. Under stable high-pressure air conditions, widespread stratus cloud may persist for several days, producing a monotonous 'anticyclonic gloom' in low-lying areas. This weather situation is not to be confused with the coastal fogs (haars) in eastern UK brought about by high-pressure easterly airflows across the North Sea in summer. Despite the anticyclonic gloom of low-lying areas in winter, upland areas above the temperature inversion may experience brilliant sunshine, high temperatures and low humidities. For example, on 31 January 1992, Manchester airport had freezing fog and a maximum temperature of 0 °C while Buxton (220 m higher and 25 km to the east) recorded 6 hours of sunshine and a temperature of 10 °C.

9.2.4 Weather types

Previous sections (4.3.1) and sections in this chapter describe the key components of the mid-latitude circulation: air masses, depressions and anticyclones. Synoptic climatology – the study of local and regional weather in relation to these large-scale atmospheric features – has now been practised for at least a century. A major impetus for this work was the need for extensive data sets for *analogue weather forecasting* (a technique involving the matching of past synoptic patterns and accompanying weather to current synoptic patterns). Analysis of the frequency of mobile atmospheric features also assists the classification of regional climates, and can signal changes in the predominance of different airflow types. Synoptic climatology is currently enjoying a renaissance thanks to its widespread applicability to regional climate scenario generation and climate model validation (see Chapter 7).

(a) Lamb 'weather types'

The most common method of weather classification involves combining aspects of prevailing wind direction (both surface and mid troposphere) with the dynamic features of pressure charts (principally depressions and anticyclones). This approach underpins the European *Grosswetterlagen*, and the closely resembling Lamb weather types (LWTs) which apply to the British Isles (50–60° N, 2° E–10° W) more specifically. The LWT classification emerged from earlier work by R.B.M. Levick and recognizes 7 'pure' weather patterns (westerly [W], north-westerly [NW], northerly [N], easterly [E], southerly [S], cyclonic [C] and anticyclonic [A]), 19 hybrid types (such as the cyclonic south-westerly [CSW] and anticyclonic north-easterly [ANE]), and one unclassifiable (U) group. The three most commonly occurring LWTs, namely the A, W and C types, account for roughly 50 per cent of all days, and the U group less than 5 per cent. Figure 9.6 shows examples of the synoptic situation associated with each of the main LWTs, and Table 9.1 summarises the typical weather conditions associated with each weather type.

Professor H.H. Lamb subjectively classified each day's weather over the British Isles from 1861 to February 1997 on the basis of surface and mid-troposphere (500 mb) pressure patterns at noon. Since 1997, the series has been updated using Jenkinson's objective version of the Lamb series. This automated classification procedure (see Box 10.1) uses daily grid-point mean sea-level pressure data to compute indices of airflow direction, strength and vorticity. These indices are then converted into LWTs using a few simple rules based on dominant direction of airflow and vorticity. Available records prior to 1997 have also been objectively reclassified to maintain internal consistency and to provide a continuous daily series from 1881 to present.

Because the British Isles are centrally placed in the mid-latitude westerly wind belt and are frequently affected by blocking of this flow, the LWT catalogue provides a good indication of atmospheric circulation over much of the northern hemisphere. In other words, the catalogue contains a disproportionately large amount of information concerning the global climate since the mid nineteenth century. The objective series is also of sufficient length and homogeneity for analyses of climatic variations over decadal timescales (see below).

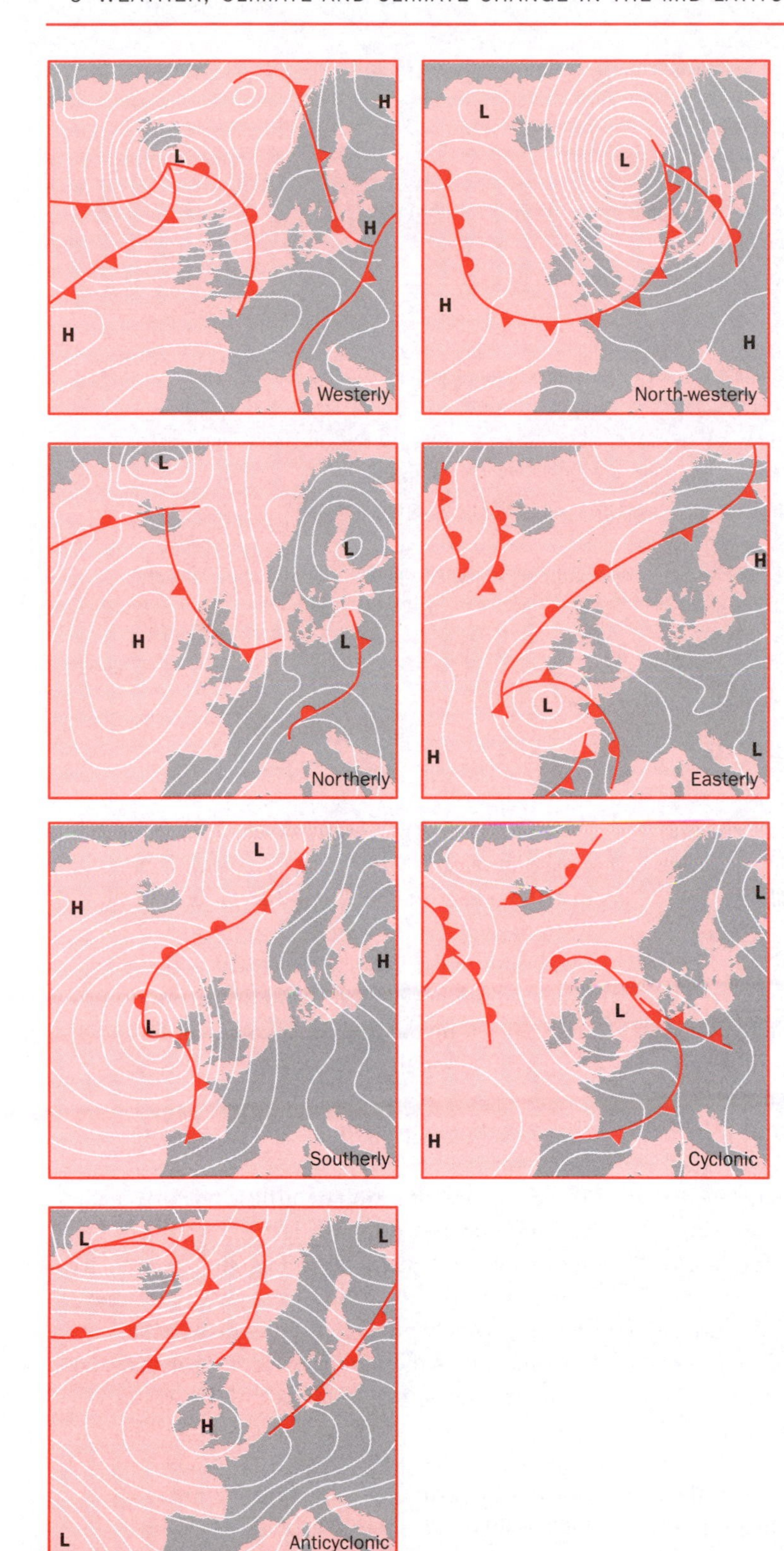

Figure 9.6 Examples of the main Lamb weather types
Source: O'Hare and Sweeney (1998)

(b) Applications

(i) Regional frequencies

The frequency with which the different regions of north-west Europe and the UK are affected by each of the LWTs varies according to their location. For instance, Stornoway in north-west Scotland receives a higher frequency of northerly (N), north-westerly (NW) and westerly (W) airflows than London in south-east England. The latter, because of its more continental position, has a higher frequency of easterly (E) and southerly (S) airflows. The anticyclonic weather type (A), with its origins in the Azores to the south-west of the UK, is also more common at London than Stornoway. By using such LWT frequency data, the weather and climate of Europe's regions can be described and classified.

(ii) Relationships with temperature

Each LWT has a characteristic temperature field (Table 9.1). Figure 9.7(a) shows that for central England, airflow type exerts a clear control over mean temperature. Weather systems moving in from the west or south help to maintain the typical mildness of the British weather, but cold winter spells result from invasions of air from the north or

Table 9.1 General weather characteristics and air masses associated with the main Lamb weather types over the British Isles

Type	Typical weather conditions
Anticyclonic	Warm and dry in summer, occasional thunderstorms (mT, cT). Cold and frosty in winter with fog, especially in autumn (cP)
Cyclonic	Rainy, unsettled conditions, often accompanied by gales and thunderstorms. May represent rapid passage of depressions across the country, or to the persistence of a single deep depression (mP, mT)
Westerly	Unsettled weather with variable wind directions as depressions cross the country. Mild and stormy in winter, generally cool and cloudy in summer (mP, mT)
North-westerly	Cool, changeable conditions. Strong winds and showers affect windward coasts especially, but the southern part of Britain may have dry, bright weather (mP, mA)
Northerly	Cold weather at all times, often associated with polar lows. Snow and sleet showers in winter, especially in the north and east (mA)
Easterly	Cold in the winter, sometimes very severe weather in the south and east with snow or sleet. Warm in summer with dry weather in the west. Occasionally thundery (cA, cP)
Southerly	Warm and thundery in summer (mT or cT). In winter, may be associated with an Atlantic low-pressure system giving mild, damp weather, especially in the south-west, or with high pressure over central Europe, giving cold and dry weather (mT or cP)

Source: Lamb (1972)

east. Our cool summers are associated with winds blowing from the west and north, while warm summer episodes are generated by airflow types from the south and east.

(iii) Relationships with precipitation

Precipitation yields associated with the LWTs can be expressed as with temperature over different space and timescales. Westerly, southerly and cyclonic circulations are responsible for the highest precipitation loadings of the seven primary LWTs. This is not surprising since they originate over the warm waters of the Atlantic before crossing the UK. When daily precipitation yields of each LWT are averaged over long periods of time (30–40 years) some interesting regional contrasts in precipitation distribution can be made. Figure 9.7(b) shows the strong west–east gradient in precipitation attributed to westerly circulations (W). In Scotland, the west coast receives between four and five times as much precipitation as the east coast (over 9 mm/day compared to 2 mm/day for westerly days). Precipitation fallout from the cyclonic weather type (C) is more uniformly distributed across the UK and Ireland than the westerly type (W). Mean daily amounts are maximum in the vicinity of the seas which provide water vapour supplies, for instance, around the Irish Sea, along the English Channel and off the North Sea along the east coast of Scotland.

(iv) Relationships with time (circulation frequency)

The daily categorisation of airflow types over the UK and Europe during the last 140 years or so allows us to calculate the average number of days when each primary airflow class prevailed. One of the most widely publicised signatures of the LWT classification is the significant long-term rise and decline in the annual frequency of the westerly type (Figure 9.7(c)). Since peaking at >100 days/year in the 1920s, the westerlies declined to <40 days/year in the early 1970s, before rising as high as 80 days/year in the 1990s. Periodic

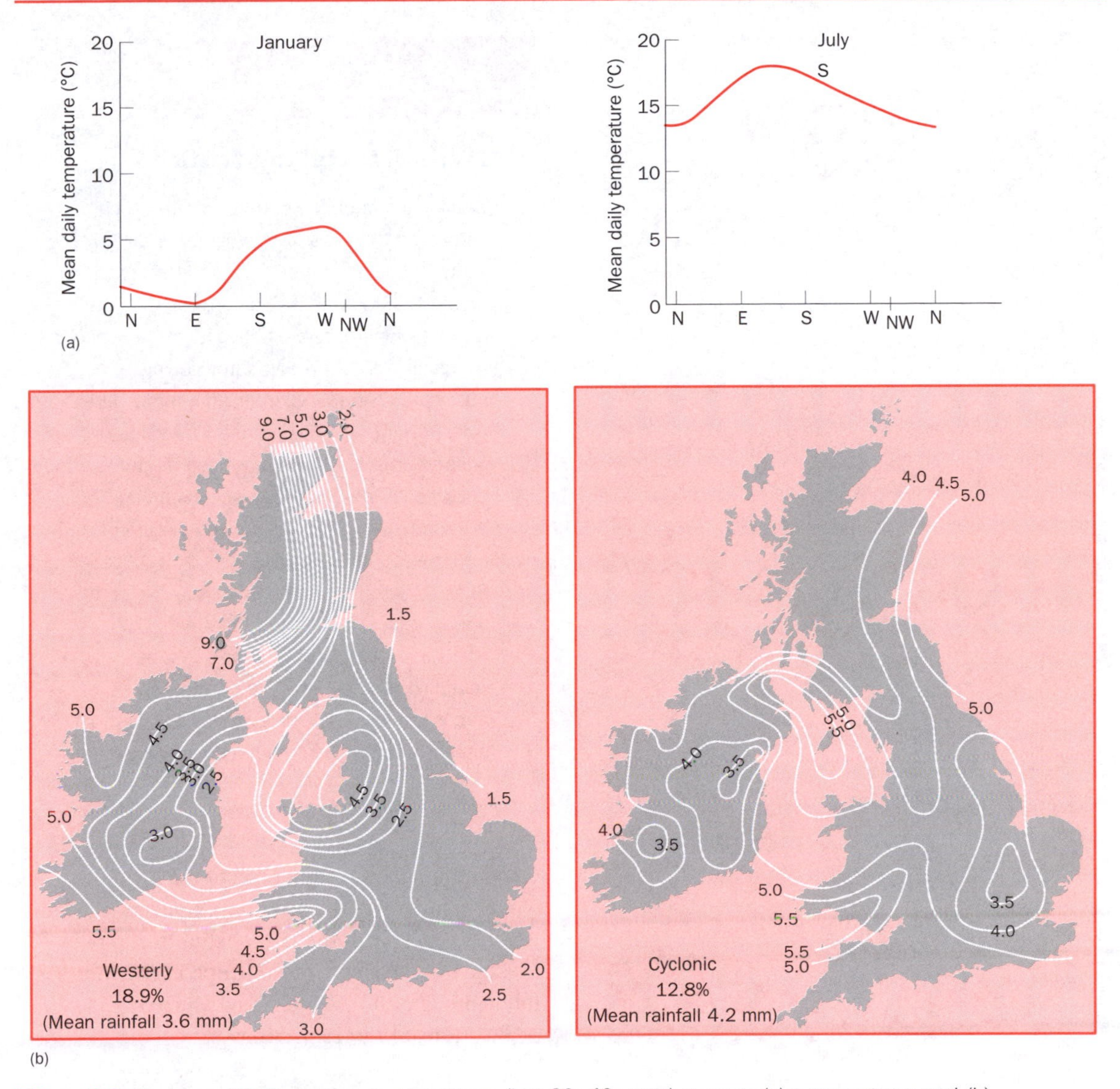

Figure 9.7 The use of LWTs in depicting long-term (last 30–40 years) average (a) temperatures and (b) precipitation conditions over the UK

continued

Source: (a) and (b) O'Hare and Sweeney (1998)

variations in the strength and dominant track of the westerlies have influenced the geographical distribution of precipitation and flooding across the British Isles, with, for example, increases in rainfall in western Scotland and northern England since the 1980s. Links have also been established between the frequency of key weather types and Central England Temperatures (see below), blizzards, frontal frequencies, air quality, precipitation chemistry, wind spectra, river water quality and geomorphological processes including river channel change. The winter frequency of the westerly type is highly correlated with the phase of the North Atlantic Oscillation Index (NAOI)

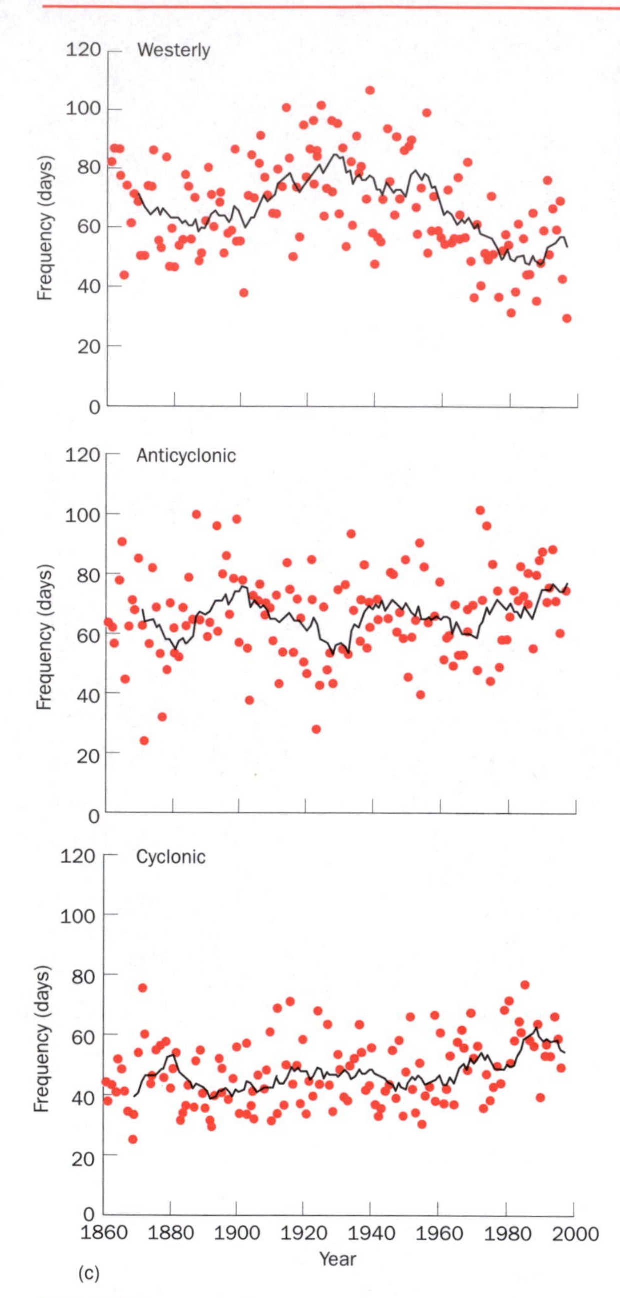

Figure 9.7 *(cont'd)* The use of LWTs in depicting long-term (last 30–40 years) average (c) annual frequency of the W, A and C Lamb weather types, 1861 to present

(see Figures 9.7(c) and 9.12), a metric of the strength of the north–south pressure gradient at the western margin of Europe. There is even tentative evidence of an El Niño Southern Oscillation (ENSO) signal (section 5.4) in the winter frequencies of the anticyclonic and cyclonic weather types, with the latter exhibiting above-average frequencies across Europe in the winters following El Niño events. See advanced applications of LWTs in Box 9.2.

BOX 9.2

THINKING FURTHER

LWT advanced applications

Catalogues of daily weather patterns such as Lamb and the *Grosswetterlagen* are useful for verifying the performance of general circulation models (GCMs), as well as providing a datum against which possible changes in atmospheric circulation might be compared. Using automated weather classification procedures (see Box 10.1), it is possible to translate gridded pressure fields produced by GCMs into daily weather types. Comparisons among historic types, GCM simulations of current and future weather types, assist in the verification of GCM climatology, and can identify systematic trends/biases in climate model behaviour. Statistical downscaling techniques (see section 7.4) may also use changes in atmospheric circulation patterns to construct high-resolution climate change scenarios.

(c) Limitations

Despite the above applications, weather classification methods have not been without their critics. It is well known that marked contrasts in weather can occur simultaneously in different parts of the British Isles under the same LWT. Moreover, two or more airflow types can influence the UK on the same day, giving markedly different weather patterns for each LWT. This notion also allows us to expect (as shown previously) that certain regions of the country, for example north-west Scotland, can expect a different frequency and assemblage of LWTs than other regions such as south-east England. Therefore, there is a strong case for more regionally specific weather type systems. However, such systems have seldom been constructed for long periods and, by definition, tend to be rather parochial in nature. Other problems include a lack of synchronisation between the calendar day used for weather classification, and the monitoring day used to record meteorological observations (e.g. in the UK precipitation totals are generally recorded for the 24 hours ending at 09:00 whereas the LWT daily classification is centred on 12:00 GMT). There may also be significant within-type variations in local properties, for example, the average precipitation yield of westerly types may not be the same in the 1920s as in the 1970s. Finally, even regional weather typing cannot resolve the meso- and microscale phenomena (such as sea breezes or convective processes) that are so often of significance to local weather.

9.2.5 Climate regions of north-west Europe

Regional climate classifications of north-west Europe reflect a hierarchy of factors. At the macroscale, Rossby waves oscillate in response to equatorward incursions of polar air, countered by the penetration of tropical air masses poleward (see Chapter 4). Seasonal variations in the balance between these two forces are reflected by the position of the Rossby waves, and hence precipitation-bearing depressions. These depressions are guided eastward by the westerlies, but their intensity and the strength of upper flow are weaker in summer than winter. The net effect of these factors is a general decline in precipitation yield from western coastal regions to the continental interiors of the east. Superimposed on these large-scale patterns are numerous meso- and microscale features. Examples include the relative degree of continentality or oceanity, the frequency and location of blocking anticyclones, thunderstorm activity and topographic effects. When these factors are combined, two broad climate regimes emerge for north-west Europe: the Mediterranean and the temperate maritime climates.

(a) Mediterranean

The Mediterranean climate is characterised by hot, dry summers and mild, wet winters and is represented by Athens, Greece, in Figure 9.8(a). The Mediterranean and temperate maritime climates of western Europe are distinguished on the basis of precipitation seasonality and, to a lesser extent, the differential in solar radiation. The Mediterranean regime is governed by the westerlies in winter and by the expanded Azores anticyclone in summer. The transition to winter conditions begins abruptly in the second half of October and is marked by a sudden drop in pressure and significant increase in the likelihood of precipitation.

The winter maximum precipitation is a consequence of high sea temperatures relative to mean air temperatures (a difference of about 2 °C in January), and resultant convective instability associated with incursions of mP air by westerly and north-westerly winds. This air mass gives rise to deep cumulus development and the formation of Mediterranean depressions under conditions of a low zonal index (marked wavelike airflow in the upper westerlies). Nearly three-quarters of all depressions develop over Europe, i.e. in the western Mediterranean to the east of the Alps and Pyrenees; relatively few (less than one in 10) originate from the Atlantic. Movement of Mediterranean depressions is predominantly eastwards over the Balkans, Turkey, the Black Sea and into north-west India, but complicated by topography and incursions of cP air from Russia. Although the winter half-year is the rainy period, the actual number of rain days is relatively low: between 6 (northern Libya) and 12 (south-east Spain) per month due to the dominance of anticyclonic circulation types over the area.

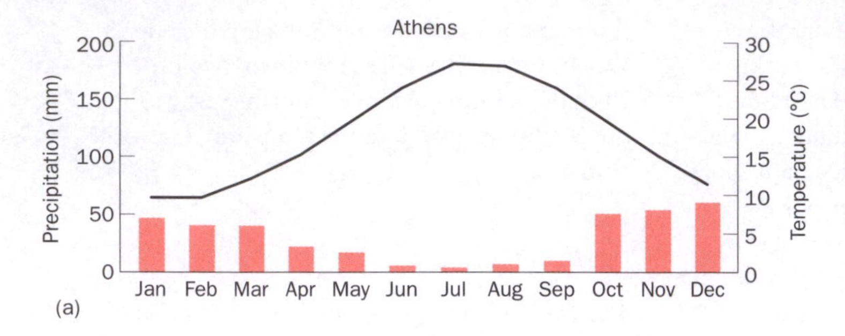

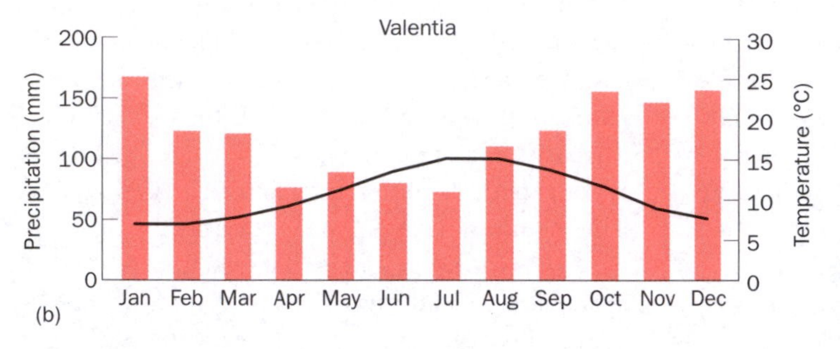

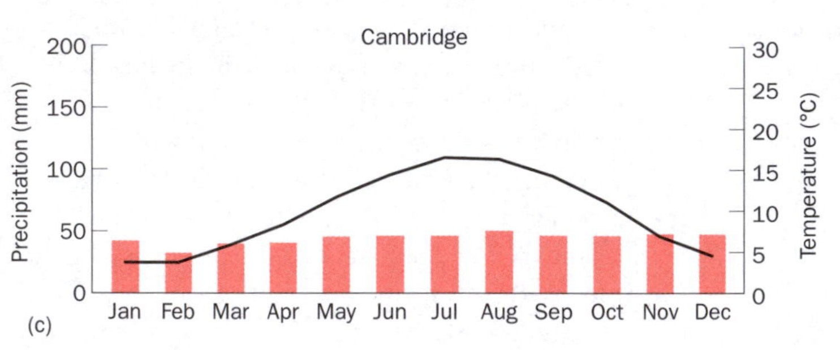

Figure 9.8 Monthly mean precipitation and temperature regimes reflecting (a) Mediterranean (Athens, Greece); (b) temperate maritime (Valentia, Ireland); and (c) near 'continental' (Cambridge, England) climates, 1961–90

Winter is followed by a long and unpredictable spring lasting from March to May. By April a weakened Eurasian high-pressure cell, combined with the expansion of the Azores anticyclone, favours the northward displacement of depressions. By this time, sea surface temperatures are relatively cool, giving rise to more stable conditions even in higher-latitude air. By mid-June an expanded Azores anticyclone, and low pressure across the Sahara, lead to northerly airflows. Subsidence in the large-scale anticyclonic region tends to weaken depressions, and the daily weather increasingly reflects local sea breezes. A number of important regional winds – such as the hot, dry and dusty scirocco of North Africa – also emerge during the summer half-year.

(b) Temperate maritime

The Alps separate the temperate maritime climates of north-west Europe from the Mediterranean climatic region to the south-east. The maritime temperate regime is represented by Valentia in south-west Ireland, and is characterised by all year round depressions and hence precipitation in all seasons (Figure 9.8(b)). The synoptic situation of the North Atlantic is dominated by the Icelandic Low, the Azores High and by the Siberian anticyclone in winter. These pressure patterns cause depressions to track towards either the Norwegian or the Mediterranean seas in winter, and along a more westerly path in summer. The summer minimum in precipitation as a consequence of less vigorous depressions is evident in the temperate coastal climate but is less marked than in the Mediterranean. Summer temperatures are also relatively low due to frequent cloudy periods. However, the most significant feature of the temperate maritime climate is the 'equable' winter climate arising from the warming influence of the North Atlantic Drift (Box 9.3).

(i) Role of relief and altitude

As already described in section 4.3, relief and altitude exert a strong control over air motions within the climate system. Topography also has

BOX 9.3

THINKING FURTHER

The North Atlantic Drift

Winter temperatures in north-west Europe are about 11 °C more than would be expected for the latitude. This anomaly is due to the moderating influence of the North Atlantic Drift – primarily a wind-driven current that originates from the Gulf Stream off Florida but is fortified by the Antilles Current. The current transports sensible and latent heat from the western Atlantic towards Europe at a speed of 16–32 km/day. When polar or Arctic air masses to the south-east of Iceland come into contact with the warm water, temperatures within the air may rise by 9 °C. In the absence of major topographic barriers in Europe, prevailing onshore winds warmed by the current are able to influence temperatures up to 1,000 km inland. With increasing distance from the coast there is a gradual reduction in precipitation amounts and an increase in the temperature range due to continentality effects. The exception is Scandinavia, where the mountain divide between Norway and Sweden produces a sharp thermal gradient from west to east, and up to 80 per cent less precipitation on the lee side.

a marked effect on the climate patterns across Europe. Mountainous regions tend to experience significantly lower temperatures, and higher wind speeds and precipitation totals than adjacent low-lying areas. Mountain barriers may also promote the development of lee depressions, formed when a westerly airflow is forced over a north–south barrier causing divergent (anticyclonic) conditions over the ridge, and convergent (cyclonic) conditions in the lee. Such phenomena are common in winter to the south of the Alps when there is a low-level, north-westerly airflow. When this occurs, cold katabatic air from the high mountains and plateaux is drawn into the lee of the passing depressions, for example the mistral of the Rhone Valley in France, the tramontana of Catalonia, Spain, and the bora of the northern Adriatic. Air can also be induced to flow from higher-pressure cells in the Mediterranean northwards towards western and central Europe when low-pressure systems prevail there. When air flows from the Mediterranean and northern Italy across the Alps, loss of moisture on the southern slopes and dry adiabatic warming on the northern side (see orographic precipitation model in section 2.3.5) cause the warm dry wind known as the föhn. As the föhn descends the northern slopes of the Alps in the Aar, Rhine and Inn valleys, temperatures may rise by 5–6 °C, with clear skies and excellent visibilities.

Although the relief of the British Isles is not nearly as marked there are still clear divides between the Atlantic-dominated climates of the north-west, and the near 'continental' climates of the south-east. The orography of upland regions can increase the amount of rainfall associated with frontal depressions and unstable air over relatively short distances. For example, and as described in section 9.2.4(b), annual precipitation totals in the western mountains of Scotland, the Lake District and Wales average more than 2,400 mm/year, compared with less than 700 mm in lowland Scotland, 250 km to the east. There is also a marked spatial gradient in the frequency of rain days and seasonality of precipitation. The north-west is more directly affected by Atlantic depressions, leading to a winter season rainfall maximum and over 230 rain days/year. Conversely, central and eastern districts as represented by Cambridge in south-east England (Figure 9.8(c)), are less influenced by depressions. They exhibit a weak summer maximum of rainfall, reflecting higher contributions from convective thunderstorms, and less than 170 rain days/year.

Key ideas

1. Air masses are mobile bodies of air whose physical properties (temperature, lapse rate, humidity, etc.) are relatively uniform over large areas and reflect the character of underlying source regions (Arctic, polar, tropical, maritime or continental).

2. Fronts are zones of separation between contrasting air masses (i.e. mT and mP), marked by characteristic patterns of pressure, clouds, precipitation and wind shear.

3. The 'conveyor belt' model is an important advance on the classic model of frontogenesis because it highlights the three-dimensional structure of frontal systems.

4. Travelling anticyclones are mobile, synoptic-scale features characterised by descending air, generally clear skies, local cloud formation and below-average rainfall in summer, often associated with air pollution episodes.

5. Weather type classifications are useful for cataloguing long-term changes in atmospheric circulation, for understanding extreme events, and for testing the realism of climate models.

6. The Mediterranean and temperate maritime climates of north-west Europe are largely differentiated by the relative degrees of continentality, and by the frequency, seasonality and location of blocking anticyclones and depression tracks.

9.3 Past climate variations

Climate variability can be detected at a range of temporal scales: from apparently random, extreme precipitation events operating over hours or days, through to periodicity in temperature at the millennial scale. The relative significance of natural versus human causes of climate change has become the subject of considerable debate.

The following sections focus on notable climatic events across north-west Europe, and examine the dominant *natural* mechanisms operating at different temporal scales. Human-induced climate variability will be discussed in section 9.4 with reference to climate change scenarios for the UK.

9.3.1 Extreme events

Our attitude to climate is largely governed by the frequency and intensity of extreme weather events. Public awareness has been heightened by the apparent volatility of weather patterns over the last decades of the twentieth century, and the possibility that such events could be precursors to adverse changes in climate over the twenty-first century. However, even in Europe, where the density, quality and homogeneity of climate records are relatively high, it is not yet possible to unambiguously isolate human-driven climate trends from natural variability using operational monitoring networks alone.

Public attitudes to climate are also driven by how different societies and environments cope with climate-related disasters. During the last two to three decades there has been an increasing impact and loss from weather-related disasters (see Figure 12.2). As previously mentioned, this may be due to an increasing frequency or intensity of extreme events. What is more clear is that the effects of climate events are being magnified by increasing populations, growing poverty, poor land-use planning and environmental exploitation. The effects of extreme weather on human society and environment are also heightened by existing mainstream rehabilitation programmes that fail to fully address the damage caused. These and related issues are more fully discussed in Chapter 12. Here, three events from the UK (the protracted 1988–92 drought, the summer 1995 heatwave/drought, and the autumn/winter 2000/1 flooding) and one from central Europe (the extensive summer 2002 flooding) are used to illustrate how persistent and devastating climate extremes can be.

(a) The 1988–92 UK drought

The extraordinary weather and river flows experienced in the UK between 1988 and 1992 may provide a useful analogue for the future climate (see section 9.4). The event was noteworthy for the exceptional persistence of rainfall deficits punctuated by abrupt phases of very wet winter weather (e.g. winter 1989/90 was the wettest in England and Wales on record). At that time, the years 1988, 1989 and 1990 were successively the warmest since records began in 1659. According to a combined hydrological/plant-stress index, the period 1989–91 witnessed the most severe drought at Kew, Surrey, since 1698. In parts of eastern England the 1988–92 runoff deficit was the most extreme for at least 150 years. As a consequence, the winter of 1991/92 produced river flows in eastern, central and southern England that were among the lowest on record, with some eastern aquifers receiving less than 30 per cent normal recharge. Rapidly falling groundwater levels were eventually arrested by very unsettled weather and intense storms at the end of September 1992, but the legacy of the drought persisted across many southern and eastern districts into 1993.

Paradoxically, the four-year period following spring 1988 was the wettest over Scotland since records began in 1869, underlining the highly regionalised nature of the drought. The origin of the north-west/south-east divide in the weather lay in a series of high-pressure systems centred over or to the south of the British Isles, culminating in the highest recorded air temperature in the British Isles (37.1 °C, measured at Cheltenham on 3 August 1990, now superseded by a temperature of 38.5 °C at Faversham in Kent on 10 August 2003). However, persistent south-westerly and westerly airflows over northernmost parts of the British Isles led to pronounced winter rain shadow effects over eastern Scotland. For example, during the winter of 1990 Fort William (in the west) had the wettest February on record, while Aberdeen (in the east) enjoyed its sunniest February of the twentieth century.

(b) Summer 1995 UK heatwave and drought

The short-lived but intense drought of summer 1995 provided the UK economy with a foretaste of possible water supply difficulties to come under future climate change (see section 9.4). The winter of 1994/95 was the wettest for Britain as a whole since 1869, giving no indication of the drought that followed. In fact, excessive amounts of precipitation between 20 and 30 January 1995 over much of western Europe were responsible for widespread flooding on the continent. For example, water levels reached record heights on the River Rhine and several of its tributaries, whereas the centre of Cologne was flooded, and an area of about 200 km^2 was inundated by the River Meuse in The Netherlands.

However, the wet winter was followed by an April–August England and Wales rainfall total that was the lowest for any 5-month period in over 200 years. During this period a northward extension of the Azores high-pressure area deflected most frontal systems and brought subtropical air masses to the British Isles. Throughout July and August, much of central and southern England recorded only a few days with showery rain, and dry spells of 30 days were common. August 1995 was also the second warmest month in the Central England Temperature series since records began in 1659 (section 9.3.3).

Like the record drought of 1976, the shorter-lived 1995 drought was notable for its large spatial extent. Across the Anglian, Severn Trent, Southern, Thames and Yorkshire Environment Agency regions, estimated return periods for the drought exceeded 200 years. Fortunately, the abundant rainfall of the preceding winter ensured groundwater levels and river flows sustained by the chalk aquifer of lowland England remained mostly within their normal range. However, small reservoirs in the impermeable catchments of Cornwall, West Yorkshire and the Lake District were seriously depleted by late summer as the rate of resource demand exceeded replenishment. The drought was terminated by the wettest September since 1918 but, because few frontal systems penetrated to the north, the water resource outlook remained fragile in parts of northern England and the Midlands well into the autumn. Perhaps the greatest legacy of the 1995 drought was to impress upon UK water resource managers that the historical rarity of warm dry summers would no longer be a reliable guide to their contemporary and future frequency. This is because the sequence of serious UK droughts in 1976, 1983, 1984, 1989, 1990, 1994 and 1995 has no modern close parallel.

(c) Autumn and winter 2000/1 UK floods

The 4-month period September to December 2000 recorded the highest rainfall total for England and Wales since records began in 1766 (Table 9.2). The resulting riverine flooding was the most extensive since the snowmelt-generated events of March 1947. Although peak flows in Sussex and Yorkshire had return periods exceeding 100 years, the 2000/1 flooding was noteworthy because of the geographical scale and duration of the inundations. For example, the winter of 2000/1 produced the highest 90-day river flow volume on record for the Thames at Teddington, but a peak daily mean discharge that ranked only fiftieth since records began in the 1880s. None the less, during the course of the event about 9,000 UK premises were flooded (some on several occasions), and the 10-year-old Aldington flood defence dam in Kent failed.

The widespread flooding began with localised, mostly urban, flooding in the English lowlands following unusually wet conditions in September. Extensive and sustained frontal rainfall in late October/early November then resulted in the appearance of floods across much of the rest of the UK (Plate 9.3). Rapidly rising groundwater compounded the problem from late November as water tables reached the surface in many permeable catchments. Following a brief respite in January, the succession of frontal systems continued until April 2001, thereby establishing a new 8-month maximum for the England and Wales rainfall series (see Table 9.2).

A study commissioned by DEFRA and the Environment Agency (EA) concluded that the autumn 2000 flooding could not in *itself* be attributed to climate change. However, there is evidence of increased winter rainfall and river flow extremes in the UK over the last 40 years, especially for longer duration events (such as those accumulated over 30- or 60-day periods). There is also some evidence that very heavy precipitation events are contributing proportionately more to

Table 9.2 Maximum 4- and 8-month rainfall totals for England and Wales, 1766–2001

Rank	4-month total (mm)	% LTA*	End month	8-month total (mm)	% LTA*	End month
1	640	178	12/2000	1033	166	04/2001
2	624	175	01/1930	1014	152	01/1853
3	599	176	10/1799	988	148	02/1961
4	591	167	11/1852	946	142	01/1873
5	563	158	01/1961	941	141	01/1769
6	548	161	10/1903	915	137	01/1840
7	546	161	10/1775	913	147	04/1877
8	538	151	01/1873	908	136	01/1928
9	534	164	02/1915	903	139	12/1872
10	527	162	02/1877	902	135	02/1904

*LTA is the long-term average (for the interval ending in the given month).
Source: Marsh (2001)

winter season totals than they did 40 years ago. This finding is consistent with climate model projections of human-induced climate change at regional scales (see below). However, the DEFRA/EA study also conceded that modelled changes in extreme rainfall frequencies since 1860 are small compared to natural variability.

(d) Summer 2002 European floods

The summer of 2002 witnessed some of the worst flooding on record in central Europe. The weather of the first few weeks of August was characterised by high pressure over northern Europe and a series of low-pressure systems advancing along a southern path across the Mediterranean and into central Europe. This synoptic situation resulted in abnormally high precipitation throughout the Mediterranean, particularly on the western coast of Greece where totals were ~500 per cent the usual for August. The disastrous floods in central Europe were caused by the passage of two distinct depressions and accompanying frontal systems between 6 and 15 August (Plate 9.4). The Czech Republic was particularly hard hit by the rainiest sectors of both depressions, which advanced very slowly northwards over western Bohemia. Parts of the Krušné hory Mountains and the Českomoravská vrchovina Highlands received over 300 mm in a single day (e.g. Cínovec station recorded 312 mm on 12 August, which is estimated to be three times the daily total that normally occurs once in 100 years).

According to the Czech Hydrometeorological Institute the storms of 6–15 August yielded ~5 km^3 of rainfall in the Vltava Basin. This produced a peak discharge of 5,160 m^3/s in Prague on 14 August at noon with an estimated recurrence interval of 500 years. However, it was difficult to obtain reliable measurements of flood volumes at such large river discharges because some gauging stations were submerged or destroyed, access to others was restricted and, where flow depths were recorded, the values were beyond the range of calibrated discharge rating curves. Nonetheless, flood marks on the River Vltava's left bank near the Charles Bridge indicate that the water level in August 2002 was 140 cm above that of September 1890, 75 cm higher than the snowmelt event of March 1845, and 55 cm above the 1784 mark.

The summer 2002 floods resulted in more than 100 fatalities across central Europe and left tens of thousands with damaged homes and property. The cost of clean-up and reconstruction has been estimated at 2 billion euros for the Czech Republic, and as high as 15 billion euros for Germany – a bill that will be met using unspent funds released from the EU's structural budget. The flooding also devastated tourism in the historic cities of Prague and Budapest, while factory and farm production ground to a halt across large parts of central Europe. It is sobering to think that the latest climate model projections suggest summer flood risk associated with storms of several days' duration is expected to increase.

9.3.2 Seasonal variability

(a) Natural seasons

The term 'singularity' is used to describe recurrent weather features around specific dates of the year. The idea that certain weather patterns tend to recur at specific times and seasons of the year has a long history founded in weather lore. Professor Lamb examined evidence for natural seasons, by identifying pronounced peaks or troughs in the annual frequency curves of the seven main LWTs (section 9.2.4). Spells of characteristic weather types exceeding 25 days in length were used to identify five main seasons: high summer (18 June–9 September), autumn (10 September–19 November), fore-winter (20 November–19 January), late winter or early spring (20 January–29 March) and spring or early summer (30 March–17 June). A number of singularities were then identified within the main seasons. For example, using daily weather charts for the 50 years 1898–1947, Lamb showed that the

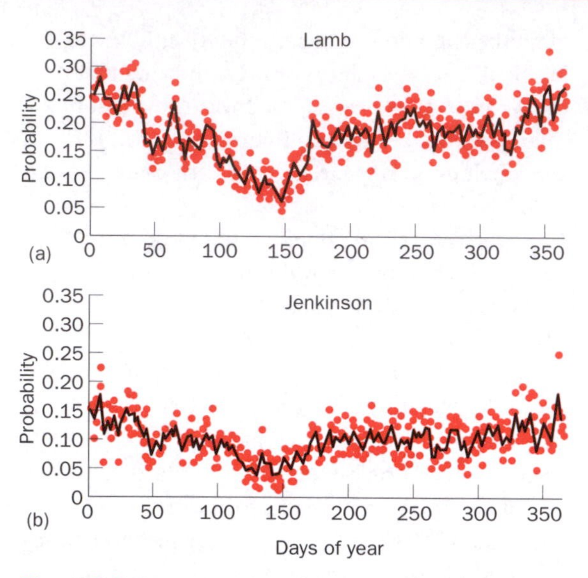

Figure 9.9 Annual cycle in the probability of a westerly weather type over the British Isles according to the Lamb catalogue 1861–1997 (upper panel), and the Jenkinson method 1881–2001 (lower panel)

westerly type occurred on about 50 per cent of days between 5 and 11 January, rising to 60 per cent of days falling on 8 January. The average frequency and timing of the westerly weather type for both the subjective and objective (Jenkinson) LWT catalogues are shown in Figure 9.9. Both data sets indicate a clear seasonal cycle in the prevalence of the westerly type, with a marked minimum likelihood of occurrence between 8 and 26 May, a period coinciding approximately with the maximum frequency of northern and central European highs. The maximum frequency of westerly types occurs on 17 December (33 per cent) and on 25 December (25 per cent) for the Lamb and Jenkinson catalogues respectively (Box 9.4).

(b) Physical basis

Several factors have been linked to the seasonal development of the large-scale circulation of the northern hemisphere. The annual sequence of events is as follows. First, the latitudinal gradient

BOX 9.4

THINKING FURTHER

Limitations of the singularity concept

Many weather spells do not stand up to rigorous statistical testing. For example, it is possible to synthesise false singularities using simple models of weather pattern sequencing. Given the raw data in Figure 9.9 it can be shown that using chance alone there is roughly a 1 per cent chance of any given date (e.g. 1 April) having more than 34 westerly days in a 137-year record. This implies that during the course of the year one would expect about 4 such days with a very high frequency of westerly weather. This is, in fact, far fewer than the actual number of dates with higher than expected numbers of westerly days in the Lamb catalogue, providing some support for the existence of winter (spring/early summer) maximum (minimum) westerly spells.

The value of the singularity for analogue forecasting hinges on a number of factors, in particular the homogeneity of the underlying data set. This is because the analogue method rests on the assumption that comparable pressure charts on different days yield comparable weather at the local scale. In other words, past pressure patterns can be used to forecast future weather based on historic transitions from one pattern to the next. Also, that similar weather tends occur at roughly the same time each year because of singularities. In practice, the atmospheric circulation patterns associated with a given singularity may yield very different weather at the same site at different times, or simultaneously at different locations. The indeterminate weather patterns that mark transitions between spells might also be legitimately classified as an 'unsettled' spell. Above all, the singularity concept rests on the assumption of a stationary climate regime, i.e. past weather episodes are representative of future episodes. In practice, anthropogenic forcing of the climate system is expected to change the frequency and intensity of weather patterns over coming decades.

of the energy balance and zonal index attain their maximum in midwinter, bringing more westerly airflows (see Figure 9.9). Second, winter maximum snow and sea-ice cover affect the position of extratropical storm tracks around February/March. Third, radiative cooling in high latitudes results in the development of the major polar anticyclone over northern Canada in March/April. Fourth, the meridional energy balance gradient and intensity of Atlantic circulation attain their minimum in May (note the minimum in westerly types at this time in Figure 9.9). Fifth, there is a northward (southward) displacement of cyclone tracks over western North America (the Atlantic–European sector) linked to contrasts in heating between landmasses and the Arctic Ocean in July/August. Sixth, the deepening of the Icelandic Low and reinvigoration of the Atlantic circulation in September. Finally, the southward displacement of cyclone tracks associated with high-latitude cooling and intensified Circumpolar Vortex from October onward.

9.3.3 Interannual variability

(a) Central England Temperature series

Long catalogues of weather patterns and airflow types, together with homogeneous instrumental records and proxy or reconstructed climate variables, are essential for detecting climate change and its impact. The Central England Temperature (CET) series is the longest instrumental record of temperature in the world. The monthly mean surface air temperatures are available for the period from 1659 to present for a region representative of the English Midlands. The daily series begins in 1772 and is currently compiled from observations at Preston, London and Bristol. Since 1974 the daily data have been corrected for urban warming.

Figure 9.10 shows the record of winter (December to February), spring (March to May), summer (June to August) and autumn (September to November) mean temperature anomalies for the CET from 1861 to 2001. During the twentieth century annual mean temperatures showed a warming of +0.6 °C, with six of the warmest years in the twentieth century occurring since 1989 in rank order: 1999, 1990, 1997, 1995, 1989 and 1998. It is worth pointing out that 1998 was the warmest for the planet as a whole. Relative to 1961–90 all these years were between 0.9 and 1.2 °C warmer than average. The years 1994 and 2000 were also unusually warm with anomalies close to +0.8 °C. Within the year, warming has been greatest from midsummer to late autumn: July (+0.8 °C), August (+1.2 °C), September (+0.9 °C), October (+1.2 °C) and November (+1.3 °C) respectively.

The daily CET record has been used to investigate the annual frequency of 'hot' (mean temperature above 20 °C) and 'cold' (mean below 0 °C) days. Since the eighteenth century, the number of cold days has fallen from around 15–20 per year to around 10 per year presently. Most of this change occurred prior to the twentieth century and is, therefore, probably unrelated to human influences on climate. At the same time, there has been an imperceptible rise in the frequency of hot days in the twentieth century, despite 1976, 1983, 1995 and 1997 returning some of the highest frequencies of such days. However, the daily CET also indicates that the thermal growing season for plants has increased by about 30 days since 1900, and that central England presently enjoys longer frost-free spells than at any time during the pre-industrial era.

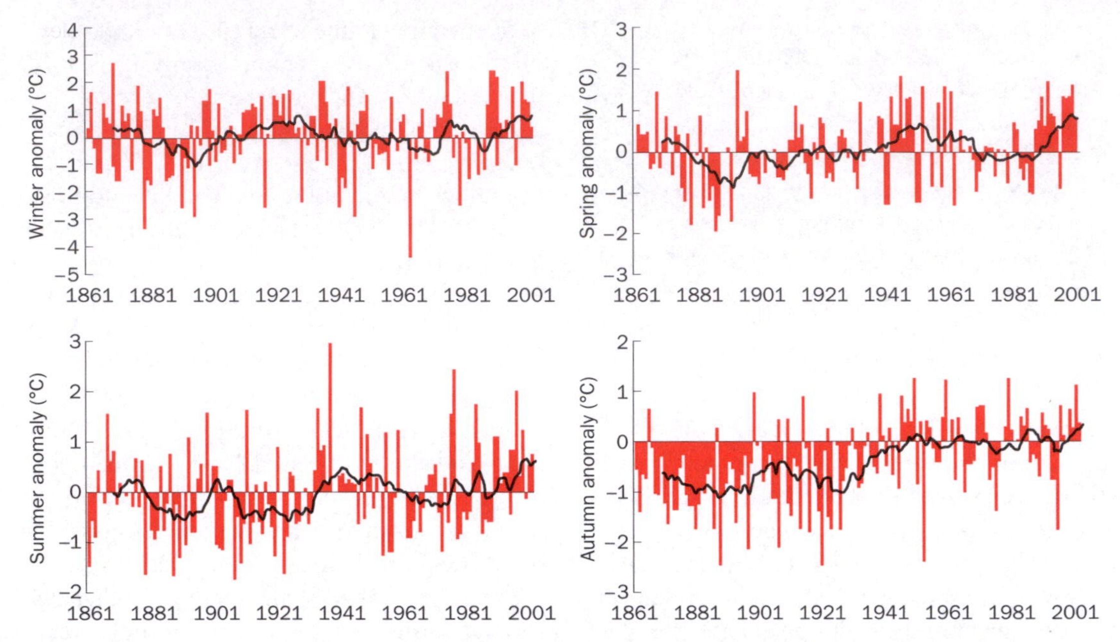

Figure 9.10 Central England Temperature (CET) record for the period 1861–2001. All seasonal anomalies are calculated with respect to the corresponding average for 1961–90 which is conventionally used as the 'baseline' condition. For example, temperatures in autumn prior to the 1940s were well below average; those in spring and summer after the 1980s were above average

(b) England and Wales Precipitation series

As with temperature, there is a good quality precipitation record for England and Wales going back to the mid eighteenth century. Using the England and Wales Precipitation (EWP) series (Figure 9.11(a)) it can be seen that there is no long-term trend in the amount of annual precipitation over England and Wales (the same also applies to Scotland). However, there is considerable variability in the annual precipitation with wet decades in the 1770s, 1870s, 1920s and 1960s, and a protracted dry period between the 1880s and 1900s, with lesser dry decades in the 1800s and 1850s. Individual extreme years also figure prominently. For example, 2000 was the wettest year in the twentieth century with 95 per cent more precipitation than in 1921, the driest (see section 9.3.1). Overall, the wettest year on record was 1872 (with 141 per cent of the 1961–90 average), and the driest was 1788 (with 67 per cent of the average).

A remarkable feature of the EWP series is the changing ratio of winter to summer precipitation (Figure 9.11(b)). The most recent 30 years have witnessed the greatest disparity between winter and summer precipitation totals since records began. Between the first and last 30 years of the record, annual precipitation increased by only 24 mm, yet winters became 55 mm wetter and summers 45 mm drier. Such trends are entirely consistent with increased flooding in England and Wales in recent years, and the latest future climate change scenarios for the UK (see section 9.4.3). The wet winters and dry summers of both 1990 and 1995 are particularly noteworthy in terms of their seasonal ratios, and provide useful analogues of the future climate.

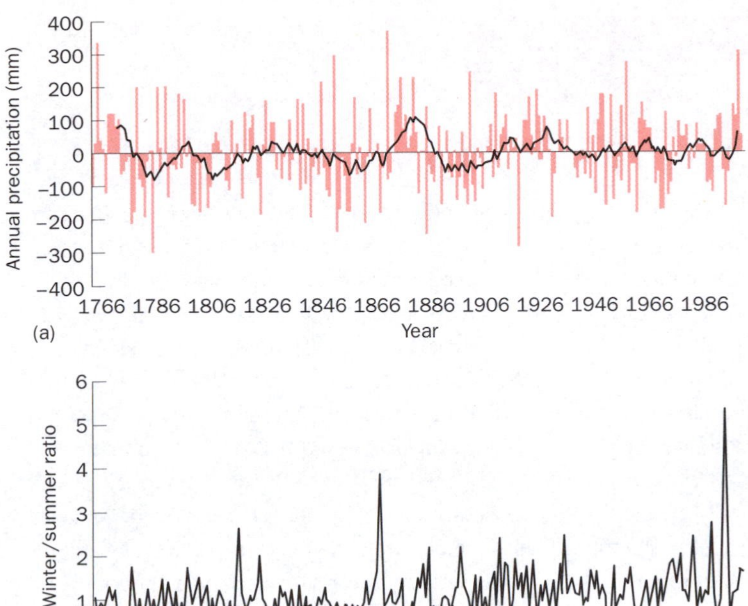

Figure 9.11 England and Wales Precipitation (EWP) (a) annual totals and (b) ratio of winter to summer precipitation, 1766–2001. Anomalies are relative to the 1961–90 average. The smooth line shows the 10-year moving average

(c) Links to the North Atlantic Oscillation

Observed seasonal changes in EWP since the 1960s have coincided with increased westerly airflows over maritime western Europe, manifested by more northerly tracks for Atlantic frontal systems. Although the Arctic Oscillation is the dominant mode of atmospheric variability in the northern hemisphere (section 8.1.3), its local expression is the North Atlantic Oscillation (NAO). The NAO reflects variations in the pressure gradient between the Azores High and Iceland Low. When the gradient is steep (shallow), the airflow across north-west Europe is more (less) westerly. The NAO is particularly important in *winter*, when it exerts a strong influence on the hydroclimatology of both sides of the Atlantic.

Strong positive phases of the NAO are associated with above-normal temperatures in the eastern USA and across northern Europe, but lower than normal temperatures in Greenland, south-east Europe and the Middle East (see similar effects for the Arctic Oscillation in Tables 8.1 and 8.2). The NAO also has a regional influence. For example, the CET is correlated with the NAO index (Box 9.5) in all seasons with the exception of summer, but the strongest correlation is in winter, with positive NAO corresponding to above-average temperatures. The NAO index is also strongly correlated with the westerly weather type and negatively correlated with the winter frequency of cyclonic weather types. As a consequence, a strongly positive NAO index generally coincides with above-average rainfall over northern Europe and Scandinavia, and drought over southern and central Europe. Again, at the regional scale, a positive NAO index is associated with higher rainfall over Scotland and north-west England, but below-average falls across southern, central and north-eastern England.

Recent trends in the NAO (Figure 9.12) have increased the spatial gradient in UK precipitation between the uplands of the north-west (intercepting the moisture-laden westerly airflows) and the lowlands of the south-east. There is also a weak association between the NAO index and the frequency of severe gales in the winter half of the year in that both increased between the 1960s and 1980s. The NAO has even been linked to the position of the Gulf Stream (see section 8.6.1). It has been suggested that a positive NAO mode and accompanying stronger westerly and trade winds give rise to a more northerly Gulf Stream 2 years later.

THINKING FURTHER

BOX 9.5

The North Atlantic Oscillation (NAO) index

The NAO index is conventionally defined using two main approaches. Method one is based on the normalised mean sea-level pressure anomaly at a station in the Azores (at Lisbon, Gibraltar or Ponta Delgada) minus that for Iceland (at Akureyri, Stykkisholmur or Reykjavik). Normalisation involves subtracting the long-term mean, and dividing by the standard deviation of the data. This gives a new series with mean equal to zero and standard deviation equal to one, allowing equal weighting to be given to the two ends of the pressure gradient.

Method two employs empirical orthogonal function (EOF) analysis of hemispheric mean sea-level pressure anomaly maps. The EOF technique reduces large data sets (in this case gridded pressure data) into a new set of composite variables that efficiently represent the original data. The main advantage of method one over two is that it can be extended back as far as 1820. Conversely, method two provides a measure of changes in the overall pressure *field* of the North Atlantic rather than the pressure *gradient* between just two stations. Figure 9.12 shows the winter (December–February) NAO index derived from sea-level pressures at Gibraltar and Ponta Delgada minus those at Reykjavik since 1820 (i.e. method one). Note the marked upward trend since the 1960s and the high year-to-year variability (see also section 5.3.1).

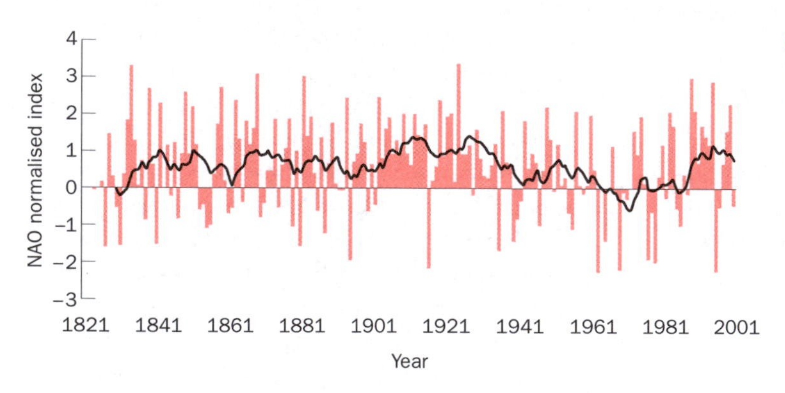

Figure 9.12 The winter North Atlantic Oscillation (NAO) index based on the difference between the normalised sea-level pressures in the Azores and Iceland, 1820–2002. The smooth line shows the 10-year moving average
Source: Climatic Research Unit

9.3.4 Longer-term climate change

As outlined in section 6.3, information on climate change prior to the seventeenth century can only be obtained from documentary sources (diaries, paintings, literary works), and environmental proxy records (tree rings, pollen analysis, ice cores, marine and lake sediments, peat deposits, corals). It can also be seen in section 6.4 that evidence from these sources has alerted us to possible global-scale climate episodes such as the Medieval Warm Period (AD 900–1200) and the Little Ice Age (AD 1350–1850). Records of ENSO and the NAO, however, are difficult to derive because no proxy translates directly into atmospheric circulation. Instead, the proxy is measuring the environmental *influence* of the circulation patterns, so there is a real danger of circular argument when relating the record to a reconstructed temperature series. Indirect cause–effect relationships (say between patterns of atmospheric pressure and tree-ring growth) are also known to vary in strength over long periods of time. It is not surprising, therefore, that there is little agreement between various indices of the NAO for periods prior to 1820. None the less, spectral analysis (a technique for identifying the cyclical patterns) of observed and reconstructed NAO indices suggests periodicities of 7 years, 19–23 years, 50–68 years (in spring), and 54–88 years (in summer). As we shall see below, the ability of climate models to replicate such behaviour is a powerful test of their physical realism.

Key ideas

1. The extraordinary drought of 1995 and winter flooding of 2000/1 provide useful climate change analogues for the UK, as did the catastrophic summer floods of 2002 for central Europe.

2. Recurrent weather patterns around specific dates (singularities) have long been used as a basis for long-range weather forecasting, but the physical explanation and persistence of such 'spells' remain uncertain.

3. Lengthy, homogeneous meteorological records (such as the Central England Temperature series) are invaluable for detecting climate change(s) and for quantifying natural levels of climate variability in the pre-industrial era.

4. Interannual variations in the North Atlantic Oscillation influence patterns of winter precipitation and temperatures across north-west Europe, and exhibited an unusually strong phase between the 1960s and 1990s.

5. Proxy climate records (based on information held in tree rings, ice cores, lake sediments, etc.) suggest that mean temperatures in the northern hemisphere are now higher than at any time in the last 1,000 years.

9.4 Climate change projections for the UK and Ireland

9.4.1 Climate change: internal and external factors

As we have seen in previous sections (3.5 and 6.5), natural climate variability results from the action of both external and internal factors. Such forcing mechanisms cause the continuous redistribution of heat between the ocean and the atmosphere, and between different states, such as vapour to liquid, or water to ice. Such exchanges occur over timescales from seconds to millennia. Globally, the two most important modes of internal climate variability are the El Niño–Southern Oscillation (section 5.4) and the Arctic Oscillation (section 8.1). Both display characteristic patterns of variability as expressed by temperature and precipitation anomalies. As noted above, the NAO correlates with many aspects of the winter climate across north-west Europe. For instance, a particularly strong trend towards the positive phase of the NAO since the 1960s has coincided with marked increases in winter precipitation and warming over the region.

Other internal factors that are responsible for climate variability include, for instance, land surface change (which affects the amount of energy reflected at the earth's surface) and modifications in ocean circulation (which redistributes heat from equatorial regions to high latitudes). Because these latter mechanisms are difficult to observe or understand, and may account for unexplained hemispheric temperature variability, they are often referred to as 'noise' factors especially for the purposes of climate modelling.

On the other hand, external factors are directly linked to variations in the receipt of solar radiation by the earth. Ice core records of the cosmogenic isotope ^{10}Be imply generally increasing solar activity during the twentieth century, a fact which has contributed to rises in global temperature.

Furthermore, a number of internal factors including both human agencies and natural events can modify the input of solar radiation. Reconstructions of sulphate aerosols injected into the stratosphere by the volcanism of the seventeenth century suggest significant reductions in downward short-wave radiation, coinciding with the climate cooling of the Little Ice Age. More recent records of electrical conductivity and sulphate deposition in ice cores from Greenland and Antarctica reveal a cluster of volcanic eruptions in the early twentieth century and a quiescent period between 1920 and 1960 (coinciding with a period of below average global temperatures). Since the Industrial Revolution, emissions of anthropogenic greenhouse gases (CO_2, methane, nitrous oxides and chlorofluorocarbons) have contributed to a net radiative forcing of the atmosphere, by reducing the amount of upward emission of long-wave radiation at the tropopause, and by increasing downward emissions from the stratosphere (section 6.5).

9.4.2 Climate model experiments

Taken in isolation or in combination, neither solar variability nor volcanism entirely explains twentieth-century hemispheric temperature change. Contention, therefore, surrounds the extent to which the missing climate variability is attributable to anthropogenic increases in greenhouse gases, or due to internal climate variability, or due to a combination of both. Such theories are now being tested using climate models to quantify the effects of natural variability, along with solar, volcanic and anthropogenic forcing (see Chapter 7).

Energy balance models (EBMs) calculate global mean temperature changes from the difference between incoming and outgoing radiation, given the planetary albedo and heat capacity, and solar influx (see section 7.2.1). More sophisticated EBMs incorporate latitudinal temperature gradients, simple representations of ocean and atmosphere layers, land surface feedbacks and thermal inertia of deep oceans. EBMs have recently been used to investigate the causes of the northern hemisphere temperature changes (see Figure 9.13), using 1,000-year long reconstructions of solar and volcanic forcing, with anthropogenic greenhouse gas forcing since 1850. Comparisons of temperature reconstructions and instrumental records for the same periods suggest that between 41 and 64 per cent of pre-1850 decadal-scale temperature variations were due to changes in solar irradiance and volcanic eruptions. The remaining variance was consistent with climate

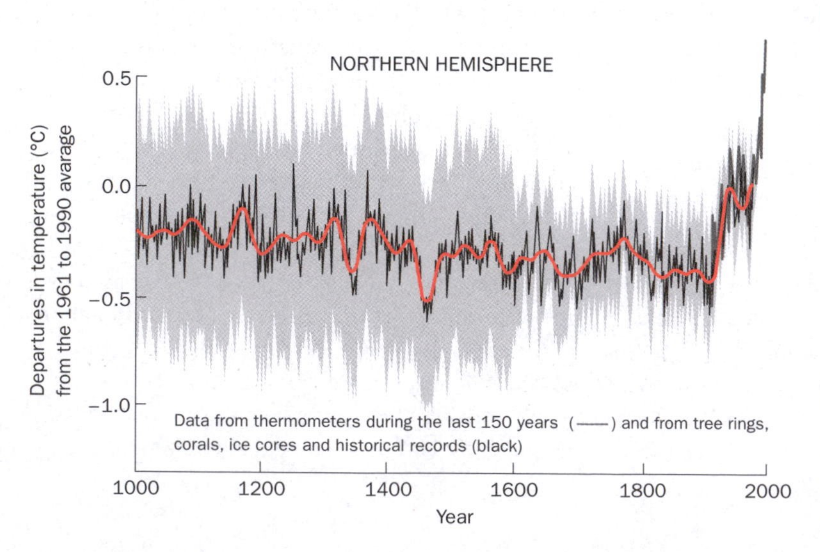

Figure 9.13 Annual (red curve) and 50 year average (black curve) variations in the mean surface temperature of the northern hemisphere for the past 1000 years. Climate trends compiled from proxy data calibrated against more recent thermometer data. Proxy data are taken from tree rings, corals, ice cores and historical records. Annual uncertainties in the proxy record are shown by the grey region. The data shows that the rate and duration of 20th century warming has been much greater than in any of the previous centuries. Source: IPCC

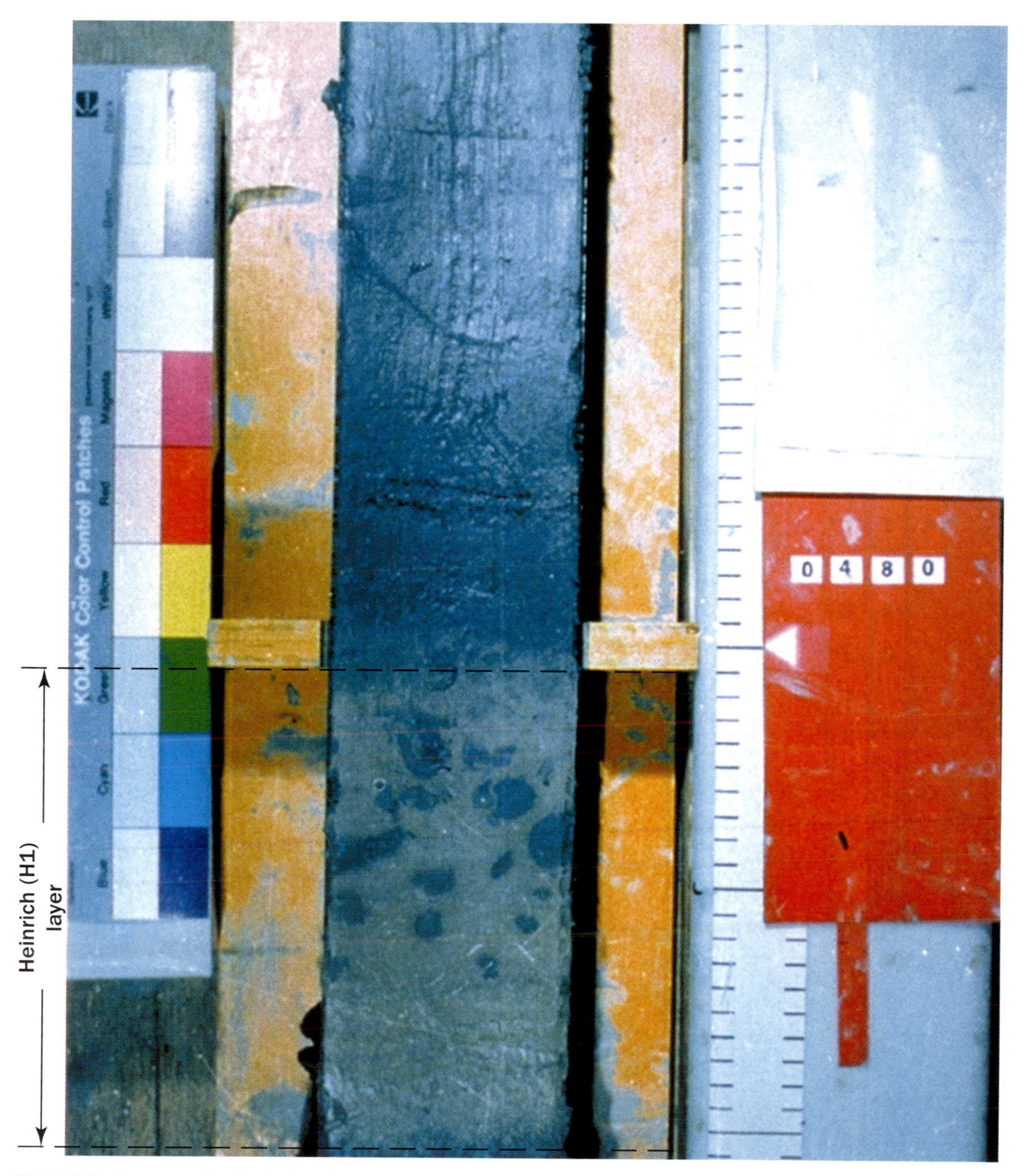

Plate 8.3 Close-up of a North Atlantic sediment core (H1) showing deposition of a 20 cm deep, light-coloured, carbonate-rich Heinrich layer. The black mottles are due to sediment churning by living organisms

Source: NOAA Paleoclimatology Program and INSTAAR, University of Colorado, Boulder, USA

Plate 9.3 Serious flooding of the River Trent valley near the M1 motorway in the English Midlands on 10 November 2000

Plate 9.4 (right) AVHRR image of one of the storms that brought severe flooding to central Europe in August 2002

Source: Dundee satellite receiving station (http://www.sat.dundee.ac.uk)

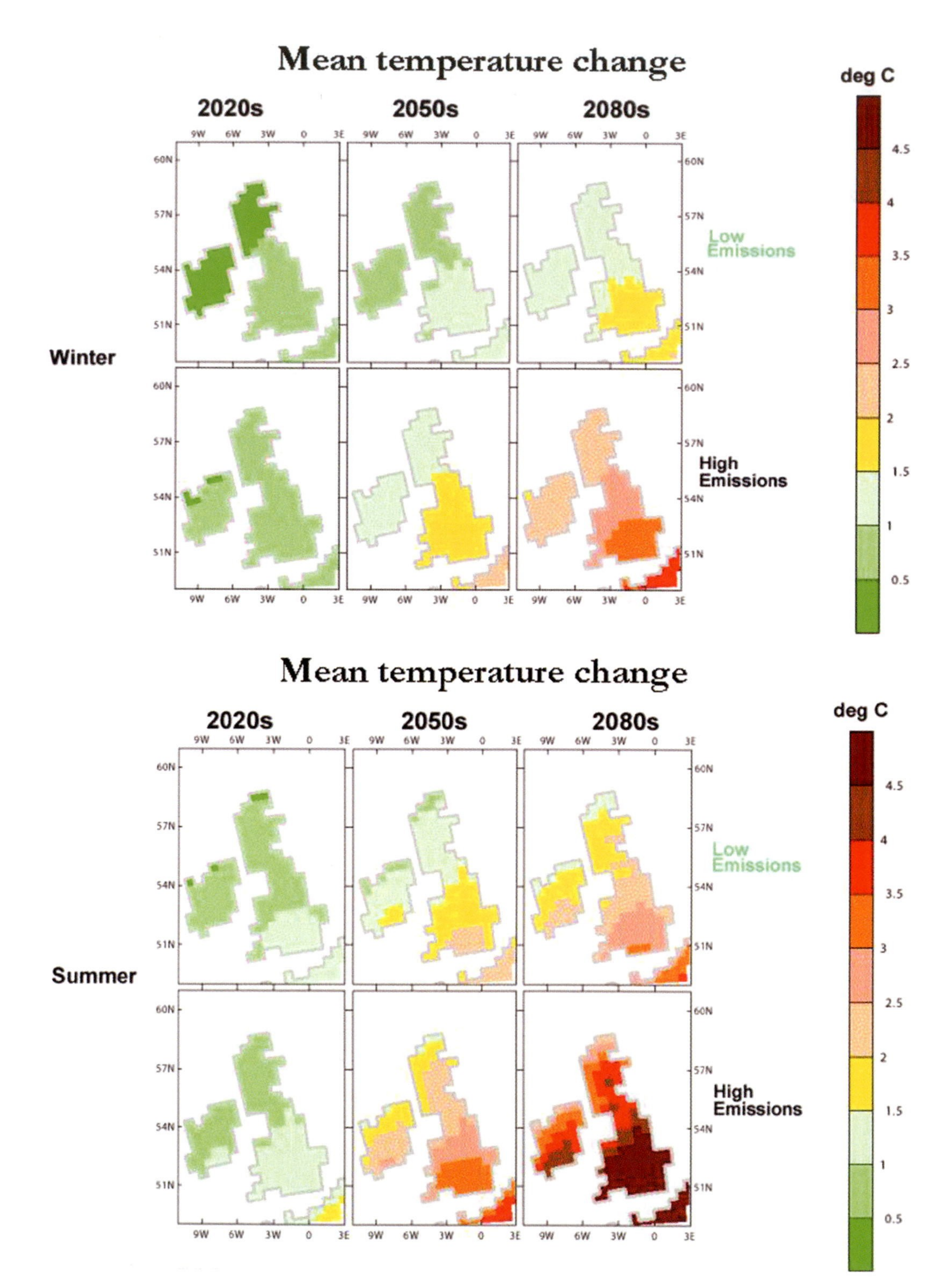

Plate 9.5 Change in average annual, winter and summer temperature for the 2020s, 2050s and 2080s for the low emissions and high emissions scenarios

Source: UKC1P02 Climate Change Scenarios (funded by DEFRA), produced by Tyndall and Hadley Centres for UKC1P

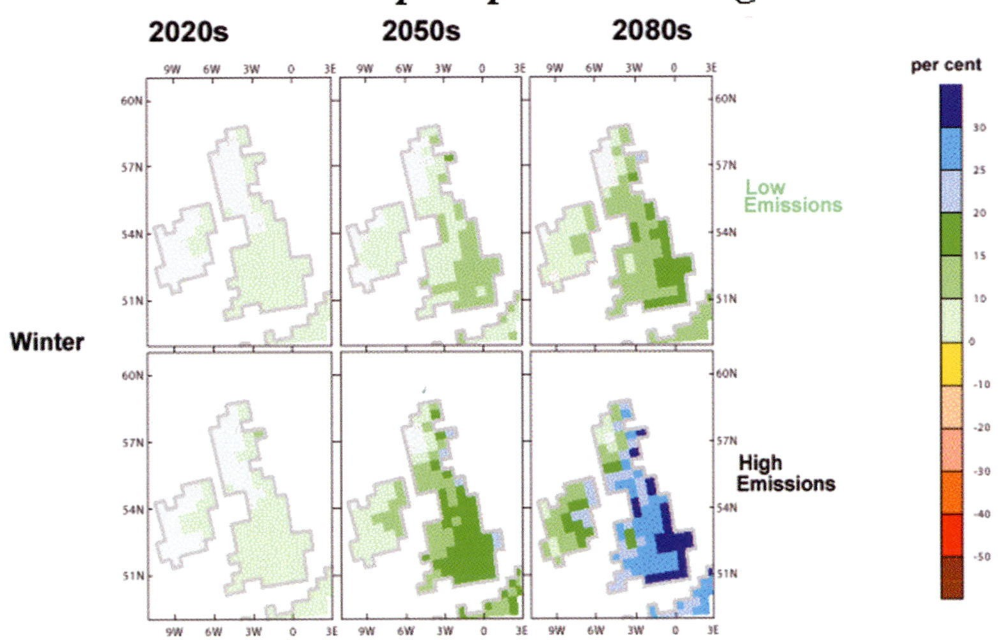

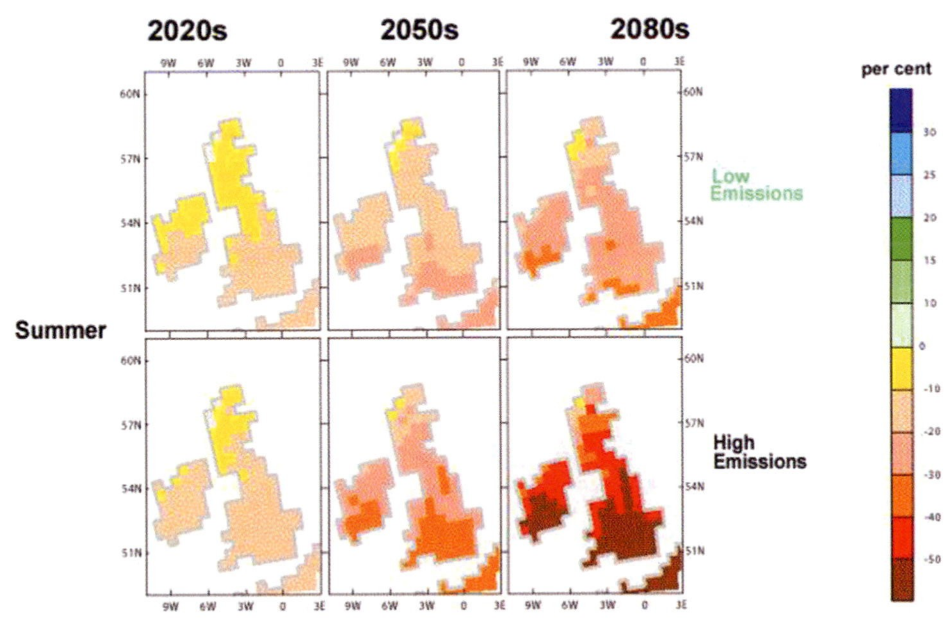

Plate 9.6 Per cent change in average annual, winter and summer precipitation for the 2020s, 2050s and 2080s for the low emissions and high emissions scenarios

Source: UKC1P02 Climate Change Scenarios (funded by DEFRA), produced by Tyndall and Hadley Centres for UKC1P

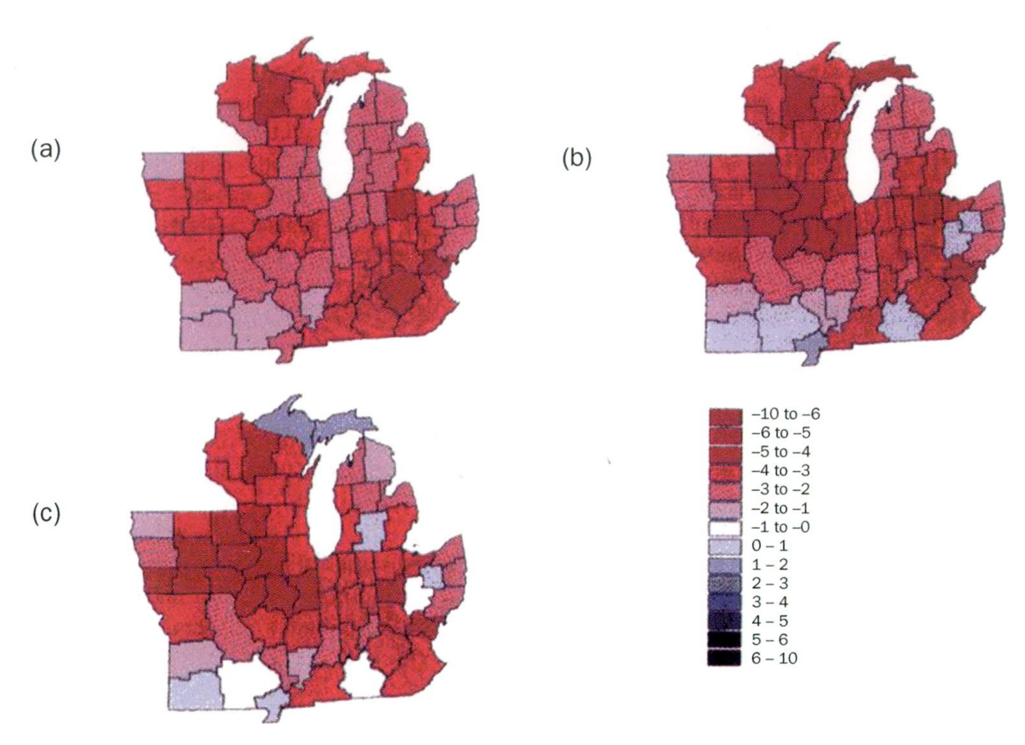

Plate 10.1 Midwest monthly Palmer Drought Severity Index for (a) June, (b) July and (c) August 1988
Source: Adegoke and Carleton (2000)

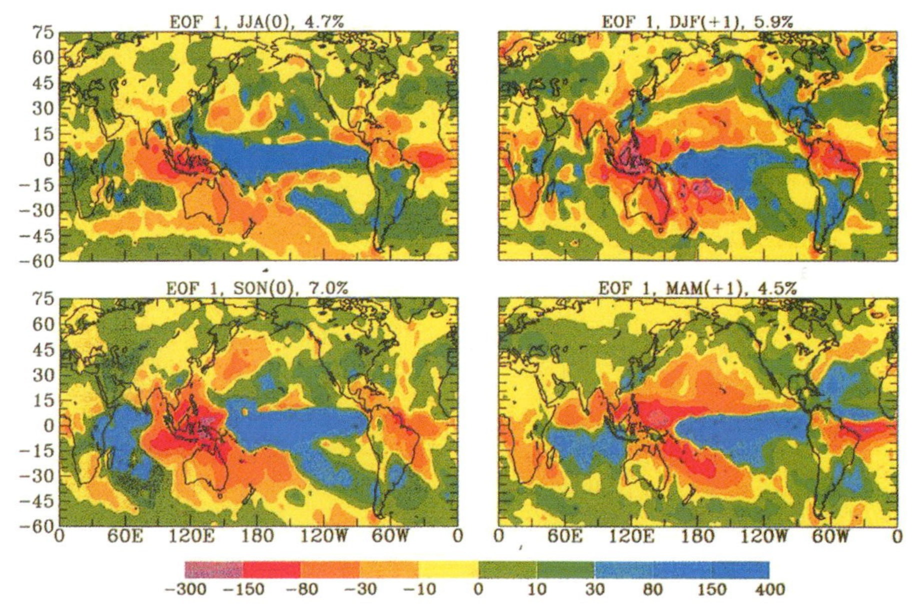

Plate 10.2 Seasonal precipitation anomalies (mm) associated with typical El Niño years
Source: Dai and Wigley (2000)

Plate 11.1 The Cherrapunji climate station in the Khasi Hills, north-east India. This station is located in one of the wettest places on earth

Plate 11.2 One of the biggest dust storms ever recorded by NASA's SeaWIFS satellite on 3 March 2000. The image shows a vast developing cloud of Saharan Desert dust blowing from north-west Africa 1,000 miles or more out over the Atlantic Ocean

Source: NASA SeaWIFS Project

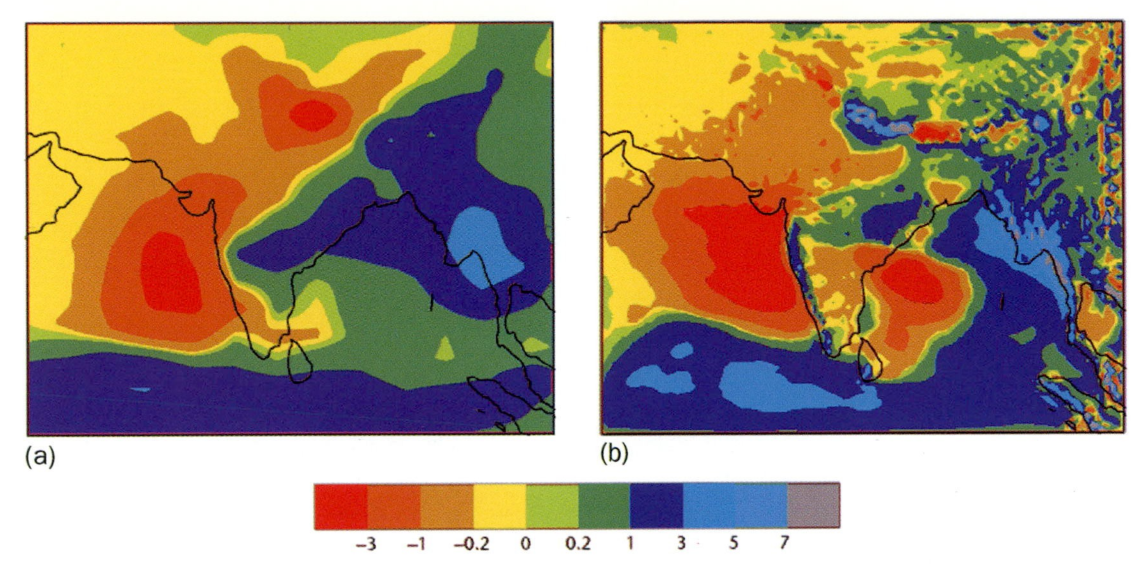

Plate 11.3 The change in summer precipitation over the Indian subcontinent (mm/day) between the middle of the nineteenth century and the 2050s predicted using (a) the Met Office's Hadley Centre GCM and (b) an RCM
Source: Hadley Centre, UK

Plate 12.1 Satellite image of Hurricane 07B moving westwards and making landfall over eastern India (Godavari delta) at midnight on 6/7 November 1996. This hurricane began as a tropical cyclonic low on 4 November, and quickly developed into a tropical storm the following day. It then energised into a severe tropical storm early in the morning on 6 November and into a hurricane by mid-morning on 6 November
Source: NOAA

Plate 12.2 Female 'low caste' rural labourers in the Godavari delta

Plate 12.3 Hurricane shelter, east Godavari

model estimates of unforced, internal climate variability. However, when the EBM was used to remove the effects of solar variability and volcanoes (but *not* greenhouse gases) after 1850, unprecedented warming was still found in the twentieth century. In this case, the temperature increase far exceeded the maximum range of variability witnessed in the previous 1,000 years. This result supports the view that the anthropogenic greenhouse effect has already established itself above the level of natural variability expected for the climate system.

Despite their simplicity EBMs can provide useful insights into climate behaviour at global scales, but they do not deliver information on *spatial* patterns of climate change due to internal/external forcing. In order to simulate the three-dimensional, time-varying, behaviour of the climate system at spatial resolutions of 200–300 km, a coupled ocean/atmosphere general circulation model (O/AGCM) is required (see section 7.2.2). When provided with observed changes in human and natural forcing, state-of-the-art coupled O/AGCMs now produce credible simulations of observed global mean temperature trends over the same period. Hence, O/AGCM experiments have become an important line of evidence in support of the claim that human activities cause global warming.

9.4.3 The UK Climate Impacts Programme (UKCIP02) scenarios

Table 9.3 summarises the most recent climate change scenarios for the UK published by the Climate Impacts Programme UKCIP02 in April 2002. Regionalised (grid box) UK temperature and precipitation projections are shown in Plates 9.5 and 9.6 based on low emissions and high emissions scenarios of future greenhouse gas concentrations. The climate projections are also shown for different seasons, and for three future periods: 2020s (years 2011 to 2040), 2050s (2041 to 2070), and 2080s (2071 to 2100). All changes are expressed with respect to the average 1961–90 climate, which itself may incorporate some climate change.

Table 9.3 A summary of the key results presented in the *UKCIP02 Scientific Report*

- The UK climate will become warmer by between 2 and 3.5 °C by the 2080s, with parts of the south-east warming by as much as 5 °C in summer
- Higher summer temperatures will become more frequent and very cold winters will become increasingly rare
- Winters will become wetter and summers may become drier everywhere
- Summer soil moisture may be reduced by 40 per cent or more over large parts of England by the 2080s
- Snowfall amounts will decrease throughout the UK
- Heavy winter precipitation (rain and snow) will become more frequent
- Relative sea level will continue to rise around most of the UK's shoreline
- Extreme sea levels will be experienced more frequently
- The Gulf Stream may weaken in the future, but it is unlikely that this weakening would lead to a cooling of the UK climate within the next 100 years

By the 2080s, annual temperatures averaged across the UK may rise by 2 °C for the low emissions and by about 3.5 °C for the high emissions scenario, with higher increases of 4.2 °C in the south-east (Plate 9.5). In general, there may be a greater warming in summer and autumn than in winter and spring, and there may be greater warming during night in winter and during day in summer. This implies that heating degree days will decrease, and that cooling degree days will increase. The likelihood of extreme temperatures is also enhanced. For example, the summer maximum temperature that has a 5 per cent chance of occurring on a given day under the current climate may increase from about 28 to 36 °C by the 2080s under the medium–high emissions.

By the 2080s, winter precipitation in the south and east may increase by 10–20 per cent for the low emissions scenario, and by between 25 and 35 per cent for the high emissions scenario (see

Plate 9.6). The pattern is reversed in summer, with a decrease in rainfall for the low emissions scenario of up to 30 per cent, and by 50 per cent or more for the high emissions scenario. The net effect on annual precipitation totals is a reduction of about 10 per cent for most of England and Wales. There will also be less snowfall over south-east England – perhaps up to 90 per cent reductions by the 2080s for the high emissions scenario, and 50–70 per cent reductions for the low emissions scenario. The frequency of heavy winter precipitation, however, is projected to increase. For example, the maximum daily precipitation amount that currently occurs once every two winters may increase in intensity by between 10 and 20 per cent for the low emissions scenario and by more than 20 per cent for the high emissions scenario. Conversely, the frequency of intense summer storms declines by more than 20 per cent over most of England and Wales.

Rates of change in mean sea level around the UK depend on natural land movements as well as on the thermal expansion of the world's oceans and melting of land glaciers. For example, by the 2080s the net sea-level rise (taking vertical land movements into account) for London may be 26 cm under the low emissions scenario and 86 cm under the high emissions scenario, relative to 1961–90. These values were derived using the low end of the low emissions scenario (9 cm global sea-level rise) and the high end of the high emissions scenario (69 cm rise), plus an assumed vertical land change of 1.5 cm/decade. However, most coastal damage is caused during storm surges. According to the Proudman Oceanographic Laboratory model, the 1 in 50-year extreme sea level increases by more than 1.1 m in the Thames Estuary by the 2080s under the medium–high emissions scenario. Unfortunately, much uncertainty is associated with this result, because the projections depend very much on the particular ocean model used.

Significant changes are expected in other climate variables. Most display a clear spatial gradient between south-east England and north-west Scotland. For example, by the 2080s, cloud cover may decrease in summer by 6 per cent over much of Scotland but by 15 per cent over the south-east for the high emissions scenario, with concomitant increases in summer sunshine. Summer relative humidities reduce by 10 per cent or more over south-east England for the high emissions scenario, with fewer fog days expected in winter. Local wind speeds are problematic to estimate from climate models, none the less the UKCIP02 scenarios suggest more frequent depressions cross the UK in winter leading to stronger winds in southern England, with little or no change in wind speeds over Scotland. (However, the climate model is known to have a southward bias in storm tracks.) Finally, average soil moisture may decrease by between 20 and 40 per cent or more under the high emissions scenario, and by less than 20 per cent for the low emissions scenario. Again, there is a clear gradient in soil moisture changes between maximum changes in the south-east and the smallest changes in the north-west.

The associated environmental and socio-economic impacts of the UKCIP02 scenarios have been assessed via regional scoping studies undertaken for much of the UK and Ireland. For example, Table 9.4 lists the most significant impacts expected for London. Follow-up studies are now being planned to develop regional adaptation strategies. For instance, research is already under way to redesign the Thames Barrier and flood defences in London to counter the climate-related threats of rising sea levels, storm surges and river flooding.

Key ideas

1. Variations in global mean temperatures since the Industrial Revolution reflect internal climate forcing (due to the continuous redistribution of heat between the ocean and atmosphere, and between stores of ice, water and vapour), and external climate forcing (due to changes in solar output, volcanic emissions of sulphate aerosols, and atmospheric greenhouse gases released by human activity).

Table 9.4 Summary of potential climate change impacts for London

Issue	Main impacts
Higher temperatures	• Intensified urban heat island, especially during summer nights • Increased demand for cooling (and thus electricity) in summer • Reduced demand for space heating in winter
Flooding	• More frequent and intense winter rainfalls leading to riverine flooding and overwhelming of urban drainage systems • Rising sea levels, storminess and tidal surges require more closures of the Thames Barrier
Water resources	• Heightened water demand in hot, dry summers • Reduced soil moisture and groundwater replenishment • River flows higher in winter and lower in summer • Water quality problems in summer associated with increased water temperatures and discharges from storm water outflows
Health	• Poorer air quality affects asthmatics and causes damage to plants and buildings • Higher mortality rates in summer due to heat stress • Lower mortality rates in winter due to reduction in cold spells
Biodiversity	• Increased competition from exotic species, spread of disease and pests, affecting both fauna and flora • Rare saltmarsh habitats threatened by sea-level rise • Increased summer droughts cause stress to wetlands and beech woodlands • Earlier springs and longer frost-free season affect dates of bird egg-laying, leaf emergence and flowering of plants
Built environment	• Increased likelihood of building subsidence on clay soils • Increased ground movement in winter affecting underground pipes and cables • Reduced comfort and productivity of workers
Transport	• Increased disruption to transport system by extreme weather • Higher temperatures and reduced passenger comfort on the London Underground • Damage to infrastructure through buckled rails and rutted roads • Reduction in cold weather-related disruption
Business and finance	• Increased exposure of insurance industry to extreme weather claims • Increased cost and difficulty for households and business of obtaining flood insurance cover • Risk management may provide significant business opportunity
Tourism and lifestyle	• Increased temperatures could attract more visitors to London • High temperatures encourage residents to leave London for more frequent holidays or breaks • Outdoor living, dining and entertainment may be more favoured • Green and open spaces will be used more intensively

Source: London Climate Change Partnership (2002)

2. Climate models suggest that the unprecedented global warming of the twentieth century lies outside the range of natural climate variability alone.

3. The UKCIP02 scenarios show summer warming of >4 °C and precipitation reductions >50 per cent in south-east England by the 2080s under a high emissions scenario.

4. The most significant impacts of climate change for London include: reduced water supply, an intensified urban heat island, more frequent air pollution episodes in summer, increased flood risk, and biodiversity changes threatening beech woodland and wetlands.

Further reading

Howe, J. and White, I. (2002) The geography of autumn 2000 floods in England and Wales. *Geography*, **87**: 116–24.

Hulme, M., Jenkins, G. J., Lu, X., Turnpenny, J. R., Mitchell, T. D., Jones, R. G., Lowe, J., Murphy, J. M., Hassell, D., Boorman, P., McDonald, R., Hill, S. (2002) *Climate Change Scenarios for the UK: The UKCIP02 Scientific Report*. Tyndall Centre for Climate Change Research, School of Environmental Sciences, University of East Anglia, Norwich, UK, 120 pp.

Marsh, T. J. (2001) Climate change and hydrological stability: a look at long-term trends in south-eastern Britain. *Weather*, **56**: 319–28.

Marsh, T. J. and Monkhouse, P. S. (1993) Drought in the UK, 1988–92. *Weather*, **46**: 365–76.

Marsh, T. J. and Turton, P. S. (1996) The 1995 drought – a water resource perspective. *Weather*, **51**: 46–53.

O'Hare, G. P. and Wilby, R. L. (1995) Ozone pollution in the United Kingdom: an analysis using Lamb circulation types. *The Geographical Journal*, **161**: 1–20.

Semple, A. T. (2003) A review and unification of conceptual models of cyclogenesis. *Meteorological Applications*, **10**: 39–59.

Ulbrich, U., Brücher, T., Fink, A. H., Leckebusch, G. C., Krüger and Pinto, J. G. (2003) The central European floods of August 2002: Part 1 – Rainfall periods and flood development. *Weather*, **58**: 371–7.

Wilby, R. L. (2001) Cold comfort. *Weather*, **56**: 213–15.

Useful websites

UK climate station data:
http://metoffice.gov.uk/climate/uk/averages/1971 2000/index.html

Historic weather events in the UK:
http://www.met-office.gov.uk/education/historic/index.html

Air masses and fronts:
http://www.met-office.gov.uk/education/training/air.html

European Climate Assessment and Dataset:
http://www.knmi.nl/samenw/eca/index.html

UK Climate Impacts Programme:
http://www.ukcip.org.uk/

Indicators of climate change in the UK:
http://www.nbu.ac.uk/iccuk/

Dundee Satellite Receiving Station:
http://www.sat.dundee.ac.uk/

Tornado and Storm Research Organisation:
http://www.torro.org.uk/

International response to climate change:
http://www.tyndall.ac.uk/

The UK National Air Quality Information Archive:
http://www.airquality.co.uk/archive/index.php

CHAPTER 10

Weather, climate and climate change in the mid-latitude continental interiors North America

10.1 Climate of the North American interior

This chapter examines the underlying mechanisms and defining character of the weather and climate of mid-latitude continental interiors with specific reference to North America. For the purposes of our discussions, the continental interior is defined as the region bounded to the west by the Rocky Mountains, to the east by the Appalachians, to the south by the Gulf of Mexico, and to the north by Hudson Bay.

All the climate regimes of this domain are strongly influenced by the physical presence of the Rocky Mountains. This is because the north–south alignment of the mountain chain, and its location at the western edge of the continent, acts as a barrier inhibiting the penetration of maritime air from the Pacific Ocean. This means that a fundamental contrast with interior Europe exists in terms of air masses. Unlike Europe, where the topography enables maritime air from the Atlantic to flow easily eastwards, and the Alps inhibit tropical air masses from flowing northwards, in North America the battleground is between two vastly different air masses. Cold dry Arctic air from northern Canada and warm moist tropical air from the Gulf of Mexico are the principal protagonists determining interior North American climate, with a much reduced role for the oceanic air masses that have such a moderating effect in Europe.

It can be seen in Figure 10.1 that, during July, Arctic and Atlantic air from the north and maritime tropical air from the south dominate the continental interior. Although the Rockies act as a barrier to west-to-east air movement, pools of mild Pacific air (as upper westerlies) can penetrate into the continental interior via the states of Washington and Oregon, and to a lesser extent from the colder north Pacific. In the south-west, continental tropical air dominates under stable high-pressure conditions (see Figure 4.12 and Table 4.1). The western mountains also tend to anchor the upper westerly waves (see Figure 4.9) producing a southward-flowing upper air pattern to the east of them, and then a northward trajectory over eastern regions. This encourages the formation of anticyclones east of the Rockies (see Figure 4.11) and results in semi-arid climate regimes in the western interior. In the eastern interior, depressions form along the rising limb of the upper westerlies (the polar front) contributing to a much wetter climate. The first climatic characteristic of interior North America is therefore its extreme variations in temperature and precipitation, even on a day-to-day basis, as conditions switch between fundamentally different controls.

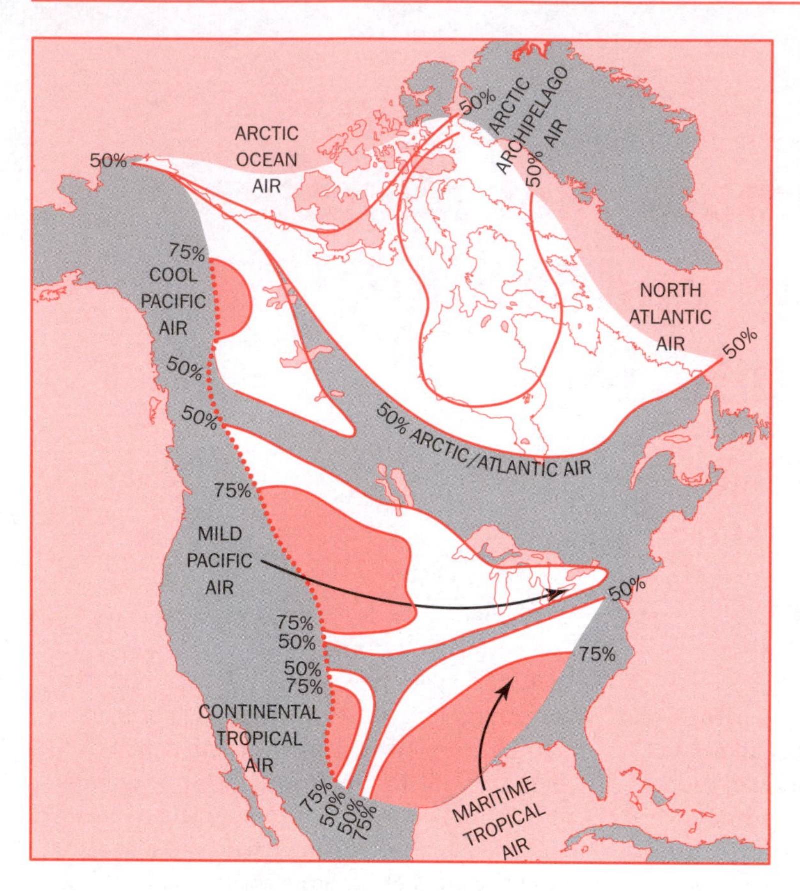

Figure 10.1 Frequency of occurrence of various air masses in summer (July) over North America east of the Rockies. Regions show dominance of air masses for more than 50 and 75 per cent of the time
Source: Bryson (1966)

Key ideas

1. The weather climate of the continental interior of North America is dominated by the physical presence of the Rocky Mountains, and by the interplay of contrasting air masses originating in the Gulf of Mexico and Arctic.

10.2 Processes and controls

The dynamic climatology of North America may be explained at different scales. At a macro scale, semi-permanent pressure systems act to drive air masses across the continent. The role and effectiveness of these vary seasonally. Second, at a meso scale, features such as depressions and convective systems exhibit distinctive seasonal and spatial occurrences as a result of the constantly readjusting balance of power at a macro scale. Finally, local factors such as topography and the moderating influence of inland water bodies such as the Great Lakes embellish further the climatic mosaic at a smaller scale. The following sections examine the most important atmospheric and physical controls, and distinguishing features of the changing weather and climate of the continental interior of North America.

10.2.1 Dynamic climatology

(a) Airflows from the Pacific

Two semi-permanent pressure systems control the flow of Pacific air into the continent. The Californian High is the subtropical high created as a result of the descending limb of the Hadley cell over the north Pacific, and becomes most pronounced in the summer season when it extends on land over the south-western USA. The offshore high with its clockwise-rotating winds generally drives maritime tropical air eastwards over the Rockies as mild Pacific air (Figure 10.1). The Aleutian Low, centred over the Gulf of Alaska, is most intense in winter and via its anticlockwise winds drives maritime polar air eastwards as cool Pacific air. Much of this air movement is in the form of west to east-moving depressions. As with the North Atlantic, frontal depressions develop at the boundary between the maritime tropical air of the Californian High and the maritime polar air of the Aleutian Low. The precise trajectory of these frontal disturbances over the Pacific and their penetration and passage across North America depend not only on seasonal factors as just described (summer–winter dominance of the two Pacific pressure systems) but also on (i) the reactivation of Pacific-derived disturbances across North America in relation to the strength and location of Arctic and polar fronts and (ii) long-term Pacific sea surface temperatures and their anomalies.

(b) Arctic and polar fronts

The two most important frontal zones affecting central North America and the passage of depressions across the continent are the Arctic and polar fronts. Although their respective positions are quite variable, both show an equatorward shift in winter, and a poleward retreat in summer (Figure 10.2). In winter, the Arctic front which is aligned across northern Canada is formed between the dry and very cold continental Arctic (cA) or cP air and Pacific maritime air from the Aleutian Low that has warmed and dried adiabatically as it has crossed the Rocky Mountains. The polar front at this time becomes aligned along the Appalachian Mountains and the eastern seaboard, separating cold interior polar continental air from maritime tropical air from the Atlantic. As shown in Figure 10.2, during the winter period the Rossby waves and their jet streams sweep in a highly sinuous fashion across the continent producing dry conditions in the west under its downward limb and wet weather in the east along its upward limb in association with frontal depressions. Latitudinal variations in the position of the polar and Arctic fronts in winter can result in marked temperature changes over short periods for the continental interior.

In summer, the polar front which separates cold polar and Arctic air masses to the north and milder air to the south moves northwards into Canada. The mild air has its source in the Pacific, entering North America via the states of Washington and Oregon (Figure 10.1). The jet stream and storm tracks associated with the polar front also move northwards to their most poleward position. The jet on average enters the west coast near Seattle, Washington, and tracks in a less sinuous route across southern Canada to Newfoundland with the storm track shifting into Canada (Figure 10.2). During summer, important large-scale low-pressure storms are rare across the interior regions.

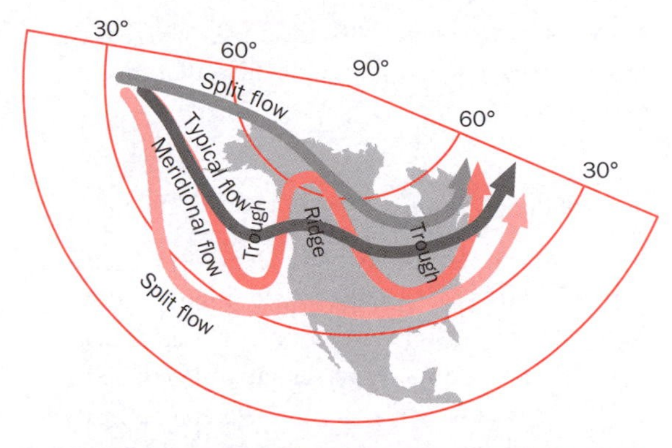

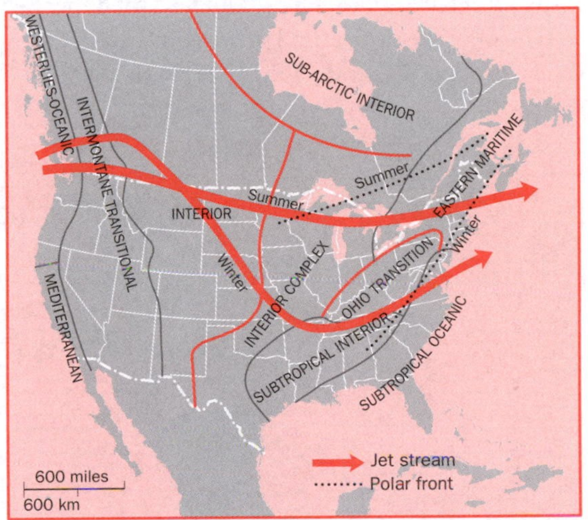

Figure 10.2 Dynamic climatology of North America showing the summer and winter positions of the polar front (as black dotted lines) and the jet stream
Source: Adapted from Henderson-Sellers and Robinson (1986) and Sheppard *et al.* (2002)

Depression frequency also diminishes; across the Great Lakes it is less than one-third that of the winter months.

(c) Pacific sea surface temperatures

The location and intensity of depression tracks across North America also reflect long- and short-term changes in the distribution of ocean-surface temperature anomalies in the north Pacific (see Pacific Decadal Oscillation in section 5.3.2). As shown in Figure 10.3, during winters when the water of the north central Pacific is relatively warm (between 1947–76 and 1890–1924) with colder coastal waters around Alaska, Canada and western USA, there is a northward displacement of the jet stream over the north Pacific. This causes a corresponding southward displacement of the jet over the western USA, with an influx of cold air from the north over central and western North America, but warmer conditions over the eastern and southern regions. Conversely, when there is an anomalous cold pool of water in the central Pacific, (since 1977 and 1925–46) with warmer water to the east, storms develop in the zone between them, leading to warm wet winters in California and Oregon, generally warm air over western USA and cold conditions in eastern USA. During such (i.e. current) winters westerly storm tracks typically enter North America over the states of Washington and Oregon, and follow a gently meandering zonal path at about 45–50° N, with limited opportunity for southward penetration by cold cP air (Figure 10.3).

Pacific-generated storm tracks can be pushed even further north. This happens when the typical high-pressure ridge over south-western USA is displaced westward into the Pacific, forming a low-pressure trough over the western USA. When the ridge (ocean)–trough (land)

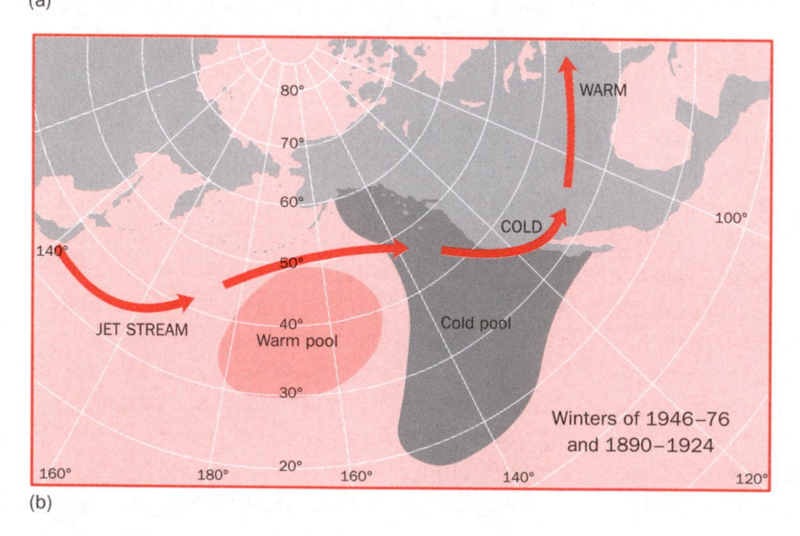

Figure 10.3 Distribution of mean ocean surface temperatures, jet stream tracks, zones of storm action and land temperatures over the north Pacific and North America. The various ocean and atmosphere patterns are shown in relation to (a) positive (cold pool) and (b) negative (warm pool) variations in the Pacific Decadal Oscillation
Source: Adapted from Wick (1973)

pattern is pronounced, a well-developed Aleutian low-pressure cell produces clearly defined storm tracks extending from Asia, across the central Pacific, and north-eastward into the Gulf of Alaska. The pattern typically yields above-average winter temperatures in the north-west and below-average temperatures in the south-eastern USA. This mode also favours the penetration of cold pools of Pacific air over Alaska and north-west Canada (see Figure 10.1). There is some evidence linking El Niño events (warm water in the eastern equatorial Pacific) with such a pronounced ridge–trough formation (called a positive Pacific–North American pattern). Conversely, a negative Pacific–North American pattern with weak ridge and trough arrangement across North America is generally associated with a La Niña (very cold water in the eastern equatorial Pacific).

(d) Summer monsoon

As outlined in the next chapter, a **monsoon** wind system is one that undergoes a marked seasonal change in direction. The North American monsoon refers to the movement, in summer, of warm moist southerly winds from the Caribbean into the central heartland of the North American continent. The term 'monsoon' as applied is somewhat of a misnomer because there is not a *consistent* pattern of wind reversal in the winter with air moving from the continental interior and south-western deserts to the Gulf of Mexico. Nevertheless, the term is commonly used in North America and will be applied in this chapter.

In summer, there is a northward shift in the position of the subtropical high-pressure cells over the Azores and north-east Pacific. As a consequence, the upper westerlies and their jet streams are typically displaced northwards following a less wavy path that crosses the Great Lakes (Figure 10.2). The onset of the North American monsoon is linked to the retreat of the upper westerlies and surface depression tracks across central North America. This allows the simultaneous advance of the Azores subtropical high-pressure cell over the south-eastern USA. At the same time intense summer heating of the continental interior creates a strong surface low-pressure cell over the central and south-western plains of the interior. Clockwise-rotating tropical maritime air on the downside limb of the Azores subtropical high is then drawn, as south-easterly and southerly monsoon winds, into the central continental low-pressure system.

(e) Mechanisms of thunderstorm formation

In summer, over the Great Plains, weather conditions are dominated by thunderstorms, larger mesoscale convective complexes (MCCs) and tornadoes. It may be recalled that the development of thunderstorms including air mass types (supercell and multicell), larger mesoscale convective complexes and squall-line forms have already been outlined in section 2.3.5. The meteorological importance of thunderstorms in the United States should not be underestimated. Over 100,000 storms form on average each year in the country. Approximately 1,000 tornadoes are released annually from such storms. Thunderstorms cause significant loss to life and property. Large hail results in nearly $1 billion in damage to property and crops, and around 10,000 forest fires are started each year in the USA by lightning, causing $100 million in annual losses.

Most thunderstorm development over the northern plains of the interior is related to squall-line types and the movement of the polar front. Ahead of the warm front, showers and thunderstorms frequently develop. Ahead of the cold front, these convective/frontal storms can organise into squall lines during the afternoon and evening hours. At night under the right conditions they can cluster into large mesoscale convective complexes covering an area of up to 100,000 km^2. These complexes form near the cold front boundary and feed off the heat of the day in the warm sector air mass. They often move many hundreds of miles during the day before dissipating. These squall lines and complexes produce rain, hail, high winds and occasionally tornadoes.

Across the southern Great Plains, frontal activity is less important in thunderstorm formation. Here, strong surface heating, low-pressure conditions and the convergence of dissimilar air masses at different levels in the atmosphere hold the key to convective instability and thunderstorm formation. Thunderstorms in the southern states require four conditions for formation. These include abundant supplies of moisture in the lower atmosphere, steep lapse rates in the middle atmosphere, cold dry air at upper levels of the atmosphere, and strong atmospheric wind shear. These conditions are supplied when three very different types of air converge over the southern plains in a particular way.

During the spring and summer months, southerly or south-easterly winds prevail across the plains. These winds from the Gulf of Mexico provide plenty of warm moist maritime tropical air at low levels in the atmosphere. When this air is forced to rise and cool, it provides the latent heat necessary to fuel thunderstorm development. A second factor is the import of colder drier air by strong west or south-west winds in the upper atmosphere. Strong vertical potential atmospheric instability is created as lapse rates increase with warm (and therefore lighter) moist air near the surface and cold heavier dry air above. If the warm moist air can be given an initial push to move upwards, the air will keep rising as latent heat is released to the storm (see Figure 2.11(b)). Third, the convergence of low-level southerly winds from the Caribbean and high-level westerly winds create wind shear in the atmosphere, helping to rotate any developing storm or tornado. The fourth factor is a layer of hot dry air between the warm moist surface air and the cool dry air aloft. This hot dry layer at middle levels in the atmosphere has its origins in winds from the Rockies which become adiabatically warmed as they move southwards over the plains. The hot layer acts as a cap and allows the warm air underneath to warm further, creating even greater potential vertical instability in the atmosphere. If this occurs conditions are favourable not only for the development of severe summer thunderstorms, but also for mesoscale convective complexes and tornadoes.

Severe thunderstorms over the Great Plains are triggered by a number of factors. Over the southern plains, daytime heating and orographic uplift (e.g. against the Rockies) are important, while across the northern plains daytime heating is allied to frontal activity to release the storms. Over the southern states especially, the lifting of air through these mechanisms helps to remove the hot mid-tropospheric cap, thereby setting the stage for explosive thunderstorm development as strong updraughts from the heated surface develop. As the rising air encounters wind shear, it may cause the ascending updraughts of the developing thunderstorm or mesoscale convective complex to rotate anticlockwise helping to release tornadoes.

(f) MCCs

Mesoscale convective complexes (MCCs) form from initially isolated cumulonimbus cells grouping into one massive convective cloud system. As pointed out in section 2.3.5, they can contribute much of the total rainfall of an area. Over the Great Plains they account for 30–70 per cent of summer (April through September) precipitation.

The development of MCCs in relation to frontal squall lines over the northern plains has already been mentioned. Over the southern plains they tend to arise from the recycling of heat and moisture on the flanks of air mass thunderstorms as follows. First, intense rainfall and hail from the thunderstorm are accompanied by cold downdraughts and evaporative cooling. The cold downdraughts can feed into adjacent growing cells displacing warmer air upwards within them, triggering latent heat release and further inflows of moist unstable air. Individual cells may cluster together into one massive organised cell covering an area >100,000 km^2 with lifetimes of about

12 hours. As with the squall-line forms, the agglomerated supercell (air mass) complex typically reaches maturity during the late evening or night-time by which time a system may have travelled from Colorado as far east as the Mississippi River or even the Great Lakes. The MCC fades as synoptic-scale features hinder further convection, leading to the cessation of rainfall and cold air production.

(g) Tornadoes

Tornadoes can often be a violent product of the new cells developing within an MCC. These phenomena occur with such regularity over the Great Plains (see above) as to be considered a significant feature of the interior's climatology, even to the point of defining a 'Tornado Alley'. Definitions of this breeding ground for violent tornadoes vary. According to the American Meteorological Society, Tornado Alley is the area of the United States in which tornadoes are most frequent: the great lowland areas of the Mississippi, the Ohio and lower Missouri River valleys (i.e. the Plains area between the Rocky Mountains and the Appalachians). Others consider Tornado Alley as the area where only the most intense killer tornadoes, the so-called 'Category-T5s' with wind speeds greater than 260 mph, have touched down (see Table 10.1). Nebraska, Kansas, Oklahoma and Texas lie at the heart of this high-risk area, but other states such as Missouri, Illinois and Indiana have also been hit hard in the past.

The spatial pattern of tornado occurrence (Figure 10.4) reflects as suggested previously the interaction of cold, dry air from the western high plateaux with warm, moist maritime tropical air from the south that penetrates northwards by means of a low-level jet. The horizontal convergence of the two air types creates wind shear and helps to rotate the developing thunderstorm or MCC in an anticlockwise direction. The wind shear spin is reinforced by the Coriolis force (section 2.2.1), especially when the storms are horizontally extensive enough (i.e. the MCCs) to rotate. As the convergent air rises, it is replaced by

Table 10.1 Tornado wind speed classification using the Fujita scale

Category	Wind speed (and possible duration)	Damage
F0 Very weak	40–72 mph (1–10 minutes)	Damage to chimneys; tree branches broken
F1 Weak	73–112 mph (1–10 minutes)	Roof surfaces removed, small trees snapped; cars pushed off road
F2 Strong	113–157 mph (20 minutes or more)	Roofs removed; large trees snapped or uprooted; house walls badly damaged
F3 Severe	158–206 mph (20 minutes or more)	Rural buildings demolished; cars lifted off ground; most trees in forest uprooted and snapped; trains overturned; block structures often levelled
F4 Devastating	207–260 mph (1 hour or more)	Well-constructed frame houses levelled; trees de-barked by small flying debris; cars lifted and rolled considerable distances
F5 Incredible	261–318 mph (1 hour or more)	Strong frame houses lifted clear off foundation; steel reinforced concrete structures badly damaged; automobile size missiles fly through air; trees debarked completely

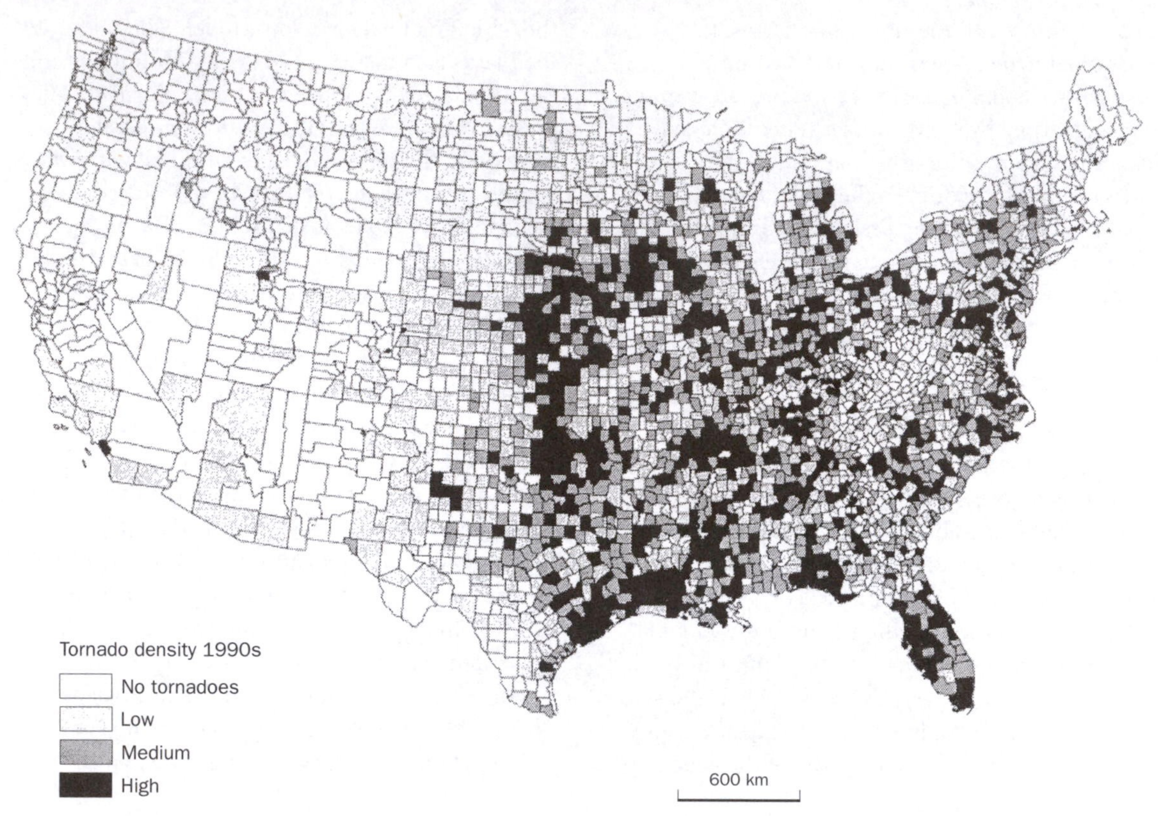

Figure 10.4 Density of tornado hazards (frequency per square mile [2.6 km^2]) by county in the 1990s
Source: Boruff *et al.* (2003)

moist air drawn from progressively lower levels, a mechanism that is often marked by a funnel cloud descending from the base of the cumulonimbus. The pressure reduction within the vortex may exceed 200–250 hPa, accompanied by wind speeds of 50–100 m/s across a swathe up to 2 km wide. The rate of horizontal movement of the tornado is determined by the strength of the low-level jet. One of the most destructive US tornadoes on record struck Xenia, Ohio, on 3 April 1974, causing 33 deaths. This was part of an outbreak of 148 tornadoes that raged through 13 states resulting in 330 fatalities and 5,500 injured in the space of 16 hours.

The number of tornadoes affecting the United States as a whole has consistently increased over the last half-century. During the 1950s, there were 4,796 events, 8,579 by the 1970s and as many as 10,696 during the 1990s. This dramatic rise, with numbers more than doubling, is undoubtedly due to better monitoring and increased detection. Since thunderstorms are primarily generated by land surface heating, part of the increase could also be a consequence of global warming. Such climate links are interesting, but in the absence of better data sets and knowledge, remain speculative at the moment. As we have demonstrated, many factors in addition to solar heating influence the formation of thunderstorms and tornadoes.

10.2.2 Physical controls

The Rocky Mountains are aligned roughly north-north-west to south-south-east and stretch for about 3,000 km from British Columbia and Alberta to New Mexico, passing though Idaho, Montana,

Wyoming, Utah and Colorado. The mountain chain, with maximum elevations exceeding 4,000 m, separates the continental climates of the interior and east from the maritime climate of the west coast. The following sections examine the unique influence of the Rocky Mountains on the temperature and precipitation regimes of the continental interior.

(a) Orographic forcing

The Rocky Mountains, together with the Tibetan Plateau, exert a major influence on the horizontal and vertical patterns of atmospheric circulation in the northern hemisphere. The most significant topographic effects on airflow include the compression and expansion of air columns, adiabatic heating and cooling due to sinking and rising air, and the release of latent heat by orographically induced precipitation (see orographic precipitation model in section 2.3.5). As a westerly airflow approaches the topographic barrier it is subject to horizontal compression and vertical stretching, resulting in convergence and cyclonic rotation. As the air rises over the summit it is both accelerated by the compression and retarded by frictional resistance with the ground, resulting in a net reduction in wind speeds compared with free moving air at the same elevation. Beyond the mountain crest the air is divergent, setting up an anticylonic motion, southward flow of air and reduced horizontal velocities (see Figure 4.11). Downwind of the barrier the descending air may form a series of lee waves often marked by lenticular (lens-shaped) clouds and turbulent (rotor) motions that result in local reversals of wind direction near the surface.

(b) Chinook

The Chinook is a local wind that develops on the lee slopes of the Rocky Mountains under conditions of strong Pacific air streams. This wind is typically dry, warm and gusty, sometimes

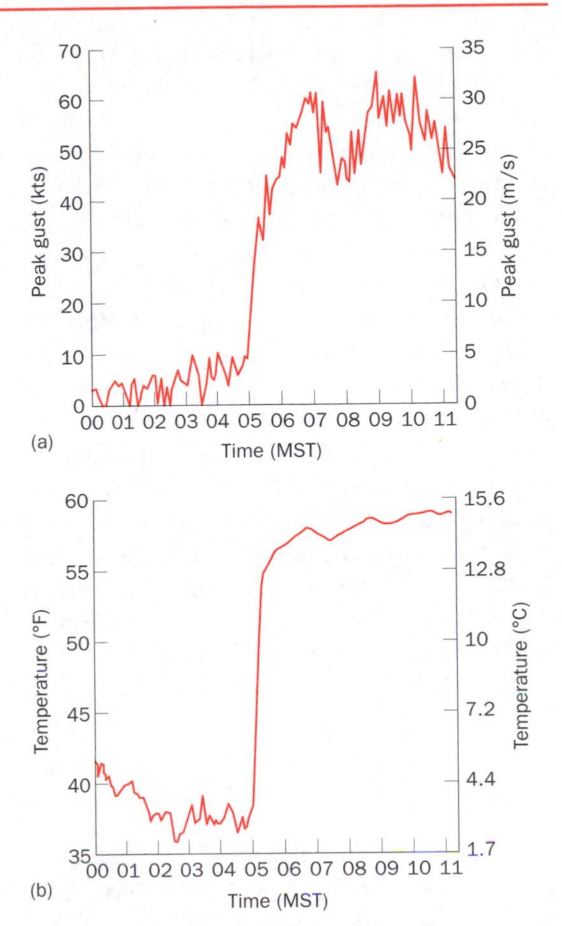

Figure 10.5 The arrival of the Chinook wind just after 5 a.m. on the eastern slopes of the Rockies on 7 December 1987: (a) wind speeds and (b) air temperatures measured at Boulder, Colorado. Kts=knots
Source: http://wxpaos09.colorado.edu/windstorms/chinook.html [accessed 13/09/02]

attaining hurricane force at the base of the mountains along a belt about 50 km wide (Figure 10.5(a)). The zone of maximum wind speeds depends on the location of a low-pressure wave trough at the base of the lee slope as air accelerates towards it. Rapid temperature rises (sometimes exceeding 20 °C in the space of several hours) may accompany the high wind speeds. This is because the air cools on the windward side of the mountains at the slower saturated adiabatic lapse rate (about 6 °C/km) but, following moisture loss through precipitation, descends at the quicker dry adiabatic lapse rate (10 °C/km) on the lee side (Figure 10.5(b)). The warming can be particularly abrupt if the chinook encounters a pool of cold air caused by night-time cooling on the lee side of the mountains, or if there is a cold air mass over the plains to the east of the Rocky Mountains. The rapid warming has been known to trigger avalanches on snow-covered slopes. On other occasions winds descending the lee slopes of the Rocky Mountains are cold. Despite adiabatic warming as the air descends, bora winds bring colder and drier winter conditions to the foothills. This happens when cold cP air masses are forced from high-elevation snowfields over the mountains, displacing warmer air in the process. Like the cold bora winds in southern Europe (e.g. the mistral and bora itself), the onset of the bora wind is usually accompanied by the passage of a cold front which draws cold air from the mountain zones behind it.

(c) Planetary waves

Mountain barriers are also known to affect the position of the Rossby waves at hemispheric scales. Climate model experiments with idealised orography and prescribed zonal airflows suggest that the Rocky Mountains would, in isolation, set up quasi-stationary planetary waves with a ridge at 40° W and a trough at 20° E. However, the actual wave pattern arises from interactions with the mountain ranges of Greenland and the Himalayas. The resultant waves typically exhibit a ridge over the northern Rocky Mountains and a trough over the Mississippi River valley, exiting the continent in a north-eastward direction. These waves move cP air deep into the continental interior and at the same time allow mT air masses originating from the Gulf of Mexico to move as far north as the Arctic Circle. Model experiments with and without mountains further indicate that the Rocky Mountains amplify the winter planetary wave, with greater blocking and diversion of the low-level flow around the mountain barrier.

(d) Continentality

The Rocky Mountains are a major factor affecting the continentality of North America. This topographic barrier severely restricts the inland penetration of maritime air masses resulting in a rapid decrease in precipitation with distance from the ocean. In fact, the line dividing the irrigated and non-irrigated regions of the western Great Plains (~600 mm precipitation/year) lies east of the Rocky Mountains in the extensive rain shadow of the western mountain ranges. Furthermore, the dominance of south-westerly airflow limits the scope for moisture advection from the Gulf to the western plains.

Although there are many measures of continentality, most are indicative of the annual temperature range corrected for latitude (on the assumption that lower latitudes have smaller annual variations in solar radiation and hence temperature). The concept may be illustrated using Gorczynski's method applied to four climate stations along a west–east axis close to 40° N (Figure 10.6), and two further stations at about 50° N. The Gorczynski index is the ratio of the annual temperature range to the sine of the station latitude.

San Francisco (38° N, 122° W) is subject to the moderating influence of the Pacific and has a relatively narrow annual temperature range (~9 °C) resulting in a continentality index of only 4. East of the Rockies, the temperature range (continentality index) for Denver (40° N, 105° W) is 24 °C (43), and for Kansas City (39° N, 95° W) is 29 °C (58). The fourth station at Richmond (37° N, 77° W) has a relatively high annual temperature range (24 °C) and continentality index (47) by virtue of the limited oceanic influence along the Atlantic seaboard. The two most northerly stations, Calgary (51° N, 114° W) and Winnipeg (50° N, 97° W), have by far the greatest temperature ranges (up to 38 °C) by virtue of the increasing dominance of cP air north of the US/Canada border. This yields a continentality index of 64 at Winnipeg.

The winter maximum precipitation exhibited at San Francisco is typical of mid-latitude west coast

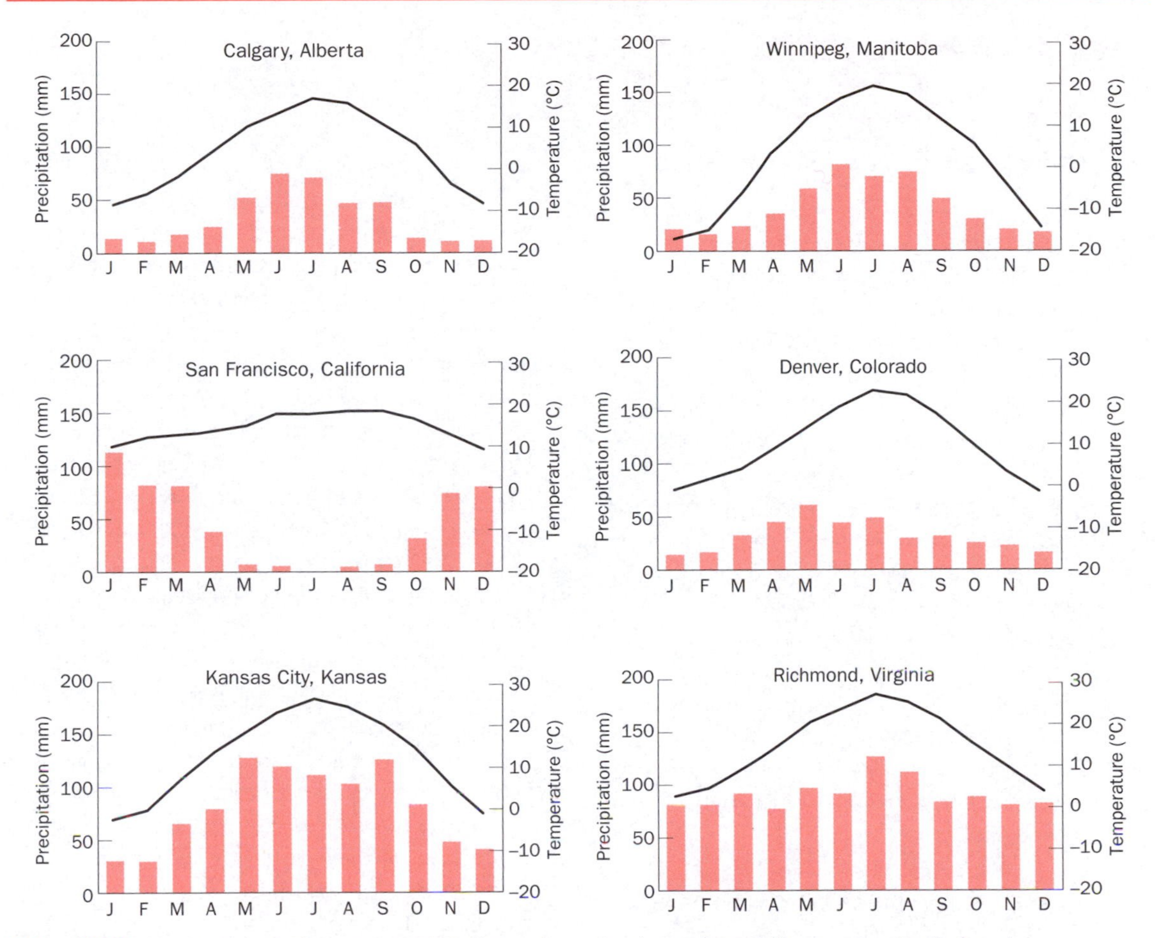

Figure 10.6 Annual temperature and precipitation regimes for six weather stations across North America

climates (see section 9.2.5). However, the similar continentality of Denver and Richmond belies the marked differences in their respective precipitation regimes – one of the limitations of temperature-dependent continentality indices. The warm season precipitation maximum for Denver is typical of much of the continental interior and is due to the summer monsoon and associated thunderstorm activity (section 10.2.1). Apart from occasional storms that manage to cross the mountain barrier from the west, winters are relatively dry over much of the western plains. Further east, the regimes are distinguished by a double maximum in May and September, as illustrated by the climatology of Kansas City. The secondary minimum in July–August is due to the temporary suppression of thunderstorm activity by an upper-level ridge of high pressure over the Mississippi Basin. Closer to the east coast, Richmond's proximity to warm and humid mT air masses from the Gulf gives rise to an even distribution of precipitation throughout the year, with a high proportion of summer precipitation originating from convective thunderstorms. The following sections will further examine these climatological features with respect to objectively defined air masses.

10.2.3 Air mass classification

Figure 10.1 shows an example of an objectively defined air mass frequency classification of North America. To date, relatively few objective weather classification schemes have been developed at the continental scale (Box 10.1). However, a spatial

BOX 10.1

THINKING FURTHER

Objective weather classification

In Chapter 9 some of the motivations, applications and limitations of weather classification were introduced with special reference to the catalogue of daily Lamb weather types. The Lamb scheme began life as a *subjective* classification of surface and mid-troposphere pressure charts for a small region centred on the British Isles. In this case, Professor Lamb's expert opinion was used to manually assign each day's weather to one of 7 'pure' circulation types or 19 hybrid types, or one unclassifiable group. Although subjective methods have been widely applied they tend to be very time-consuming to implement, suffer a lack of consistency between analysts (because no two 'experts' will always agree as to the correct classification), or display internal inconsistencies (because even a professional analyst will have difficulty in correctly classifying weather when pressure patterns are weak or transitional between states).

The advent of globally gridded data sets of sea-level pressure and geopotential height fields together with powerful desktop computing have paved the way for numerical classifications of daily weather data. The following sections describe a range of *automated* (computer-based) procedures for classifying continental-scale weather systems such as those affecting North America. Three families of technique are introduced: pattern correlation, data reduction and non-linear methods.

Pattern correlation

The basic principle underlying pattern correlation methods is that a relatively small number of 'key' weather states can be extracted from a large sample of pressure maps; all other days are then assumed to congregate around one or other of the key states. Identification of the key days may rest on expert judgement, or on their extraction via statistical means. The most straightforward automated approach is to specify an arbitrary correlation threshold (typically $r > 0.8$) and then to count the number of days with strongly correlated pressure patterns using all possible combinations of days. The pressure pattern with the largest number of days correlated at or above the threshold level is designated Type A. These days are then removed from the training data set and the procedure repeated, this time identifying the next most popular pattern, Type B, and so on. The procedure ends when all days have been removed or when the remaining pressure maps are considered uncorrelated at the threshold level.

Despite its simplicity, the so-called Lund technique has a number of shortcomings. First, the results depend on the threshold for similarity. If it is too stringent a relatively small fraction of days will be classified; if it is too relaxed the number of types will become unworkably large and the discrimination between types obscured. Second, the 'optimum' number of weather types depends on the season and region of interest; the test is the effectiveness of the scheme at describing the regional climatology. Third, some researchers have suggested that the correlation coefficient is a rather poor measure of the similarity between weather maps and that the sum of the squares of pressure differences (or Kirchhofer metric) is preferable. Fourth, the results of correlation-based classifications can depend on the sample size and period of record available for training, and on the grid spacing of the pressure fields. Finally, extreme events are often associated with unusual weather conditions which, by definition, are poorly correlated with key types and therefore often judged as 'unclassifiable' by objective schemes.

Data reduction methods

Classifications based on data reduction assume that large numbers of weather variables (e.g. grid values of surface pressure, humidity, temperature) can be collapsed into relatively few composite terms that succinctly describe most behaviour in the data set. A widely employed technique, principal components analysis (PCA), uses linear combinations of variables to explain dominant modes of behaviour in the atmospheric data. A key advantage of PCA is that the resultant composite terms (principal components or PCs) are mutually uncorrelated and the amount of variance explained is additive. However, the amount of explained variance decreases with each successive component so care must be taken in the interpretation of later PCs because they progressively explain more noise (or local variability) in data.

Having extracted the first few PCs, the information is typically used in one of two ways. First, the daily PC values derived from, for example, the geopotential height field over the target region are statistically related to the variable of interest, such as maximum daily temperature. In this approach, the weather pattern is being used to 'downscale' large-scale information to the local-scale phenomenon directly (see section 7.2.4). Alternatively, the PCs may provide the basis for further data processing in order to identify clusters of similar synoptic weather types. In this case, nearest-neighbour approaches may be used to identify assemblages of weather types (or PC scores) that maximise inter-group differences while simultaneously minimising *within*-group variability. In some recent approaches, the boundaries are 'fuzzy' rather than discrete, enabling any given day to have membership of more than one weather state in a probabilistic sense. As in the case of the pattern correlation methods, the key problem remains the justification of the number of weather types, and the criteria used to judge similarity between sets of patterns.

Non-linear methods

The final suite of automated weather classification procedures acknowledges that correlation-based and linear data reduction methods are poorly equipped to capture non-linear or multi-scale behaviour in weather data. Self-organising maps (SOMs) and artificial neural networks (ANNs) are growing in popularity for such tasks and are now widely employed across the environmental sciences. SOMs attempt to identify a limited number of nodes among the 'cloud' of observations around which nearby values tend to congregate. A major advantage of SOMs is that they are unsupervised classifiers, that is, they can extract characteristic weather types purely on the basis of interrelationships hidden within data. Furthermore, the resultant weather classifications are readily visualised and display smooth transitions from one type to another.

ANNs mimic the function of the human brain by establishing non-linear relationships between a suite of data inputs (by analogy our senses) and outputs (our responses to those stimuli). This is accomplished through a trial and error approach to pattern recognition by which the ANN 'learns' to produce the same response to a given set of inputs. Just like the human brain, information is processed by the ANN in a hidden layer comprising of a network of interconnected neurons. By adjusting sets of weights between the inputs, the connecting neurons and the output, the ANN input–output response is 'hard-wired'. For example, the ANN may be trained to relate patterns of surface pressure and humidity to patterns of rainfall.

Opponents of SOMs and ANNs argue that the procedures are overly complex and that comparable results can be obtained from far simpler methods such as multiple linear regression. Much skill is also required when training the models to ensure that they are not over-fitted to training data, thereby losing their ability to generalise for unseen cases. Once again, the choice of input variables, the number of weather states, and their applicability to extreme weather, are significant issues affecting results from non-linear classification methods.

synoptic classification has recently been produced for winter weather across the USA, east of the Rockies for the period 1961–90. The method was based on the initial identification of six air mass types and their typical meteorological conditions (from specified ranges of afternoon surface temperature, dew point, dew point depression, wind speed, wind direction, cloud cover and diurnal temperature range) at each station in the network. Representative 'seed days' were then used to train a statistical model that classifies previously unseen days into moist and dry variants of polar, temperate and tropical air, given local meteorological conditions.

The resultant daily classification provides a means of examining spatial gradients in the occurrence of each air mass across the interior and eastern part of the continent. For example, Figure 10.7(a) shows the wintertime frequency of the contrasting dry temperate (DM) and moist tropical (MT) air masses. As expected, the frequency of MT air decreases rapidly with

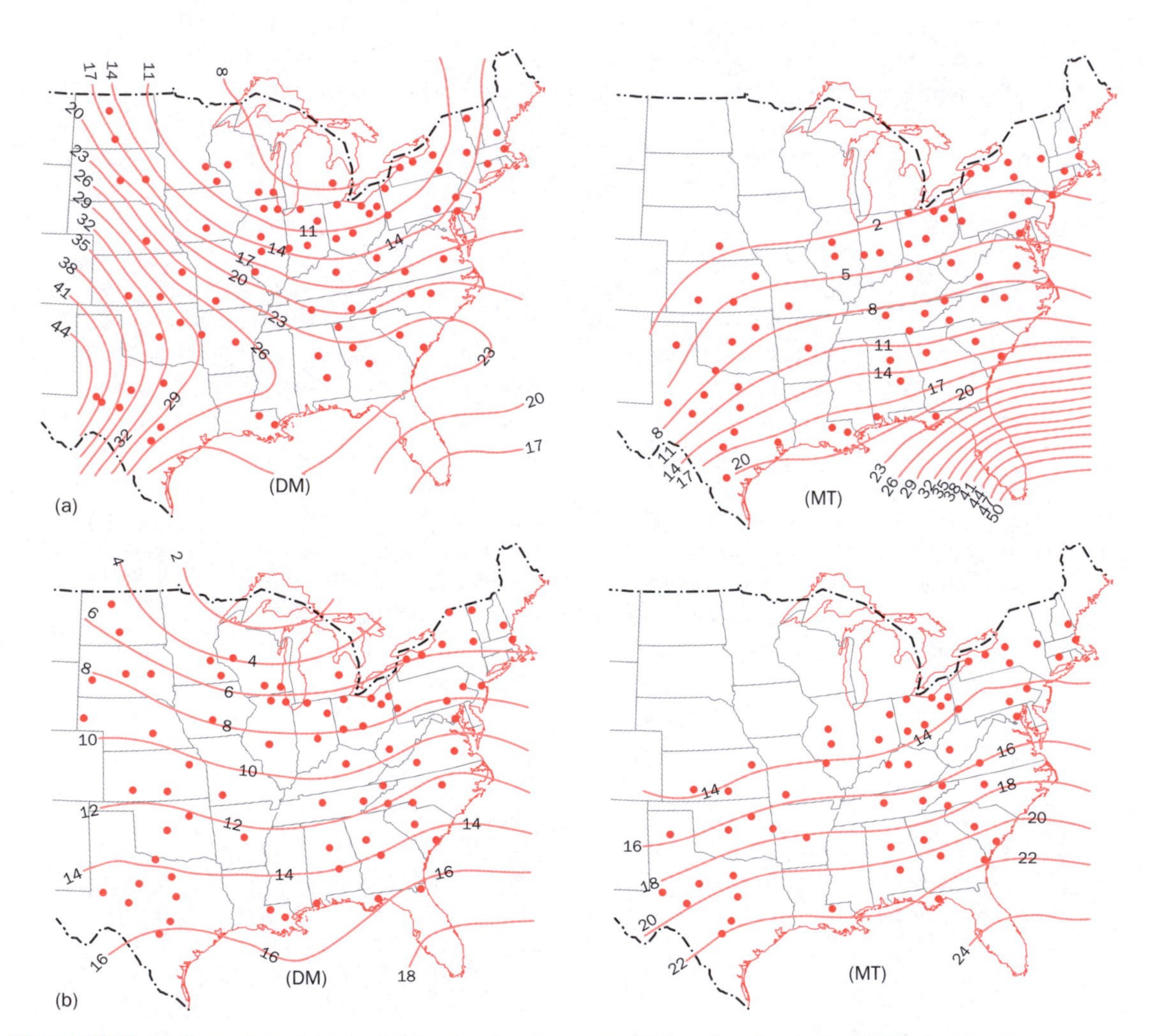

Figure 10.7 (a) Percentage frequencies for the dry temperate (DM) and moist tropical (MT) air masses; (b) corresponding afternoon temperatures (°C) 1961–90

Source: Kalkstein *et al.* (1996)

distance northward from Florida, whereas DM is indicative of the downslope motion of air masses crossing the Rockies during zonal flow yielding maximum rates of occurrence over the southern Great Plains. Associated patterns of mean temperature show modifications with increasing distance from their respective source regions (Figure 10.7(b)). For example, the temperature of MT air decreases by about 0.5 °C per 100 km near Florida, and by approximately 1 °C per 100 km further north where the air masses increasingly come into contact with snow cover. Spatial variations in dew point temperatures within DM air (not shown) suggest little west–east change in moisture content as this air moves away from the source region in the lee of the Rockies.

Developers of synoptic weather classifications point to a number of potential applications of their research. For example, there are strong links between certain air masses at particular locations and patterns of weather-related mortality, especially during summer months. This arises because of the association between air mass type and high concentrations of air pollutants such as sulphates, particulates and ozone (see Box 9.1). If an air mass with life-threatening properties is forecasted, a 'health watch' can be issued for high-risk groups and locations. In the following sections it will be shown how an appreciation of air mass properties also contributes to a deeper understanding of extreme weather events affecting the continental interior.

Key ideas

1. The polar front marks the sharp temperature divide between moist tropical maritime (mT) air in the south-east and dry continental polar (cP) air to the north.

2. Atlantic air masses are the major source of moisture for the continental interior east of the Rockies.

3. The location and intensity of depression tracks across North America largely reflect the changing distribution of ocean surface temperature anomalies in the north Pacific.

4. The North American monsoon is fed by moisture advection from the Gulfs of Mexico and California, and is characterised by the onset of thunderstorm activity in June/July over the western USA, often resulting in torrential rainfall and flash flooding.

5. Mesoscale convective complexes and tornado activity are typically triggered by cold, dry eastward-moving air overlying unstable mT at the near surface.

6. The Chinook is a dry, warm and gusty wind that develops in the lee of the Rocky Mountains under conditions of strong Pacific air streams.

7. The Rocky Mountains severely restrict the inland penetration of mT air masses, resulting in rapid decreases in precipitation and increased annual temperature range with distance from the ocean (termed continentality).

10.3 Past climate variations

Recent evidence suggests that precipitation has increased by about 10 per cent across the contiguous United States since 1910. Over half of the total increase is accounted for by heavy and extreme daily precipitation events (defined as >50 mm/day). It has also been estimated that flood-related fatalities total about 100 per year, with annual flood losses rising from $1 billion in the 1940s to about $5 billion in the 1990s. Although climate change plays an important part in explaining the increasing social and economic costs of extreme events, rising damage is partly due to increased population, wealth and exposure of society to natural hazards (such as the bias towards new development in flood-prone coastal regions). The following sections will focus on notable climatic events affecting the continental interior of North America, and examine the role of atmospheric *and* terrestrial forcing of these

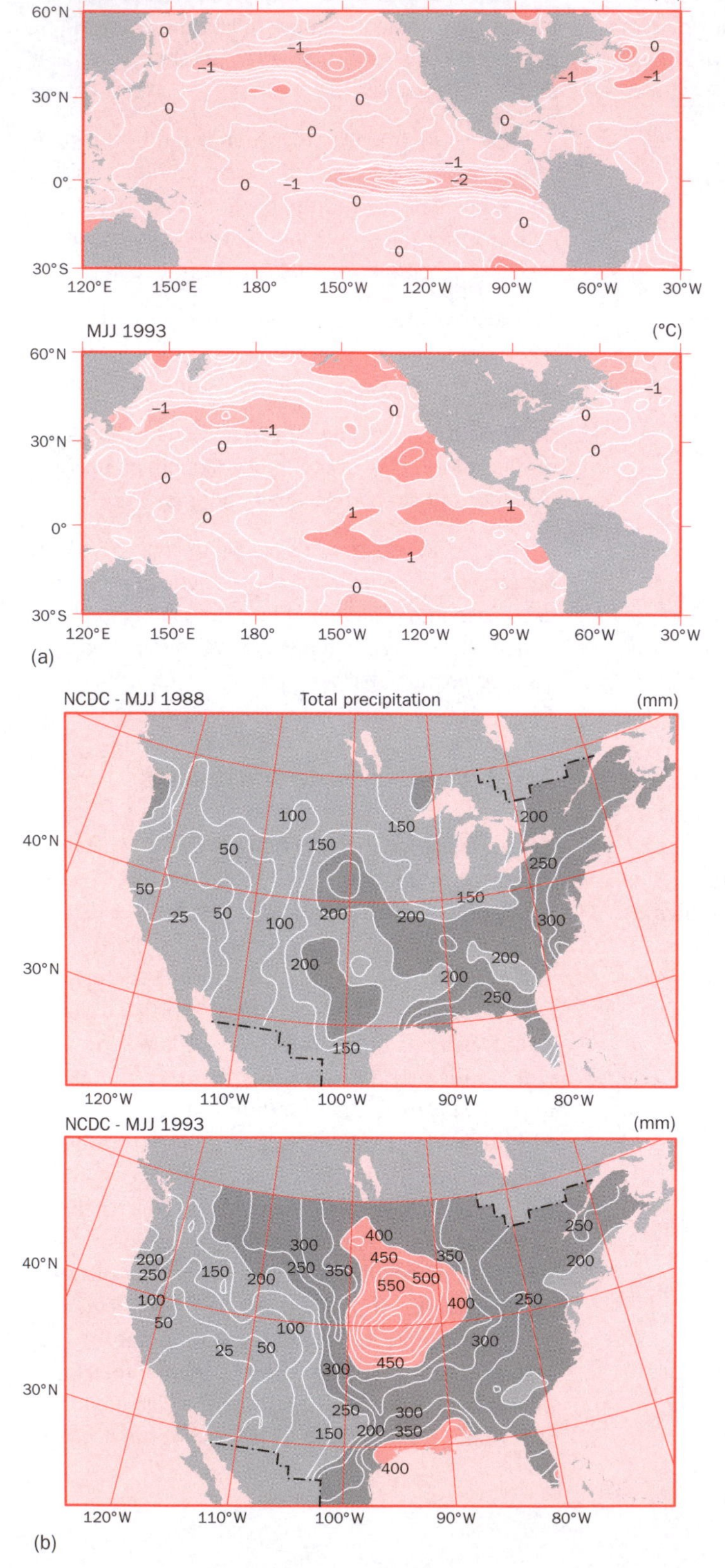

Figure 10.8 (a) Anomalous SST fields for May–July 1988 and 1993, relative to 1951–79. The contour interval is 0.5 °C, positive values exceeding 1 °C are stippled, and negative values <–1 °C are hatched. (b) Accumulated rainfall totals for May–July 1988 and 1993. The contour interval is 50 mm except for an additional 25 mm contour. Values greater than 200 and 400 mm are stippled with increasing density
Source: Trenberth and Guillemot (1996)

phenomena. This provides a platform from which to assess potential human-induced climate change impacts in section 10.4.

10.3.1 Extreme events

There have been two notable events affecting the central United States in recent decades. The spring and summer drought of 1988 and associated heatwaves resulted in the loss of some 5,000–10,000 lives and several tens of billions of dollars. The devastating summer floods of 1993 resulted in comparable economic damage and cost over 50 lives in the Midwest. A common feature of both events was the anomalous and persistent atmospheric conditions throughout North America. By comparing and contrasting the two extremes, some insight can be gained as to the links between large sea surface temperature anomalies in the tropical Pacific and the positioning of the subtropical jet stream (and hence storm tracks) across North America. As we shall see, terrestrial feedback mechanisms involving soil moisture also played an important part in perpetuating the anomalous circulation patterns once set in place.

(a) The 1988 North American drought

It is widely accepted that the origins of the 1988 drought lay in the large sea surface temperature (SST) anomalies in the tropical Pacific Ocean (Figure 10.8(a)). By May 1988, SSTs along the equator at 110° W were more than 4 °C cooler than normal in association with the strongest La Niña in over 20 years. As a consequence, patterns of tropical atmospheric heating were modified, displacing an intensified ITCZ north of normal. This in turn forced a Rossby wave-like pattern of high- and low-pressure anomalies across North America, with a deeper than normal upper-level trough along the west coast and strong (dry) anticyclonic conditions over the north central continental interior. Under normal circumstances, the summer jet stream is located at about 48° N, but in 1988 the jet stream was displaced further north into Canada. East of the Rockies the resulting storm track was shifted northwards and accompanying eddy activity was suppressed by the strong anticyclonic circulation over the interior, weakening potential links to the moisture source of the Gulf.

By the latter half of July 1988, over 40 per cent of the area of the contiguous United States was experiencing severe or extreme drought as defined by the Palmer Drought Severity Index (Box 10.2). The spatial extent of the drought had been exceeded by just four earlier events in 1934, 1936, 1954 and 1956. Figure 10.8(b) shows that the accumulated summer rainfall totals east of the Rockies were relatively uniformly distributed in the range 150–200 mm. With precipitation totals less than 10 per cent of normal, parts of Wisconsin, Illinois, Indiana and Ohio experienced the driest April to June since 1895. The accompanying clear skies and soil moisture deficits led to heatwaves across much of the interior as more solar heating went into increasing temperatures and less into evaporating moisture. Furthermore, the absence of soil moisture discouraged local evaporation and moisture recycling through precipitation, thereby perpetuating the drought. Experiments with climate models also confirm that the anomalous Pacific SSTs and displaced ITCZ established the unusual pattern of storm tracks with their associated drought over the central plains up to June 1988, but thereafter land surface feedbacks probably contributed to the severity and persistence of the drought.

(b) The 1993 Mississippi flooding

The onset of the 1993 summer flooding over North America was preceded by mature El Niño warm phase conditions in the spring. Contrary to the La Niña cold phase situation in 1988, positive tropical SST anomalies (Figure 10.8(a)) resulted in a southward displacement of the Pacific ITCZ and the establishment of a pronounced and persistent Pacific–North American (PNA) teleconnection pattern. It may be recalled that with this pattern, a high-pressure ridge extends over the eastern Pacific with a low pressure prevailing over the land area of the west and south-west. As a consequence the jet stream and associated storm paths followed a more spring-like, southerly track across the entire United States (see Figure 10.2). A persistent area of high pressure over the south-east also greatly enhanced the poleward flux of moisture out of the Gulf of Mexico. This strong ‘river’ of atmospheric moisture fed a constant succession of storms over the Midwest in June and July 1993.

The abnormally persistent weather patterns conspired to produce the most devastating flood in US history. Parts of North Dakota, Kansas and Iowa received more than double normal rainfall amounts, with accumulations of 300–700 mm in the upper Mississippi Basin (Figure 10.8(b)). By the end of September 1993, the River Mississippi at St Louis, Missouri, had been above flood stage for 144 days, and approximately 3 billion m^3 of river water had inundated the floodplain downstream of the city. The flooding damaged 56,000 homes and covered 44,000 km^2 of land across an area

BOX 10.2

THINKING FURTHER

The Palmer Drought Severity Index (PDSI)

The Palmer Drought Severity Index (PDSI) is one of the most widely applied measures of regional drought. The PDSI uses monthly mean precipitation and temperature to estimate the monthly water balance in an idealised, two-layer soil model. Several important processes are involved, such as potential and actual evapotranspiration, soil water storage and recharge, runoff and water loss from the soil. Among the model parameters are soil water storage capacities and a weighting factor used to scale soil moisture anomalies by local conditions. The model also defines local hydrological normals relative to the temperature and precipitation averaged over a predefined calibration period. In this respect, the PDSI does not represent drought in absolute terms but, rather, the soil moisture status of different regions using a standardised scale. This ranges from less than −6 (extreme drought), to −4 (severe drought) to −2 (moderate drought). Values between −0.5 and 0.5 are considered near normal, whereas larger positive values are indicative of moisture excess. As an example, Plate 10.1 shows the pattern of PDSI for the central Midwest in August 1988 when the majority of values lay between −4 and −7.

Although the PDSI quantifies two of the most elusive properties of droughts, namely their intensity and duration, the index has several limitations. First, monthly resolutions can be too coarse to capture the complex interplay between precipitation and potential evaporation, particularly in late summer. Second, the default soil moisture capacities in the model are somewhat arbitrary, restricting the applicability of the original model to North American climate regimes. Third, no lag is incorporated in the model to account for the delay between the generation of excess water and its appearance as runoff. Fourth, the thresholds used to partition rainfall into runoff or soil moisture often tend to underestimate recharge during the summer and early autumn. Fifth, no allowance is made for snowmelt or frozen ground. Sixth, the designation of the drought severity classes is arbitrary. Seventh, drought magnitudes (but not duration) are very sensitive to the choice of the regional weighting factors. And finally, the number of months designated as moderate to extreme drought depends on the choice of base period used to estimate model parameters.

Despite these limitations, the PDSI remains a popular measure of climatological drought severity for North America and has been used extensively to quantify past, present and future drought severity across the continent.

spanning nine states. The enormous volume of fresh water discharged into the northern Gulf of Mexico led to widespread algal growth and a vast sediment plume (7,000 km^2) that extended as far east as the Florida continental shelf.

Although anomalous tropical SSTs and convection ultimately set up the planetary waves, jet stream location and associated storm track activity, the heavy summer rainfall and accompanying latent heat release are believed to have perpetuated the situation by favouring cyclonic conditions locally. This in turn allowed the large-scale advection of moisture from the Gulf. At the same time, analyses of the continental moisture budget strongly suggest that local evaporation and moisture recycling were significant sources of precipitation, thereby perpetuating the wet conditions (see also the central Amazon, Box 11.2). As in the summer of 1988, land surface feedbacks served to amplify and prolong the weather triggered by anomalous SSTs in the tropical Pacific.

10.3.2 Interannual variability

The El Niño–Southern Oscillation (ENSO), incorporating both warm phase El Niños and cool phase La Niñas, is the single most important driver of multi-year variability in global patterns of precipitation, droughts and floods (see sections 5.4 and 5.5). Even so, ENSO explains at best ~7 per cent of the total variance in global precipitation in autumn (September to October) and, during a typical El Niño year, enhances global mean precipitation by just ~0.2 per cent. However, the timing, direction and impact of the teleconnection is highly region dependent (Figures 5.8 and 5.9 and Table 5.2). The main areas where precipitation is known to be strongly affected (positively or negatively) by ENSO include Indonesia–Australia, the equatorial Pacific and Atlantic (Plate 10.2). Over these regions ENSO causes shifts in rain-belts and explains ~30–60 per cent of the annual variability in precipitation.

(a) North American teleconnections with ENSO

ENSO-related climate anomalies (section 5.5.2) have also been reported for parts of North America. During the northern winter, the El Niño signal consists of an anomalously deep Aleutian Low extending south-east along the west coast of North America, and an amplification of the northward branch of the tropospheric Rossby wave over North America (see Figure 10.2). This results in higher temperatures and lower precipitation over the Pacific north-west, but cool, wet winters over the south and south-eastern United States. As discussed in section 5.5.3, warm El Niño events have also been linked to a reduction in the number of hurricanes and tropical cyclones in the western North Atlantic–Caribbean Sea (5.4 per year compared with 9.1 in non-El Niño years). Other responses include decreased montane snowpack and annual runoff in the north-west as storm systems are deflected north towards Alaska, and increases in the south-west due to the entrainment or transfer of more moisture from the Pacific on the poleward flank of the subtropical jet.

The mid-latitude responses to La Niña events generally oppose but are not perfectly symmetrical to those of El Niño. This is because negative anomalies in tropical convection and precipitation are not as strong or extensive as those during El Niño events. Thus, a positive sea-level pressure anomaly in the north Pacific replaces the Aleutian Low, but extends to the Bering Sea and along the west coast of North America. As a consequence, there is a weakening of the subtropical jet accompanied by a dampening of the amplitude of the wave train over North America. This results in a southward shift in the Pacific storm path leading to average increases in precipitation over the Pacific north-west (decreases with El Niño), and reductions over the south-west (increases with El Niño).

The typical El Niño and La Niña responses have been derived from composite analyses of regional climate, drought and streamflow patterns. This technique involves *averaging* weather anomalies associated with opposing phases of large-scale indices (such as ENSO) and is widely used to characterise global teleconnection patterns. A major limitation of the approach is that there is often large variability among the responses of years subject to the same (i.e. La Niña or El Niño) ENSO phase. This is a significant confounding force that reduces the long-range predictability of ENSO-related weather anomalies across North America. However, there is growing evidence to suggest that the more slowly evolving Pacific Decadal Oscillation (PDO) (see section 5.3.2) exerts a modulating effect on ENSO teleconnections. As illustrated in Figure 10.3, positive (negative) PDO phases are characterised by an anomalously deep Aleutian Low (high), cold (warm) waters in the western and central north Pacific, and warm (cold) waters in the eastern coastal, central and eastern tropical Pacific.

Two-way classifications of positive and negative PDO with warm and cold ENSO events suggest that typical El Niño patterns (less wet north Pacific coast, cooler wetter southern Pacific coast) are most pronounced and consistent when they coincide with the positive phase of the PDO. It has been hypothesised that the deeper Aleutian Low of the positive PDO favours a more southerly storm track into which El Niño feeds enhanced eastern tropical moisture for the storms. Conversely, typical precipitation patterns during La Niña winters (wetter north Pacific coast with warmer drier south Pacific coast) are most consistent when accompanied by a negative PDO phase. Under these circumstances cyclonic storms are steered further north away from the eastern tropical Pacific moisture source. This leaves the negative PDO–El Niño and positive PDO–La Niña combinations which yield relatively weak and spatially incoherent climatic anomalies.

(b) Trends in climate

A growing body of evidence suggests that the climate of North America has become more extreme in recent decades. Such observations raise the perennial questions as to how much change (if any) may be attributed to natural climate variability, and how much to direct and/or indirect forcing by human agency (e.g. greenhouse gas emissions, land surface change)? Analyses of homogeneous, gridded data sets with global coverage for monthly precipitation, temperature, Palmer Drought Severity Index (PDSI), and streamflow have gone some way to addressing these important issues.

Across the country as a whole, the United States warmed by about two-thirds of a degree Celsius during the twentieth century (Figure 10.9(a)). However, the most rapid warming was in the three months March to May over the north-west and north-east (noting that much of the south-east cooled by ~1 °C over the same period). Some of the warming may be attributed to the marked increase of night-time minimum temperatures in large urban areas. However, declines in Arctic ice thickness, less snow cover, earlier

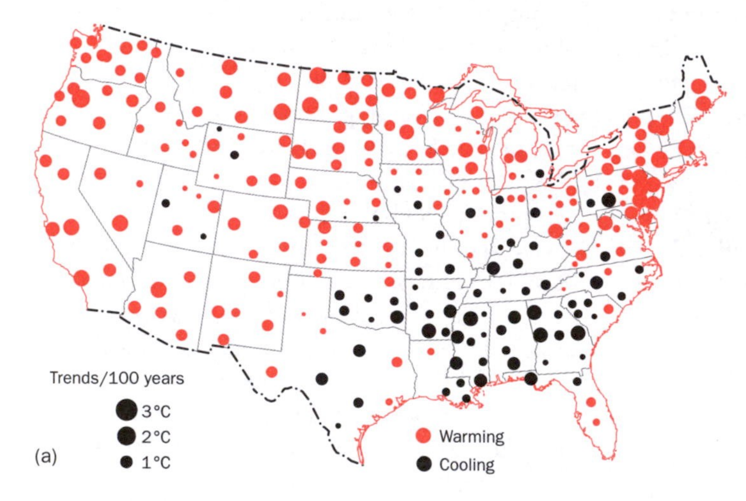

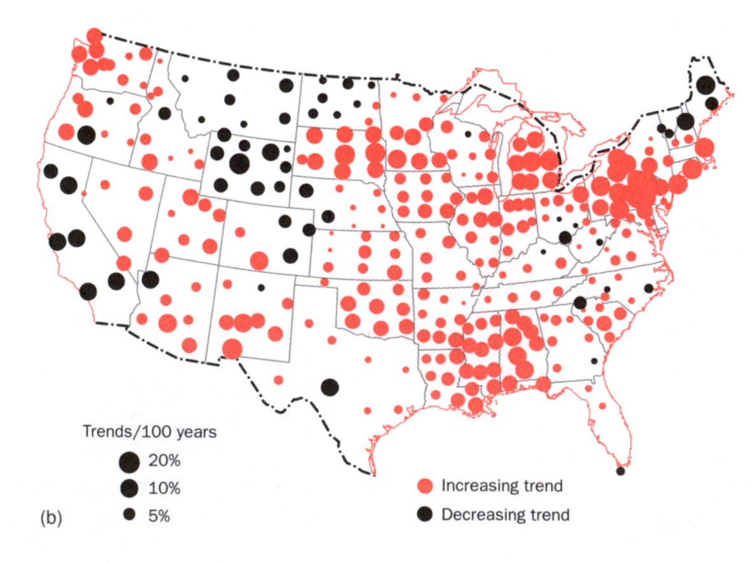

Figure 10.9 Changes in (a) annual mean temperatures and (b) annual total precipitation across the contiguous United States, 1900–94
Source: Karl and Knight (1998)

snowmelt-related floods in western Canada and the western USA, thawing of large areas of Alaskan permafrost, and the melting of glaciers at unprecedented rates, all support the impression of a warming continent (section 8.1.4). Additionally, since the 1980s, the growing season between 45 and 70° N increased by up to 12 days, and plants have bloomed about a week earlier.

Far greater uncertainty surrounds the detection of trends in precipitation because of the quality of available data and high natural variability (Figure 10.9(b)). None the less, the percentage of land areas defined as 'wet' by the PDSI has increased from ~12 to >24 per cent since the 1970s, and the percentage of 'dry' areas has decreased by a similar amount since the 1940s. The frequency of wet days and the intensity of the most intense rainstorms have both increased. For example, in the Midwest, the number of precipitation events of 50 mm/day or more increased by ~20 per cent between 1901 and 1994. Trends in rainfall are broadly reflected by continental-scale streamflow changes. For example, a number of recent studies highlight a distinct concentration of stations in the upper Mississippi and Ohio River valleys with greater winter and spring streamflow. Much of the increase appears to have occurred in the lower half of the streamflow distribution, between the annual minimum and the annual median flow.

Recent research indicates that ENSO forcing may account for some of the increases in temperature and in the distribution of severe wet and dry areas across North America since the late 1970s. In particular, a shift towards more warm phases after the mid 1970s has been reflected in changes to the PDSI over ENSO-sensitive regions. There have also been significantly higher El Niño-induced PDSI anomalies during 1979–95 than would have been expected from pre-1978 data. In other words, the wet area associated with ENSO is now much greater than for earlier warm phases. One explanation is that greenhouse-gas induced climate changes in the last decades of the twentieth century have favoured a more vigorous hydrological cycle, manifested by increased moisture surplus (i.e. large positive PDSI) over much of the continent.

10.3.3 Decadal variability of drought

Great droughts in the 1930s, 1950s and 1980s caused immense economic and social hardship to the communities of the Great Plains. Due to the brevity of the instrumental record, estimates of the return period of the infamous 1930s ('Dust Bowl') droughts, however, are highly uncertain (ranging between once in 75 and once in 3,000 years). None the less, it is important to understand how rare the severe droughts of the twentieth century really were, and whether or not *even more* extreme events are possible. Such questions can only be addressed through reference to the full range of past drought variability contained in historical, archaeological and palaeoclimatic records. For instance, newspaper reports from Kansas, and records of increased wind-blown (aeolian) sand, suggest that the 1860 drought may have matched the severity of the 1930s drought. Furthermore, multiple proxy sources such as tree rings, aeolian and lake sediments suggest that even the most severe droughts of the nineteenth and twentieth centuries were eclipsed by several droughts in the last 2,000 years, but most recently in the late sixteenth century.

(a) Evidence from tree rings

In section 6.3.3, environmental proxy records including tree rings, pollen and sediments were described and evaluated. Tree ring-based reconstructions back to the seventeenth century for sites across the North American corn belt of Iowa and Illinois suggest that four decades were drier than the 1930s (respectively 1816–25, 1735–44, 1696–1705 and 1664–73). Tree-ring data from Virginia further indicate that the lost colony of Roanoke Island disappeared during the most extreme drought in the last 800 years (1587–89), while the near abandonment of the Jamestown colony occurred during the driest 7-year episode in

770 years (1606–12). A reconstruction of flows in the Colorado River supports the contention that the period 1579–98 was the longest and most severe drought in the record. Archaeological evidence from abandoned Anasazi settlements in the south-west points to another 'Great Drought' in the last quarter of the thirteenth century. Such records provide collective evidence that the multidecadal droughts of the late thirteenth century and sixteenth century were more severe than any event experienced in the twentieth century. The three centuries between these two major drought-rich periods were, however, largely free of widespread drought.

As we look further back in time, the temporal resolution and number of palaeo-records for the continental interior diminish. Fortunately, modern instrumental data indicate that conditions in the interior are closely related to conditions in the west and south-west of the continent where proxy records are more abundant. For example, analyses of relict tree stumps suggest that the Sierra Nevada, California, experienced extremely severe drought conditions for more than two centuries before AD 1112 and for more than 140 years prior to AD 1350. During these periods, the Sierran runoff was significantly lower than during any persistent droughts in the instrumental record of the last 140 years. Summer temperatures were also higher than at any time with the exception of the most recent decades of the twentieth century.

Although longer proxy records have lower temporal detail or resolution (decadal rather than annual), several suggest a regime change around AD 1200. Prior to this time droughts appear to be characterised by even greater intensity and frequency than those of the thirteenth and/or sixteenth centuries. Drought-sensitive tree-ring chronologies point to the existence of at least four 'mega-droughts' in the Great Plains between AD 1 and 1200. For example, tree rings from the White Mountains, Four Corners region and Colorado Front Range document widespread drought around mid AD 1100. The next major drought ~ AD 950 coincided with the onset of aeolian activity in the western Great Plains, and the third major episode between 700 and 900 is corroborated by archaeological evidence from the Four Corners area and lake sediments in North Dakota (see below). The earliest of the mega-droughts began in the middle of the third century AD and, according to drought-sensitive giant sequoia, may have lasted up to three centuries.

(b) Evidence from lake sediments

For millennia as opposed to centuries, it is necessary to reconstruct environmental conditions using techniques other than tree rings. For example, pollen and diatom assemblages deposited in lake sediments are widely used to reconstruct histories of vegetation change and lake-water salinity respectively. Diatoms are members of the algal flora whose microscopic, unicellular forms, abundance and community structure are highly related to lake-water quality and temperature. By establishing statistical relationships between diatom abundance/type and water quality for current lake conditions, it is possible to infer past salinity/lake levels (and hence climate) from counts of diatoms preserved at different depths in lake sediments. The age of sediments is typically obtained by means of radiocarbon and lead-210 dating.

One such proxy record from Moon Lake, North Dakota, supports the view that droughts of greater intensity than those of the 1930s were more frequent in the north Great Plains before 800 BP (Figure 10.10(a)). Furthermore, drought-rich periods persisted for centuries rather than decades, being most noticeable during AD 200–370, AD 700–850 and AD 1000–1200 (see above). Supporting evidence of multi-decadal drought is provided by giant sequoia growing in the San Joaquin Valley, California. The sequoia tree rings are suggestive of persistently dry periods during AD 250–350 and AD 700–850, and therefore, of a common source of continental-scale climate forcing. By modern analogy, the drought epochs were probably due to more frequent patterns of

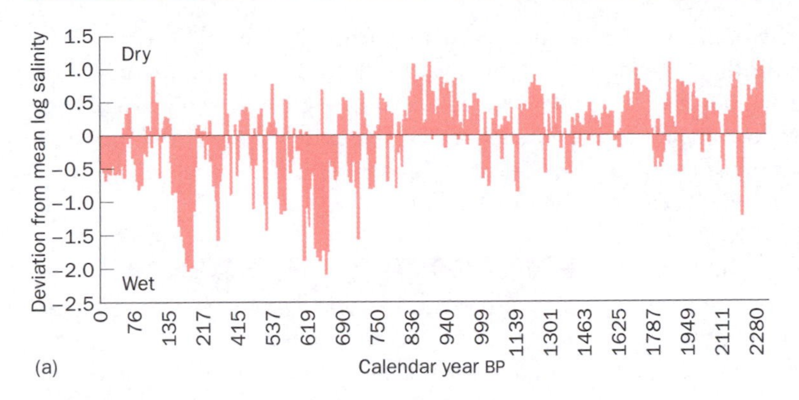

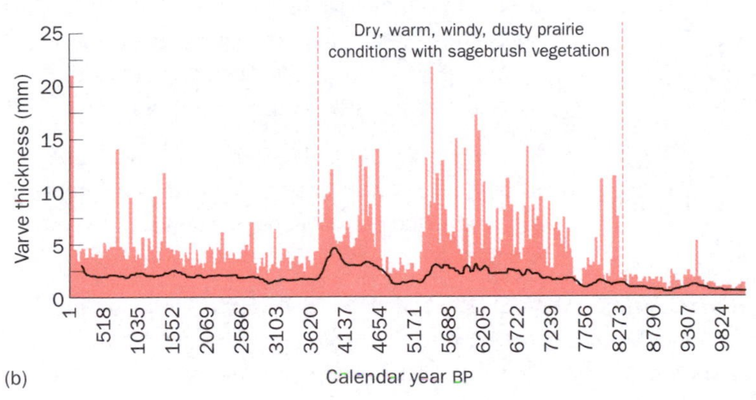

Figure 10.10 Reconstruction of droughts in the Great Plains using palaeo-environmental sources: (a) North Dakota Moon Lake salinity record for the past 2,300 years. (b) Upper Mississippi Elk Lake varve thickness for the past 10,000 years
Source: (a) http://www.ngdc.noaa.gov/paleo/drought/drght_laird96.html (b) http://www.ngdc.noaa.gov/paleo/drought/drght_dean.html. NOAA Paleoclimatology Program

persistent upper-level high-pressure ridges over central North America, than typically occur today.

High-resolution records of the physiochemical properties of lake sediments (such as grain size, layering and isotope concentrations) have also been used to reconstruct past drought and climate variability. One such record from Elk Lake, in the headwaters region of the upper Mississippi Basin (north-eastern Minnesota), provides a history of climate and other environmental conditions over the last 10,000 years (Figure 10.10(b)). The lake sediments were deposited in annual layers (or varves) whose thickness and mineralogy depend on the relative abundance of wind-blown quartz-rich silt and clay. Because thicker varves are indicative of drier, warmer, windier and dustier prairie conditions, it is evident that there have been two major drought phases. The first occurred between 6,000 and 8,000 years ago, and was followed by a wet interlude, before the onset of the second phase between 4,300 and 4,800 years ago.

(c) Palaeo-drought mechanisms

Palaeoclimatic data provide evidence of drought phases in the pre-instrumental era that would have had devastating consequences for the region if they occurred today. Furthermore, the situation would be exacerbated by anticipated human-induced changes in climate over the course of the twenty-first century (see below). It is, therefore, imperative to identify the physical mechanisms leading to severe, continental-scale droughts. Although the causes of droughts over seasonal timescales (e.g. 1988) are generally well known (section 10.3.1), the origins of droughts spanning years, decades or even centuries (e.g. the 1930s and palaeo-droughts) are poorly understood.

It has been speculated that major droughts of decadal duration or more are a consequence of persistent anomalies in boundary conditions such as the thermal properties of the Pacific and Atlantic oceans. For example, the lake levels of Mono Lake, California – inferred from the abundance of ^{18}O in sediments dated between 34,500 and 12,900 years BP – show four low stands each lasting between 1,000 and 2,000 years that coincide with changes in the rate and location of the thermohaline circulation in the North Atlantic.

Ocean temperatures in this region have a major impact on precipitation patterns across the Great Plains. This is because they affect changes in the position of the Azores high-pressure cell which in turn influence the amount of Gulf moisture penetrating the interior. For instance, it has been suggested that dune reactivation about 1,000 years ago coincided with a slight easterly shift in the Azores High and an associated reduction in the continental moisture influx. It is worth noting here that such an easterly shift in the Azores High could have pushed warm tropical air (mT) further north over Europe helping to create the Medieval Warm Period of the tenth to twelfth centuries (see section 6.4.3).

There is also growing evidence to suggest that ENSO varies in intensity at the century timescale as a consequence of internal variability, external forcing or a combination of both (see section 10.4.1). Furthermore, the Pacific Decadal Oscillation (PDO) is known to exert a modulating influence on ENSO-related precipitation anomalies over North America (see above). The droughts of the 1930s to 1950s, for example, coincided with a period of low ENSO variability. A more controversial hypothesis attributes the decadal-scale precipitation variability evident in many tree-ring and sediment records to the ~20-year solar–lunar cycle. However, until a physical explanation is found for these associations, the complex interplay of low-frequency variations in Atlantic and Pacific SST forcing remains a more plausible explanation for observed decadal patterns of drought.

Finally, as suggested above, it is tempting to suggest that North American records of warm/dry weather may have coincided with the European Medieval Warm Period of the ninth to the fourteenth centuries. This could provide evidence of drought forcing on a global scale. Unfortunately, regional drought episodes are not strongly synchronous (e.g. evidence for the south-east United States does not show unusually warm weather at that time). In fact, long-term palaeo-temperature records from around the world exhibit substantial decadal to multidecadal variability over the last millennium with only a few decades in the seventeenth to nineteenth centuries AD (i.e. the Little Ice Age) exhibiting a consistent (cold) temperature bias of truly global extent.

Key ideas

1. The origins of the severe summer drought of 1988 and summer flooding of 1993 both lay in large sea surface temperature anomalies (of contrasting signs) in the tropical Pacific, leading to unusual displacements of storm tracks over North America.

2. The severity of the 1988 drought (and 1993 flooding) was exacerbated by land surface feedbacks involving restricted (enhanced) soil moisture recycling to precipitation.

3. El Niño events are generally associated with above average temperatures, less precipitation and snowpack accumulation over the north-west; cool, wet winters over the south-east; and fewer cyclones in the Caribbean.

4. During the course of the twentieth century the USA warmed by approximately two-thirds of a degree Celsius, the frequency and intensity of rain storms increased, and the land area defined as 'wet' has doubled since the 1970s.

5. Newspaper reports from the time suggest that a drought in 1860 may have matched the severity of the infamous 'Dust Bowl' drought of the 1930s.

6. Evidence of past climate conditions from tree rings and archaeological data point to four 'mega-droughts' of multidecadal duration prior to AD 1200 that have no modern precedent.

7. Pollen and diatom records from lake sediments support the view that the continental interior experienced two major drought phases in 4300–4800 and 6000–8000 BP, linked to persistent changes in ocean temperatures in the North Atlantic and/or tropical Pacific.

10.4 Climate change projections for North America

There is compelling evidence from palaeoclimate data and current general circulation models that climate variability and change could pose serious challenges to the production of food and energy, as well as human and ecosystem health, and the functioning and properties of the hydrologic cycle across continental North America. The following section describes the major findings of the National Assessment of Potential Consequences of Climate Variability and Change for the United States ('the National Assessment') undertaken in the year 2000. The focus will be on future changes in extreme events, with particular reference to changes in high-intensity precipitation events, droughts and associated water sector impacts.

10.4.1 The US National Assessment

The National Assessment was initiated by the 1990 Global Change Research Act, mandated by Congress to undertake a 'comprehensive and integrated United States research program which will assist the Nation and the world to understand, assess, predict, and respond to human-induced and natural processes of global change'. Under the US Global Change Research Program (USGCRP) Federal research agencies such as the US Geological Survey and Department of the Interior were instructed to cooperate with a view to analysing and evaluating the potential consequences of climate variability and change with respect to broad public, environmental and resource concerns. The study had three major components: (i) a national synthesis of findings emerging from (ii) regional analyses of potential consequences of climate variability, and (iii) sectoral analyses to address potential impacts specific to agriculture, forests, human health, coastal areas, marine resources and water. The following paragraphs present a summary of key changes in seasonal climate identified by the national synthesis team.

(a) Climate model experiments

The National Assessment was based largely upon the results of two climate models: the Canadian Global Coupled Model (CGCM1) and the UK Meteorological Office's Hadley Centre Coupled Model (HadCM2). In both cases, future climate experiments were derived from a 'business as usual' scenario in which there is a compound 1 per cent per year increase in CO_2 emissions with a doubling of sulphur emissions by 2100. According to the second Coupled Model Intercomparison Project (CMIP2), CGCM1 is considered a relatively 'warm' model and HadCM2 a relatively 'cool' model (in terms of their respective transient climate responses).

Both the CGCM1 and HadCM2 produce future climate scenarios at relatively coarse spatial resolutions – in the case of HadCM2, a grid spacing of 2.5° latitude by 3.75° longitude. The second phase of the Vegetation–Ecosystem Modeling and Analysis Project (VEMPAP2) involved the development of higher-resolution, topographically adjusted monthly climate variables (temperature, precipitation, solar radiation, humidity and wind speed). These were produced, first, for historic data from 1895 to 1993, and then interpolated from the output of the two climate models to a 0.5° grid from 1994 to 2100. These data were generated for the conterminous United States and have been an important resource underpinning many impact studies conducted within the National Assessment framework.

(b) Temperature

The CGCM1 and HadCM2 climate models suggest an average annual warming of between 3 and 6 °C across the continental United States by the 2090s. However, this range conceals large variations in warming at seasonal timescales and regional spatial scales (Table 10.2). For instance, CGCM1 shows strong winter warming exceeding 9 °C over much of Canada and the USA, with more conservative winter warming of up to 5 °C projected by

Table 10.2 Changes in annual mean temperature (°C) projected by the CGCM1 and HadCM2 climate models by 2030 and 2095

Region	2030		2095	
	CGCM1	HadCM2	CGCM1	HadCM2
North-west	1.8	1.7	4.9	4.1
South-west/California/Rockies	2.0	1.8	5.5	4.0
Great Plains	2.2	1.6	6.3	3.6
Great Lakes/Midwest	2.4	1.1	6.1	2.7
South-east	1.8	1.0	5.5	2.3
North-east	1.8	1.0	5.6	2.7
United States	2.1	1.4	5.8	3.3

Source: US National Assessment

HadCM2. The CGCM1 suggests that the most rapid (slowest) warming will occur in the Great Plains (north-west), whereas HadCM2 points to the north-west (south-east). Neither result can be taken as 'truth' because of differences in each model's structure and parameterisations. However, there is a very high degree of confidence in a scenario of rising temperatures across the continent, but confidence in the regional projections of any given model remains low.

(c) Precipitation

There is a very high degree of confidence among climate modellers that future global mean precipitation will increase. However, future changes in regional precipitation are far less certain. This is because precipitation mechanisms occur at spatial scales smaller than those currently resolved by climate models (see Chapter 7). Each climate model has slightly different schemes for representing subgrid-scale processes such as clouds, convective precipitation, soil moisture, evaporation and runoff. As a consequence, each model displays characteristic biases in the representation of hydrological components at regional scales. For example, both the CGCM1 and HadCM2 models tend to be too wet in the west and north-eastern USA in spring and summer, and too dry over the south-east and Midwest in winter and summer.

The changes shown in Table 10.3 are, therefore, only indicative of future precipitation patterns across the continent. Even so, it is interesting to note that both models signal large increases in annual and winter precipitation across the south-western USA, California and Rocky Mountains (a pattern as shown previously that is currently associated with warm ENSO events). These precipitation changes reflect a greater number and intensity of storms in the Gulf of Alaska rather than changes in the position of storm tracks. Conversely, there is a decrease in the Atlantic storm tracks of the east coast, and neither model replicates the frontal development (cyclogenesis) on the lee side of the Rockies.

Summer precipitation changes are less pronounced: both models show increases across southern California but CGCM1 suggests future drying of the south-eastern USA, whereas HadCM2 shows the same area becoming wetter. The increased precipitation in the south-west arises from an expansion of the zone of convective precipitation associated with the warming of SSTs along the Pacific coast. Precipitation reductions throughout the Rockies and the southern region of North America correspond to areas of decreased soil moisture. Enhanced warming over land leads

Table 10.3 Changes in annual mean precipitation (%) projected by the CGCM1 and HadCM2 climate models by 2030 and 2095

	2030		**2095**	
Region	**CGCM1**	**HadCM2**	**CGCM1**	**HadCM2**
North-west	+8	+11	+31	+13
South-west/California/Rockies	+16	+8	+67	+27
Great Plains	−2	+6	+13	+16
Great Lakes/Midwest	−2	+9	+20	+27
South-east	−19	+3	−13	+22
North-east	−6	+8	0	+24
United States	−4	+6	+17	+23

Source: US National Assessment

to drying of soils, reduced evaporation and precipitable water, fewer clouds, less precipitation, and more energy for surface heating – a positive feedback loop of the type that perpetuated the drought of 1988 (section 10.3.1). Differences in the details of summer precipitation change probably reflect differences in the respective land surface schemes in CGCM1 and HadCM2.

(d) Evapotranspiration

The estimation of future rates of evapotranspiration is highly problematic. This is because the two components of evapotranspiration – evaporation from land and water surfaces and transpiration from vegetation – are regulated by a wide range of meteorological factors. The potential for evaporation increases with greater levels of solar radiation and wind speed, but is limited at higher atmospheric water contents and/or by the actual availability of water held on plant and soil surfaces. A greater proportion of precipitation may be evaporated from water intercepted by vegetation in regions of frequent rainfall compared to sites exposed to fewer but more intense precipitation events.

Transpiration is affected by both meteorological and physiological factors, including plant architecture, stomatal behaviour and response to rising atmospheric CO_2 concentrations. Some laboratory and field studies have shown that certain plants have lower rates of water loss at higher CO_2 levels because of higher stomatal resistance (i.e. guard cells around the stomata impede the vapour flux from plant to atmosphere). Other studies highlight the control exerted by water or nutrient limiting of plant growth and hence changes in leaf area (a proxy for the number of stomata) that may either increase or decrease water-use efficiency depending on local resources. In any event, the extrapolation of global changes in evapotranspiration from laboratory findings so-called 'upscaling') is an area of active research for hydrologists, soil scientists and plant physiologists. Current estimates suggest that global average potential evaporation could increase by up to 15 per cent for a doubling of atmospheric CO_2 concentration, depending on the amount of warming at that time. Projected evaporation changes over the USA broadly correlate with precipitation changes, but warming can lead to lower actual evaporation if soils are dried and the potential for moisture recycling diminishes.

10.4.2 Future changes in extreme events

Some of the most profound impacts of climate variability and change arise through costs associated with flood and drought events. Unfortunately there have been relatively few credible studies of riverine flood risk associated with climate change because of the difficulties associated with adequately modelling high-intensity precipitation at sub-daily timescales and/or finer resolutions than individual climate model grid scales. The simplest approach is to infer future flood risk from projected changes in storm frequency or magnitude of precipitation events. This method is illustrated below along with an assessment of future drought scenarios for the North American continent.

(a) High-intensity precipitation

Precipitation changed significantly over the twentieth century across many regions of the world and it was noted earlier that precipitation has increased by about 10 per cent over the contiguous United States since 1910. Comparable changes have been reported for Russia, parts of Scandinavia, Australia and Canada. A common feature in all such regions has been the disproportionate increase in heavy daily precipitation events relative to the mean. This observation has been supported by climate model experiments. For example, an earlier version of the Canadian Climate Model shows a 4 per cent increase in the mean daily precipitation amount, but an 11 per cent increase in the size of the 20-year extreme precipitation event over North America. Other models have consistently shown statistically significant increases in the fraction of annual precipitation contributed by extreme events, with qualitatively similar changes observed in the UK, Australia and elsewhere.

Many climate models project further changes in the frequency distribution of daily precipitation amounts. For example, both HadCM2 and the National Center for Atmospheric Research Climate System Model (CSM) show disproportionate increases to extreme (>50 mm/day), heavy (25–50 mm/day) and moderate (2.5–25 mm/day) precipitation categories, at the expense of light events (<2.5 mm/day). Although days with extreme precipitation contribute less than 1 per cent of the winter total, this category has by far the largest percentage gain across North America. For example, extreme winter precipitation events are projected to increase by over 230 per cent in CSM and by 65 per cent in HadCM2 by the end of the twenty-first century. In summer there is a greater contribution to total precipitation from moderate events than in winter, and generally less from other categories. However, as in winter, there is again a disproportionate increase in the future contributions from extreme and heavy events (although these are much rarer events in CSM than in HadCM2). Light and moderate events change little in CSM, while the light category decreases slightly in HadCM2.

The most striking changes in modelled average daily precipitation amounts occur over the tropical to mid-latitude Pacific. These may reflect differences in the responses of the models' El Niños to anthropogenic forcing, or to the degree to which each model shows an El Niño-like temperature-change pattern in the Pacific (i.e. enhanced warming in the east relative to the west). Indeed, it has been speculated that the frequency and intensity of El Niño events may increase due to greenhouse warming, with significant consequences for patterns of precipitation and flooding across the continent. However, the very large precipitation changes shown over the Pacific by HadCM2 in summer probably reflect the high sensitivity of this model's precipitation to changes in atmospheric moisture content.

(b) Drought: how to define it?

The major droughts of the nineteenth and twentieth centuries led to mass migrations and considerable social hardship, so water managers are highly concerned about the risk of severe droughts in the future. However, determining changes in drought frequency or intensity is complicated by the fact that no single definition of drought is universally applicable. For example, droughts may be defined and quantified according to precipitation or soil moisture deficits (meteorological definition), changes in reservoir yields or low flows (hydrological), reduced crop production and/or soil losses due to dust storms (agricultural), the loss of hydro-power production or industrial output (economic), increased water demand, mortality and discomfort to human populations (social), or adverse consequences for fisheries, wetlands and natural resources (ecological drought). For practical purposes the hydro-meteorological approach is the most straightforward, using ideas related to the water balance (i.e. precipitation, evaporation and runoff) to evaluate change.

(c) Hydro-meteorological drought

Projected decreases in total precipitation, more frequent dry spells, and enhanced temperatures imply more frequent and severe droughts. Unfortunately, there is very little consensus among different climate models about drought scenarios at the regional level – a direct consequence of the large model uncertainties attached to precipitation at this scale. For example, CGCM1 suggests that runoff will decline by the 2030s in all regions except California (Figure 10.11); by the 2090s there is further drying in the east but additional

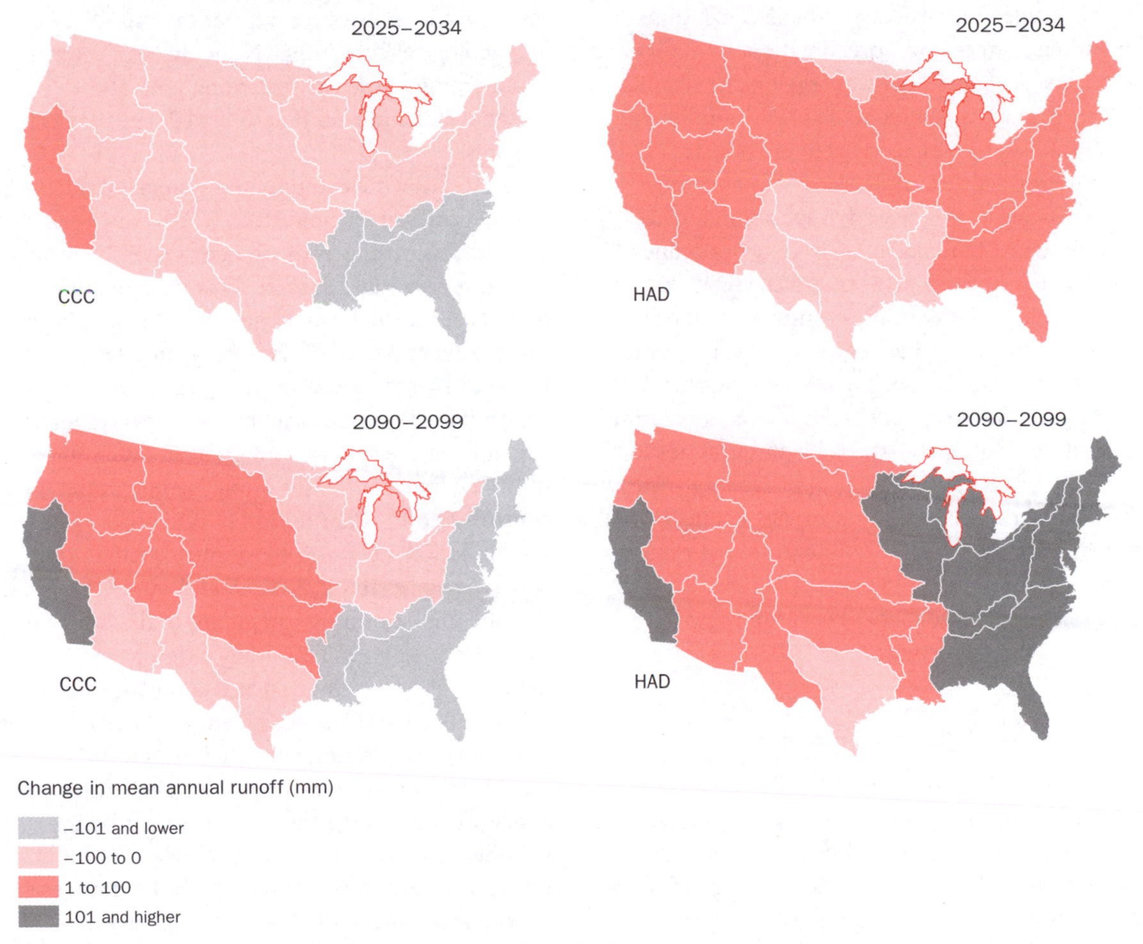

Figure 10.11 Changes in runoff by US hydrologic region as projected by the CGCM1 (CCC) and HadCM2 (HAD) climate models
Source: Wolock and McCabe (1999)

increases to water supplies in the west. In contrast, HadCM2 projects increases in water resources across most of the USA with the exception of the lower Mississippi, Souris–Red–Rainy region, Texas, and Rio Grande. By the 2090s, the HadCM2 model indicates wetter conditions in all regions except the Texas–Gulf area. The most substantial resource increases are projected for the nation's arid and semi-arid regions (e.g. California and the Great Basin). Inter-model differences highlight the difficulties associated with quantifying future risk of drought, and underline the need to regard model scenarios as sensitivity studies.

Like runoff, lake levels are known to respond to a wide range of hydro-meteorological conditions including temperature, precipitation, humidity and wind speeds. Large open (exorheic) lake systems are particularly sensitive to changes in inflows and outflows, as well as rates of open-water evaporation. For example, most climate model scenarios suggest increased evaporation due to warmer water temperatures, leading to declines in the levels of the North American Great Lakes and associated reductions in outflows to the St Lawrence River. Higher temperatures also point to less extensive lake ice formation in winter, and deeper thermal stratification of the water column in summer. Both processes have implications for nutrient cycling and ecological productivity within affected lakes. Adverse water quality changes (such as increased water temperature, salinity or deoxygenation due to enhanced algal growth) represent further challenges confronting water managers during droughts.

10.4.3 Hydrological impacts

The IPCC Third Assessment Report asserts that climate change may lead to substantial changes in total runoff, the seasonality of river flows, and probabilities of very high- or low-flow conditions across North America. Climate variability and change will also have many indirect impacts on water-related infrastructure and future patterns of water demand. Furthermore, water sector impacts depend on a host of non-climatic factors such as the technologies and policies chosen to address future water demands, as well as global environmental change.

(a) River flow regimes

Projections of changes in annual runoff are highly uncertain at the continental scale mainly because of inter-model differences in future patterns of precipitation change (see Figure 10.11). This problem is compounded still further at the river catchment scale where climatic factors are mediated by land surface properties such as topography, snow cover, vegetation, aspect, soil and geological conditions. None the less, there is an emerging consensus regarding possible seasonal shifts in runoff and related hydrological impacts.

In the snowmelt-dominated systems of the mountainous areas of western North America, enhanced warming leads to increases in the ratio of rain to snow and shorter snow accumulation periods at intermediate elevations. Other responses include earlier onset of the spring-melt season followed by reductions in spring and summer runoff. However, the exact hydrological response depends on the complex interplay between the future position of the snow line (temperature control) and the distribution of potential accumulation sites with elevation (physiographic control). For example, hydrological models suggest that the high-elevation catchment of the Merced River, California, will experience significant increases in winter snowpack accumulation (Figure 10.12(a)) as a consequence of large increases in winter precipitation projected by HadCM2 for this region (see Plate 7.1). Also evident is a forward shift in the timing of peak snowpack volume from March presently to February in the future, and later onset of snowpack development (November instead of October). Higher winter temperatures also increase the likelihood of liquid precipitation falling on snowpack resulting in a marked increase in the magnitude of winter runoff

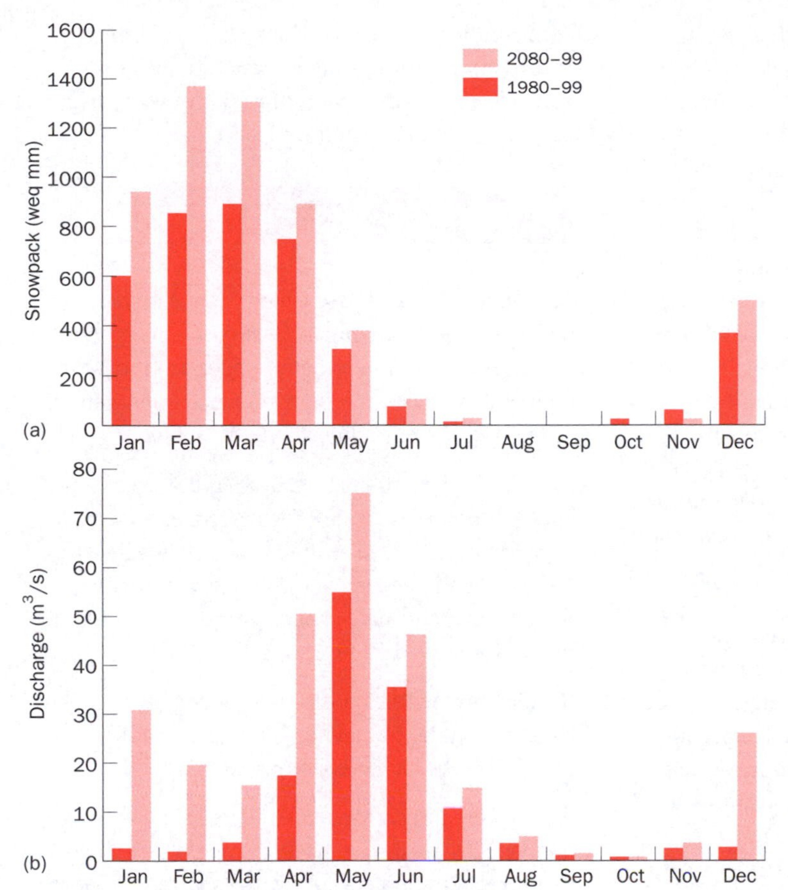

Figure 10.12 Changes in monthly (a) snowpack accumulation and (b) runoff between 1980–99 and 2080–99 for the Merced River, California.

(Figure 10.12(b)). At the same time there is approximately a doubling of the peak daily discharge (not shown) arising from a combination of more intense precipitation events falling on relatively impermeable snow-covered areas. The total runoff accumulated throughout the year is higher because of enhanced winter precipitation and moisture storage in snowpack.

In comparison, the future hydrological regimes of low-elevation or semi-arid sites are far more responsive to projected changes in the timing and magnitude of individual precipitation events. Arid environments exhibit highly non-linear runoff responses to precipitation so are particularly sensitive to any changes in the temperature and rainfall regimen. Unless there are deep soils or geology favouring water retention, precipitation is routed directly to the drainage network with only limited opportunities for water storage within the catchment. As a consequence, scenarios of increasing aridity in the midwest USA and Canadian Prairies translate into falling groundwater levels, declines in summer river flow, and increased likelihood of severe droughts. Rivers that have their headwaters in mountainous regions and descend to the semi-arid zone will, however, respond in more complex ways due to the dual influence of the snowmelt- and rainfall-dominated components of the flow regime.

(b) Non-climatic factors

The impacts of anthropogenic climate change on the water sector must be viewed alongside the historic legacies of river regulation and groundwater exploitation. For instance, the Ogallala aquifer supports the agricultural activities of the western part of the Great Plains and is probably one of the most heavily 'mined' fossil aquifers in the world. In 1914 there were just 139 irrigation wells in all of west Texas, by 1971 there were 66,144, and by 1975 the overdraft from the Ogallala region was the equivalent to the entire discharge of the Colorado River. As a consequence, the water table at Dallas–Fort Worth has fallen 120 m since the 1960s.

Some climate model scenarios suggest that future droughts may be more frequent (see

Figure 10.11) and accompanied by higher temperatures, leading to further drawdown of the aquifer. The long-term ramifications for irrigated agriculture and the regional economy are clear, but there are even wider implications for global food production and US foreign exchange. Some commentators have suggested that a second Dust Bowl era is in the making since all that holds the soil in place is crops and water, both of which cannot last under present rates of water mining. Others have a more optimistic outlook, believing that irrigation depletions have already peaked and that there has been a shift towards greater civil consumption and water efficiency since the 1980s. In either case, it is evident that unsustainable water resource practices have far-reaching consequences, and may profoundly affect a region's susceptibility to future meteorological droughts.

Finally, it is increasingly recognised that local land-use practices influence regional climate, vegetation and runoff regimes in adjacent natural areas. For example, results from the Colorado State University Regional Atmospheric Modelling System (RAMS) indicate that historic land-use changes have caused summer cooling of about 0.6 °C to the east of the Colorado Rockies relative to the climate associated with natural vegetation cover. Regional cooling is thought to arise from the replacement of dry, natural biomes by irrigated vegetation and cropland. This is because human-modified landscapes have lower albedos and higher surface roughness and soil moisture than natural surfaces. The net effect of these physical changes is to partition a larger proportion of solar energy into latent heat (associated with evapotranspiration) and less into sensible heating of the overlying atmosphere. Increased cloudiness due to elevated moisture fluxes and atmospheric instability further reduces daytime temperatures.

Thus, a complete understanding of local climate trends requires consideration of both regional and global anthropogenic changes to the land surface as well as to atmospheric chemistry. Furthermore, historic land surface changes may have had a confounding influence on surface temperature records that – alongside natural variability – complicates the detection of radiative warming due to increased greenhouse gas concentrations.

Key ideas

1. The US National Assessment was charged with undertaking a comprehensive and integrated research programme to better understand, assess, predict and respond to human-induced and natural processes of global change.

2. The two climate models used in the assessment (CGCM1 and HadCM2) suggest average annual warming of 3–6 °C over the continent, large winter precipitation increases in the south-west, and reduced summer precipitation across the southern tier and Rockies by the 2090s.

3. Climate models consistently project significant increases in the fraction of annual precipitation contributed by extreme storm events.

4. Projected decreases in total precipitation, more frequent dry spells and higher temperatures imply more frequent and severe droughts in the Midwest and Canadian Prairies.

5. Most climate models show increased evaporation due to warmer water temperatures, leading to declines in the levels of the North American Great Lakes and associated reductions in outflows to the St Lawrence River.

6. In the snowmelt-dominated river catchments of the Rockies, enhanced warming leads to a greater proportion of liquid precipitation (i.e. less snowfall), shorter accumulation periods for snowpack, earlier onset of the spring-melt season, higher peak flows in winter, and reduced river flows in summer.

7. The historic legacies of land-use changes by agriculture and the depletion of major aquifers such as the Ogallala by irrigation, could increase the susceptibility of the continental interior to future droughts.

Further reading

Biondi, F., Gershunov, A. and Cayan, D. R. (2001) North Pacific decadal variability since 1661. *Journal of Climate*, 14: 5–10.

Boruff, B. J., Easoz, J. A., Jones, S. D., Landry, H. R., Mitchem, J. D. and Cutter, S. L. (2003) Tornado hazards in the United States. *Climate Research*, **24**: 103–17.

Chase, T. N., Pielke, R. A., Kittel, T. G. F., Zhao, M., Pitman, A. J., Running, S. W. and Nemani, R. R. (2001) Relative climate effects of landcover change and elevated carbon dioxide combined with aerosols: a comparison of model results and observations. *Journal of Geophysical Research-Atmospheres*, **106**: 31685–91.

Dai, A. and Wigley, T. M. L. (2000) Global patterns of ENSO-induced precipitation. *Geophysical Research Letters*, **27**: 1283–6.

Felzer, B. and Heard, P. (1999) Precipitation differences amongst GCMs used for the U.S. National Assessment. *Journal of the American Water Resources Association*, **35**: 1327–39.

Kalkstein, L. S., Nichols, M. C., Barthel, C. D. and Greene, J. S. (1996) A new spatial synoptic classification: application to air-mass analysis. *International Journal of Climatology*, **16**: 983–1004.

Trenberth, K. E. and Guillemot, C. J. (1996) Physical processes involved in the 1988 drought and 1993 floods in North America. *Journal of Climate*, **9**: 1288–98.

Wolock, D. M. and McCabe, G. J. (1999) Estimates of runoff using water balance and atmospheric general circulation models. *Journal of the American Water Resources Association*, **35**: 1341–50.

Woodhouse, C. A. and Overpeck, J. T. (1998) 2000 years of drought variability in the Central United States. *Bulletin of the American Meteorological Society*, **79**: 2693–714.

Useful websites

US weather and climate disasters:
http://www-libraries.colorado.edu/ps/gov/us/climate.htm

Billion dollar weather disasters in the USA:
http://lwf.ncdc.noaa.gov/oa/reports/billionz.html

Global/US weather and climate data bases:
http://www.cdc.noaa.gov/index.html

Canadian climate bases and climate change scenarios:
http://www.cics.uvic.ca/index.cgi?/About_Us/Canadian_Institute_for_Climate_Change_Studies

Records and methods of reconstructing former climate:
http://www.ngdc.noaa.gov/paleo/paleo.html

Modelling climate change using vegetation/ecosystem indices:
http://www.cgd.ucar.edu/vemap/

Climate system interactions in global climate change:
http://www.nacc.usgcrp.gov/

Climate subsystems and teleconnections (e.g. North Atlantic Oscillation, North Pacific Circulation):
http://www.cpc.ncep.noaa.gov/data/teledoc/telecontents.html

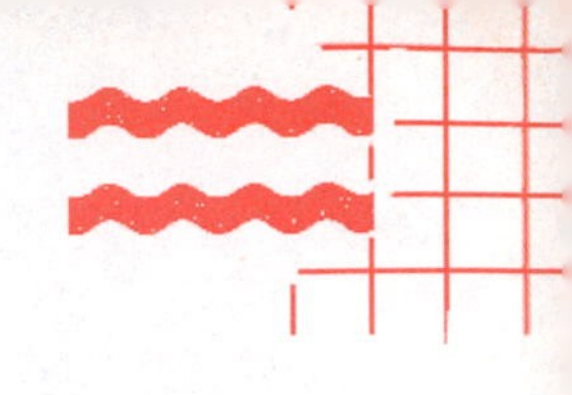

CHAPTER 11

Weather, climate and climate change in the low latitudes

11.1 Tropical climate

Tropical climates are found over a large pan-global zone lying between about 30° N and 30° S of the equator. All tropical climate is dominated to a greater or lesser extent by the great convective/convergent circulation of the Hadley cell (see section 4.1 and Figure 11.1) and its migration as it follows the sun north and south of the equator between the seasons. The seasonal migration of the Inter-Tropical Convergence Zone (ITCZ, the surface low-pressure region of the

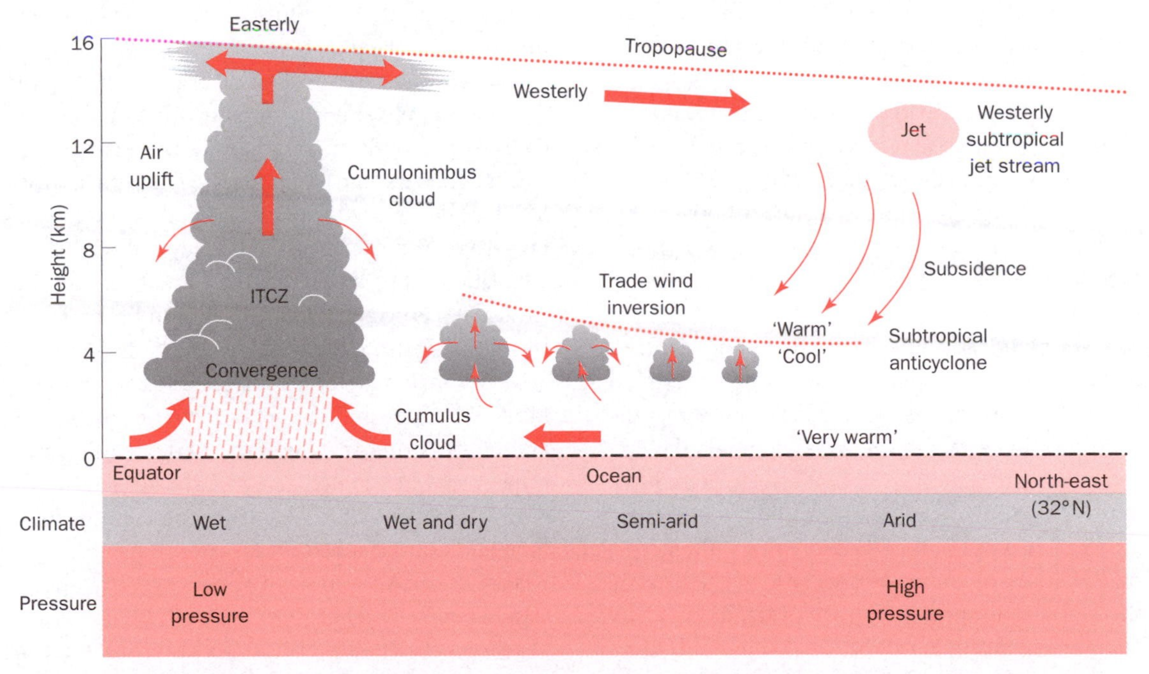

Figure 11.1 The Hadley cell showing air uplift (wet) at the equator and air descent (dry) at the subtropics. Much of the region between the equator and the subtropics is also relatively dry because air ascent is limited by the TWI
Source: Adapted from Robinson and Henderson-Sellers (1999, p. 125)

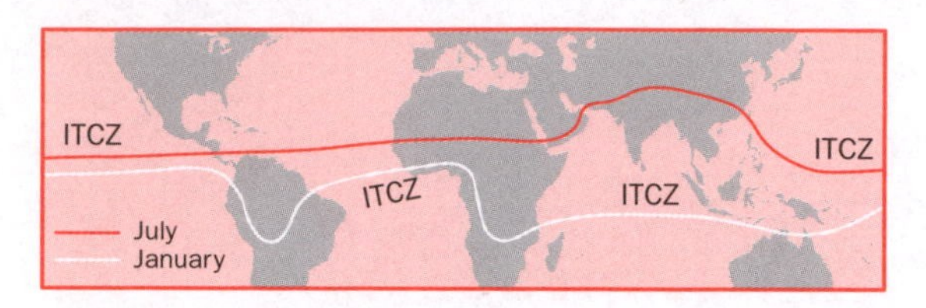

Figure 11.2 The July and January position of the ITCZ

Hadley cell) between July and January can be seen in Figure 11.2.

The various tropical climates are most effectively differentiated on the basis of moisture patterns, specifically on the amount and distribution of rainfall rather than by temperature condition. Tropical climates are by definition warm and can be defined as climates where the mean temperature of the coldest month does not fall below 18 °C (64 °F). On the other hand, rainfall patterns are extremely varied in time and space. For instance, moist tropical climates with a good distribution of rainfall throughout the year are found in the equatorial zone which coincides with the unstable air of the ITCZ. In the stable air of the subtropical highs (20–30° N and S of the equator) where air streams are descending are found the tropical arid climates (i.e. tropical dry climates with no real wet season). Located between the rainfall control areas of the ITCZ and the subtropical high-pressure zone and influenced by both are found climates with a distinct wet (high sun summer period) and dry (low sun winter period) season.

Other types of tropical climate can be identified including those influenced by a strongly modified Hadley cell. Examples include (a) tropical regions (and certain extratropical areas) dominated by the monsoon circulation such as West Africa and South and South-East Asia, and (b) equatorial/ tropical climates in the Pacific region influenced by the El Niño–Southern Oscillation (ENSO). All the aforementioned climates are examined in this chapter except ENSO-dominated climates in the Pacific which are dealt with in Chapter 5. In view of the great amount of energy and wide range of moisture stored in tropical atmospheres, tropical climates include locations with some of the wettest (Cherrapunji, Assam, Plate 11.1) and driest (Saharan Desert, Plate 11.2) climates on earth. Tropical climates are also associated with many of the most severe weather events on the planet (e.g. intense tropical rainfalls, hurricanes, thunderstorms, droughts).

11.1.1 The Hadley cell: processes and controls

Figure 11.1 shows a detailed model of air circulation within the Hadley cell (northern hemisphere only). Air flows equatorwards from the subtropics around 20–30° N at low elevations within the north-east trade wind belt, but is forced to rise in areas of low pressure in the great convergence/convection zone of the equatorial trough (ITCZ). It can be seen that cloud formation and rainfall are highly abundant near the equator where air uplift is encouraged. The latent heat released to the air on air ascent provides much of the energy needed to continue the whole cellular motion of the Hadley cell.

The air that ascends then diverges at the top of the troposphere and flows poleward. As it flows it cools by loss of thermal radiation, increases in density and descends especially in the subtropical latitudes between 20 and 30° N. On descent, the cool high-level tropical air of the Hadley cell warms by adiabatic heating so that by the time it reaches the surface it has become a warm dry cloudless air mass. It is not surprising then that in this high-pressure zone of subsiding air and surface air divergence, rainfall is scarce and the great deserts of the world are found. Moreover, the warmed descending air traps somewhat cooler air at the surface so that semi-permanent subsidence inversions occur in the region of the subtropical surface high-pressure system. It can be seen from Figure 11.1 that the area between the ITCZ and the subtropical high-pressure belt (20–30° N) is also subject to more or less persistent subsidence – if less vigorous than at the subtropics. This inversion, known as the trade wind inversion (TWI), is highest and weakest near the equator and strongest and closest to the surface at the poleward limits of the Hadley cell. The persistence of the TWI throughout much of the tropics explains the relative scarcity of precipitation over much of the subtropics.

11.1.2 Seasonal movement of the ITCZ

It has already been demonstrated that the low-pressure zone of the ITCZ is the principal rainfall-generating system of the tropics. Rainfall distribution patterns in the tropics are largely determined by the ITCZ and its seasonal movement north and south of the equator. The ITCZ tends to follow the sun in its annual migration between the Tropic of Cancer (sun overhead on 21 June) and the Tropic of Capricorn (sun overhead on 21 December). Figure 11.2 shows the transtropical migration of the ITCZ (also called equatorial trough, ET) between July and January. As the whole general circulation of the atmosphere moves in sympathy with the ITCZ, the various zones of vertical air uplift (wet areas) and downdraught (dry areas) migrate towards the hemisphere where the summer prevails.

11.1.3 Zonal model of tropical precipitation

The seasonal movement of the ITCZ and its associated areas of rising (wet) and descending (dry) air within the tropics can be used to demonstrate four main tropical climates based on the amount and distribution of their moisture supplies. These climates are namely tropical moist, tropical wet and dry, tropical semi-arid and tropical arid. Their spatial relationship in relation to the Hadley cell is best illustrated along a transect from the equator to the subtropical highs (20–30° N) in West Africa (Figure 11.1 and Box 11.1, Figure 11.3). It can be noted here that as with the polar and mid-latitude regions, the role of air masses is important in conferring the defining characteristics of the tropical climate. As we shall see, the different tropical climates can be described using air mass characteristics (Table 4.1 and Figure 4.12) and frequency. However, although varying greatly in moisture, the various tropical and equatorial air masses are relatively homogeneous in terms of temperature. As a consequence, tropical weather and climate are not dominated by frontal activity in the same way as for example the mid latitudes are.

BOX 11.1

CASE STUDY

Adjustment to a wet and dry climate

Figure 11.3 illustrates, for West Africa, a relationship between prevailing climatic type (especially rainfall amount and occurrence), vegetation formation and agricultural pattern. It can be seen that as rainfall diminishes in amount and duration (monthly frequency), the main vegetation zones alter first from tall tropical rainforest (TRF) to less abundant monsoon forest, then to wet and dry savanna grassland and eventually to thorn-bush scrub and desert vegetation. A corresponding pattern reflecting the rainfall regime can be seen in changes in the agricultural use of the West African region, first from rubber production and forest timber, to coffee and oil palm, then to yams and maize and eventually to cotton and groundnuts. It needs to be kept in mind that such relationships between climate, vegetation and agricultural activity reflect only broad-scale general patterns. The application of local irrigation schemes, for instance, can modify the aforementioned climate–land-use associations, so that rice production can be found in many parts of West Africa including those with wet and dry and semi-arid regimes (e.g. The Gambia). Also, short-term changes in climate from year to year and especially longer-term climate change over many years can radically alter climate–land-use patterns.

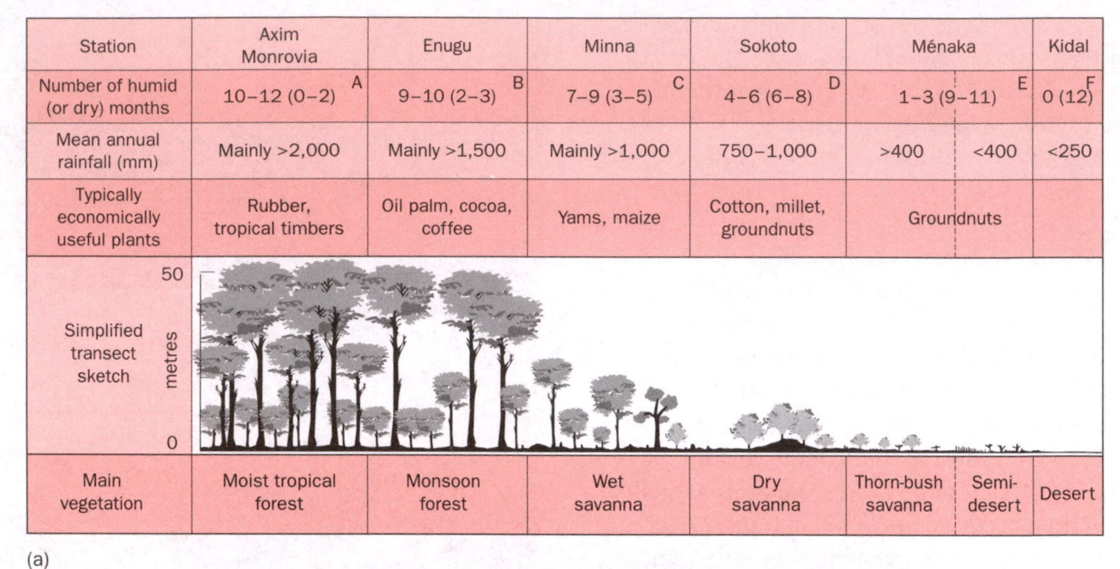

Station	Axim Monrovia	Enugu	Minna	Sokoto	Ménaka	Kidal
Number of humid (or dry) months	10–12 (0–2) A	9–10 (2–3) B	7–9 (3–5) C	4–6 (6–8) D	1–3 (9–11) E	0 (12) F
Mean annual rainfall (mm)	Mainly >2,000	Mainly >1,500	Mainly >1,000	750–1,000	>400 / <400	<250
Typically economically useful plants	Rubber, tropical timbers	Oil palm, cocoa, coffee	Yams, maize	Cotton, millet, groundnuts	Groundnuts	
Simplified transect sketch (metres, 0–50)						
Main vegetation	Moist tropical forest	Monsoon forest	Wet savanna	Dry savanna	Thorn-bush savanna / Semi-desert	Desert

(a)

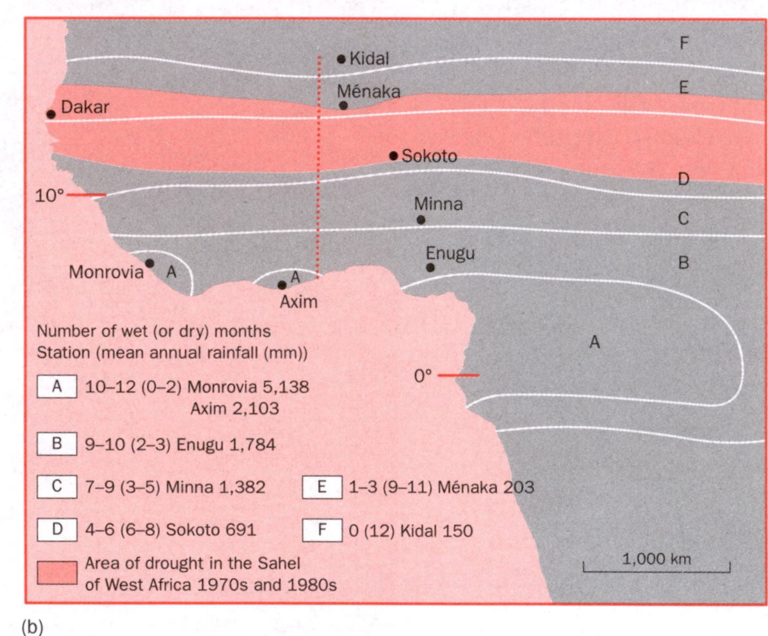

(b)

Figure 11.3 (a) Profile and (b) plan view of West Africa, showing the relationship between rainfall distribution (climate type), vegetation formation and cropping pattern between the equator and the tropic of Cancer
Source: Manshard (1979). Taken from O'Hare and Sweeney (1986, p. 136)

(a) Tropical moist climate of the equatorial zone

The tropical moist climate is found in lowland areas in the Amazon Basin of Brazil, the Zaire Basin in central Africa, coastal regions of West Africa and in Indonesia. Here, maritime equatorial air (mE) dominates, giving abundant rain throughout the year with no real dry season. This is due to the permanence of the ITCZ and its rising airflows. Rainfall totals are often around 2,000 mm/year and can exceed 4,000 mm/year on the windward side of mountainous coasts. As shown in Figure 11.4(a) for Eala in the Zaire Basin in central Africa, there are often two peak periods of rainfall related to the passage (twice) of the sun overhead during the year. Although there is variation between the monthly rainfalls, only two months have rainfall values less than about 100 mm/month, a threshold indicating rainfall deficiency in warm tropical areas. This is because water loss by evaporation and plant transpiration can be very high in the tropics (Box 11.2). Such evapotranspiration losses have to be set against the rainfall inputs, leaving water that can be effectively used for agriculture (irrigation), industrial and domestic uses. Influenced by equatorial maritime air the whole year round, temperatures are uniformly and constantly high in the tropical moist climate, often in the range 25–27 °C. They do not exceed those in the dry tropics, however, because of the abundance of clouds in the equatorial zone (see Figure 3.5). Seasonal temperature variations are very low, usually less than 3 °C, because the noonday sun is always high in the sky and day-lengths do not change much during the year. The annual temperature range at Belem in the Amazon, for instance, is 1.3 °C. Temperature ranges are in fact greater on a daily basis between the warm sultry days (mean maxima about 28–30 °C) and the cooler nights (mean minima about 22 °C).

Almost every day in the moist tropics, strong surface heating under the convergent low-pressure conditions of the ITCZ encourages convective–convergent updraughts to punch through the weak trade wind boundary. Towering cumulus clouds up to 10 km in diameter and sometimes reaching 15–20 km in height begin to build and produce heavy localised showers/thunderstorms by early afternoon. As the evening approaches the surface

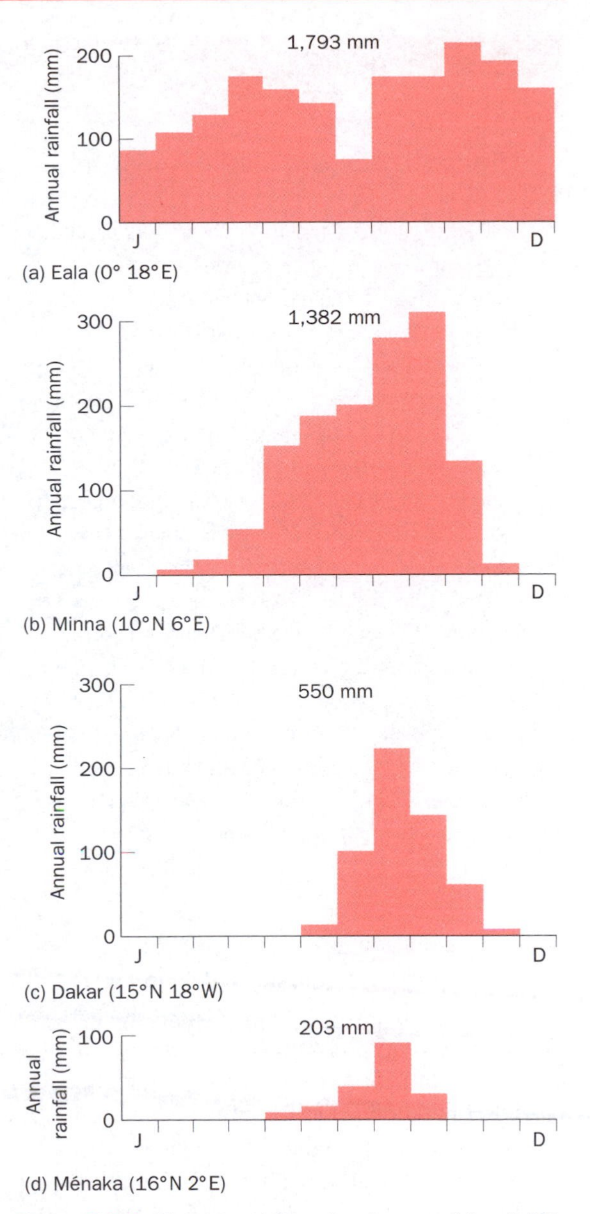

Figure 11.4 Climate regimes, showing monthly rainfall distribution for (a) the tropical moist, (b) tropical wet and dry, (c) tropical semi-arid and (d) tropical arid climates

BOX 11.2

THINKING FURTHER

Rainfall recycling in the moist tropics

Moisture recycling mechanisms are fundamental in the moist tropics for the maintenance of high and well-distributed rainfall over the year, especially in interior continental regions like the Amazon and central Africa. Almost every day in the convergent–convective zone of the ITCZ, towering cumulus clouds begin to build and produce heavy localised showers by early afternoon. Surface heating during the morning under clear tropical skies, an abundance of moist air and prevailing low pressure encourage strong convective updraughts and the building of the tall cumulonimbus towers. After several hours of heavy afternoon rain, the surface cools, convection declines and air stability levels are increased. Most of the heavy rain that falls from the cumulonimbus clouds remains on or near the surface (wet surface soil, moist plant surfaces, rapid overland flow feeding large rivers). As the surface heats the following day this wet surface and subsurface provide an ideal source for evaporation to restart the whole sequence of convective moist air ascent, air cooling and condensation and rainfall production. Modelling studies have shown that as much as 75 per cent of the moisture that fuels the rain in the Amazon Basin has its origins in the Atlantic Ocean, being initially transferred to the more coastal parts of the basin by winds. Once over the Amazon Basin, the moisture is efficiently recycled from the atmosphere by condensation and rainfall processes to the region's soils, plants and surface water bodies and then back again by strong evapotranspiration to the atmosphere. Winds are able to push the recycled moisture in the atmosphere further in towards the interior to start the whole process of recycling again. These processes explain why interior moist tropical Amazon stations like Manaus and Iquitos (2,740 mm rainfall per year) receive such heavy daily rainfall.

cools, the clouds dissipate and the skies begin to clear. Precipitation frequency is therefore high, occurring on more than 50 per cent of the days. Duitenzorg in Java, Indonesia, averages 322 days a year with thunderstorms, which are intense and brief. Rainfall intensities for hour-long periods are not significantly higher in the tropics than in the mid latitudes, although hourly rainfalls of 95–120 mm (3.7–4.7 inches) have been recorded in Indonesia. *Sustained* periods of heavy rainfall are much more common, however, in tropical areas than elsewhere. Under certain conditions individual cumulonimbus clouds can become grouped together into mesoscale convective complexes (MCCs) up to 100–200 km across. Several MCCs may gather together to comprise a massive cloud cluster from 100 km to as much as 750 km across (sections 2.3.5 and 10.2.1).

(b) The wet and dry tropics

The wet and dry tropics are found in a zone immediately poleward (north and south depending on hemisphere) of the equatorial region. These latitudes are influenced by the ITCZ with its rain in the high sun period (summer) and by the subtropical anticyclones with their trade wind inversions and dry conditions in the low sun period (winter). The tropical wet and dry climate is thus dominated by tropical maritime (mT) air in summer and tropical continental air (cT) in winter. Rainfall is usually less therefore than the moist tropics, i.e. more than 1000 mm but usually less than 2000 mm. The distinguishing feature of this climate as shown by Minna (10° N, 6° E) in West Africa is its high summer rainfall during a fairly long wet season (up to 6–8 months) and scanty rainfall in winter (Figure 11.4(b)). The reason for the monthly differences in the wet season relates to station location. Areas of the wet and dry climate nearest to the equator will have lengthier wet season rainfall than those regions close to the subtropical highs. The summer rainfall comes mostly from low-pressure disturbances including line squalls and thunderstorms and is more irregular than in the moist tropics with mean annual totals fluctuating widely (15–20 per cent) from year to year. Intra-seasonal rainfall is also irregular so heavy floods can be followed by serious droughts during the wet season.

There is also a greater thermal range than in the moist tropical climates. The annual mean temperature range is typically between about 22 °C during the winter and up to 32 °C in the warm to hot summer. Daily maximum summer temperatures can reach over 40 °C so potential evapotranspiration rates can easily match and exceed the rainfall totals. Potential evapotranspiration is the amount of water that would be lost from the soil by evaporation and from plants by transpiration *given no theoretical restriction in the water supply*. This means that the water supply in the wet and dry tropics is not always assured or in surplus. One consequence of the reduction in the usable or effective rainfall is that tropical rainforest gives way in the tropical wet and dry climates to tropical deciduous forest (the trees drop their leaves during the dry winter). Much of this forest has, however, been cleared by human activity (cutting and burning) and maintained as tropical grassland by cattle ranching and arable cultivation.

(c) Tropical semi-arid regions

Moving polewards from the tropical wet and dry climate on either side of the equator, rainfall continues to decline and the dry season becomes more pervasive. This is because here the rain-bearing ITCZ is at the limits of its migration north and south of the equator, only bringing maritime Equatorial air (mE) and rain to the region for a few (3–5) months each year. For the rest of the time, the climate is dominated by the subtropical highs with their tropical continental air (cT) and strong trade wind inversion. Annual rainfall totals range from around 800 mm in the wetter parts on the equatorial margins of the semi-arid zone to less than 300 mm on its poleward fringes. At Dakar (15° N, 18° W) in West Africa much of the annual rainfall of around 500 mm falls during a short wet season lasting about 4 months and very little during a long dry season lasting up to 8 months (Figure 11.4(c)).

The relatively low mean annual rainfall of the tropical semi-arid climate is subject to marked annual variation. While annual rainfall variability can range between 20 and 35 per cent in the wetter parts of the zone, yearly rainfall variation from the mean can reach as much as 50–75 per cent in the drier parts of the area. When the annual rains are reduced in the tropical semi-arid zone, it is usually because of the failure of the ITCZ to move far enough (north or south depending on hemisphere) into the region. This seems to have happened in the Sahel zone of sub-Saharan Africa where persistent droughts occurred between 1965 and 1990, and in southern Africa where serious rainfall deficits have been a feature of the last 20 years (see section 12.3.3). With summer maximum daytime temperatures commonly in the range 35–40 °C, the more cloud-free tropical semi-arid climates also experience greater heat and thus higher rates of potential evapotranspiration in the wet summer months compared to the wet and dry tropics. This means that the effective rainfall is much less than the given rainfall supplies suggest. One result of this is that many tropical semi-arid areas have a thorn scrub vegetation rather than a tropical woodland or tall savanna grassland cover. The region has also proved fragile under human use and has been subject over many years to the process of desertification. Desertification can be defined as a deterioration in the quality of soils and vegetation under human use (excessive woodcutting, overcultivation and overgrazing) so that desert surfaces are created. Even today with parts of the Sahel showing some recovery from the former drought (especially in 1974 and 1984/85) soil fertility decline, high population growth rates and increasing food imports remain a feature of the region.

(d) Tropical arid climate

Poleward of the tropical semi-arid zones are the tropical arid regions. They are located in the subtropical belt 20–30° N and S of the equator. These regions are associated with year-long dominance of the subtropical highs with their tropical continental air (cT), descending air streams, strong TWIs and permanent dryness (see Figure 11.1). The tropical arid climate is usually defined as one with less than 250 mm of rain per annum. Ménaka (16° N, 2° E) in West Africa (Figures 11.3 and 11.4(d)) is representative of the wetter parts of the tropical arid zone. Its annual rainfall of around 200 mm per year is not only low but falls during a very brief summer period.

The highest average annual temperatures on the planet are found in tropical arid climates. They vary between 29 and 35 °C. At Lugh Ferrandi in Somalia, the mean annual temperature is 31 °C. Winter temperatures dip below those in other parts of the tropics as loss of thermal radiation at night under clear desert skies rapidly cools these areas. Averages are as low as 15–20 °C. The lower winter temperatures give the tropical arid climates the highest annual range found in the tropical regions. Aswan, Egypt, has an annual range of 19 °C. Daytime temperatures in the summer can rise to 50 °C and fall well below 20 °C during the evening hours. One of the highest official temperatures ever registered is 58 °C recorded on 13 September 1922, at El Aziza, near Tripoli in North Africa. Summer temperatures dictate that potential evaporation rates are extremely high, being many times that of the annual rainfall (Figure 11.4(d)).

Precipitation occurs sporadically in time and space. Heavy downpours sometimes occur from convective storms and other disturbances including low-pressure waves and squalls. One rain event may bring 125–150 mm of precipitation then no precipitation will fall for several years. One station in the Thar Desert, north-west India, where rainfall averages 100 mm, received 850 mm in 2 days.

(i) Dust- and sandstorms

Tropical disturbances can generate weather in the desert even when there is no rainfall involved. These weather events include dust- and sandstorms (Box 11.3). The dust in these storms is the result of deflation. The desert wind has long shifted the larger sand particles from many areas into specific sand dune regions helping to keep most of the surface swept clean. Trade winds blowing out of the deserts are often dust-laden. On the south side of the Sahara, they call this dry dusty wind the *harmattan*. In the sandy areas of the desert (the Sahara has only 30 per cent of its surface with a sandy cover) sandstorms can be driven by intense surface convective heating and associated air turbulence. It is not unusual for sand temperatures at the surface to reach 85 °C. These convectional sandstorms are common during the daytime and in the hottest months. The second type of sand/duststorm results from low-pressure disturbances in the TWI passing across the area. These storms are often stronger and last longer than the diurnal driven ones. One sandstorm in the Nafud Desert (central Sahara) lasted more than 40 hours, during which time the sand blew constantly. This kind of regional sand/duststorm halted the American advance on Baghdad during the Iraqi War in the late spring of 2003. Some of the most intense low-pressure disturbances can shift millions of tonnes of dust many miles from their origin over neighbouring seas and oceans and are spectacular when seen from space (Box 11.3; Plate 11.2).

(ii) No rain but plenty of fog

The trade winds of the subtropical deserts are reinforced in the vicinity of cold ocean currents that help to stabilise the air even further. The most well-known examples of this is the Atacama Desert

BOX 11.3

CASE STUDY

Huge African duststorms

Atmospheric disturbances and major winds blowing across the arid and semi-arid areas of the world can send millions of tonnes of dust thousands of kilometres from their source of origin. The dust itself is composed of mineral elements such as iron, phosphorus and sulphates and may act to fertilise ecosystems when it is deposited at the surface. The mineral dust also carries a whole army of micro-organisms including bacteria, fungi and viruses, some of which may be harmful to humans and other living creatures. Plate 11.2 shows a dramatic satellite image taken on 26 February 2003, of a vast cloud of Saharan dust blowing from north-west Africa (lower right) a thousand miles or more out over the Atlantic Ocean. By 4 March, the thinning cloud of dust had reached the north-east coast of Brazil via the trade winds. Such dust events show a cyclical pattern based on the movement of the trade wind belt. From February to April, the dust descends on the Amazon Basin where some evidence suggests it may help to add scarce mineral nutrients to the Amazon rainforests. The trade winds then shift north, and from June to October the Caribbean and North and Central America bear the impact of the storms which take 5–7 days to make the trip. The same type of dust that fertilises the Amazon Basin may also stimulate phytoplankton growth in tropical waters. A correlation, for instance, has been shown between dust events and algal blooms like the 'red tides' of red-coloured algae that occur in Florida's coastal waters.

Figure 11.5 shows the overall increase in Saharan dust reaching the Caribbean island of Barbados between 1965 and 1992. The general increase in dust concentration has been linked to desertification and the expanding Saharan/Sahel desert region in Africa. The peak episodes of 1983 and 1987, however, may relate to El Niño events in the Pacific which are correlated with greater disturbance and higher wind shear in the tropical atmosphere over the Atlantic. There are suggestions by scientists that potential disease organisms carried by the dust could be causally linked to enhanced rates of coral deaths in the Caribbean (Figure 11.5). Other factors, however, are thought to be implicated in coral decline, including coral bleaching under globally warmed ocean waters and excessive herbivorous attack. Besides the potential microbe and fungi hazard, the dust grains themselves are known to cause respiratory and allergic reactions. The arrival of African dust in Florida for example is associated with heightened pollution alerts. One study in Barbados revealed a 17-fold increase in the incidence of asthma attacks since 1973 and a correlation between such attacks and periods of heavy dust fallout. Asthma attacks are, however, related to a whole complex of factors including local tree and grass pollen, local air pollution levels, lowered human immune systems under a generally cleaner environment and the concentration of dust mites in the home.

along the west coast of South America in northern Chile and Peru, and the Namibian Desert in south-western Africa (Figure 11.6(a)). These coastal deserts are often reported to be the driest places on earth with rainfall occurring only once or twice every decade or so. At Iquique in the Atacama Desert, northern Chile, the rainfall averages 28 mm/year but during a 20-year period no measurable rainfall fell at all. The driest location on earth, however, is actually in the mountain

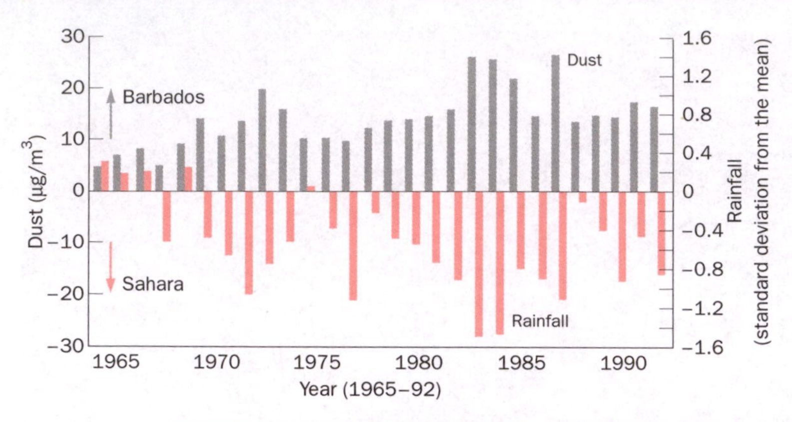

Figure 11.5 Mineral dust concentrations at (top) Barbados and (bottom) rainfall departures from the mean in sub-Saharan Africa, 1965–92

Source: Dr Joe Prospero, NOAA Institute for Marine and Atmospheric Sciences, University of Miami

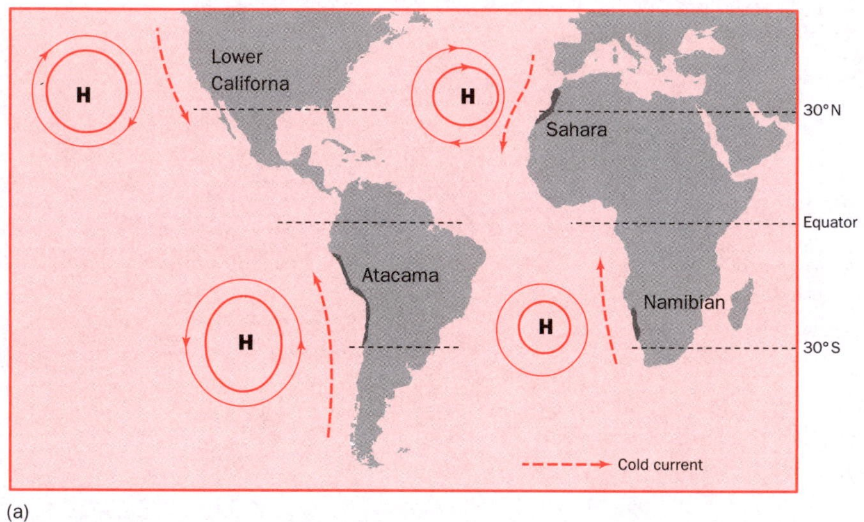

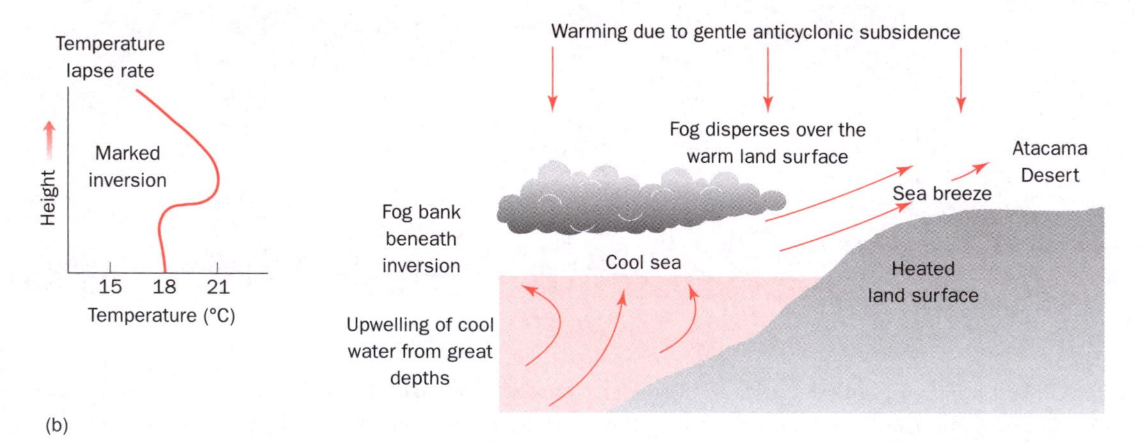

Figure 11.6 (a) The location of the subtropical coastal deserts in relation to semi permanent high pressure and the movement of cold ocean currents, and (b) the influence of the cold Peruvian (Humboldt) current on air stability and moisture conditions in the Atacama Desert

Source: (a) Various; (b) Flohn (1969). Taken from O'Hare and Sweeney (1986, p. 126)

valleys of central Antarctica where extremely cold air and 'rain' shadow effects combine to reduce rainfall (snowfall) occurrence to once every several hundred years!

The trade winds of the tropics can pick up a good deal of moisture when they pass over a large expanse of warm tropical ocean. Because of the TWI, however, they are normally unable to release it as precipitation, so that many areas of the tropics though dry can be very humid, often in the 80–90 per cent range. At Iquique, the relative humidity averages 81 per cent in winter (August). The moisture-laden trade winds can be brought to saturation, however, by a small amount of cooling. As shown in Figure 11.6(b), warm moist trade wind air passing towards the west coast of South America from the high-pressure cell over the central eastern Pacific (Figure 11.6(a)) is cooled to dew point from below, over the cold Humboldt current. This surface advection cooling causes a strengthening of the TWI in the coastal regions (see Figure 11.6(b)). The resultant stability causes a persistent fog regime below the inversion (section 2.3.3) and generally prevents convective rain cells developing. On moving inland, however, over a warm surface such as the Atacama Desert of northern Chile, such fogs quickly disperse as the (visible) water droplets of the fog quickly evaporate to (invisible) water vapour.

Key ideas

1. Tropical climates, where mean monthly temperatures do not fall below 18 °C, are predominantly found between the latitudes of 30° N and 30° S of the equator.

2. Tropical climates show a great variety of rainfall regimes and are most usefully differentiated on the basis of moisture.

3. A simple zonal climate model based on the Hadley cell and the trade wind inversion explains the rapid decline of annual rainfall and the length of the wet and dry season between the equator and the subtropics.

4. The tropical moist climate of the equatorial zone, with rain well distributed throughout the year, is dominated by the rain-bearing Hadley cell.

5. In contrast, tropical dry climates with year-long aridity and found 20–30° N and S of the equator are dominated by the tropical high-pressure cells and a strong trade wind inversion.

6. Sandwiched between the tropical moist and arid climates (12–20° N and S) are climates that are wet in summer (when dominated by the Hadley cell) and dry in winter (when dominated by the tropical highs).

7. These include the tropical wet and dry climate when the dry season is relatively short (closer to the equator) and the tropical semi-arid climate (closer to the subtropical high) when the dry season is long.

8. While tall cumulonimbus clouds with their high-intensity rainfall are a frequent feature of the moist tropics, dust and sandstorms are a frequent event in the tropical arid zone.

9. Cold ocean currents intensify the aridity of tropical desert climates and cause frequent fog along their coastal margins.

11.2 Indian monsoon circulation

The term 'monsoon' derives from the Arabic word *mausin* which means season. A monsoon wind system is one that undergoes a marked seasonal change of direction. Many regions including West and East Africa, South and South-East Asia, China, Indonesia, north and south Australia and much of the west coast of North America come under the influence of monsoon-type circulations when the wind shift between the seasons is between 120 and 180°. In its ideal form there is a complete 180° reversal of wind direction between the seasons, blowing from one direction in summer and from the opposite direction in winter. The Indian monsoon is often regarded as the 'classic' monsoon where a combination of land and sea distribution, the alignment of coasts and mountain ranges, and high elevation (the Himalayas and Tibet) make it the most intense and varied of all monsoon climates. A detailed analysis of the Indian monsoon will now be given.

11.2.1 The low-level thermal cell

(a) A gigantic land and sea breeze

It was the English scientist, Edmund Halley, who in 1686 first proposed the thermal concept of the origin of the Asiatic monsoon. According to this model, monsoons are perceived as gigantic convection systems produced by differential seasonal heating of continental and oceanic areas (i.e. a continental-scale land and sea breeze effect).

(b) Differential (land–ocean) sensible heating

Differential heating is the result of the contrasting energy responses of land and sea to incoming solar radiation as the sun migrates back and forth across the equator. Water has a higher specific heat capacity than dry soil, so heats up much more slowly and is better able to distribute and disperse heat away from its surface layers (chiefly by wind-induced turbulence). During spring and summer, the air over the hot continental landmass of north-west India becomes much warmer than the air over the cooler water of the Indian Ocean. As shown in Figure 11.7(a), the heated air over land expands and rises vertically upwards in the atmosphere producing a region of low atmospheric pressure at the surface. In contrast, air over the colder seas absorbs less solar energy from the oceans and remains much cooler. Less heat is conveyed upwards through the atmosphere which compared with that over the land sinks closer to the surface. As a result of this sinking, a region of high pressure develops over the surface of the ocean. While low pressure exists at the surface over the warm land, a zone of *relative* high pressure is found above it. Relative air pressure is always high away from the surface in an expanded column of warm air because of the greater volume and weight of expanded atmosphere above. This high pressure is best observed at around 6–7 km elevation in the atmosphere. Because of the shallower depth of the cooler atmosphere over the ocean, a zone of *relative* low pressure is located in the atmosphere over the ocean at the same elevation as that over the land.

Figure 11.7(a) shows the circuit of winds within the thermal cell created by the heating differences between land and sea. Despite having

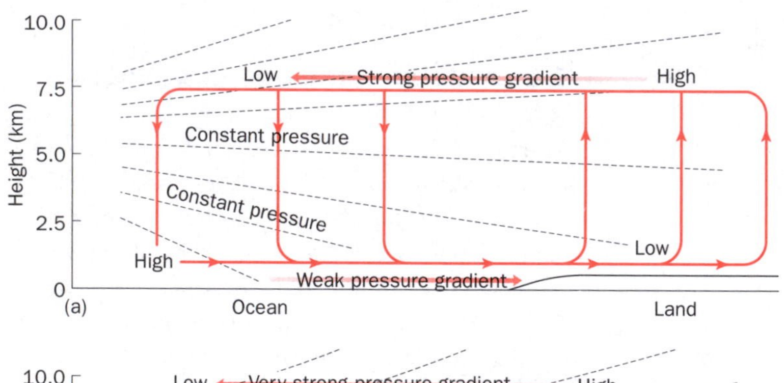

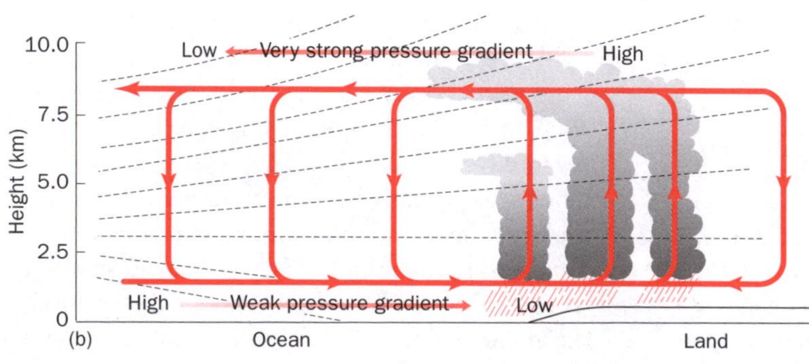

Figure 11.7 The Indian summer monsoon showing surface winds from ocean to land: (a) the low-level dry thermal cell, and (b) the higher and more vigorous moist thermal cell. The moist cell is characterised on its landward side by condensation, cloud formation and rainfall, and gains extra energy from latent heat release
Source: Webster (1987)

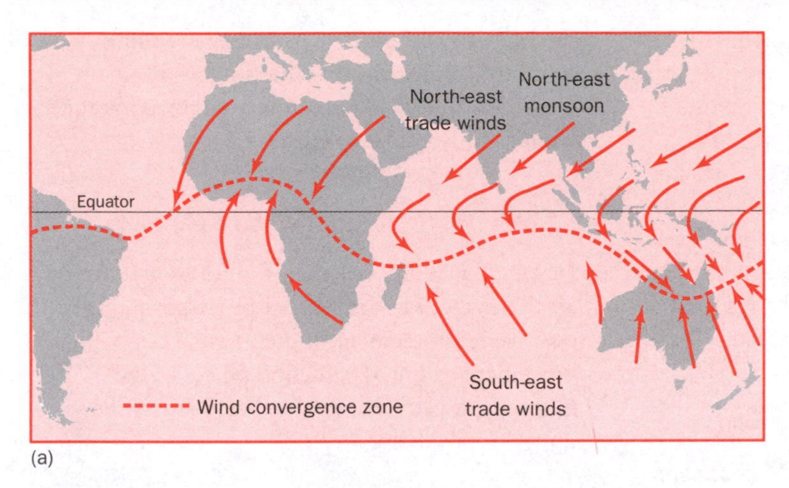

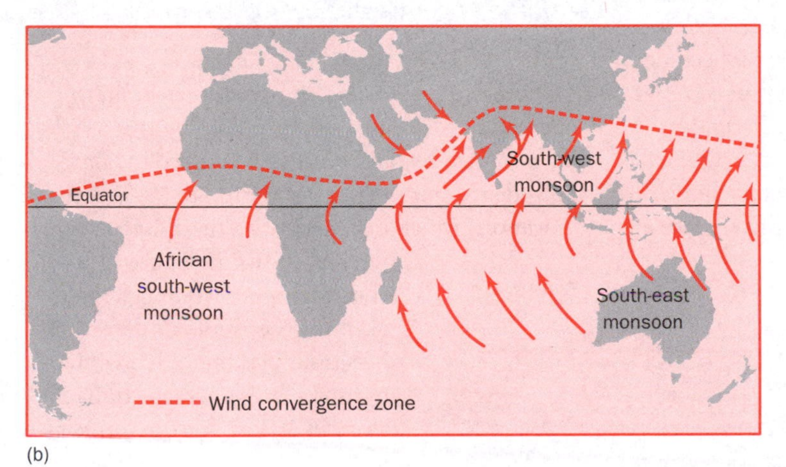

Figure 11.8 The regional pattern of surface monsoon winds over India and other regions of the world (a) north-east in winter (January) over India and (b) south-west over India in summer (July). The winds are shown in relation to the ITCZ

a vertical depth of about 6–7 km, the thermal direct cell so described is classified as a *lower air or lower tropospheric* circulation. This is because the thermal cell occupies only about half the lower atmosphere (troposphere) which over India has a vertical extent of about 14 km. The surface limb of the circulation shown, i.e. that from sea to land, constitutes the familiar rainy summer monsoon winds over India. As shown in Figures 11.8(b) and 4.14, this monsoon air stream is deflected by the Coriolis force, first, to the left of its path in the Indian Ocean (as south-easterlies), and then to the right of its path (as south-westerlies) as it crosses over the equator to the Indian subcontinent.

In the northern hemisphere winter, because the land cools down much more than the oceans, a complete reversal of the low-level thermal cell shown in Figure 11.7(a) takes place. High pressure builds over the land surface while low pressure is created over the surface of the ocean. In winter, subsiding air from the land anticyclone moves out at the land surface over the oceans to be replaced by a counterflow of air at higher levels (3–4 km) from ocean to land. The surface limb of the thermal cell from land to sea represents the dry winter monsoon winds over India. As shown in Figure 11.8(a), this monsoon air stream is deflected by the Coriolis force first to the right of its path (as north-easterlies) over India and then to the left of its path (as north-westerlies) as it crosses the equator.

(c) Latent heat collection and transfer

During the summer monsoon, condensation takes place in the atmosphere as air ascends and cools over the land (Figure 11.7(b)). The latent heat liberated by condensation is now experienced as sensible heat and makes the rising air warmer and thus more unstable and buoyant (the warmer the air the faster it will rise). As the air undergoes further ascent, more cooling, condensation and heat release takes place, reinforcing the whole system of buoyant ascent (see Figure 2.11(b)). Due to latent heat release, the column of the warm moist air over the land is now higher and more vigorous than it would be if it remained dry (Figure 11.7(b)). Latent heat release thus results in an overall higher and stronger monsoon, with a more powerful inflow of moist air at low levels from ocean to land. Its effect is greatest in the late summer when the temperature of the sea surface and the overlying air are at their highest.

(d) The migration of the planetary winds

In January, the position of the ITCZ lies at about 10° S over the Indian Ocean, but migrates because of strong continental heating as far as 25° N over the Indian subcontinent in July where it becomes known as the monsoon trough (MT). In summer, the surface low of the MT is aligned along the Ganges Valley from Bengal in the east to the Punjab in the north-west. Although very similar to the thermal direct cell, the surface convergence at the MT can be regarded as driving the summer monsoon, and when over the southern Indian Ocean (as the ITCZ) in January as powering the winter monsoon (see Figure 11.2).

11.2.2 High-altitude winds

Despite the additional energy given to the lower air thermal cell by latent heat processes and the positive reinforcement of the thermal cell by the movement of the MT, the full power of the monsoon is provided by wind currents in the upper troposphere.

(a) The upper westerlies and the winter monsoon

During winter, air descent over India within the low-level thermal cell (3–4 km high) is surmounted by, and positively coupled to, air subsidence from the upper westerlies (5–10 km elevation) which flow across northern India at this time. As shown in Figure 11.9(a), when the upper westerly jet is taken into account, air descent in the subsiding limb of the winter monsoon over northern India becomes 10 km deep and much

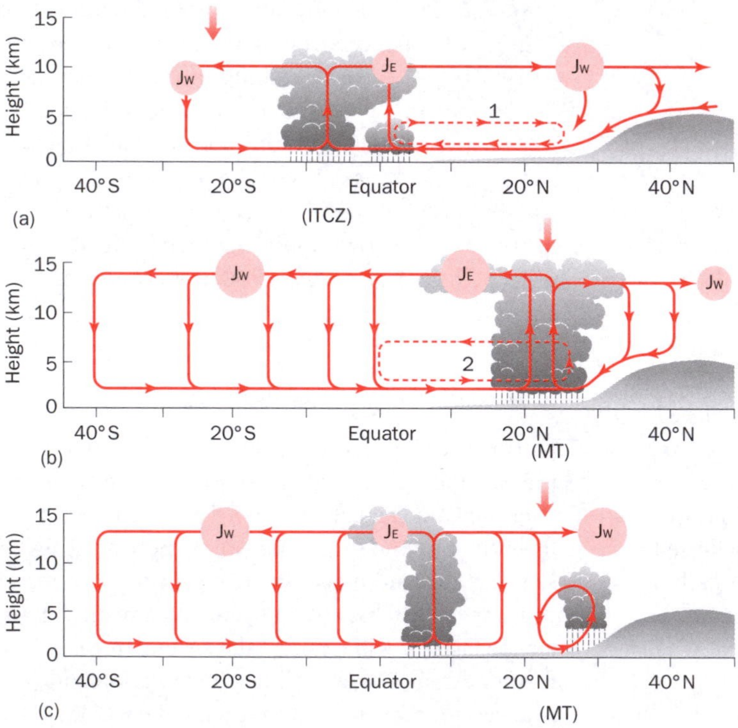

Figure 11.9 North to south vertical transect of the India monsoon circulation from Tibet in the north to the southern Indian Ocean south of the equator. The westerly jet stream (Jw) moves out of the paper eastwards across northern India, while the easterly jet stream (Je) moves 'into paper' and westwards across India. Three phases are shown: (a) the winter monsoon (Dec.–Feb.); (b) the active summer monsoon (June–Aug.); and (c) summer (June–Aug.) 'break' monsoon. The low-level winter and summer thermal cells (1 and 2 respectively) are shown in relation to the upper airflows
Source: Webster (1987)

more vigorous than that described by the simple 3–4 km deep thermal cell. A 10 km high jet-induced monsoon circulation is eventually established between northern India and the southern Indian Ocean. The upper westerlies and westerly jet thus serve to *deepen, broaden and strengthen* the dry outblowing north-easterly winds of the winter monsoon.

(b) The upper easterlies and the summer monsoon

During late May and early June, continued heating of the northern hemisphere weakens the upper atmospheric low-pressure centre that sits over the high-pressure cell at the polar surface. As the upper-level low drives the upper westerlies, its decline signals a weakening of these high-level winds over northern India, which become intermittent. The MT that is associated with scattered but strong convective showers, pushes northwards over southern India with each weakening of the upper westerlies south of Tibet. More importantly, summertime heating of the Tibetan Plateau eventually produces strong upper easterly winds over India (Figure 11.9(b)). These *upper easterlies*, with a strong jet core at their centre (i.e. the subtropical easterly jet) encourage air ascent. By doing so they help to activate the full force of the south-west summer monsoon.

(c) The Tibetan Plateau

The Tibetan Plateau plays a crucial role in the development of upper air jet streams which first trigger then drive the summer monsoon circulation over India. During the summer months the Tibetan Plateau acts as a powerful upper-level sensible heat source. Intense solar energy absorption by the 4 km high plateau surface heats the air above it sufficiently to produce warm air at high levels in the atmosphere. Moreover, abundant latent heat release over northern India associated with rising and condensing air currents from the surface heat low (see Figure 11.7(b)), spills over the southern flanks of the plateau and adds further heat to the middle and upper troposphere. By early June, air in the upper troposphere over Tibet becomes warmer than that at similar altitudes over the equator. An upper tropospheric anticylone develops over Tibet whose clockwise-rotating winds become upper easterlies on its southern limb over northern India.

(d) Summer monsoon mechanisms in action

Between late May and the beginning of June, the westerly jet weakens over northern India and moves north of Tibet. As it migrates, the easterly jet (elevation 10–15 km) suddenly appears about 15° N and then spreads out at high levels over much of India during the rest of the summer. Unlike the westerly jet, the easterly jet acts like an exhaust system to air in the lower troposphere over India by encouraging surface air *ascent* on its northern flank (Figure 11.9(b)). Air rising from the lower thermal cell in late May and early June particularly from the summer heat low over northern India (see Figure 11.7) is drawn upwards by the upper easterlies to near the top of the atmosphere (14–15 km). The strength of the jet-induced 14–15 km deep monsoon is now such that it begins to draw moisture-laden low-level winds from the Indian Ocean south of the equator over India (see Figure 11.8). It achieves this via a low-level (1.5 km elevation) jet stream called the Somali jet. This jet forms at the centre of the surface south-west monsoon winds where wind speeds are highest, and is an important mechanism in transferring heat and moisture into the subcontinent. The easterly jet thus *deepens, broadens and strengthens* the low-level south-west monsoon. It is only when the westerly jet disappears from northern India, and is replaced by the easterly jet, that the summer monsoon finally switches into full gear and becomes 'active'. The speed of jet stream replacement is so quick, that it causes the monsoon to suddenly 'burst' over India bringing copious supplies of rainfall.

Occasionally, however, during the south-west summer monsoon, for periods lasting over several days or several weeks, the upper easterly jet weakens and moves south, while the MT migrates north against the Himalayas (Figure 11.9(c)). Low pressure and rising air currents associated with these two systems encourage convective rainfall over the extreme south and north of the country. On the other hand, surface pressure rises over most of central and north India as these areas come under the descending branches of the two thermally forced convective circulations. With subsiding air and rising surface pressure, reduced rainfall prevails over much of the subcontinent. This condition, which literally reverses the normal summer situation bringing dry conditions, is termed the 'break' monsoon.

Key ideas

1. A monsoon climate has a marked seasonal (winter–summer) change in wind direction.

2. The Indian monsoon with warm moist south-westerly winds blowing from ocean to land in summer (June–September) and cool dry north-easterly winds blowing from the land to ocean in winter (Dec.–March) is regarded as the 'classic monsoon'.

3. The Indian monsoon (both summer and winter) is driven first by a lower atmospheric thermal cell that is created by seasonal temperature differences between land and sea.

4. Moisture processes in the atmosphere including strong evaporation over the southern Indian Ocean and condensation of that moisture with the subsequent release of latent heat, add energy and strength to the summer thermal monsoon cell.

5. Another factor driving the Indian monsoon is the seasonal migration of the ITCZ or monsoon trough (MT) from the southern Indian ocean in winter (around 10° S) to northern India (25° N) in summer.

6. The full dynamic of the Indian monsoon, i.e. its width, depth and strength, is governed by the action of upper air motion, chiefly by upper westerly winds in winter (air subsidence) and upper easterly winds (air uplift) in summer.

7. Air subsidence by upper westerlies confirms the dryness of the winter monsoon while air elevation by upper easterly winds reinforces the summer wet season.

8. The Tibetan Plateau plays a pivotal role in determining the track and intensity of the upper westerlies and the upper easterlies over northern India between the seasons.

9. Interruption in the summer rains called 'break' periods can happen at any time during the monsoon and are usually caused by a temporary weakening of the upper easterlies.

10. Break periods whether they occur at the start, middle or end of the monsoon, can cause serious drought and be associated with major food shortages.

11.3 Monsoon rains

While early definitions of the Indian monsoon were based on seasonal wind reversal, today the monsoon means the south-west monsoon and even more popularly 'the rainy season'. Unlike the West where the year is divided into four seasons, the Indian calendar consists of a triad: the cold season from October to December, the hot season from January to May and the rains of the summer monsoon from June to September. The monsoon rains play a critical role in industrial and agricultural production. Industrial performance can be hampered by poor monsoons when there is insufficient water for a range of manufacturing processes and hydro-power production. With over 650 million people living in the rural areas of India and deriving much of their income from agriculture, the main impact of monsoon failure is on food production and supply. If the rains are delayed or withdraw early, or if they are low or fail

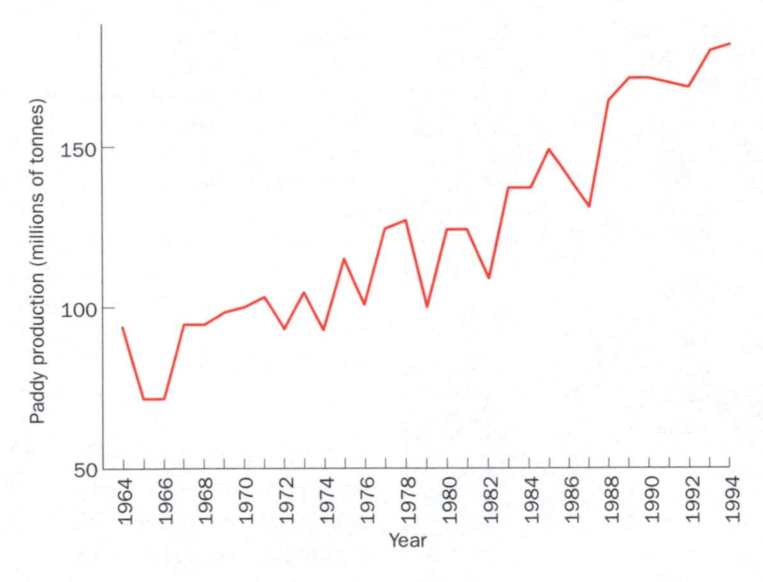

Figure 11.10 Annual wetland rice production, 1964–94. The overall trend is upwards because of the Green Revolution, but years with lower output can be linked to poor monsoons with deficit rainfall amounts

during a break monsoon period, crop output is reduced. Figure 11.10 shows a clear link between annual rice (paddy) production from 1964 to 1994 and monsoon rainfall expressed as a percentage departure from the long-term mean. The overall production trend is upward as a result of the Green Revolution with its higher inputs of technology, irrigation, seeds and fertilisers but dips in production relate to years with poor monsoons. With reductions in crop production, widespread hunger and economic ruin can result. Some of India's worst famines have been associated with times of monsoon failure, although famine in India as elsewhere (see Chapter 12) is also related to other factors. Major famines in India (1877, 1899, 1918 and 1972) have occurred during years when total food grain stocks in the country have been adequate, suggesting that political attitudes and food distribution also play a part in the creation of famine.

11.3.1 Mean annual distribution

(a) Orographic enhancement and shadowing

The most dramatic control on Indian rainfall is exercised by surface elevation, a relationship best observed in the mean annual distribution (Figure 11.11). Because elevated barriers encourage air convergence and vertical ascent (section 2.3.5), the highest rainfall yields with over 2,500 mm are experienced in mountain areas that lie directly in the path of the south-west monsoon. These areas include the Western Ghats, the foothills of the eastern Himalayas and above all in the Khasi Hills in Meghalaya, north-east India. The rainfall station at Cherrapunji (1,313 m) on the southern slopes of the Khasi Hills and one of the wettest places on earth (Plate 11.1) illustrates the role of altitude, land configuration and air circulation in rainfall enhancement (Figure 11.12(b)). Since the 1970s, when automatic rainfall monitoring was introduced, the minimum annual rainfall at Cherrapunji was registered in 1976 at 9,019 mm (over 9 m). The maximum annual rainfall was recorded in 1974 at 24,555 mm (over 24.5 m!) – a world record. During one day (16 June 1995) 1,563 mm or 1.5 m of rain (i.e. more than twice the rainfall of lowland Britain) was received – another world first. The same day also notched up the world record for rainfall amount during 1 hour – 420 mm. The reasons for the spectacularly high rainfall amounts at Cherrapunji are threefold.

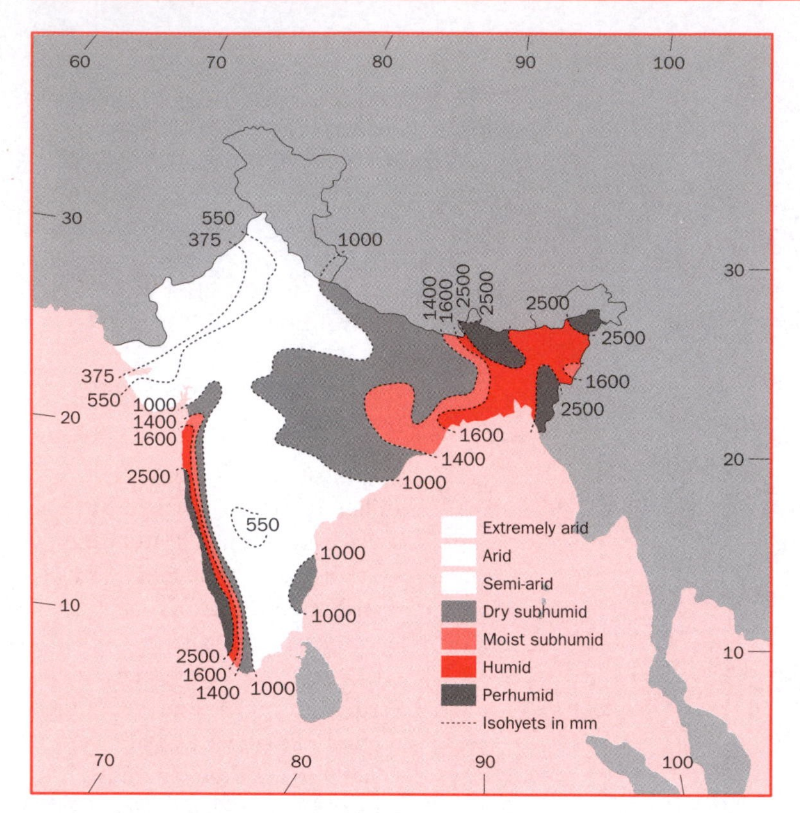

Figure 11.11 Mean annual rainfall distribution in India. The annual isohyets (lines of equal rainfall) when used in conjunction with rates of evapotranspiration mark boundaries of different climatic zones. For example, the mean 375 mm isohyet delineates the extremely arid from the arid, while the 1,400 mm isohyet separates the dry subhumid from the moist subhumid

Source: Singh *et al.* (1991)

First, the warm moist winds of the northward-moving Bay of Bengal branch of the summer south-west monsoon (see Figure 11.8(b)) occupy an extensive area over the southern Bay but dramatically converge towards the north over the Khasi Hills. Second, the Khasi Hills produce significant orographic lifting of the monsoon winds because of their east–west alignment lying directly in the path of the Bay of Bengal branch of the summer south-west monsoon. Third, in summer, orographically lifted air over the Khasi Hills is continually drawn up by vigorous upper tropospheric easterly winds and evacuated at high levels across India.

In direct contrast to the high rainfall of elevated regions, the driest parts of the subcontinent, with less than 200 mm, are located in lowland areas including the western Thar Desert in the extreme north-west, and in lowland areas in the rain shadow of the Western Ghats. Rainfall is also low over north-west India not only because of low altitude, but as a result of strong regional air subsidence in this area which prevents surface air uplift by convective and convergent means. Air subsidence over north-west India produces semi-desert conditions even on mountain ranges such as the Aravalli Hills which lie in the path of the south-west monsoon.

(b) Tropical cyclone paths

Orographic rainfall enhancement is not the only factor explaining the annual distribution of rainfall in India. Much of the south-west monsoon air that is forced up by the Western Ghats, for instance, is deflected by them to form large offshore tropical cyclones (low-pressure systems) lasting 2–3 days and bringing 100 mm of rain in 24 hours along the western coastal belt of the peninsula. Cyclonic weather disturbances also contribute much of the relatively high summer monsoon rainfall (over 1,000 mm) recorded in the low northern plains of India (Figure 11.11). The main source of supply for this rainfall comes from tropical cyclones that move in over the northern plains from the Bay of Bengal in summer. Most rainfall in India is actually produced by such regional-scale weather systems embedded in the monsoon winds. Thus, a second important factor explaining the distribution of summer monsoon rainfall distribution in India is the distribution and alignment of the most favoured tropical cyclone tracks, i.e. those crossing over the Western Ghats and those aligned north-westwards along the Ganges Valley from the Bay of Bengal.

11.3.2 Seasonal variations

The south-west monsoon dominates India's rainfall supply to such an extent that it produces over 80 per cent of the total annual rainfall amount. As a consequence, the summer seasonal distribution resembles that for the year as a whole (Figure 11.11).

(a) Monsoon duration period

Figure 11.12(a) shows the average times of the onset and withdrawal of the summer monsoon. The average date of arrival is 20 May in the extreme south of the country, but it takes till 5 July (over 6 weeks) for the south-west monsoon to migrate northwards across India to Rajasthan and the Punjab in the north-west. Withdrawal times are somewhat longer. The south-west monsoon winds begin to retreat from the extreme north-west on 1 September and it takes till around 15 November (10 weeks later) for the southward-withdrawing monsoon to clear the southern tip of India.
As a result, duration periods for the summer monsoon vary across the country. It is clear from Figure 11.12(b), which shows a five-zone model of monsoon duration times, that the period June to September, normally considered as the period of monsoon rains for the whole of the country, is found only in the central zone. It is shorter in the north-west where the summer monsoon season lasts 2 months and longer in the south where it can last up to 6 months.

(b) Seasonal rainfall regimes

Figure 11.13 shows the mean annual and monthly rainfall for a range of climate stations in India. These stations are widely distributed across the

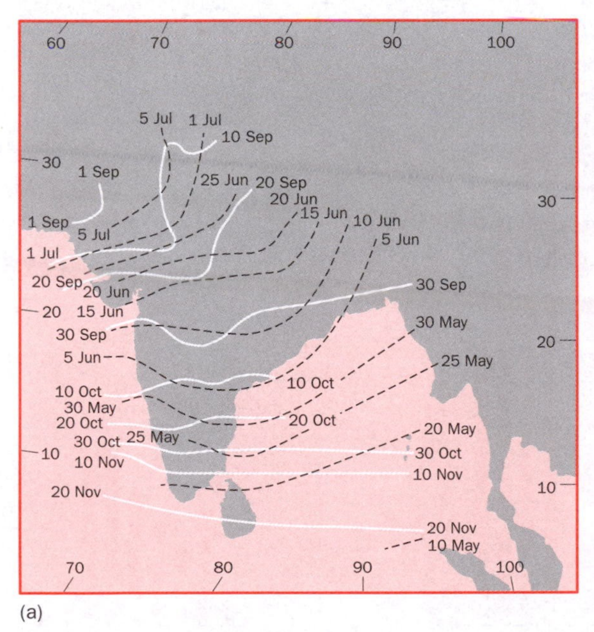

(a)

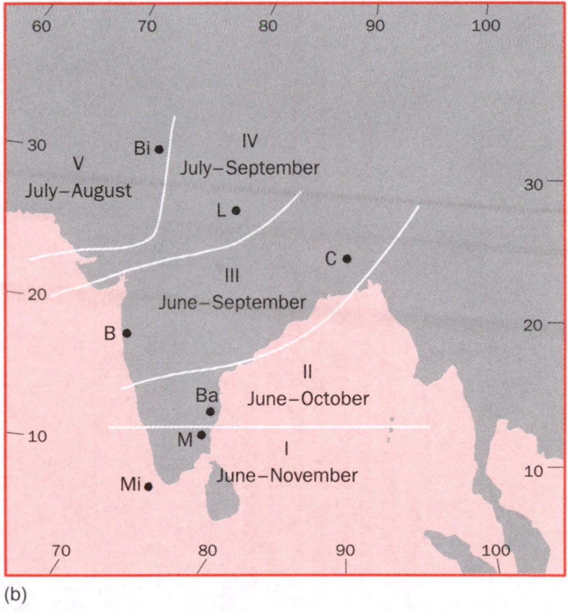

(b)

Figure 11.12 (a) Average onset (dashed lines) and withdrawal (solid lines) dates for the summer south-west monsoon; (b) duration times of the summer monsoon with selected climate stations: (Bi) Bikaner, (L) Lucknow, (C) Cherrapunji, (B) Bombay, (Ba) Bangalore, (M) Madras, (Mi) (Minicoy)
Source of dates: Subbaramayya and Naidu (1995, pp. 159–66)

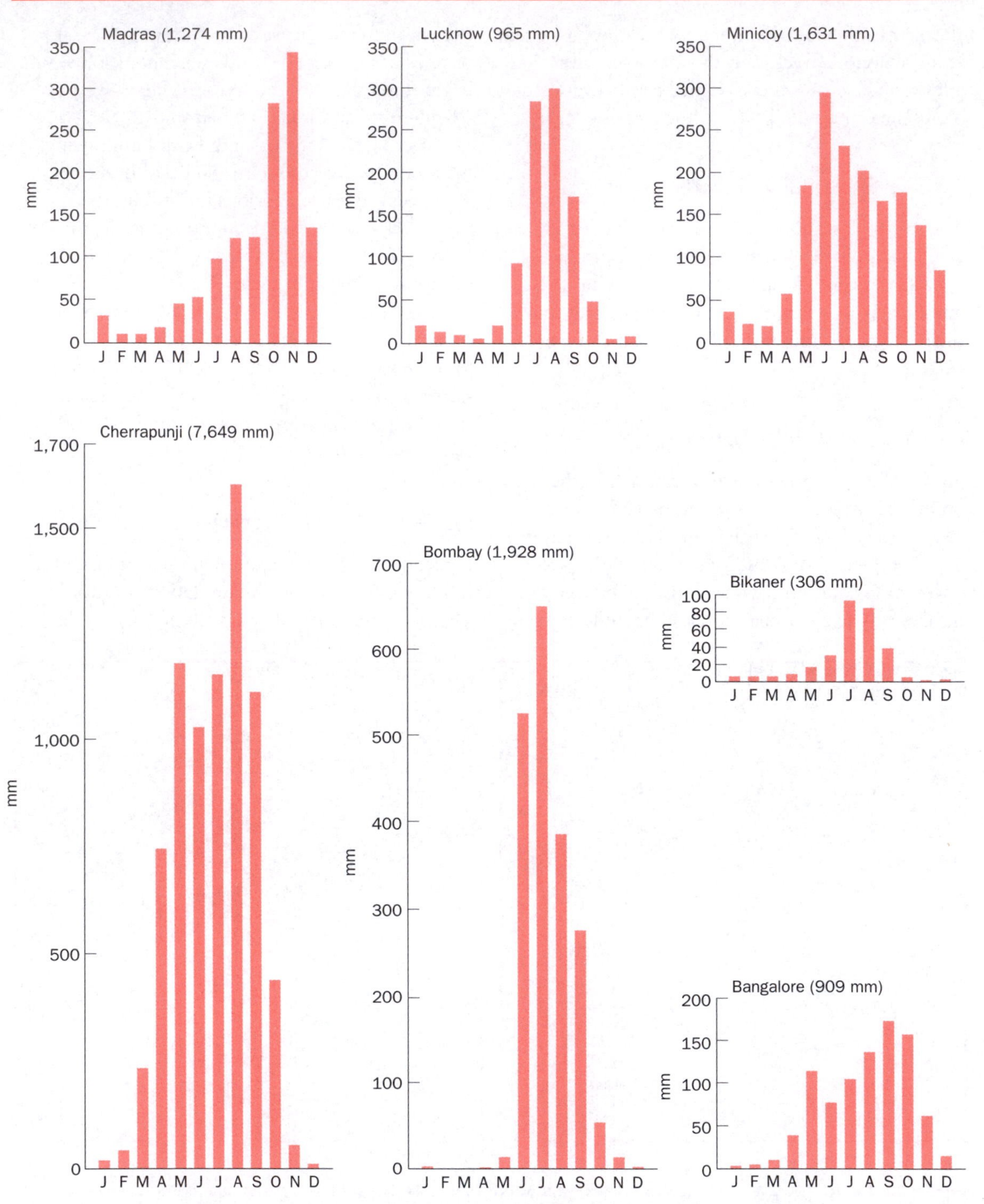

Figure 11.13 Mean annual and monthly rainfall for selected climate stations in India. Note conformity with stations shown in Figure 11.12(b)

Table 11.1 Seasonal rainfall regimes for selected stations in India (based on long-term twentieth-century rainfall records)

Zone 1 (6 months from June to November)
1. *Minicoy*: This zone 1 station has significant precipitation from June (maximum) to November. Low rainfall between December and April from the dry north-east monsoon. Convective showers contribute to rainfall in May and November. These showers occur just before the onset and just after the withdrawal of the south-west monsoon. They are associated with the passage of the low-pressure MT as it migrates north and south over India following the overhead sun

2. *Madras*: Low rainfall between June and September from south-west monsoon because it lies in rain shadow of the Ghats mountains. Most rainfall in October and November from cyclonic disturbances which develop over the Bay of Bengal during the transition period between the summer and winter monsoons. North-east monsoon winds in Dec.–Feb. are a poor source of moisture despite trajectory over Bay of Bengal

Zone 2 (5 months from June to October)
Bangalore: Stormy convective rainfall in May and September (maximum) when this station comes under the influence of the migrating MT. Bangalore also lies in the rain shadow of the Western Ghats, and compared with Bombay receives very low monthly rainfall amounts between June and October from the south-west monsoon

Zone 3 (4 months from June to September)
1. *Bombay*: Classic rainfall profile of the south-west monsoon. Ninety per cent of Bombay's rainfall coincides with the months from June to September. High rainfall associated with orographic enhancement. Rapid arrival of the rains (i.e. the monsoon 'burst') witnessed in the contrasting rainfall totals of May (dry) and June (very wet)

2. *Cherrapunji*: The enormous rainfall yield at this station is attributed to very rapid and sustained air uplift of the northward-moving south-west monsoon against the east–west aligned mountain range of the Khasi Hills. Most rainfall in August when the south-west monsoon is at its peak over northern India. High rainfall (more than 100 mm) is extended to the months of April, May and October by convective thunderstorms. Early spring rains known as mango showers; these nourish the spring rice crop in Bengal and bring first flushes of green leaves to the tea plantations of Assam

Zone 4 (3 months from July to September)
Lucknow: Classic monsoon rainfall profile with most rain falling between July and September. A lowland station without the orographic enhancement of Bombay

Zone 5 (2 months from July to August)
Bikaner: Extreme north-western station with low rainfall amounts (less than 100 mm) during July and August. During the winter months this station intercepts small amounts of rainfall from eastward-moving depressions which are attracted by the upper westerly jet over northern India. Little rain falls during the rest of the year as region is dominated by subsiding airflows

country with at least one climate station in each of the five monsoon duration zones. Table 11.1 also provides a concise rainfall description for each station.

The zonal duration model in Figure 11.12(b) is generally well supported by each station's rainfall regime. While Minicoy (zone 1) experiences the monsoon rains for 6 months between June and November (a rainfall month is one with more than 100 mm of rain), Bangalore (zone 2) has monsoon rainfall for 5 months between June and October (June has less than 100 mm, but see below). Bombay (zone 3) shows the classic 4-month monsoon season with rains between June and September. Lucknow clearly represents zone 4 with a 3-month main rainfall period between July to

September, while Bikaner, a dry zone 5 station, has monsoon rainfall for only 2 months (July and August). No month at Bikaner has more than 100 mm of rain, however.

Other factors help to explain more fully the rainfall patterns at each station. The very high monthly rainfall totals at Bombay and especially at Cherrapunji in the Khasi Hills give evidence of strong orographic enhancement at these stations. In contrast, Bangalore and Madras have much smaller rainfall amounts during the rainy season because of orographic shadowing effects as they lie in the lee of the Western Ghats. At Lucknow, a lowland station in northern India (zone 4) where rain may be expected to be slight between July and September, relatively high summer totals are received from tropical cyclones. Cyclonic rainfall associated with the retreat of the south-west monsoon or the onset of the north-east monsoon (with moist winds blowing over the Bay of Bengal) encourage high rainfall in October, November and December at Madras. The rainfall season is also extended by convective thunderstorms just before the onset of the monsoon at Minicoy and Bangalore (in May) and notably at Cherrapunji (in March, April and May). Convective showers also supplement the rainfall supply at several stations during the months following the retreat of the monsoon, as at Cherrapunji (October), Bangalore (October) and Minicoy (November, December).

(c) Variations in monsoon duration

Over the last 100 years there has been 46 days' difference between the dates for the earliest (7 May) and latest (22 June) arrival of the monsoon over Kerala, south India. One of the most serious delays in the onset of the monsoon occurred in 1972 when it was late by 18 days. When long lasting (i.e. 2 to even three weeks) the monsoon 'break' with its diminished rainfall can also be responsible for serious drought. It causes extensive crop damage and loss because it often occurs during the middle of the summer growing season when water demand and crop productivity are at their highest. Serious break conditions occurred in 1972, when the monsoon as mentioned above was delayed by 18 days. The result was severe drought and a colossal loss of about one-third of the country's food crop (see Figure 11.10). Widespread famine occurred in western and central India. Finally, an early departure of the south-west monsoon is catastrophic since it dries up crops which are reliant on a continuance of the rains. In 1883, the monsoon left Bengal a month too early, ruining the rice crop and producing a serious famine.

11.3.3 Individual weather systems

(a) Tropical disturbances and rainfall intensity

Most of India's rain is intermittent and localised, lasting over a period of several hours to several days. Much of it comes from relatively frequent but short-lived weather disturbances embedded in the general monsoon airflow. These disturbances include (a) synoptic or regional-scale *tropical cyclones* from 150 to 1,000 km across and (b) local (10–50 km) and mesoscale (100–500 km) *convective thunderstorms*. A characteristic feature of tropical cyclones and thunderstorms is their violent and intense rainfall (25–75 mm/hour). However, as shown below, these weather systems are also associated with long periods of rain of much lower intensity.

Studies have shown that around one-half of India's summer rainfall is contributed in the majority (80–90 per cent) of rain days from falls of low intensity. In other words, the most long-lasting and persistent rainfall during the summer monsoon *is similar to moderate to heavy rainfall in the UK*. Most dramatically, however, the other half of India's monsoon rain is of very high intensity and is delivered over a relatively small number (10–20 per cent) of rain days. Rainfall intensities are at their lowest over peninsula India, east of the Western Ghats (20–30 mm/day). Over most of northern India, 50 per cent of total rainfall is from amounts around 40–50 mm/day, while

Table 11.2 A four-fold classification of tropical cyclones in India. The weather systems form a continuum. During the summer monsoon, tropical lows (8 m/s) often develop into the stronger tropical depression (8–17 m/s) and occasionally into a tropical storm (17–23 m/s). During the pre- and post-monsoon periods when there is little wind shear, tropical depressions can sometimes develop into severe tropical storms (24–32 m/s) and eventually into hurricanes (>32 m/s)

Type of disturbance	Speed (m/s)	Vertical depth (km)	Duration (days)	Horizontal dimension (km)	Frequency	Rainfall total (cm)
A. Cold core systems						
Tropical low	<8	2–4	1–3 (l/s)	150–300	Frequent	5–10
Tropical depression	8–17	4–8	2–5 (s)	250–500	Common	10–20
B. Warm core systems						
Tropical storm	17–32	8–10	3–10 (s)	300–600	Occasional	20–50
Hurricane	>32	8–12	5–7 (s)	400–1,000	Rare	50–150

(s) = origin over the sea; (l/s) origin over land or sea.

along the west coast intensities reach as high as 50–60 mm/day.

(b) Four types of tropical cyclone

As suggested previously, the majority of India's rainfall comes from regional-scale tropical cyclones. Table 11.2 shows the four main types of tropical cyclone which dominate India's rainfall supply. They are classified primarily on the basis of their wind speed, but physical size, frequency of occurrence and duration, rainfall supply and core temperatures are also important parameters. The tropical cyclonic low (wind speed less than 8 m/s) is the weakest system, followed by the tropical cyclonic depression (8–17 m/s). Next is the tropical cyclonic storm (17–32 m/s), and finally the most powerful system, the hurricane (see section 5.2 and Table 5.1) which is defined as a very severe cyclonic storm with hurricane force wind speeds (more than 32 m/s) at its core.

In section 5.2.2 it was suggested that if conditions are right, it may take only 2 or 3 days for a weak tropical cyclone to develop into a severe storm with hurricane force winds (see also Plate 12.1). Since the various types of tropical cyclone shown in Table 11.2 are linked during the process of cyclogenesis, it is not surprising that they share a number of common features. They are all identified by low pressure at their centre, encirclement by anticlockwise (in the northern hemisphere) converging winds (as in a vortex), and rising buoyant air across an extensive area. For all types of tropical cyclone, rain falls from deep nimbostratus (extensive rain-bearing layered clouds) and, particularly for storms and hurricanes, from embedded cumulonimbus (rain-bearing clouds of very deep vertical extent). Skies remain predominantly overcast, though the intensity of rainfall may vary considerably. Most tropical cyclones that make landfall over India originate over the surrounding seas. This is because they derive their energy from intense evaporation over warm seas. Moisture evaporated over the oceans is drawn into the developing vortex. When this moisture (water vapour) is converted into cloud (water droplets) by cooling and condensation, latent heat is released maintaining the energy of the system.

(i) Tropical lows and depressions

The weakest types of tropical cyclone, the tropical low and the tropical depression, occur relatively frequently over India during the monsoon (every 4–5 days) and produce rains of intermediate to high intensity (Table 11.2). As a result they are responsible for most of India's rainfall. These two weather systems are fairly shallow being less than 4 and 8 km deep, respectively, and have, typically, at their centre, a cold column of air that decreases in temperature with height. Being shallow systems they develop when wind speeds change rapidly with height or direction in the atmosphere (i.e. when there is large vertical wind shear) and when there is rapid air convergence at the surface. These conditions are fulfilled in the wet summer monsoon months (June–Sept.) when there are surface westerlies but strong easterly winds aloft (see Figure 11.9). They are most frequent at the height of the monsoon in July and August when sea and air temperatures are high. It is worth pointing out that the tropical low and depression which gain their energy from evaporation over the sea, are very different from their mid-latitude counterparts that obtain their dynamic energy from the contrast in (potential) energy between air masses of different temperature and moisture.

Many tropical lows and depressions originate in the Arabian Sea to the west of the Western Ghats before passing over them on the south-west monsoon current. The majority, however, develop over the ocean at the head of the Bay of Bengal. A typical Bay depression is shown in Figure 11.14. These systems (as well as the less powerful tropical lows) are pushed inland by the recurving winds of the south-west monsoon which are attracted to the summer heat low over north-west India (see Figure 11.7(a)). As they migrate, they often follow the southern edge of the MT that is typically aligned over the northern plains from Bengal to the Punjab during the height of the summer. As these systems progress inland, they are accompanied, especially on their southern side, by heavy rains. Figure 11.15 shows the rainfall yield associated with a typical tropical depression moving west-north-westwards across the Gangetic Plain over 3 consecutive days between 8 and 10 July 1973. It can be seen that large regions receive more than 50 mm/day of rain and significant areas receive more than 100 mm on any consecutive day. It is important to note that the highly productive rice-growing districts of the Ganges Valley with their associated large agricultural populations are directly supported by, and critically dependent on, such rains.

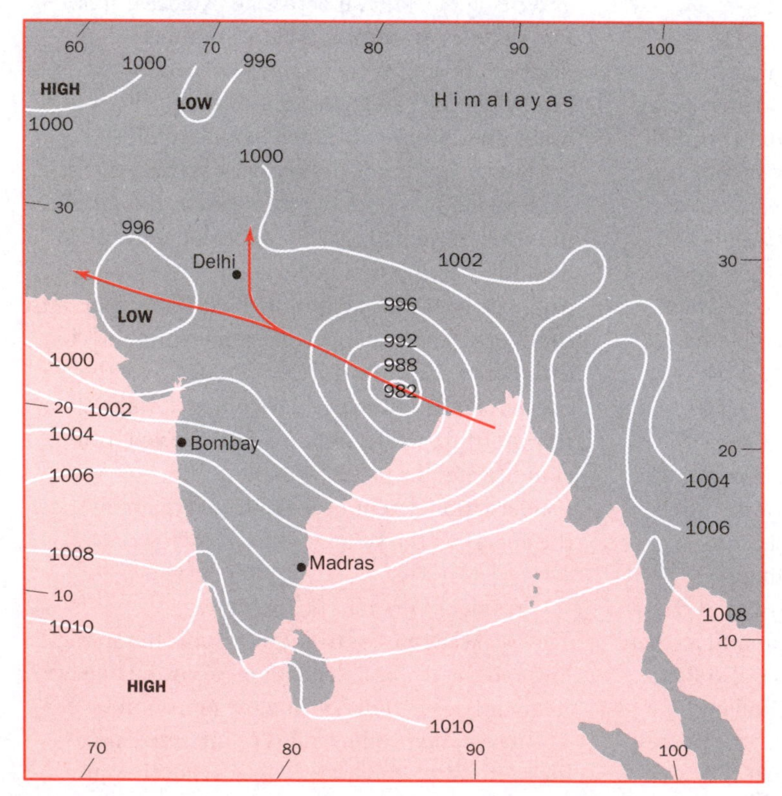

Figure 11.14 Surface weather chart depicting typical 'active' summer monsoon conditions. Low pressure is widespread with a tropical depression moving in over north-east India from the Bay of Bengal. Favoured depression tracks shown by arrows
Source: Das (1987)

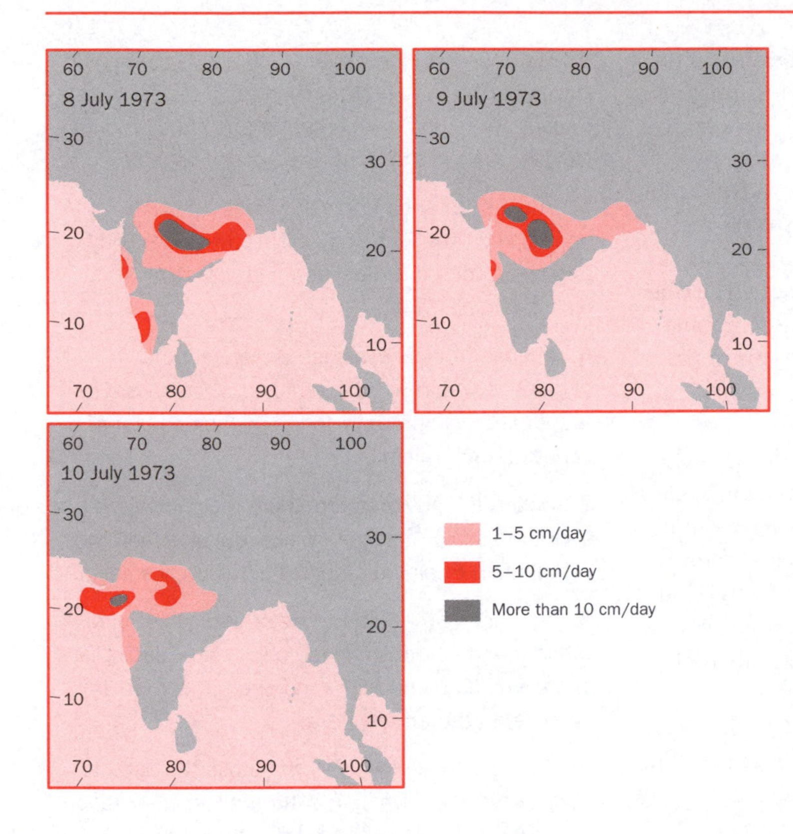

Figure 11.15 The distribution of daily rainfall (in cm) associated with a Bay of Bengal depression moving east to west across the northern plains of India
Source: Webster (1981, pp. 70–80)

(ii) Tropical storms and hurricanes

Though capable of producing the most rain in a single event, tropical cyclonic storms including those with hurricane wind speeds at their centre are relatively infrequent. On average only one or two occur over the Arabian Sea while six to eight develop in the Bay of Bengal each year. They therefore contribute little aggregate rainfall on a seasonal or annual basis. Compared to the less powerful cyclones, severe tropical storms and hurricanes have great vertical extent (8–12 km) and have a warm core with descending air at their centre (the eye of the storm/hurricane). Because they are vertically deep systems with hurricanes reaching the top of the troposphere, they cannot develop when there is a large change in wind speed/direction with height through the atmosphere (see factors in hurricane formation in section 5.2.2). They require weak winds aloft with little wind shear in the surrounding atmosphere so that they are not toppled over. As a consequence, they originate over the warm seas around India during the transition periods *between* the winter and summer phases of the monsoon, from late April to early June and late September to early December. It may be recalled that during these periods, the upper easterly and westerly jet streams are relatively weak with little wind shear in the lower atmosphere. As the most intense of the tropical cyclones, hurricanes are capable of enormous damage. The most destructive cyclones defy imagination. Three of the largest Bay of Bengal cyclones which devastated large areas of Bengal and Bangladesh include the earliest known killer cyclone of 7 October 1737 which killed or drowned over 300,000 people, and the Bangladesh hurricanes of 13 November 1970 (up to 500,000 dead) and 29 April 1991 (139,000 dead).

(c) Convective thunderstorms

Convective thunderstorms differ in a number of ways from tropical cyclones. Unlike tropical cyclones they tend to develop over *land* in areas of strong surface heat convection (sections 2.3.5 and 10.2.1). Moreover, they occur when vertical wind shear and lower tropospheric convergence are both high. They are associated not with the monsoon *rains* which come from tropical cyclones, but with *showers* which fall from scattered towering cumulus or cumulonimbus cloud. Though frequent, they are smaller in scale and more localised than tropical cyclones. As a result, the contribution of convective thunderstorms to all-India rainfall is relatively small. This situation contrasts with that in southern USA where rainfall from thunderstorms and mesoscale convective complexes (MCCs) contributes most to regional rainfall totals. Nevertheless, thunderstorms are often associated with very intense rainfall and can cause severe damage. This is particularly true when individual convective cells occasionally combine into larger MCCs that can be comparable in size to the smallest tropical cyclone (i.e. 100–500 km across).

The factors that lead to the formation of organised convection within a thunderstorm (clear skies, hot land surface, high wind shear) are thus very different from those that are necessary for tropical cyclone formation (warm seas, low wind shear). As a consequence, the two weather systems tend to occur at different times and in different places. Cyclonic disturbances develop most frequently in association with the MT at the height of the summer monsoon in July and August, while convective thunderstorms tend to occur under clear skies (a) in April/May and September/October at the onset and withdrawal of the summer monsoon, (b) near the MT during breaks in the summer rains and (c) in the rain shadow of mountain ranges during the monsoon rains.

Key ideas

1. The mean annual distribution of rainfall in India is related to surface elevation and the main tracks taken by tropical cyclones.

2. The highest mean rainfall is in the mountains of north-east India and the Western Ghats (over 2,500 mm), while the lowest rainfall (less than 550 mm) is found in the Thar Desert to the north-west and in rain shadow regions east of the Ghats mountains.

3. There is marked variation in the duration of the monsoon over India because of different arrival and departure times.

4. The monsoon lasts over 6 months (June to November) in the south, 4 months (the classic June–September period) in the centre and only 2 months (July–August) in the north-west of the country.

5. The less intense tropical cyclones (lows < 8 m/s and depressions 8–17 m/s respectively) are fairly frequent during the monsoon and deliver most of India's rainfall.

6. Intense tropical cyclones (severe storms (17–32 m/s) and hurricanes (over 33 m/s)) are rare events and account only for a small proportion of the country's total rainfall.

7. Convective thunderstorms are more localised and smaller in scale than the tropical cyclones and their contribution to all-India rainfall is relatively small.

8. The factors that lead to organised convection within thunderstorms include clear skies, strong land surface heating and high wind shear (near jet streams) in the atmosphere.

9. They are thus different from those that lead to tropical cyclone formation (warm ocean evaporation and low wind shear with weak jet streams).

10. Tropical cyclones thus differ from mid-latitude cyclones (depressions) which depend on the energy contrast between the temperature and moisture content of converging air masses and high wind shear (upper jet streams).

11.4 Monsoon variability and change

11.4.1 Interannual and interdecadal variability

There is considerable variability from one year to the next in the amount of rainfall delivered by the summer monsoon. Figure 11.16 shows one index of the interannual variability in all-India rainfall for the 130-year period between 1871 and 2000. The index shows the annual departure in summer monsoon rainfall (JJAS) from the long-term 1961–90 mean. When the all-India rainfall

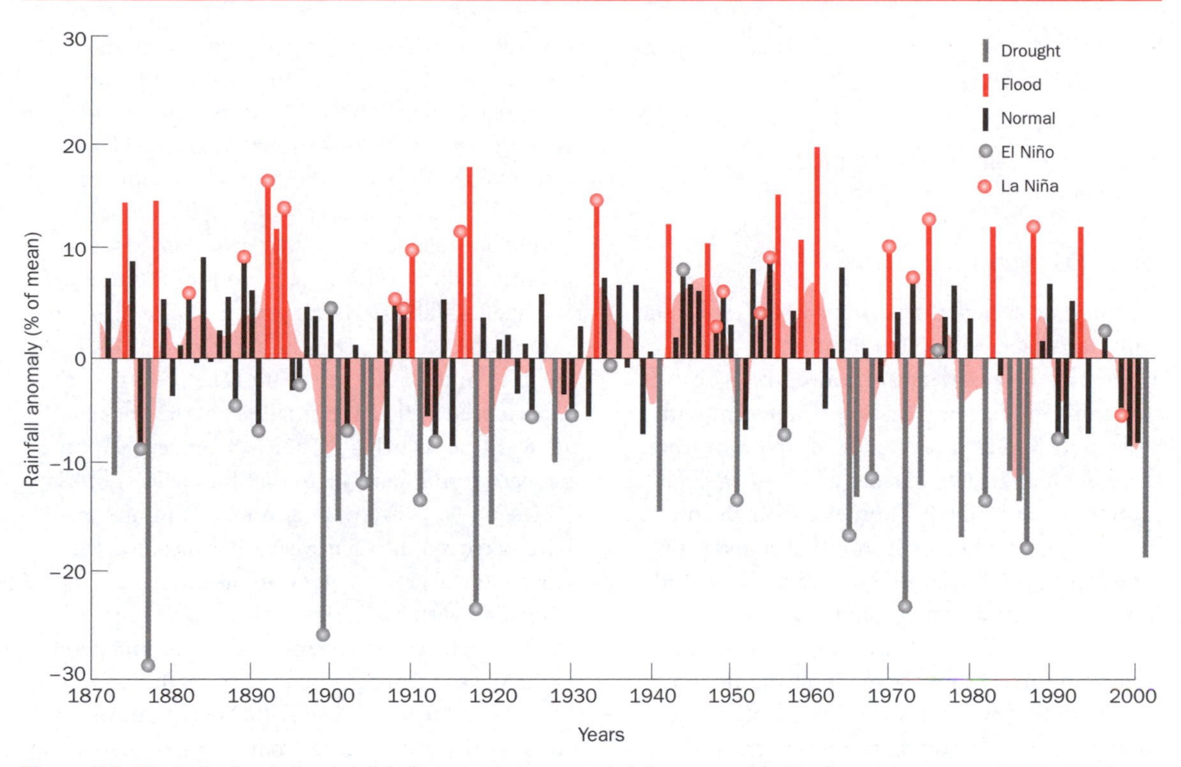

Figure 11.16 Annual variation in all-India mean summer rainfall compared to the long-term mean, 1871–2000. El Niño and La Niña years are shown in relation to the rainfall variation. Rainfall data developed by the Indian Institute of Tropical Meteorology

departure is more than about +/–10 per cent, we can expect significant environmental impact: serious drought occurs in certain parts of the country when the all-India rainfall is more than 10 per cent deficient, and serious flooding takes place when the overall rainfall is more than 10 per cent in excess. Particularly dry summer monsoons occurred in the years 1877, 1899, 1918 and 1972 when national rainfall deficits of more than 20 per cent were achieved; while the years 1892, 1917, 1956 and 1961 saw some of the wettest monsoons where national rainfall totals were more than 15 per cent of the all-India mean. The year-to-year rainfall variability is so marked that there are only a few instances when two wet or two dry years run consecutively, i.e. the wet years of 1892/3/4, 1916/17 and the dry spells of 1904/5, 1965/66 and 1985/86/87. Despite the high interannual variability shown, the overall pattern in Figure 11.16 could be argued to be relatively stable. There is no apparent long-term trend of increasing or decreasing rainfall. Some studies, however, suggest that there is a discernible decadal fluctuation or cycle in the all-India rainfall series, with the periods 1880–1900 and 1931–60 being wetter than the overall average, and the periods 1901–30 and 1961–90 being drier than average.

Another important facet of the summer monsoon rains is that they are also very differently and very unevenly distributed from year to year across the landmass of India. This means that when the rains fail in one part of the country, they are often adequate or even in excess in other parts of the country. For instance, India suffered a serious drought – the worst for 12 years – in the summer of 2002. While central and northern India were the worst regions hit by scanty and deficient rainfall, excessive rainfall with serious flooding and loss of life occurred in north-east India and in the west of the country.

11.4.2 Explaining interannual monsoon variation

(a) ENSO events

The marked annual fluctuations in the monsoon rains have been linked to a number of factors. There have been statistical connections with variations in solar output, where wet years with large-scale floods have a tendency to occur when there is a high degree of solar activity (e.g. high numbers of sunspots). Anomalously high amounts of winter/spring Eurasian snow cover and snow mass have been linked to weak rainfall over India in the following summer. Snow may be an important factor in the external forcing of the monsoon since a snow cover cools the air above it by reflecting a high proportion of incoming sunlight. Large amounts of snow cover over the Tibetan Plateau may retard the heating of the upper atmosphere. This in turn may restrict the development of an upper air anticyclone over the Tibetan region and thus delay the appearance of the easterly jet over India that eventually triggers the onset of the monsoon. The strongest control, however, governing the interannual behaviour of the Indian monsoon is ENSO (El Niño–Southern Oscillation) events.

As discussed in section 5.4, ENSO events describe periodic, global-scale changes in the ocean–atmosphere circulation in the Pacific Basin. The opposing air (and ocean) circulations during an El Niño and a La Niña create worldwide changes in global air movement. Some of the strongest links (atmospheric teleconnections) set up by El Niño/La Niña events are with regions immediately adjacent to the Pacific Basin (see also section 10.3.2). One such neighbouring region is the Tibetan Plateau where upper air pressure systems and wind movements are so crucial to the onset and development of the monsoon. Many studies have shown that El Niño events are generally associated with poor monsoons and La Niña events with good monsoons over India. The great majority of the El Niños generated between 1871 and 2000 shown in Figure 11.16 are associated with poor (dry) monsoons. Conversely, most of the La Niña events during the same period are related to good (wet) monsoons. Strong El Niños falling within the periods of less than average rainfall (1901–30, 1961–90) appear to be responsible for the greatest droughts, e.g. 1877 when as much as 64 per cent of the country experienced deficient rainfall, 1899 (65 per cent), 1918 (71 per cent) and 1972 (48 per cent). It can be seen from Figure 11.16 that the weakest link between ENSO events and monsoon rainfall may have occurred in recent years. For instance, the reasonably average to good monsoon in 1997 coincided with the very strong El Niño event of 1997/98. Moreover, the relatively poor monsoon of 2000 was seemingly unaffected by the La Niña of 1998–2000. These two ENSO events, however, were fairly protracted and this may have had some impact on the development of the monsoon.

Finally, it needs to be recognised that as well as being affected by global weather systems like ENSO events, the Indian monsoon itself is powerful enough to influence weather and climate in other regions of the world. For instance, while India experienced one of its worst droughts for many years in the summer of 2002, some of the coldest and wettest weather on record occurred in southern and central Europe. This very wet episode caused hundreds of people to lose their lives and billions of pounds of damage (see section 9.3.1). The atmospheric teleconnection between these two regions is not difficult to explain. As we have shown, as the summer monsoon season arrives, huge volumes of air begin to rise over the Indian landmass. These air streams push north and west, allowing other columns of air to descend over the Mediterranean region, causing stable high-pressure systems to develop. These stable surface anticyclones block the movement of unstable rain-bearing weather systems (i.e. mid-latitude depressions) moving into the Mediterranean from the Atlantic. Thus, when the Indian summer monsoon is very weak (as it was in the year 2002) its control over southern Europe is released, so that there is no stable high-pressure system to prevent the entry of the unstable rain-bearing mid-latitude depressions.

11.4.3 Longer-term monsoon change

As previously outlined, the most important factor behind the monsoon climate is solar heating. Variations in solar heating between the seasons (especially summer–winter) and between land and sea are responsible for driving the main monsoon circulations. We have also seen from section 8.3 that earth orbital variations around the sun exert a strong control on the amount and distribution of solar radiation impinging on the earth. These solar variations have been linked to the development of glacial advance and retreat in the high polar latitudes. It is also possible that changing solar radiation amounts could control the strength of monsoons over orbital timescales of 10,000–100,000 years. The theory that monsoon performance could be linked to orbital variations in solar heating is called the orbital monsoon hypothesis (Box 11.4).

In the low and middle latitudes, where monsoon circulations are strongest, changes in the amount of incoming solar radiation that follow the 23,000-year cycle of orbital precession or 'wobble' (section 6.5.2) have been shown to have a measurable effect on monsoon strength and intensity (Figure 11.17). The northern limit of today's monsoon summer rainfall coincides approximately with the 100 mm annual rainfall isohyet. Northwards of this limit, annual rainfall dips below 100 mm, and active sand deserts develop. During the last glacial maximum (LGM) about 1,800 years ago, precessional solar inputs were low and monsoon circulation was weaker than today. In response to these changes, the southern limit of active sand deserts moved southwards over the northern continents by about

BOX 11.4

THINKING FURTHER

The orbital monsoon hypothesis

Palaeoclimatic evidence from ocean, lake and sand dune deposits show that monsoon circulations have been strongly influenced by orbital-scale variations in solar energy receipt. Insolation variations linked to the 23,000-year precessional or 'wobble' cycle (section 6.5.2) have a particularly marked effect on the extent and intensity of the monsoon, especially over South Asia and North Africa.

1. Increases in aridity during the LGM

During the last glacial maximum (LGM) about 18,000 years ago, precessional inputs of solar radiation in the low to mid latitudes of the northern hemisphere were low. With less land heating, not only was the Indian summer south-west monsoon relatively weak (Figure 11.17) but there was also a much stronger dry north-easterly monsoon in winter. Stronger north-easterlies with winds blowing from the Indian subcontinent over the Arabian Sea increased aridity in the region as indicated by an increase in sand input into the Arabian Sea, and an increase in dune activity in Arabia.

2. Warmer coastal Arabian seas

With weaker south-westerlies and stronger north-easterlies 20,000 years ago, relatively warm surface water was pushed up against the coast of Saudi Arabia, so that studies of ocean plankton (Foraminifera species) show warmer sea surface temperatures (SSTs) along the Arabian coast during the LGM than at present.

3. Stronger upwelling along the coasts of Arabia and West Africa

In accordance with the precessional cycle, summer insolation values at the Holocene Optimum, i.e. 8,000–10,000 years ago (one half-cycle), were much higher (by 8 per cent) than today. With greater land heating, the summer monsoon intensified and extended its range (Figure 11.17). With a stronger south-west summer monsoon over the Arabian Sea more warm surface water was dragged away from the Arabian coast towards India. The result was that cold water upwelling was intensified along the coast of Arabia as identified in planktonic foraminifera assemblages.

4. Higher lake levels in North Africa

A stronger south-west monsoon 8,000–10,000 years ago also brought much higher rainfall amounts to extended parts of tropical and subtropical North Africa and India. One important consequence of this was that many lakes in Africa that are dry today were filled to high levels at this time. Today's lower summer levels of insolation and weak monsoon are apparently insufficient to bring most North African lakes into existence.

5. Freshwater lake diatoms in the tropical Atlantic

Further evidence that the size of North African lakes fluctuates at the 23,000-year tempo of orbital precession can be found in sediment cores from the North Atlantic. Distinct layers of freshwater diatoms from North African lakes appear in the *ocean* sediments of the North Atlantic every 23,000 years or so. These layers of diatoms must mark times in the past when the dry north-easterly monsoon in Africa was strong. The dry north-east monsoon with its origins over the Sahara would encourage lakes to dry out across North Africa exposing their muddy beds to the atmosphere. As the lake sediments dried out (every 23,000 years), strong winter north-easterly winds would have carried the exhumed diatoms southwards and westwards across North Africa to the Atlantic Ocean (not too dissimilar from dust storms driven by the harmattan wind today).

6. 'Stinky muds' in the Mediterranean

Higher rainfall levels from stronger summer monsoons (10,000–11,000 years ago and every 23,000 years since then) in the Ethiopian Highlands also gave rise to greater runoff and massive flooding by the River Nile. The plume of fresh water that resulted from the River Nile floods would have spread at the surface across the Mediterranean because it is lighter (i.e. less dense) than the salty waters of the Mediterranean Sea itself. The low-salinity freshwater layer which effectively put a cap over the Mediterranean Sea would have encouraged stratification of the surface layers and stagnation of the bottom waters. Stronger river flow may have at the same time brought more nutrients into the Mediterranean so that surface plankton productivity could have increased. This high productivity at the surface would have continually sent organic-rich remains of dead plankton towards the sea floor. Continued sinking and decay of this organic carbon soon depleted oxygen levels in the deep Mediterranean muds on the seafloor. A lack of oxygen in the bottom water prevented the normal decay of further organic additions to inorganic mineral form so that the muds became highly concentrated in black semi-decayed organic remains. A lack of oxygen also led to the deposition of iron sulphides giving the black muds a stinky (rotten-egg) odour. These distinct layers of black organic-rich muds are called sapropels. The sapropel layers can be traced back every 23,000 years or so in the Mediterranean. The most recent sapropel layer can be traced back in the eastern Mediterranean to about 8,000–10,000 years ago, the same time interval when African lakes were high and the African monsoon was strong.

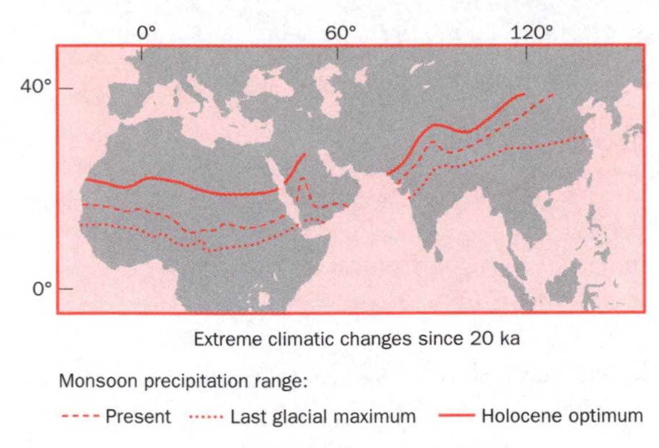

Figure 11.17 Long-term changes in the limits of monsoon rains during the last glacial maximum (LGM) about 18,000–20,000 years ago and the Holocene climatic optimum about 8,000 years ago
Source: Wilson *et al.* (2000, p. 27)

300–400 km. In contrast, 8,000–10,000 years ago (i.e. back one-half precessional cycle) solar radiation forcing was at a period high. As a result, monsoon circulations became much stronger with rain belts extending much further inland. So much so that during this period, active sand dune desert formation (100 mm annual rainfall isohyet) was pushed almost 1,000 km further north compared with today.

11.4.4 Monsoon futures

(a) The case of India

The South Asian monsoon is central to the Indian economy and predicting future changes to it is a priority. Whilst general circulation models (GCMs) like the Hadley Centre HadCM3 is able to reproduce the mean large-scale airflow over the country fairly well, rainfall is only simulated adequately over the relatively flat terrain of central India. Regional circulation models (RCMs), on the other hand (see section 7.2.3), such as the one devised by the Hadley Centre in association with the Indian Institute of Technology, are better at simulating the observed rainfall over the south-east region. As a result, scenarios of future rainfall amounts during the south-west monsoon for the year 2050 differ between the GCMs and RCMs (Plates 11.3(a) and (b)). It can be seen that over the western coastal mountains (the Western Ghats) the RCM predicts large increases in summer rainfall that do not appear in the GCM. The RCM also signals the likelihood of less rainfall over southern India east of the Western Ghats, and higher rainfall amounts over the central region of the Himalayas in northern India. Despite their recognised uncertainty, such modelled scenarios suggest better monsoons in the future for the rice- and wheat-growing areas of the northern Ganges Valley but the greater likelihood of droughts in the rice-growing deltas of southern India.

Human populations are affected not only by changes in long-term means in the climate but also by extreme weather. For instance, the flooding damage caused by storm surges (tidal waves) from tropical cyclones in the Bay of Bengal can lead to considerable loss of life and property (see section 12.2). The behaviour of these storm surges is particularly pronounced in shallow regions, such as on the Bay's continental shelf regions. RCMs are therefore being employed to analyse how frequently storm surges occur, and to predict how this might change in the future. The surge scenarios are critically dependent on the ability of the RCM to accurately predict cyclone frequency, track and strength. This is no easy task, however, since modelled forecasts of the track positions of actual hurricanes are often well short of the mark (see Box 5.1).

Key ideas

1. The Indian summer monsoon shows considerable annual variation in rainfall amount.

2. Poor monsoons with significant rainfall deficits across the country are strongly linked to warm El Niño events in the Pacific Ocean.

3. Good monsoons with high rainfall totals are strongly associated with cold La Niña events in the Pacific Ocean.

4. Longer-term decadal variations in the monsoon have been identified, including wet cycles occurring in 1880–1900 and 1930–60 and drier episodes in 1901–30 and 1961–90.

5. The driest summer monsoons are associated with severe but localised famine, the wettest with serious and localised flooding.

6. One theory (the orbital-monsoon hypothesis) suggests that long-term (over a 23,000-year cycle) variations in monsoon intensity and coverage are linked to orbital variations of the earth around the sun.

7. Regional circulation models (RCMs) suggest, by the year 2050, higher rainfall amounts in the rice-growing areas of the northern Ganges Valley but more rainfall deficits in the rice-growing deltas of south-east India.

Further reading

Barry, R. G. and Chorley, R. J. (1992) *Atmosphere Weather and Climate*. Routledge, London, 392 pp.

McGregor, G. R. and Nieuwolt, S. (1998) *Tropical Climatology*. Wiley, London, 339 pp.

O'Hare, G. (1997) The Indian Monsoon: Part 1 The wind system. *Geography*, **82**(3): 218–30.

O'Hare, G. (1997) The Indian Monsoon: Part 2 The rains. *Geography*, **82**(4): 335–52.

Pant, G. B. and Kumar, K. R. (1997) *Climates of South Asia*. Wiley, London, 320 pp.

Robinson, J. and Henderson-Sellers, A. (1999) *Contemporary Climatology*. Longman, 317 pp.

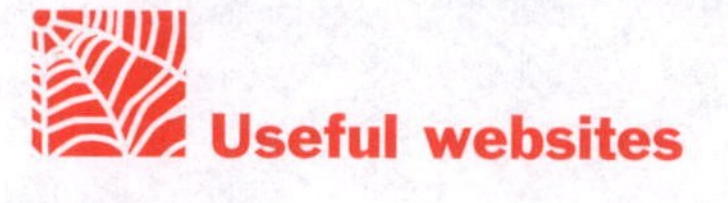

Useful websites

Indian Institute of Tropical Meteorology:
http://www.tropmet.res.in/

Indian ocean sites:
http://indianocean.free.fr/links.htm#monsoon

Monsoon On Line:
http://www.tropmet.res.in/~kolli/MOL/Monsoon/Historical/air.html

NASA:
http://rossby.metr.ou.edu/~spark/AMON/v1_n2/Yang/Yang2.html

NASA:
http://antwrp.gsfc.nasa.gov/apod/ap000303.html

NOAA:
www.pmel.noaa.gov/tao/elnino/impacts.html

NOAA:
http://www.osei.noaa.gov/Events/Dust/

NOAA:
http://soman.tropmet.res.in/paleo/forcing.html#volcanic

CHAPTER 12

Humans and climate change

The subject of weather and climate impacts on society and environment has been a common and regular feature of this book. This chapter continues the theme but defines it further by putting more emphasis on the human adjustment side of the equation. We have already seen that climate change occurs when the climate moves beyond certain previously defined boundaries or limits. These limits may be, for instance, mean temperatures or the number of annual storms defined over a 20–30 year period. How quickly and in what way these boundaries might be crossed have been discussed in section 6.2.1. A theoretical discussion is given in this chapter of how climate change might affect the future frequency and intensity of extreme weather events (climate hazards). Also considered are the main principles and tenets that underline human–climate hazard relationships, including the concepts of human adaptive capacity, vulnerability and risk. A number of detailed case studies are also used from the developing and developed countries alike to illustrate the various types of climate impact and how human societies adjust or adapt to them. Finally, as the well-being of the human species is closely dependent on the world's plants and animals, an analysis is given at the end of the chapter on the possible (harmful) effects of climate change on species survival and extinction.

12.1 Extreme weather and climate change

12.1.1 Extreme weather: how is it predicted?

There is, at the time of writing, a general lack of good reliable data linking global warming to changes in extreme weather events. Because of this, the exact relationship between global warming and extreme weather (storm, flood, drought, extreme heat and cold) has not been fully worked out, either in theory or in practice. Climate scientists actually use fairly simple mathematical and statistical models to predict the intensity and frequency of extreme weather events in the future (Figure 12.1, Box 12.1). It is generally presumed by climate scientists that in a globally warmed world, increased heat will put more energy into the world's weather systems, triggering an increase in the frequency and intensity of extreme climate events. Nevertheless, this hypothesis will be challenged to some extent by a reduction in the differences in rising temperatures between the poles and the tropics, because the coldest poleward parts of the world are heating up the fastest. A best-guess scenario or guessario from the UK Meteorological Office is that with global warming, there will be more storms but they will not necessarily be more violent.

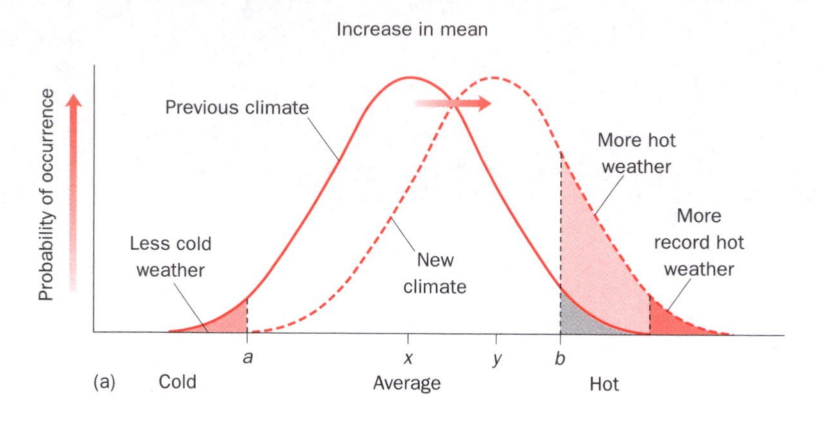

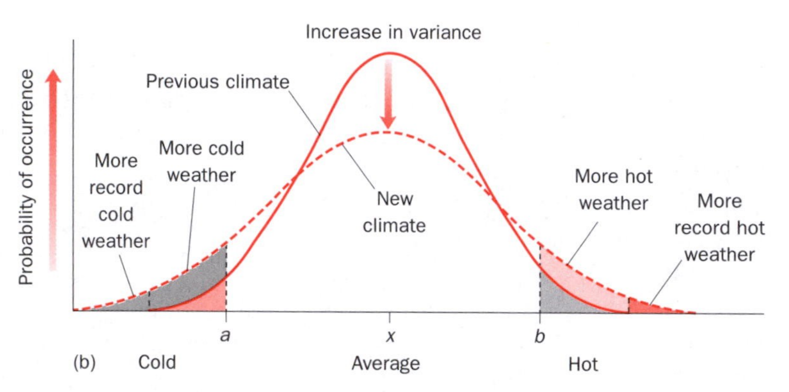

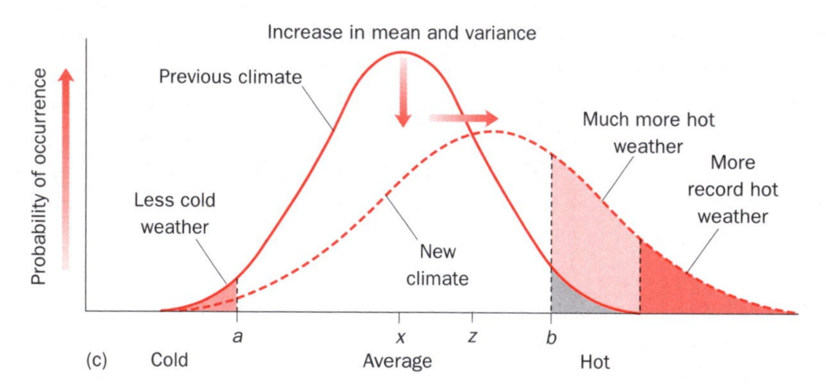

Figure 12.1 Model showing the effect on extreme temperatures when (a) mean temperatures increase, i.e. from *x* to *y*, (b) variance increases around an unchanged mean at *x*, and (c) when mean increases (*x* to *z*) and variance increases for a normal distribution of temperature
Source: McCarthy, J. J. *et al.* (2001, p. 423)

(a) Climate hazards: mean and the extreme values

For most systems and communities, changes in the mean climate condition are more easily adapted to than changes in extremes. Many economic and social systems including agriculture, forestry, settlements, industry, transportation, human health and water resource management have evolved to accommodate some changes in average conditions, particularly if the changes are gradual. This capacity of systems to adjust to general climate change from year to year is known as the coping range. Changes in the mean commonly fall within the coping range. An example of a coping range for a given climate is shown in Figure 12.1 in Box 12.1. Economic and social systems can adjust and manage within the coping range. In contrast, communities are less adaptable, i.e. more vulnerable, to conditions that fall outside the coping range. This means changes in the frequency and/or magnitude of climate extremes, which fall beyond the coping range. The term 'climatic hazard' is often used to capture those climate extremes, in addition to changes in annual averages, to which a system under study is vulnerable.

(b) Increasing climate hazards

Climate-related disasters (e.g. storm, flood, drought) cause more deaths and property loss than other natural hazards (earthquake, volcano). Recent years have shown that while the relative significance of the different types of climate hazard has altered, the global impact of severe climate events taken together is increasing. Flooding is

BOX 12.1

THINKING FURTHER

Predicting extreme weather events

We have already looked at various definitions and models of climate change in this book. Climate scientists use specific mathematical/statistical models in order to represent and understand the implications of climate change. Figure 12.1 shows one way a given climate's temperatures (or rainfall or storms) can be modelled. In the case of temperatures, most temperature recordings (for example daily temperatures) cluster around the mean condition (at *x* in Figure 12.1(a)) with the frequency of recording falling away as the more extreme values are reached. Thus, a typical climate's temperatures (or rainfall or storms) show a normal distribution or bell-shaped curve around the mean value. Climate scientists use these types of statistical representation in order to forecast the way in which the frequency and intensity of extreme weather events (e.g. hot episodes, wet periods, storms) will alter with a change in the mean weather condition (e.g. temperature, rainfall, wind speed respectively). It can be seen in Figure 12.1(a) as the mean temperature increases from *x* to *y*, the probability of extreme weather events, i.e. hot and record hot episodes, increases markedly, while the frequency of cold weather events decreases. This particular relationship is revealed because the variance, or in this case, the normal distribution or bell-shaped curve of temperature occurrences around the mean *has been kept constant*.

Using this model, the probability of observing future temperature conditions, in any region, can be worked out. For instance, the current (1961–90) mean temperature of central England is 15.3 °C. Under a rise of mean temperature of 1.6 °C by the 2050s (a scenario predicted by climate scientists), the probability of observing very warm temperatures similar to that of 1995 increases from 13 per thousand to 333 per thousand. Under the same scenario, a very mild winter of the type observed in 1988–89 would recur on average every 5 years, compared to three or four times a century at present. By the 2050s decade the chance of a very cold winter in England such as that in 1947 and 1963 is extremely low.

It needs to be borne in mind, however, that this model depicts only one possible relationship in a globally warmed world between the mean climate condition and extreme events. Another situation can be envisaged (Figure 12.1(b)) where the mean condition is maintained while the amount of variance around it is increased, so that the frequency of both hot/record hot and cold/record cold events is increased. It is quite likely that in a globally warmed world, both the mean and the variance about that mean will increase as shown in Figure 12.1(c), so that the number of hot and record hot episodes will increase dramatically. Another possibility (not shown) is that a change in mean will be accompanied by a complex change in variance, resulting in the bell-shaped curve of distribution becoming skewed or asymmetrical around the mean. If the skew is to the right of the new mean in Figure 12.1(c), then the frequency of extreme events (hot/record hot) could rise even further.

rapidly becoming one of the most prevalent natural disasters globally. Until the early 1980s, more people died as a result of drought than flood. From the early 1980s, floods overtook droughts as a global hazard. Around 140 million people globally are affected by floods, and 40 million people are affected directly by drought.

Real economic costs from weather-related global natural hazards, particularly in the more developed world, have spiralled in recent years rising from less

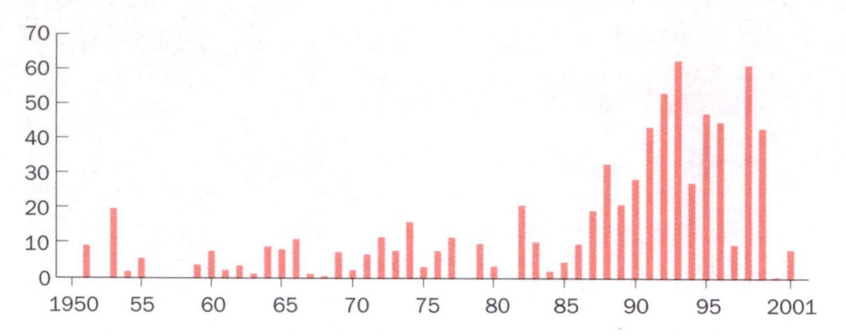

Figure 12.2 Financial costs of weather-related natural disasters, 1950–2001. Costs normalised for inflation in $ billions (2001 prices)
Source: *The Economist*, 6 July 2002, p. 16. Munich Re

than $5–10 billion in the 1950s to more than $50 billion in the 1990s (Figure 12.2). These economic costs, allied to unacceptably high human losses in the less developed countries (1,000 deaths per hazard in 1947–67 increasing to over 2,000 deaths per hazard in 1969–89), have alerted us to the growing vulnerability of communities and regions to the threat of environmental hazards. The increasing impact and loss from climate disasters are a response to three factors (see section 9.3.1). First, there is a realisation that in a globally warmed world, the frequency and intensity of climate extremes (storm, flood, drought) may be increasing, although good statistical evidence for this is still not available. Second, in recent years, increasing populations, growing poverty, poor land-use planning and environmental disturbance, including deforestation, have conspired to exacerbate the effects of the climatic hazard the world over. Third, there is increasing realisation that current mainstream approaches to hazard management have rarely been successful. As a result, there is a real need to seek new approaches to hazard mitigation and to programmes of sustainable development.

12.1.2 To cope or not to cope

(a) Adaptive capacity

The adaptive capacity of a system refers to the potential or capability of that system to adapt to, and thus cope with, the effects of environmental change. In relation to climate change impacts, the adaptive capacity of a system depends on

1. the severity of the climate change in relation to the mean and the extremes;
2. the vulnerability of the system: this is a reflection of the inherent ability of the system to cope with the climate hazard; and
3. the technical resources available to raise the system's ability to cope with the effects of the climate hazard.

(b) Vulnerable systems

The coping range or vulnerability of a system depends on the inherent ability of the system to respond to and recover from climate change. Vulnerability is usefully defined as the characteristics of a person or group in terms of their capacity to anticipate, cope with, resist and recover from the impact of a natural hazard. Countries and communities in the developing world with limited economic resources, low levels of technology, poor information and skills, minimal infrastructure, unstable or weak institutions, low levels of empowerment and access to resources have little capacity to adapt and are highly vulnerable to climate change. Specific sets of vulnerable people based on livelihoods include the following: (i) rural smallholder agriculturalists with limited land and labour resources, especially when farming marginal lands; (ii) pastoralists in the semi-arid and subhumid areas; (iii) rural wage labourers especially in remote or marginal agricultural lands; (iv) urban poor and informal sector workers; (v) refugees and displaced people and (vi) destitute groups.

In addition to livelihood groups, individuals may be particularly vulnerable because of their

Table 12.1 Vulnerable people and groups in society

- Low-caste communities
- Ethnic minorities
- Women, especially those who may be widowed or deserted
- Aged men and women
- Children, particularly female
- The disabled
- People dependent on low or daily incomes
- Those in debt
- Those most isolated from infrastructures, e.g. transport, health services

Source: Dina Abbot, University of Derby.

social status and access to food and resources. As shown in Table 12.1, they include the very old and young, the disabled, marginalised groups and those living in extreme poverty. These individuals have low adaptive capacity because they are unable to prepare in advance for, and protect themselves from, an adverse climate event, and to respond or cope with its effects. Such vulnerable people intersecting with livelihood groups, form populations at the highest risk to climate change and other natural and human hazards.

(c) Risk

Vulnerable groups are said to be at **risk** from climate extremes because definitions of risk involve concepts of vulnerability. Risk is defined as the probability of occurrence of a climate hazard multiplied by the vulnerability (expected losses) of the target group. Thus, people are more at risk when the likelihood or frequency of the hazard increases and/or the vulnerability of the target group increases. Many sectors and groups that are vulnerable to climate change are also under pressure from other forces such as population growth and resource depletion. In vulnerable communities exposed to hazardous climate events, coping abilities actually decline and risks increase with increases in poverty. Risk may also be said to increase when, for instance, there is an increase in population or settlement within the hazardous zone (vulnerability and expected losses increase) or when people migrate into a hazardous region (here the probability of the hazard increases for the group).

12.1.3 Strategies in raising adaptive capacities

Activities that seek to raise the adaptive capacity of communities and regions to climate change are equivalent to those promoting sustainable development. Sustainable development for the poorer groups in society can be defined as the wise or optimal use of local resources by a community so that its long-term standard of living can be raised and sustained. Sustainable development and enhancement of adaptive capacity is a necessary condition in reducing vulnerability, particularly for the most sensitive regions, nations and socio-economic groups. Vulnerability can be reduced and the coping range expanded with new or modified adaptations within a system, for example improvements in socio-economic and technical provision including better infrastructure, as well as local community action and regional institutions to deal with a crisis. Two very different perspectives, however, concerned with enhancing the coping abilities of regions and communities to environmental hazards exist.

(a) The traditional viewpoint: technological solutions

This approach emphasises a homeostatic or self-regulating balance between environmental hazards and society. It infers that an acceptable balance is eventually achieved between environmental disturbance and society since people and institutions are uniformly and unambiguously committed to removing known manageable risks from life, and only fail to do so when the hazard is highly uncertain. This perspective emphasises a 'top-down' approach to the disaster problem, and sees solutions (since the Industrial Revolution and more particularly since the 1950s) through the application of physical measuring and monitoring techniques, and the use of structural management programmes involving large engineering works and architectural design.

Such dominant technocratic/techno-fix approaches have been transferred to developing countries, often through aid/development schemes. However, from the late 1970s, they have come in for increasing criticism for being inappropriate (environmentally, socially, economically) in many development contexts. They have also been criticised for reinforcing the dependency or reliance of recipient (less developed countries) upon donor (more developed countries), leading to the development of further poverty and underdevelopment. One of the main reasons why externally imposed approaches to disaster management tend not to work is because nature and disasters are social constructions, in the sense that meanings given to them are different in different societies. People and communities develop their own view of the world and of nature and their response to disaster is governed by such personal views of reality. Top-down or external influences can override these feelings and can escalate vulnerability to loss, perpetuating and even exacerbating the disaster issue. Western governments are now reassessing their approach to disaster management not least because of the escalating costs of traditional technological approaches to mitigation.

(b) Modern views: self-empowerment and community action

The second perspective, increasingly promoted from the mid 1980s, envisages a more open-ended positive feedback process between society and hazard where the cumulative process of rising populations and settlement expansion (especially in poor societies) inevitably leads to greater vulnerability and growing potential for disaster. This more people-centred or 'bottom-up' perspective calls for greater community awareness and participation in vulnerability reduction. It claims that vulnerability and impacts should be considered more in terms of what people do, and greater consideration should be given to the effects of disasters on social and community groups. These changes in perspective mirror the general trend in overseas aid/development programmes towards empowering people to help themselves – building on the ideas of Schumacher (appropriate technology, self-reliance) in the late 1960s and early 1970s. In the face of global environmental change, societies are thus increasingly being encouraged to learn to live with nature rather than fight against it with ever bigger technology, necessitating a greater community awareness of, and involvement in, environmental processes and their mitigation. With this view, non-structural solutions to disasters are recommended, including careful land-use planning, risk assessment and management, and where possible, support and insurance schemes.

Key ideas

1. Climate scientists use mathematical or statistical models to predict the frequency and magnitude of extreme weather events under future climate change.

2. The validity of models that assume no change in climate variance with increases in the mean value are highly uncertain in their prediction of future extremes.

3. A climate hazard is a climate event whose incidence/frequency or magnitude or persistence is such that it falls beyond a system's (region or community) ability to cope with it.

4. The poorest and most marginalised groups in society, particularly in the developing world, have the lowest coping abilities or adaptive capacities in face of extreme climate hazards, and thus remain the most vulnerable to them.

5. Risk is defined as the probability of occurrence of a climate hazard multiplied by the vulnerability (expected losses) of the target group.

6. Societies in the developing world are becoming more vulnerable to climate hazards in response to (i) possible increases in hazard frequency and magnitude under climate change, (ii) growing and more impoverished populations and (iii) inadequate/inappropriate hazard management policies.

7. Large technically based and externally imposed 'top-down' hazard management schemes tend to alienate local communities and rarely offer a solution to the hazard problem.

8. 'Bottom-up' local community schemes which help communities to help themselves (i.e. those based on self-empowerment) are often seen as more appropriate and successful in face of the climate hazard.

12.2 Hurricanes and human survival

12.2.1 Hurricane impacts: rich and poor countries

The amount of damage from any given hurricane will be influenced by its severity and by the number of people living in the vicinity of its track. Equally important are living standards of the affected area. For instance, there will be much greater potential for financial loss if a hurricane passes Miami, Florida, where real estate values are high, compared with one which crosses Havana, Cuba, or Bangladesh where property values are much less. The most destructive storms in the USA (Hurricane Andrew) can result in property damage of \$30–\$40bn, i.e. at least 5–10 times that of the largest storms to affect countries in the less developed world, for example Hurricane Mitch in Central America (\$5bn) and the great storms of Bangladesh. General real estate losses caused by hurricanes in the USA have increased over time (Figure 12.3(a)). This is a reflection of the greater affluence and the rising cost of real estate as well as the increase in numbers of people, buildings and associated infrastructures in the affected coastal areas of the Gulf of Mexico. In Florida alone, population grew by 37 per cent to 10.5 million in the period from 1980 to 1993. Figure 12.3(b) shows the trend in costs attributable to the 30 most devastating and costliest hurricanes when conditions are normalised to the present time in terms of inflation, wealth and population. This

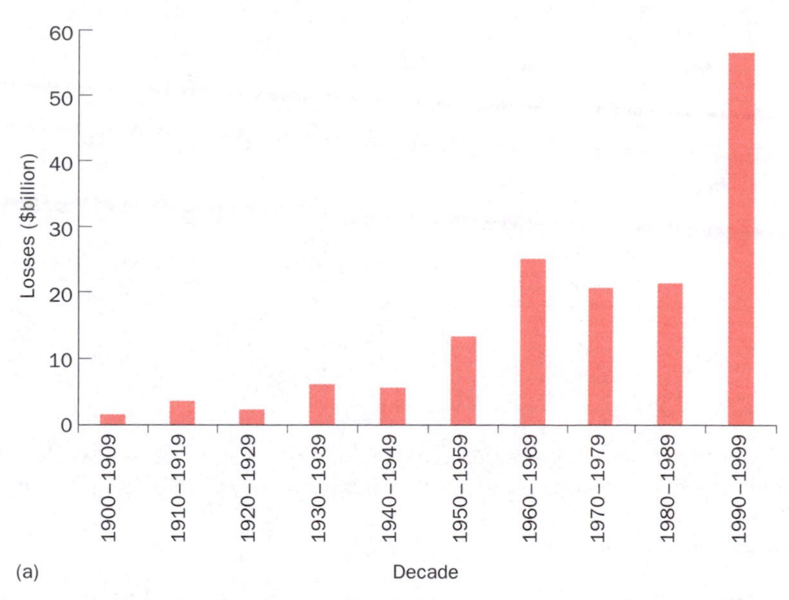

Figure 12.3 Hurricane impacts in the USA: (a) US property losses by decade, 1900–99 in \$bn (real \$ prices for year 2000)

continued

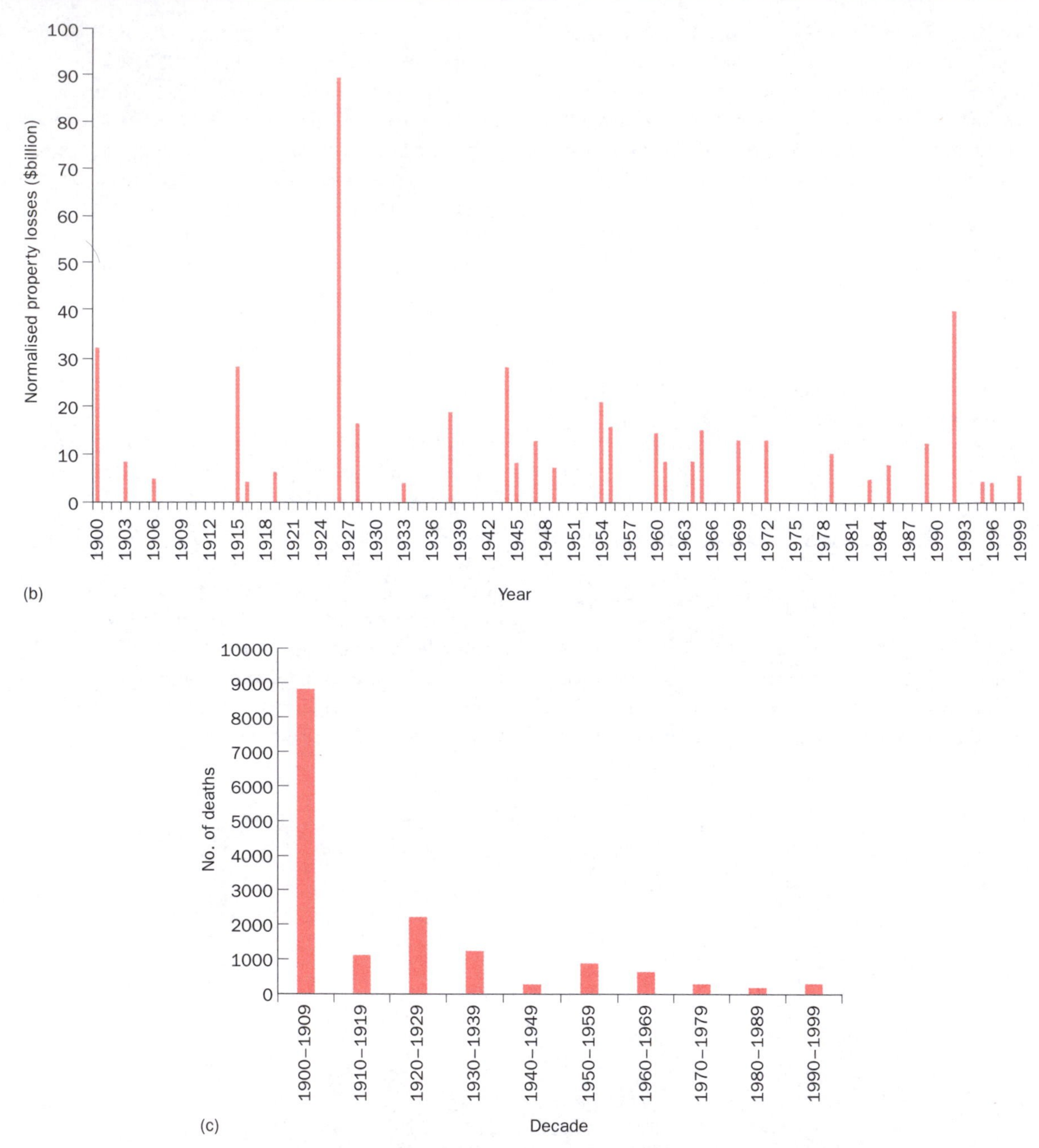

Figure 12.3 (*cont'd*) Hurricane impacts in the USA: (b) annual property losses, 1900–99 in $bn (costs shown are normalised for inflation and rising population and wealth (real $ prices for year 2000); (c) US hurricane deaths by decade, 1900–99

Source for (b) Pielke and Landsea (1998)

method shows that the hurricane of 1926, for instance, if it happened today with present population and wealth distribution, would have cost almost $90 billion. The normalised costs show that in common with hurricane frequency in the Caribbean (see Figure 5.4(b)) general financial losses due to hurricanes have *not* increased over time. Normalised costs were greater from the mid 1940s to 1970 when the frequency of the most intense and destructive storms (category 3–5) was relatively high, while normalised costs were lower during the relatively quiet years from the 1970s to the early 1990s. There has been an increase, however, in the most intense tropical storms since 1995 to the present. Some scientists believe that there may be a decadal cycle of the most intense hurricanes so that their frequency may increase in the next several decades. If this happens, massive real estate losses will be incurred in the Caribbean and south-east USA.

While the real costs of hurricane damage (when not corrected for population and individual wealth) in the USA may be said to have increased over time, the number of hurricane deaths has decreased (Figure 12.3(c)). This is a direct reflection of a decrease in the number of the most severe hurricanes between 1970 and 1995 (see Figure 5.4(c)), and better forecasting of possible hurricane track locations (see Box 5.1). With improved communications such forecasts can be converted into broadcast warnings and the affected population evacuated. In contrast, hurricane death tolls in the less developed countries can be very high (Table 12.2). Bangladesh has experienced some of most intense tropical cyclones on record. Noteworthy in this respect are the Bangladesh cyclones of 13 November 1970, which killed or drowned almost 500,000 people and that of 29 April 1991 that caused 139,000 deaths. Recent death counts (11,000) from the largest tropical storms in the Caribbean such as Hurricane Mitch (October 1998) in Guatemala and Honduras are much less than those in Bangladesh, but are nevertheless significant. High death rates in the less developed countries are a result of hurricanes impacting over coastal regions with poor and rapidly growing rural populations. Although storm tracks may be predicted, the facilities for the dissemination of early warnings are often much poorer than in the developed world. Even when

Table 12.2 Number of deaths caused by the intense hurricanes in the Caribbean and Bay of Bengal, India

Hurricane	Date	Impact zone	Deaths
Caribbean			
The Great Hurricane	10–16 Oct. 1780	Martinique, Barbados	22,000
Martinique	6 Sept. 1776	Point Petre Bay	>6,000
Galveston	8 Sept. 1900	Galveston Island	8,000
Dominican Republic	1–6 Sept. 1930	Dominican Republic	8,000
Hurricane Flora	30 Sept.–8 Oct. 1963	Haiti, Cuba	7,200
Hurricane Fifi	14–19 Sept. 1974	Honduras	8,000
Hurricane Mitch	26 Oct.–4 Nov. 1998	Honduras, Nicaragua	11,000+
Bay of Bengal			
Andhra Pradesh	Oct. 1679	Andhra Pradesh	20,000
Bangladesh	7 Oct. 1737	Bangladesh	300,000
Bangladesh	13 Nov. 1970	Bangladesh	500,000
Andhra Pradesh	26 Nov. 1977	Andhra Pradesh	>10,000
Bangladesh	29 April 1991	West Bengal	140,000

Source: Various

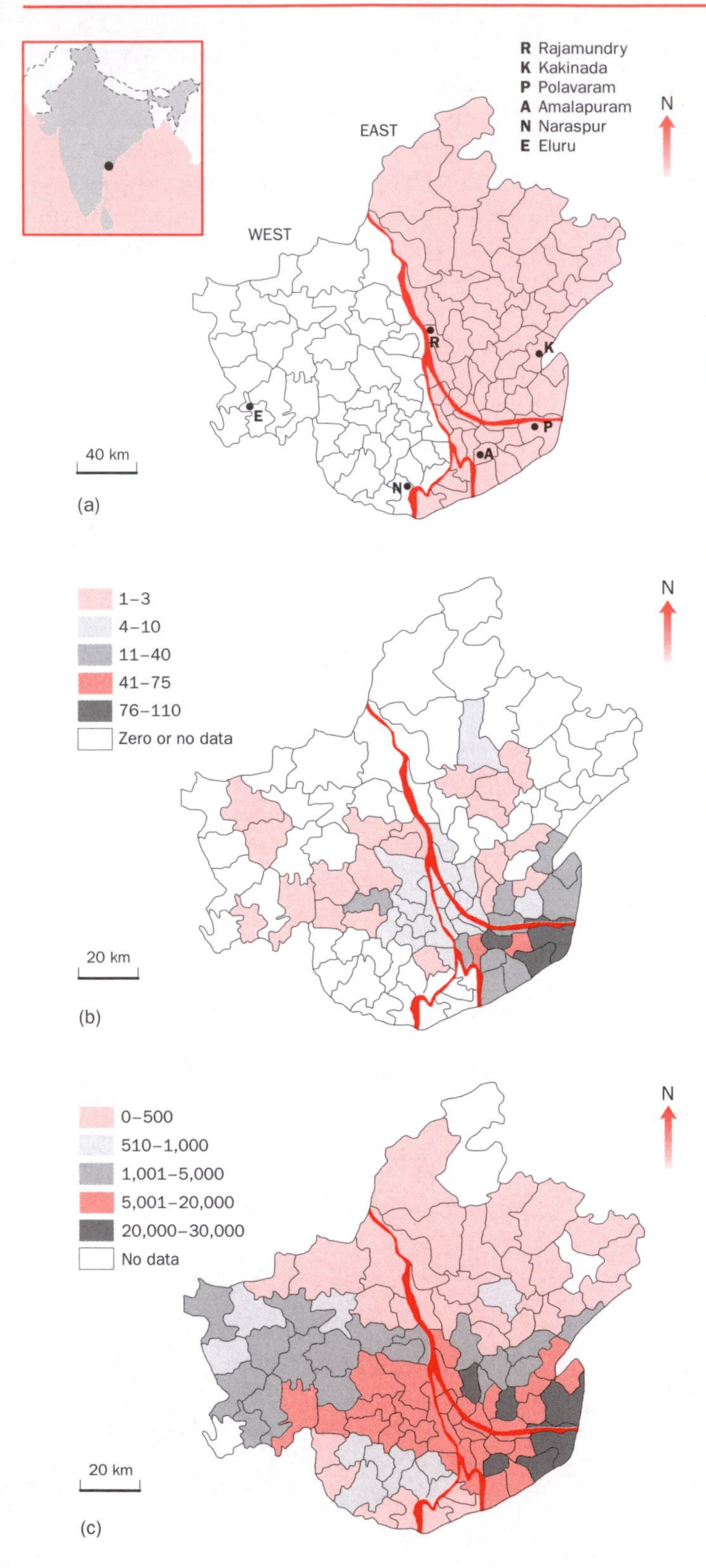

storm warnings are given many people do not have the facilities (television, telephones, telexes) to receive them. Evacuation is also difficult in the time available because of the lack of good all-weather roads and other transport infrastructure.

12.2.2 Hurricane 07B in the Bay of Bengal

A detailed analysis of a hurricane in the Bay of Bengal making landfall over eastern India is now given. The investigation explores the impact of, and the human response to, Hurricane 07B which made landfall over the Godavari delta region of central Andhra Pradesh (Figure 12.4(a)) during 6/7 November 1996. This category 4 hurricane (see Plate 12.1) with core wind speeds of up to 175–220 km/hr (about 110–135 mph) had a severe effect on the delta region, being responsible for over 1,000 deaths and incurring heavy economic losses.

(a) The Godavari delta

The Godavari delta is centrally located on the eastern coast of India (Figure 12.4(a)). It is a fertile agricultural area with extensive paddy fields, coconut gardens, banana groves, sugar-cane fields and horticultural crops. The region has a population of 10 million growing at a fairly high rate of 2 per cent per annum.

Figure 12.4 (a) The Godavari delta, Andhra Pradesh, eastern India, showing East and West Godavari and the distribution of administrative units across the region. Damage from Hurricane 07b, 6/7 November 1996; (b) number of confirmed deaths by district; and (c) number of seriously damaged houses by district
Source: O'Hare (2001, p. 32)

Most people live in villages and small towns in the southern half of the delta close to the coast where population densities can reach over 1,250 persons/km. The population is mostly rural with more than 67 per cent of the active population engaged in agriculture or agricultural processing.

The hurricane developed from a low-pressure disturbance which originated in the Gulf of Thailand and moved westwards across the Malay peninsula into the Bay of Bengal, where it rapidly intensified to hurricane strength (wind speeds >34 m/s). It then moved quickly and crossed the Godavari delta region about 50 km south of Kakinada at about 5 p.m. on 6 November. The system then raged for 6 hours over the southern central part of the delta before dissipating into a severe cyclonic storm by midnight of 6 November and weakening into a deep cyclonic depression (wind speeds less than 17 m/s) by 3.00 a.m. on 7 November some 150–200 miles inland. During the hurricane event, most of the southern half of East and West Godavari districts received over 100 mm of rainfall and the area around Kakinada over 200 mm of rainfall.

(b) Spatial incidence in hurricane damage

(i) Death tolls by storm surge

Most serious impact of all was the high death toll with more than 1,000 killed. Most of the dead were recorded from the 40 or so low-lying coastal villages in East Godavari. The devastating impact of the storm surge (2–2.5 m high) which swept 5–6 km inland, can be clearly identified in the high concentration of deaths in the coastal mandals (village districts), especially those immediately south of the mouth of the main channel where 07B made landfall (Figure 12.4(b)).

(ii) House damage and high winds

Across the delta, around 314,000 houses were seriously damaged and a further 332,000 partially damaged by severe wind speeds. When the number of damaged houses per mandal is mapped, a very dramatic imprint of the storm's devastation can be found (Figure 12.4(c)) especially when the destruction is normalised by dividing the damaged house number of each mandal by the population of each mandal. Figure 12.4(c) shows that most house destruction occurred within a narrow 40–50 km wide zone across the delta, clearly indicating the west–north-west trajectory of the storm. The progressive dissipation of the storm over land from hurricane (>34 m/s) to tropical cyclonic storm (17–33 m/s) is also evident in the stepwise landward reduction in house damage inwards across the delta.

(iii) Agricultural damage and flooding

Despite widespread damage to housing with reconstruction costs calculated at around $150 million, agricultural losses were far greater at almost $750 million. Hurricane 07B came at the most inappropriate time when paddy rice, the main crop of the region, was ready for harvest. Around one-third of a million hectares were severely damaged due to 3–4 days' flooding and submersion in the southern part of the delta. The loss in paddy yield for the region was estimated at around 0.5 million tonnes and just over $500 million. The destruction of other commercial crops including bananas, coffee, vegetables and other fruits brought the total loss under the agricultural sector to around $750 million. Agricultural losses were more widespread than deaths and house damage occurring across the whole of the southern part of East and West Godavari as a result of widespread flooding.

12.2.3 People at risk in the delta

(a) Coping strategies: rich farmers and female labourers

A small number of landowning farmers in the delta were bankrupted by the severe agricultural losses they suffered. However, the great majority were able to rely on savings and other resources to tide them over to the next harvest. Much more affected were the rural poor, especially landless agricultural

Table 12.3 Coping strategies adopted by poor people and groups (in India) when affected by disaster

(i)	Diversifying sources of income, including seasonal migration
(ii)	Drawing upon communal resources
(iii)	Drawing on social relationships, including patronage, kinship, friendship and informal credit networks
(iv)	Drawing upon household stores, food, fuel, etc. and adjusting current consumption patterns
(v)	Drawing upon assets

Source: Agarwal (1990)

labourers reliant on a meagre daily wage. One of the most vulnerable groups in this sector were migrant, scheduled (low) caste women from the state of Orissa who perform most of the agricultural work in the rice fields of East and West Godavari (Plate 12.2). Such migrants do not have a permanent home or stable social environment and are often found living in flimsy shanty dwellings on the edge of villages and towns. Dependent on a meagre daily wage, such female migrants are often unable to build up food reserves or other assets with which to survive in times of disaster. The destitution wrought on these women after the destruction of the rice crop was revealed in the range of mostly individual strategies they had to adopt in order to survive the disaster (Table 12.3). Thus, while some scheduled caste women could rely on handouts of food from considerate landlords, most had to subsist by other means, by begging, or by selling what possessions (mostly jewellery) they had. Others entered domestic service in the towns and villages where they could find such work, while others migrated temporarily either to neighbouring agricultural districts in southern Andhra Pradesh or even back to their home village. The plight of these migrant women worsened even further if they were on their own (many were unmarried or with their husband gone or working elsewhere) and with young children. It can be seen that in relation to the five main coping strategies shown in Table 12.3, the migrant female labourers of the delta tended to adopt non-collective, individually based survival methods, i.e. (i) diversifying sources of income, including (seasonal) migration, and (v) drawing on assets where possible.

(b) Fishing communities

The social group most seriously affected by the hurricane were the fishermen and their families. We have already seen that the greatest loss of life attributed to the hurricane was overwhelmingly concentrated in the low-lying fishing villages along the coast of East Godavari (Figure 12.4(b)). In addition, many livelihoods were seriously damaged in the coastal villages as many fishing boats were destroyed or lost in the storm together with a good deal of ancillary fishing tackle.

Strategies adopted by fishermen and their families in the face of disaster were similarly fatalistic, but more collective and less diverse geographically/economically, than those taken up by farmers or the female migrant rural labourers. This is almost certainly a reflection of the strong sense of social cohesion and community spirit found in the coastal villages, where fishing communities are regarded as being the most fiercely independent and closely knit in the whole delta. For instance, despite considerable losses, most people did not move but remained in the same fishing village as before and tried to recover their lives. This is also related to the fact that fishing activity though greatly reduced in the coastal villages did not cease altogether in the aftermath of the hurricane. Some fishermen who lost almost everything, simply carried on their trade by going to work for other fishermen who were more fortunate in terms of disaster losses. Many of the worst affected victims of the hurricane relied on highly developed community support strategies with families and individuals helping others with house rebuilding and food sharing. Thus, it is clear in Table 12.3 that the fishing communities tended to adopt Agarwal's (1990) more collective strategies, i.e. (ii) drawing on communal resources, i.e. ocean fish stocks,

(iii) drawing on social relations, and (iv) drawing on (where possible) surviving household stores while also adjusting consumption patterns.

12.2.4 Mitigation strategies

Three largely 'top-down' institutional (i.e. not community-based) approaches have been employed to lessen the destructive impact of the tropical cyclone, especially along the coast of eastern India.

(a) Physical approaches

Physical methods comprise the construction of stronger and more fortified defences and embankments, and surge-proof buildings capable of withstanding the hurricane force winds and storm floods. Hundreds of hurricane shelters built in high cyclone/flood risk locations along the coast of eastern India are an example of this technical approach. Hurricane shelters (many of them built over 50 years ago) can be found in most villages along the east coast of India (Plate 12.3). When local communities want to take advantage of the hurricane shelters, they are often too small to house all the people who want to use them. In other areas, residents do not always take refuge in them, partly because they are not provisioned with food, water and other services, and partly because they are afraid that their dilapidated disused structures might collapse. With many isolated ethnic and cultural groups along the coast, they may not be used simply because such externally introduced technically based mitigation measures are viewed with some suspicion.

(b) Early warning and evacuation procedures

Human avoidance systems represent another approach and include the installation of satellite-based cyclone forecasting and evacuation procedures. The implementation of INSAT-based Cyclone Warning Dissemination Systems (CWDS) at several main coastal towns in the delta in 1978 has generally been credited with reducing the death toll from major hurricanes along the Andhra coast. Nevertheless, a number of criticisms have been levelled at storm warnings given out by the Indian Meteorological Department in Andhra Pradesh, especially the inappropriateness of applying advanced technical solutions in simple developing societies. Warning bulletins given up to 24 hours ahead of the expected time of the cyclone's arrival by phone, teleprinter, telex and TV are beyond the reach of the poor fishermen in the southern part of the delta. So too are the hourly warnings given by coastal radio stations at the onset of the storm. It has been found that only 30 per cent of the delta's fishermen carry radios, and many of these are either tuned to a music station, not working or are too weak to receive the warning signals beyond about 4 km distance from the shore. Reception of messages over the radio is usually disturbed in stormy weather; other complaints are that the messages are read too fast and need to be given in a language closer to the one spoken locally.

It is also a problem removing poor people from an area because of poor communications and transport systems. Nearly half the roads in the Godavari delta region are not all-weather surfaced, and as many as 20 per cent of villages are not connected by road. Such poor communications also of course slow down central government, state and non-government aid getting to villages. Nevertheless, much of the loss of human life from cyclone 07B could have been avoided if timely efforts had been made by the state government and the district authorities to evacuate people from the low-lying coastal villages, and at least to attempt to restrain fishermen from venturing out to sea.

Social structures (or the Indian government's reluctance to understand them) in the coastal villages and fishing coves have also been identified as a problem for early warning cyclone systems where the village leader is often the only person the community pays heed to. It is not surprising that 80 per cent of educationists, farmers, fishermen and other town traders in the southern delta region have voiced the need for educating people more about cyclones and their potential impact. Most people of the region were actually unaware that early warnings were issued by the Indian Meteorological Department.

(c) Government rehabilitation policies

With respect to state-organised disaster mitigation, some basic relief was provided to the inhabitants of the delta with the provision of 25 kg rice and 3 litres of kerosene to each affected family. However, with poor roads and other communications, rice donations took from several days to as much as 2 weeks to arrive in some remote coastal fishing villages. According to many people in the coastal villages, most other elements of the initial government rehabilitation programme have not been met or only partially realised, including new house construction and promises of cash relief to families damaged by the storm.

Key ideas

1. Many communities remain vulnerable to tropical hurricanes, an example of an extreme climate hazard.

2. In general, while high real estate losses are found in the developed countries as a result of hurricanes, great loss of life occurs in the developing countries.

3. Most deaths attributed to hurricanes are the result of the storm surge while other property losses occur from high wind speeds (e.g. houses) and flooding (e.g. agricultural losses).

4. People most at risk from hurricane damage, including the poor and the old, are unable both to prepare in advance of the hurricane event, and to recover from its devastating effects.

5. Nevertheless, vulnerable people affected by hurricane impact develop individual and collective coping strategies that reflect their personal, social and economic status.

6. Many institutional programmes in the developing world designed to mitigate the effects of hurricanes including (1) the government construction of physical shelters, (2) early warning and evacuation procedures and (3) basic rehabilitation measures are insufficient to cope with the destruction and hardship caused by hurricanes.

12.3 Drought and human survival

12.3.1 Adjustment to drought

Drought is a complex, multifaceted concept that is difficult to define precisely (see section 10.4.2). It implies, however, an absence of precipitation (climate drought) for a period long enough for rivers, lakes, etc. to run low (hydrological drought) and for serious moisture deficits to develop in the soil (soil/agricultural drought). Soil moisture deficits are a result of water loss by evapotranspiration, groundwater and stream flow. Soil drought, however, is not synonymous with soil moisture deficiency although the two concepts are related. During the dry season for instance in the wet and dry tropics, there may be severe soil moisture deficiency, but a state of drought is not perceived by the inhabitants (see Box 11.1). This is because the dry season is a regular event that does not disrupt normal biological and human activity. In other words, the dry season is expected (in the winter in the case of the tropical wet and dry climate) and has been adapted to.

In contrast, drought is related to an *unexpected* failure of the rains at a particular time (during the summer in the tropics or in the UK for instance), since most activities using water will be geared to that which is normally available. In the tropical semi-arid regions, the risk of drought is greatest when a succession of good years with above average rainfall is followed by a long drought. The years of favourable rain encourage the inhabitants of the area to expand the cultivated zone, to build up livestock numbers and to increase their own population. When the rains fail during the ensuing

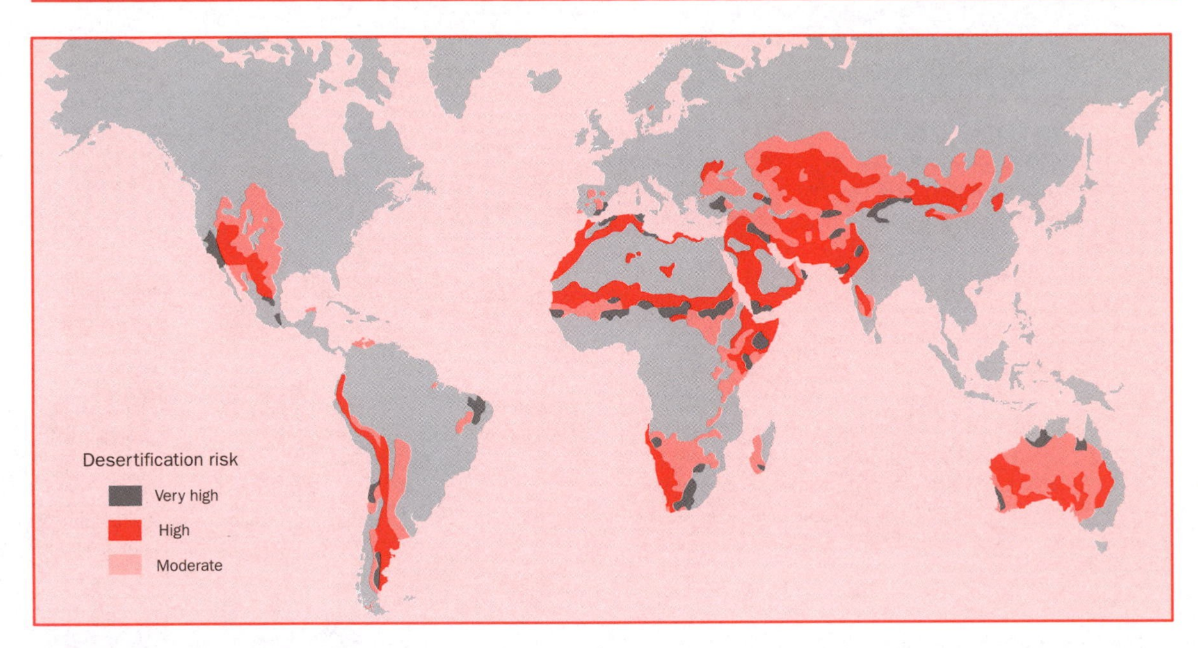

Figure 12.5 Desertification risk in the semi-arid regions of the world
Source: Thomas and Middleton (1994)

drought, the carrying capacity of the land to support such activity is quickly exceeded. Soils and vegetation, already under stress from lack of rainfall, are impoverished further by overgrazing and compaction by livestock, woodland clearance for firewood and overcultivation. Thus, according to this model, both climate drought and human exploitation of a region's resources can lead to large areas of arable and grazing land being ruined and converted to desert each year.

Desertification is the term given to this process of desert creation by drought and human exploitation. Figure 12.5 shows that desertification is a global phenomenon, with extensive distributions not only in the semi-arid and subhumid regions of the tropical world but also in many extratropical semi-arid regions, for example central Asia and southern South America. Because of widespread poverty, and soil/resource exploitation in the semi-arid and subhumid regions, the prevention and reversal of desertification are not an easy option. Nevertheless, when conditions allow, there are a number of methods and procedures that can help reverse the process. These include good land-use planning and management, for example cultivation when there is adequate rainfall, lowered livestock numbers in keeping with the carrying capacity of the land in the driest years and maintenance of woodland where possible. In addition to these techniques are a number of social, cultural and economic controls including economic development, population planning, education, reduction of grazing animal herd size and population resettlement.

(a) Famine and food security

Soil and hydrological drought are not synonymous with the condition of human famine although the concepts again are often connected. Famine is usually defined by a prolonged food deficit that eventually leads to widespread and substantial increases in morbidity and mortality. Drought can sometimes lead to famine if certain *other* factors exacerbate the event. These factors may include a combination of the circumstances shown in

Table 12.4 Factors that can exacerbate the effects of drought and other climate hazards leading to famine

1. There are prolonged rainfall deficits
2. Farmers cannot save their crops or plant again if their crops have died in a drought
3. Families have no food left over from last season
4. Families have no money to buy local food which is often expensive
5. Markets have no food to sell
6. Local food-growing areas are used for the export of cash crops
7. Local and national governments are unable or unwilling to redistribute existing food surpluses and bring in food supplies
8. War, conflict and political corruption interferes with people's ability to grow or buy food

Source: Various

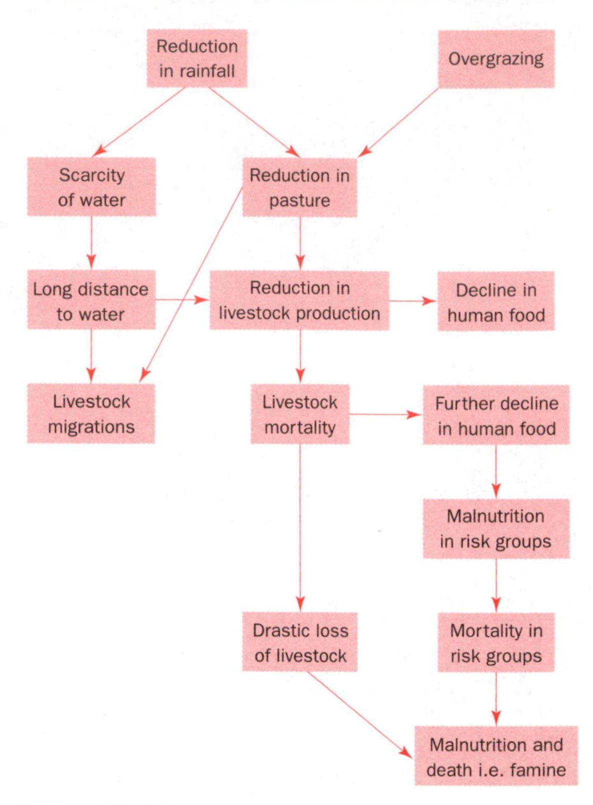

Figure 12.6 Model showing how a combination of drought and overgrazing can lead to famine within pastoral groups in the tropical semi-arid and subhumid regions

Table 12.4. Drought may not necessarily be a precondition for famine, however. Many of the great famines that have affected different regions of India at different times in the past bear little connection with periods of serious drought. Food supplies were often available but for political reasons were never distributed to the famine areas or the victims. Most if not all famines therefore result from a sequence of human-induced processes and events that reduces food availability and food entitlements.

Food security indicates an assured level in the availability of food, while hunger and famine refer to the effects of food insecurity, i.e. to the poor availability of the food supply. Chronic or long-lasting food insecurity leads to a high degree of vulnerability to hunger and famine. Vulnerable populations can reach a stage of famine with slight abnormalities in the food production–consumption system. Figure 12.6 shows how a combination of drought and overgrazing can lead to a state of famine in vulnerable pastoralist groups in the tropical arid region. Chronic food deficits have become the most visible symbol of inadequate agricultural performance and falling food production per capita in a large and growing number of African countries. Nearly half of sub-Saharan Africa's 45 independent states face serious food emergencies.

12.3.2 Drought in the Sahel

Drought, food security and famine have also been a feature of the Sahel region of Africa from the mid 1960s. The Sahel region occupies a narrow band of West Africa (see Figure 11.3) extending from Senegal on the west coast right across to Mali and Niger. On average, the Sahel receives during a short but active wet season from late June to September about 350 mm of rainfall annually. This amount varies from about 200 mm (e.g. at Menaka) on its northern edge rising to about 550 mm on its southern margins (north of Sokoto). The region

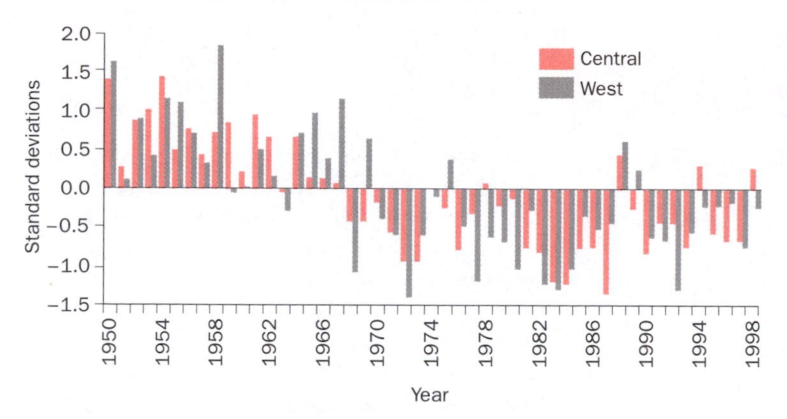

Figure 12.7 Rainfall fluctuations in the Sahel, 1901–98, expressed as regionally averaged standard deviation (long-term mean divided by the standard deviation)
Source: R. Washington, Keble College, Oxford (pers.comm.)

thus has a semi-arid climate with short grassland, and contrasts with the desert of the Sahara to the north, and the more humid wet and dry climates with their savanna grasslands to the south.

As a semi-arid region, the Sahel experiences high interannual variation in rainfall. Figure 12.7 shows, as expected, a wide percentage departure in annual rainfall from the 1901–98 mean. The region enjoyed a notable wet decade in the 1950s, leading farmers to expand the cultivated area and to build up livestock numbers. This wet period was, however, followed by long-term rainfall deficiency and serious drought from the mid 1960s to the end of the 1980s. The factors responsible for such a sustained decline in rainfall have long been debated, involving both human and natural processes (Box 12.2). As suggested in the previous desertification model, the carrying capacity of the land to support the wider areas of cultivation and greater livestock numbers was quickly exceeded, leading to increases in soil nutrient mining, crop failures and livestock deaths. The drought and human overuse of the land encouraged the Sahara Desert to move south, and destroyed farmland (rangeland and arable land) in many countries including Nigeria, Niger and Mali. In association with other socially disruptive factors such as political instability and war, it has been implicated in causing widespread famine and hardship. Serious famine for instance between 1968 and 1975 is estimated to have been linked to the deaths of as many as 200,000 people.

Some recovery in rainfall since 1990 has been achieved with rainfall levels actually above the long-term mean in 1994. Claims have been made in recent years that land productivity in a number of areas has been more or less restored. Nevertheless, questions remain of how the people of the Sahel can continue to cope with the long-term lack of water resources, continuing falls in food production per head, soil resource depletion and desertification. Coping strategies for farmers in sub-Saharan Africa as in India depend on personal circumstances. In times of serious drought, farmers will rely on savings if they have any and/or will seek alternative sources of income. Many poor farmers in the Sahel have sought other sources of income generation through migration and have moved in increasing numbers with their families to the towns and cities of the region.

12.3.3 Drought in southern Africa

Southern Africa comprises the states of South Africa, Zimbabwe, Botswana, Mozambique, Namibia, Lesotho and Swaziland. Much of the region has a subtropical semi-arid climate with rainfall varying between 300 and 600 mm per annum. As in the Sahel, it is not surprising that large parts of the region are subject to wide fluctuations in interannual rainfall, drought episodes and moderate to serious desertification (see Figure 12.5). It is claimed that two distinct types of drought have affected southern Africa in

BOX 12.2

THINKING FURTHER

Causes of Sahel drought

The 25-year period of intense desiccation between 1965 and 1990 is interesting for a number of reasons. It represents the most substantial and sustained change in rainfall for any region of the world within the period of instrumental measurements. There has been much scientific debate about the causes of drought in the Sahel, particularly concerning the role of human activity on the one hand and natural climate variation on the other.

(a) Land albedo cooling hypothesis

The traditional view has been that drought (rainfall decline) in the Sahel is the result of humans over exploiting the natural resources of the region causing desertification (see Figure 12.5). It has been argued by a number of scientists that by removing vegetation in the Sahel, the reflectivity of the ground surface (i.e. surface albedos) would increase (Figure 12.8). With higher amounts of incoming solar radiation reflected from the ground back to space, cooling would take place at the surface and in the lower atmosphere, eventually restricting the possibilities of rainfall by cloud-convective activity. As rainfall declines and soils dry out, plants die back increasing the albedo of the surface further. Later studies have countered the above argument by showing that removing vegetation in semi-arid areas probably causes the local surface to dry and eventually warm rather than to cool.

(b) Atlantic cooling hypothesis

Recent work by the UK Meteorological Office using advanced computer modelling techniques has shown that drought in the Sahel may be the result of a natural climate cycle involving changing ocean temperatures and associated loss of natural vegetation. There seems to be a strong connection between the northwards movement of the ITCZ and SSTs in the Atlantic across the equator. When SSTs decline in the Atlantic north of the equator and increase in the south (Figure 12.9) as they have done in recent decades, drought conditions are promoted in the Sahel. The reasons for Atlantic cooling are still not clear, however. Causal mechanisms have been sought in the El Niño in the Pacific, the northern Atlantic conveyor belt and even air pollution.

(i) El Niño
It has been found that when SSTs in the eastern Pacific warm (causing a warm phase El Niño), the Atlantic cools and air circulation over the Atlantic changes. These atmospheric changes limit the northwards movement of the rain-bearing ITCZ over the Sahel in the summer season and restrict the region's rainfall supply.

(ii) Ocean conveyor
Ocean cooling in the Atlantic north of the equator may reflect a reduction in the northern transport of heat in the North Atlantic. This could be caused by a weakening of the North Atlantic conveyor belt (see section 4.4.1 and Box 4.5) as a result of the freshening of surface waters by ice melt in the northern North Atlantic.

(iii) Air pollution
There is some speculative evidence to suggest that Sahelian drought may be linked to air pollution and global warming. Increases in air pollution by sulphur dioxide has increased the concentration of tiny atmospheric particles known as sulphate aerosols. They act as hygroscopic nuclei and assist in cloud formation. However, they make cloud droplets smaller. This makes clouds brighter and longer lasting, so they reflect more sunlight to space, cooling the earth's surface below. Most of the sulphur dioxide pollution is generated in the northern hemisphere so most of the surface cooling can be expected to take place there. As a result, the tropical rain belt, which migrates northwards and southwards with the seasonal movement of the sun (ITCZ), is weakened by cooling in the northern hemisphere and does not move as far north. The main impact of restricting the northern movement of the ITCZ rain belt is to generate drought in the Sahel.

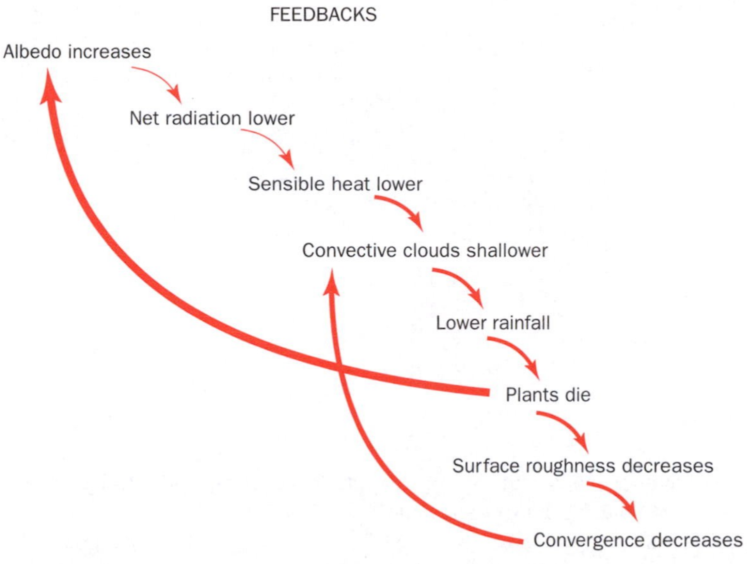

Figure 12.8 Increases in albedo and associated feedbacks in the semi-arid Sahel region
Source: R. Washington, Oxford University

the last 20 years or so. Type 1 drought occurred in the late 1980s and mid 1990s (Figure 12.10) and affected most of the northern part of southern Africa from northern Namibia to north-west Mozambique. Type 2, a more severe drought, affected a southern core region comprising southern Zimbabwe, southern Mozambique and north-east South Africa and held sway during the early/mid 1980s and early 1990s. Recent droughts in southern Africa have a complex origin and seasonality but have been linked to variations in ENSO events. In general southern African rainfall is inclined to be drier than average during an El Niño event, and wetter than average during a La Niña. The clearest link is between the occurrence of type 2 drought in the south-eastern African region and the El Niños of 1982/83 and 1991/92.

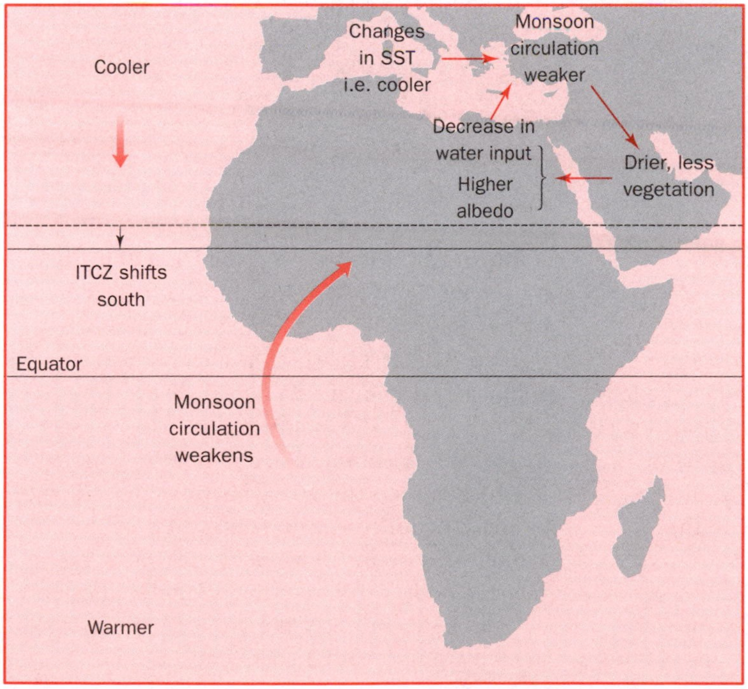

(a) Food security and adaptation

Like the Sahel region, southern Africa has suffered serious drought *and* food insecurity in recent years. Table 12.5 shows that in 2003, almost 18 million people in southern Africa were at risk from famine. The reasons are the same as in the Sahel – a varying cocktail of crises (depending on country)

Figure 12.9 The impact of cooler sea surface temperatures (SSTs) north of the equator and warmer SSTs south of the equator on rainfall in the Sahel
Source: R. Washington, Oxford University

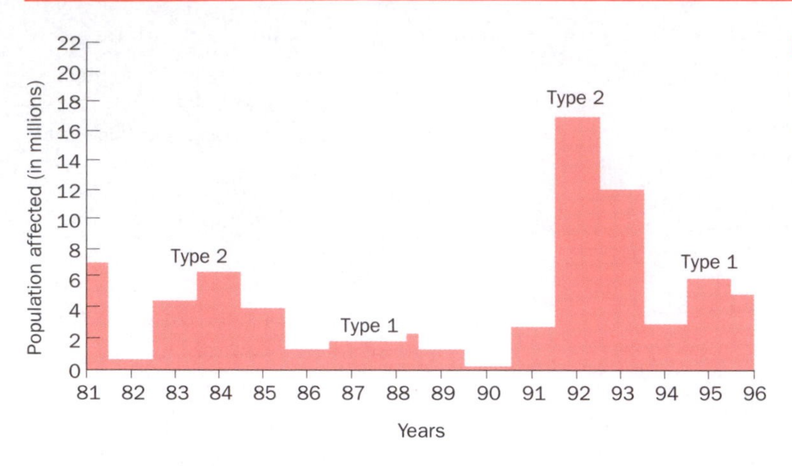

Figure 12.10 Population affected by drought in southern Africa, 1981–96
Source: Anyamba *et al.* (1998). Clarks Lab. USA

Table 12.5 People at risk from drought, flood and famine in southern Africa in 2003. Other factors including political corruption, war and ethnic divisions exacerbate the drought and flood risk

Country	Population	Climate hazard	At risk	% Population
Lesotho	2.2 million	Heavy rains Drought	760,000 in need of food assistance	34.5
Malawi	11 million	Floods Drought	3.6 million	32.7
Mozambique	17.5 million	Floods Drought	655,000	3.7
Swaziland	1 million	Erratic rainfall	297,000	29.7
Zambia	10 million	Flood Drought	2.6 million	26
Zimbabwe	12 million	Drought	7.2 million	60
Total	53.7 million		16 million (rounded down)	27.9

Source: United Nations, World Food Programme

involving political corruption, warfare, poor agricultural management and AIDS, all compounding the effects of climate drought. Lesotho, an impoverished mountain kingdom in southern Africa, provides a good example of a land and people in crisis. This small country which used to export food has been devastated by years of poor government and misguided agricultural practices which have left it vulnerable to the vagaries of drought. In 2003, there were serious food shortages in 8 out of 10 districts and 4 in every 10 children under 5 showed signs of malnutrition and stunted growth. Like many other southern Africans, a good proportion of its inhabitants have had to live on wild fruit and berries and by killing or selling their livestock – a desperate measure which will leave them completely dependent on outside help.

Figure 12.11 shows how individual groups including rich commercial and poor communal livestock farmers in the dry north-western district of South Africa cope with recurrent drought. Interestingly about 10 per cent of farmers, both rich and poor, carry the perception that they never have drought and the same proportion of farmers in each group (about 25 per cent) have no

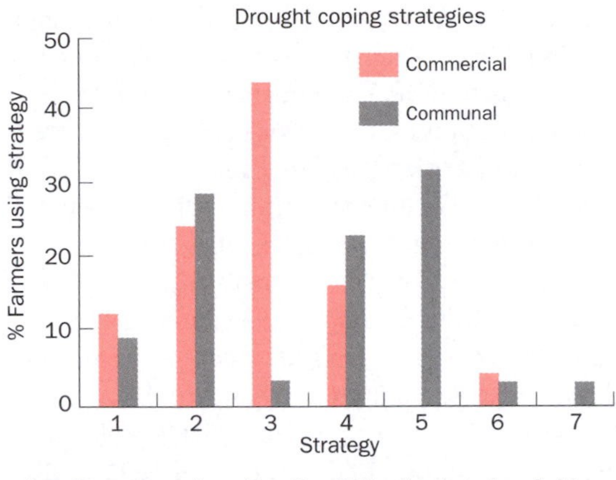

1 = Never have drought
2 = No drought strategy
3 = Sell animals
4 = Buy fodder
5 = Sell animals to buy fodder
6 = Rent pasture
7 = Feed cash crops to animals

Figure 12.11 Coping strategies by commercial and communal farmers in the semi-arid region of north-west South Africa
Source: Hudson (2001)

established drought strategy. However, the commercial farmer has many more choices than his poorer communal counterpart when drought threatens. The former can simply cash in and sell his animals, whereas this choice is not really open to the poor farmer who often regards his animals more in terms of the social status they confer rather than in terms of a cash return. In extreme drought conditions, about 30 per cent of the poor farmers questioned sold their animals for fodder so that some of their herd could be saved. A small number of commercial farmers also adopt the strategy of feeding cash crops to livestock, an option not open to the communal farmer.

(b) Raising adaptive strategies: early warning systems

A good deal of famine relief in sub-Saharan Africa today depends on food aid from international agencies like the United Nations' World Food Programme (WFP) as well as foreign government and non-governmental organisation assistance (e.g. Live Aid). The impact of drought can be lessened for societies and countries, however, with the use of better drought preparedness schemes. Unlike hurricanes and most flooding events, droughts build up gradually so that possibilities of preparing for drought are potentially higher for the latter than the former types of hazard. Nevertheless, the ability to estimate the probability of drought occurrence and the various risks associated with drought is a crucial component of drought preparedness. Some advance in drought prediction can be made in parts of southern Africa because of the link between periods of strong warm and cold ENSO phases with large-scale rainfall anomalies. Although the links are by no means perfect, ENSO signals have for example been detected in rainfall time series in the Sahel region, eastern Africa, Ethiopia and southern Africa. The potential role of ENSO events, as well as other factors, in the prediction and early warning of impending seasonal rainfall anomalies in Africa and South Asia needs to be examined and developed. However, the lack of ability of poor developing nations to make long-range forecasts of drought and to reach their rural populations remains a limiting factor.

Key ideas

1. Drought involves an unexpected failure of the rains and involves an absence of precipitation for a period long enough for serious moisture deficits to develop in the soil and for rivers and aquifers to run dry.

2. Famine is a prolonged food deficit that can lead to widespread increases in ill health and mortality.

3. Drought can be a precursor of famine but other factors like war, political ideologies and poor food distribution systems usually drive the famine process.

4. Serious drought in the Sahel region of sub-Saharan Africa between 1965 and 1990 helped to create, along with other non-climate factors, persistent famine in the region.

5. The Sahel drought has been related to land and ocean cooling processes that have respectively prevented the development of (i) convectional rainfall and (ii) the migration of the rain-bearing ITCZ across the region.

6. In contrast, drought in southern Africa during the last 20 years has been related mostly to El Niño events in the Pacific.

7. Drought in southern Africa, exacerbated by political corruption and a poor food distribution network, has resulted in widespread famine.

8. Coping strategies by the poor in Africa and India in the face of serious drought are often limited when government mitigation and support systems are weak.

12.4 Floods and human survival

12.4.1 Rising costs: huge flood events

As suggested previously, floods now represent the single greatest climate hazard facing society with upwards of 140 million people affected globally. Figure 12.12 indicates the yearly damage resulting from floods in the USA for the period 1903–99.

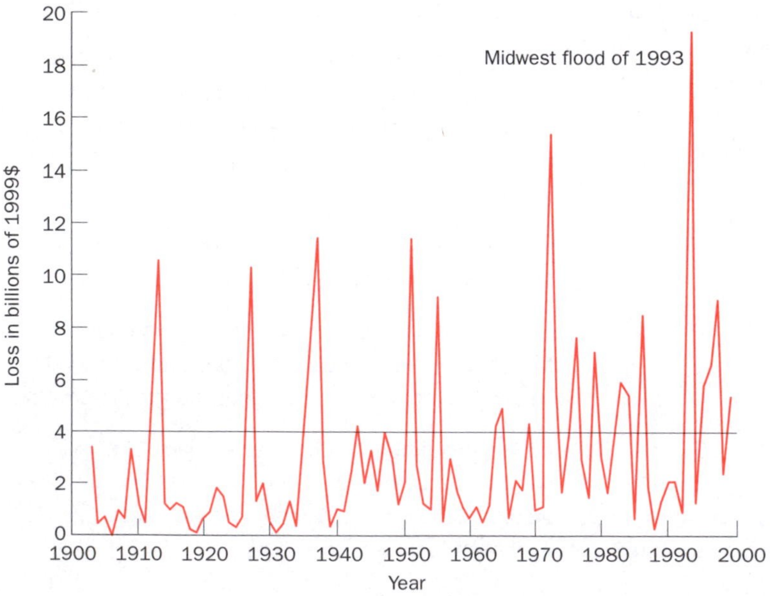

The increase in loss over the period is evident and, like hurricane-related damage, is probably due to the growth of population and wealth in the country. But unlike the hurricane impacts, there may be an additional external component. It is possible that the intensification of the hydrological cycle caused by global warming, has led to an increasing frequency of wet spells and droughts. The great Midwest flood of 1993, costing almost \$20 billion (1999 dollars) stands out in terms of its duration, extent and intensity as the US flood of the twentieth century (section 10.3.1). Prolonged heavy rains starting in the autumn of 1992, and continuing through to the summer of 1993, together with a lack of natural flood and overspill areas in the Mississippi Basin (removed in a highly engineered river system) led to the flood event. In the summer months of 1993 some midwestern states received more than 1.2 m of rain. The flood waters remained in many areas for nearly 200 days, a very unusual event for the USA. Nine states with more than 15 per cent of the area of the contiguous USA were catastrophically impacted.

While the great Mississippi flood of 1993 caused vast damage, exceeding by \$5 billion the total costs due to the 2002 floods of central Europe (\$15 billion), it is dwarfed by the 1998 summer floods in the Yangtze River in northern China. Chinese economists estimate the loss from the Yangtze disaster at more than \$35 billion. The losses were accentuated by a high population density and rapid economic growth in the region leading to an increased concentration in wealth.

Figure 12.12 Estimated annual damage resulting from floods in the USA, 1903–99 (\$1999 prices)
Source: National Water Service, NOAA, US Government

12.4.2 Recent flooding in England and Wales

(a) Agency: winter storms and heavy rainfall

Many people living in the floodplains of England and Wales have, in recent years, found themselves increasingly vulnerable to flood impacts. An unusually high level of rainfall occurred over England and Wales throughout the autumn and winter of 2000/1 as part of the wettest 12-month period in the region since records began in 1776 (section 9.3.1). Highlights of this spell are the wettest April on record with 143 mm (234 per cent of the 1961–90 mean) and the wettest autumn (Sept.–Dec.) on record with 640 mm (178 per cent of the 1961–90 mean), the latter resulting in severe flooding in many parts of England and Wales. The associated flooding was the most extreme fluvial event (max. 30-day flow at 1,108 m^3/s) since the snowmelt-generated floods of March 1947 (1,279 m^3/s) and greater than the next highest flow (1,008 m^3/s) in 1960. With extremely high runoff rates, soil erosion and landslides were a common feature, and impacts of the flooding on the community, agriculture and transport systems were very pervasive. Many impermeable catchments especially in south-east England and the Humber–Trent valley were subjected to multiple flood events, and extensive sheets of flood water became a prominent, if temporary, feature of the landscape (Plate 9.3). For instance, on 12 October 2000, there were 5 severe flood warnings (where large numbers of people and property are at risk, and there is imminent danger to life) and 47 flood warnings where evacuation preparation is advised. Some estimates put the cost of flood damage during the autumn–winter of 2000–1 as high as £2 billion, with more than 10,000 homes and businesses seriously affected.

Much of the flooding has been related to the exceptionally high rainfall totals of the period and to the possible effects of global warming which suggest that in the next 20–50 years, the UK (including southern England and Wales) will become a wetter place, particularly in winter. Although, there is no clear statistical evidence of significant increases in total rainfall over England and Wales in the last 20 years, daily precipitation intensities over the UK seem to have increased in winter over the same period. Moreover, several studies on changes in storminess over the north-east Atlantic, reveal that storminess has increased in the last few decades, although storm intensities are no higher than in the early part of the twentieth century. As indicated in section 8.1.3, the increasing frequency of gales and storms over England and Wales can be directly linked to changes in the Arctic Oscillation (AO) and North Atlantic Oscillation (NAO) index. The latter index is a measure of mean sea-level pressure gradients over the North Atlantic between south-west Iceland and Gibraltar. When the NAO index is positive as during the last 20 years or so, the flow of westerly winds across the North Atlantic and Europe is enhanced, bringing milder and possibly stormier and wetter weather. When the NAO index is negative as in the 1960s, the frequency and strength of the mild wet westerlies decline, allowing a greater incursion of different airflows and weather types over the UK (e.g. dry, cold/warm anticyclones, dry cold/warm easterlies and cold northerlies).

(b) Vulnerability and adaptive capacities

The increasing vulnerability of people to flood risk in England and Wales has more to do with *poor land-use planning and property development* than to any increases in storm frequency and intensity. This is despite the record wet winter of 2000/1, and the return of some serious flooding in the winter of 2002. A failure by local authorities to keep drainage ditches and channels clear – leaving them to be filled by sediment and clogged by vegetation – resulted in excessive rainwater and localised flooding. Also implicated in the alarming rise of wet muddy flows that often engulfed houses, were the inappropriate widespread cultivation of winter cereals – a practice that keeps the fields bare in early autumn and, therefore, susceptible to fast muddy flows, even during normal rainfall events. Increases in urban development within river catchments also helped increase the impermeability of land surfaces and accelerated surface runoff into river valleys.

The most significant factor in flood damage, however, has been the increased development of property in flood-prone areas. In recent years, major new housing areas have been allowed to expand into the floodplains of the major rivers. Twenty-five years ago just over 2 million people in 1 million homes lived in the flood-risk areas in England and Wales: today the figure is close to 5 million people in 2 million properties.

(c) Raising adaptive capacities

While local authorities and developers must be held more responsible for increasing vulnerability to flood damage, solutions to the flood problem are not easy. Floodplain management is a complex and multidimensional problem which has been ill served in the past by short-sighted non-sustainable technical engineering strategies resulting in a bias towards structural solutions (large flood defences) to flood hazards. This bias could be lessened with the adoption of more appropriate and adaptive strategies. These would include the following:

1. more local flood defences;
2. better environmental management (cultivation techniques);
3. more sensible land-use planning (not developing in vulnerable places);
4. sympathetic infra-structural design (waterproofing, avoidance of building bungalows);
5. educating and preparing people better for damage (flood proofing and attitude changes); and
6. providing economic incentives (sensible insurance costs).

These developments will go some way towards creating a more comprehensive and sustainable floodplain management system. The key to these developments lies in government incentives and its willingness to address the growing problem of flooding in the UK. However, with growing public funds being invested in European and UK water management schemes, the UK Environment Agency now has some real power to provide real long-term solutions.

Given the assumption that heavy winter rains in the UK are likely to increase in intensity and/or frequency in a globally warmed future (see section 7.1.3), a failure to develop and enforce a suitable management system will increase the vulnerability of floodplain communities in England and Wales. For some people living in the most flood-prone areas, the pursuit of a 'sustainable lifestyle' is already becoming difficult with falling house prices and increasing difficulties of obtaining fairly priced insurance cover.

Key ideas

1. A greater number of people in the world from the developed and developing countries alike are affected more by floods than by any other climate hazard.

2. The autumn and winter of 2000/1 was one of the wettest 12-month periods on record in England and Wales, and led to some of the region's worst flooding events.

3. Despite the high rainfall totals, much of the damage caused by the flooding was the result of poor land-use planning and property development.

4. For instance, the number of people living on the flood-prone river valleys of England has risen from 2 to 5 million during the last 25 years.

5. In view of climate change predictions and the likelihood of more UK flooding in the future, a government-led management plan to raise the coping abilities of people faced with the flood hazard is necessary.

6. This plan will involve more attention paid to local flood defences, better cultivation and drainage systems, more appropriate land-use planning with less house building on floodplains, improvements in house design (waterproofing and avoidance of building bungalows) and sensible insurance costs for people at risk.

12.5 Overview: climate hazards and sustainable development

12.5.1 Absolute poverty and human risk

For many of the world's poorest people, especially those who live in the developing countries of the tropics, daily access to an adequate food supply and clean drinking water remains a problem. They possess few economic resources, including low levels of income and technology, limited information and skills, meagre access to physical and social infrastructures, and minimal capacities of self-empowerment. For the poorest groups in society in the less developed countries, the realisation of the concept of sustainable development, or the idea that their living standards can somehow be improved and sustained at an adequate level is not an especially useful one. This is because the more developed world has not shown the political will or the economic commitment to really help them. The numbers of people at risk from poverty and hunger keep on increasing. This is despite recent attempts by the World Bank to convince us otherwise using seriously flawed methods of estimating their true numbers. With their roll call at present around 1 billion, there are more people living in absolute poverty today than ever before.

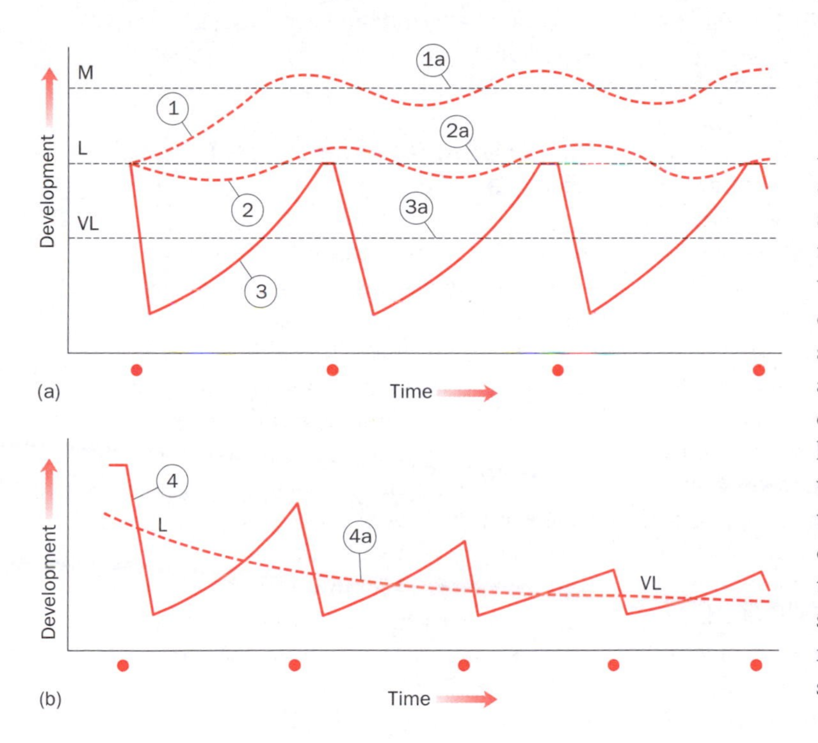

KEY
● Occurrence of climate hazard
L = Low
M = Medium } development level
VL = Very low
1 Poor society with raised and sustainable development
1a Mean development status of society 1
2 Poor society without assistance and sustainable development
2a Mean development status of society 2
3 Development path of a poor society subject to regular climate hazards
3a Mean development level of society 3
4 Development path of poor society subject to frequent climate hazards (i.e. with low return period)
4a Mean development (i.e. declining) level of society 4

12.5.2 Modelling unsustainability

A hypothetical situation of sustainable development is shown in Figure 12.13 (path 2 to path 1), where communities with initially low levels of development have their living standards raised and sustained at a higher, medium level of development. This situation is, however, unlikely for most of the world's poorest societies because the more developed countries continue to lack the political will to implement long-lasting and sustainable poverty alleviation measures to raise their living standards.

Figure 12.13 Hypothetical model showing for regions and communities in the developing world: (a) different levels of sustainable development under moderately frequent climate impacts; and (b) unsustainable levels of development in relation to highly frequent climate impacts
Source: O'Hare (2002, p. 235).

(a) Climate change

Sustaining existing standards of living is difficult when poor communities and societies are faced with the effects of climate change. This is even more so when climate change is associated with the generation of more frequent or prolonged climate hazards. Figure 12.13 (path 3) shows that it is possible for communities to cope with a recurring severe climate situation, and for recovery to be achieved. However, the important point is that recovery is usually slow so that if the climate event recurs (say over 5, 10 or 20 years) overall average development levels (path 3a) are maintained at a very low level. Associated development levels will be lower than they would be in the absence of the climate extreme (corresponding to path 2a). Another even worse situation can be envisaged if the frequency, intensity or duration of an extreme climate event increases. Examples of extreme climate events increasing in (a) frequency/intensity and (b) duration include respectively recurrent droughts and floods caused by El Niño (sections 5.5.2 and 10.3.1), and the hazard impact of persistent drought in the Sahel and southern Africa (Figures 12.7 and 12.10). If severe weather events increase in this way then long-term development in a region may not be sustained at all and show a continuous downward spiral of decline (Figure 12.13 path 4).

It is appropriate to stress again that the ability of societies to cope with climate change/hazards very much concerns the availability of physical, economic and institutional resources employed to combat the effects of the climate hazard. When these are available (e.g. food rehabilitation programmes, house insurance or rebuild programmes for damaged houses, local education systems to help communities to cope better with the climate hazard), they help reduce the vulnerability of high-risk communities and groups. When these resources are lacking, the adaptive capacity or ability of local communities to cope with the effects of severe climate impacts declines.

Key ideas

1. Improving and sustaining a community's level of development would increase the adaptive capacity of that community (i.e. lower its vulnerability) to climate change.

2. Communities faced with recurrent, increasingly frequent or more intense climate hazards suffer long-term reductions in their standards of living.

3. The chance of improving and sustaining living conditions for the world's poor, especially those faced with recurrent climate hazards, looks unlikely because of a historic and continuing lack of political will by the West.

12.6 Climate change and global biodiversity

Many of the previous case studies in this book involving climate hazards and their impact on societies and regions have examined a range of fairly direct effects of climate change. Just as important will be how human societies are affected by many indirect influences of a changing climate. First among these will be the impact of future climate change on the world's land plant and animal species. As the survival of human society ultimately depends on green plants and the biological communities they contain, a brief analysis of the possible effects of global warming on living communities is worth addressing.

12.6.1 Vegetation types

A world warmed by increasing concentrations of greenhouse gases will be very different from the one we know today. Contemporary climate change effects (section 6.4.4) and future projections of anthropogenic changes in climate (section 7.1) inform us of a rapidly changing world. In our future world, modelled projections suggest that glaciers will continue to retreat, global snow cover will gradually diminish, overall precipitation will increase, especially in the mid and high latitudes, sea levels will continue to rise, and climate variability and extreme weather will become more common. One of the most fundamental effects of climate change will be on the abundance and diversity of plants and animals, since they depend so directly on climate. Modelling studies have shown for instance that the distribution of the world's main vegetation zones will alter with a doubling of CO_2 levels. Although the wholesale migration of global vegetation will not take place, some types of vegetation will move as they try to adjust to the warmer climate while others will not. Studies have indicated that in a globally warmed world some vegetation types will be damaged more than others. It is thought that modelled projections of global warming from the Intergovernmental Panel on Climate Change (IPCC) Third Assessment Report (a rise of 0.5–3.0 °C in the mean annual global temperature by the year 2050) will seriously damage some types of vegetation particularly in the northern hemisphere. The greatest change will be in the far north where the vast treeless tundra and polar deserts will shrink. Boreal coniferous forest (spruce, fir, pine and deciduous larch) on the southern side of the tundra will spread north into the tundra as the climate and soils (permafrost) warm. Boreal forest in turn may be reduced with the spread of broadleaf forests (beech, oak, sycamore) located on their southern margins into warmer northern regions. As winter temperatures are expected to increase more than summer ones, many broadleaf evergreen species of tree such as the oaks of the Mediterranean will be able to survive the relatively mild winters of temperate latitudes, even as far north as the UK. Any spread of these species would in turn be at the expense of western European broadleaf deciduous species. The potential geographical range of the tropical rainforests of the equatorial regions will change little and might even expand given the possibilities of future higher rainfall and temperature there. On the other hand, vegetation areas of southern Europe, eastern Asia, North and South America and South Africa which are classed as shrubland will shrink, with sparse or desert-like vegetation expanding to replace it.

12.6.2 Change in species

Perhaps more important than the possible gross changes in the distribution of the main vegetation zones will be severe life-threatening dislocations to the animals and plants they contain. The loss of tundra, for instance, will affect the animals that are adapted to survive its harsh conditions, or which migrate there in summer to breed, taking advantage of the brief summer period of food availability. Several species, including the polar bear, will probably not survive, or if they do, their populations will shrink and move further north into the colder remaining ice zones.

(a) Species response

Changes in species extinction and abundance as a result of global heating are therefore likely to be much more complex and dramatic than the simple movement of the major vegetation types. At the species level, four major biogeographical effects will be seen. First, there will be changes in species phenology or the timing of seasonal activity. This is well demonstrated by the early (6 week) arrival from Africa of swallows and house martins in southern England in February 2004 during a notably mild spell during that month. Such phenological change can be damaging for some species. For instance, early flowering plants and nesting birds can be catastrophically hit by the return of severe weather. Additionally, premature plant growth and flowering might be induced to take place before animal communities can respond. This would apply for instance to where plant food reserves were made available to birds before they had a chance to breed and feed their young, or to plants flowering early and crucially before their pollinating insects were on the scene.

Second, there will be changes in the geographic distribution of species: some species will expand while others will shrink, some will see shifts in their distributions north while others will fail to move and become extinct. It is now recognised that global warming over the last 100 years and more particularly over the last 30 years has produced numerous shifts in the distribution and abundance of species. Range shifts in areas with regional warming trends have been reported in alpine plants, butterflies, birds, marine invertebrates and mosquitoes. For instance, in studies of European non-migratory butterfly species, many have shown ranges that have shifted to the (cooler) north by 35–240 km during the past century, with relatively few shifting to the (warmer) south. The range shifts parallel the 0.8 °C warming over Europe during the last century (see Figure 6.3), which has moved the climatic isotherms northwards by an average of 120 km.

Because some species will become extinct or greatly alter their distribution and abundance, a third consequence of global warming on species behaviour will involve complex changes in the composition of existing plant and animal communities. As a consequence of such altered community assemblages, there will be, fourth, new dynamic changes in species interactions.

(b) A case for species extinction

Many species, however, as with whole vegetation formations, are unable to migrate and change their distributions easily when subjected to external or internal change. For instance, deliberate habitat destruction as a result of direct human intervention, for example forest clearance, the substitution of grassland by cropland or the loss of farmland to urbanisation, has already reduced and fragmented the distribution of many species so that some are now on the verge of extinction. Habitat destruction can also occur as a result of indirect human intervention through climate change. This is because climate change, as with direct habitat loss, will alter the physical and biological conditions of habitats and thus the environmental conditions necessary to sustain various plants and animals, for example temperature conditions, water supplies, vegetation and ground conditions for living and breeding and food reserves.

Thus, one of the greatest threats of anthropogenic climate change in the twenty-first century may relate to the loss of global habitats and the animals and plants they contain. A recent major scientific study (Thomas *et al.*, 2004) has estimated that a rise of temperature of between 0.5 and 3 °C in the next 50 years could wipe out a quarter of the world's known animals and plants, more than 1 million species. It is claimed that global warming will represent a threat on par with the greatest loss of species today – the already well-documented destruction of natural habitats around the world. The study investigated over 1,100 species of plants, mammals, birds, reptiles, frogs, butterflies and other insects living in six survey areas – Europe, South Africa, Australia, Brazil, Mexico and Costa Rica. The effect of rising temperatures on each species was calculated using the future scenarios of temperature change proposed by the UN's Intergovernmental Panel on Climate Change. The IPCC has predicted minimum, mid-range and maximum global average temperature rises of between 0.5 and 3 °C by 2050 (see section 7.1.2 and Figure 7.1(b)). A warmer world would push most species towards the poles or higher up mountains, but for many the practicalities of dispersing to new areas is problematic. Many species are incapable of moving the large distances required, and those that can often fail to find large enough suitable habitats for growth and reproduction.

The survey points out that under the various global warming scenarios, more than half of Australia's 400 butterfly species and 25 per cent of bird species in Europe could be wiped out. In excess of 2,000 species in Brazil's unique tropical grassland and many species in Mexico's deserts also face extinction. Moreover, in the species-rich

South African Cape region many unique flowering plants could perish, as well as most species from the high-altitude cloud forests in Costa Rica.

When these results are extrapolated across the world, considerable losses in species are projected. Using the IPCC lowest rate of future warming, 18 per cent of global species are committed to extinction, the mid range of warming suggests a 24 per cent loss while around 35 per cent of species may disappear under the maximum warming scenario. These losses were exhibited regionally, however, with the highest percentage loss of species in tropical shrubland (28.9 per cent), temperate deciduous forest (24.2 per cent), and temperate mixed (deciduous and coniferous trees) forest (19.2 per cent). Lower percentage losses are expected from tropical ecosystems such as tropical rainforest (4 per cent) and tropical grassland (8 per cent), with the smallest percentage losses from high-latitude ecosystems including boreal forest (0.9 per cent) and tundra (1.0 per cent). Given that by far the highest proportion of the world's species resides in tropical forests, the actual number of species lost there will be considerable.

It needs to be kept in mind that calculating potential species extinction rates for the planet and its regions under various global warming scenarios is a difficult task. Many factors other than climate affect species distribution and abundance, and the scale and interactions of these factors among themselves and with climate are difficult to assess or predict accurately. With such uncertainties in mind, the numbers of species facing or committed to extinction as a result of climate change in the future can only be projected in an approximate way. Nevertheless, the numbers of species facing extinction could be far greater than anything previously realised, and be at least on par with the current losses due to the direct destruction of natural habitats. Evidence supporting the claim that climate change will be a significant threat to global biodiversity this century demands our attention. Only a rapid switch to green technologies and the active removal of CO_2 from the atmosphere could avert such ecological disaster.

Key ideas

1. Climate change in the twenty-first century will have major effects on the world's living communities of plants and animals.

2. There will be important changes in the distribution of major vegetation types with most movement and reductions occurring in the northern hemisphere in the tundra and boreal forests.

3. Even greater dislocations will take place in the distribution and abundance of individual land plant and animal species.

4. Climate change will alter the seasonal response of plant and animal behaviour, alter species distributions, change the species composition of communities and produce new species interactions.

5. One of the most serious consequences of climate change will be a large loss of species through extinction.

Further reading

Downing, T. E. (ed.) (1996) *Climate Change and World Food Security*. Springer-Verlag, Heidelberg.

McCarthy, J. J. *et al.* (2001) *Climate Change 2001: Impacts, adaptation and vulnerability*. IPCC Third Assessment Report. Cambridge, Cambridge University Press.

O'Hare, G. (2002) Climate change and the temple of sustainable development. *Geography*, **87**(3): 234–46.

Parry, M. and Carter, T. (1998) *Climate Impact and Adaptation Assessment*. Earthscan, 166 pp.

Thomas, C. D. *et al.* (2004) Extinction risk from climate change. *Nature*, **247**: 145–8.

Williams, M. A. J. and Bolling, R. C. Jr (1996) *Interactions of Desertification and Climate*. WMO and UNEP, Arnold, London, 270 pp.

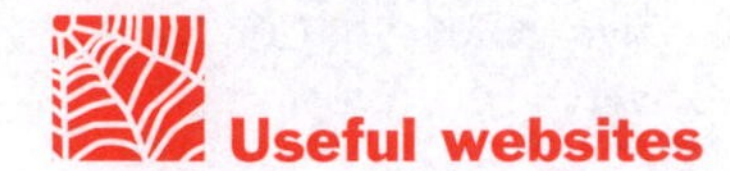

Useful websites

Disaster relief:
http://www.disasterrelief.org/index.html

Famine Early Warning Systems Network:
http://www.fews.net/about/

IPCC:
http://www.grida.no/climate/ipcc_tar/wg2/

NASA:
htpp://www.earthobservatory.nasa.gov/NaturalHazards/

NOAA:
http://www.ncdc.noaa.gov/oa/reports/weather-events.html
http://www.ncdc.noaa.gov/paleo/drought/drght_alleve.html

World Disasters:
http://www.em-dat.net

REFERENCES

Adegoke, J. and Carleton, A. M. (2000) Warm season land surface-climate interactions in the US Midwest from mesoscale observations. In McLaren, S. J. and Kniveton, D. R. (eds), *Linking Climate Change to Land Surface Change.* Kluwer Academic Publishers, Netherlands.

Agarwal, B. (1990) Social security and the family in rural India: coping with seasonality and calamity. *Journal of Peasant Studies*, **17**: 341–412.

Ahrens, C. D. (1993) *Essentials of Meteorology*. West Publishing Co., New York.

Ahrens, C. D. (2000) *Meteorology Today*, 6th edn. Brooks/Cole, Pacific Grove, Calif.

Anyamba, A., Eastman, J. R. and Tucker, C. J. (1998) *Warm Event of 1997/98: NDVI precursors and drought pattern prediction for southern Africa.* Preliminary Report to the Southern African Regional Climate Forum. 12–15 May 1998, Pilanesberg, South Africa.

Barry, R. G. and Chorley, R. J. (2003) *Atmosphere, Weather and Climate.* Routledge, London.

Belasco, J. E. (1952) *Characteristics of Air Masses over the British Isles.* Geophysical Memoirs 87 (M.O. 530b), 33–4. HMSO. Meteorological Office FitzRoy Road Exeter Devon EX1 3PB United Kingdom.

Blunier, T. and Brook, E. J. (2001) Timing of millenial-scale climate change in Antartica and Greenland during the last glacial period, *Science*, **292** (5 Jan.): 109–14.

Bond, G. C., Broecker, W., Johnsen, S., McManus, J., Labeyrie, J. and Bonani, G. (1993) Correlations between climate records from North Atlantic sediments and Greenland ice. *Nature*, **365**: 143–7.

Boruff, B. J., Easoz, J. A., Jones, S. D., Landry, H. R., Mitchem, J. D. and Cutter, S. L. (2003) Tornado hazards in the United States. *Climate Research*, **24**: 103–17.

Bradshaw, M. and Weaver, R. (1993) *Physical Geography: An Introduction to Earth Environments*, Mosby, St Louis.

Broecker, W. S. (1995) *The Glacial World According to Wally*. Eldigio Press, Palisades, 344 pp.

Bryson, R. A. (1966) Air masses, streamlines and the boreal forest. *Geography Bulletin*, **8**: 228–69.

Buckle, C. (1996) *Climate and Weather in Africa.* Longman, London.

Budyko, M. I. (1974) *Climate and Life*, Academic Press, New York.

Christopher, R. W. (1995) *Elemental Geosystems*, Prentice-Hall, Englewood Cliffs, NJ.

Critchfield, H. J. (1983) *General Climatology*, 4th edn. Prentice Hall.

Costanza, R. *et al.* (1997) The value of the world's ecosystem services and natural capital. *Nature*, **387**: 253–60.

Dai, A. and Wigley, T. M. L. (2000) Global patterns of ENSO-induced precipitation. *Geophysical Research Letters*, **27**: 1283–6.

Das, P. K. (1987) Short and long range monsoon prediction in India. In Fein, J. S. and Stephens, P. L. *Monsoons*, Wiley, London, pp. 549–78.

Dawson, A. and O'Hare, G. (2000) Ocean atmosphere circulation and global climate. *Geography*, **85**(3): 193–208.

Domato, F. *et al.* (2003) A remote-sensing study of the inland penetration of sea-breeze fronts from the English Channel. *Weather*, **58**: 219–26.

Drake, F. (2000) *Global Warming: The science of climate change.* Arnold, London.

Flohn, H. (1969) *Climate and Weather*. Weidenfeld and Nicolson, London.

Friss-Christensen, E. and Lassen, K. (1991) Length of the solar cycle: an indicator of solar activity closely associated with climate. *Science*, **254**: 698–700.

Fritts, H. C. (1976) *Tree Rings and Climate*. Academic Press, London.

Gibson, P. and Power, C. (2000) *Introductory Remote Sensing: Digital image processing and applications.* Routledge, London.

Giorgi, F. and Mearns, L. O. (1999) Regional climate modeling revisited. An introduction to the special issue. *Journal of Geophysical Research*, **104**: 6335–52.

Hansen, J., Russell, G., Rind, D., Stone, P., Lacis, A., Lebedeff, S., Ruedy, R. and Travis, L. (1983) Efficient three-dimensional models for climate studies: Models I and II. *Monthly Weather Review*, **111**: 609–62.

Hardy, J. T. (2003) *Climate Change: Causes, effects and solutions*. Wiley, Chichester.

Harvey, L. D. D. (2000) *Climate and Global Environmental Change*. Pearson, Harlow.

Henderson-Sellers, A. and McGuffie, K. (1987) *A Climate Modelling Primer*. Wiley, Chichester.

Houghton, J. (1997) *Global Warming: The complete briefing*, 2nd edn. Cambridge University Press, Cambridge.

Houghton, J. T., Ding, Y., Griggs, D. J., Noguer, M., van der Linden, P. J. and Xiaosu, D. (eds) (2001) *Climate Change 2001: The scientific basis*. Contribution of Working Group I to the Third Assessment Report of the IPCC. Cambridge University Press.

Hoyt, D. V. and Schatten, K. H. (1997) *The Role of the Sun in Climate Change*. Oxford University Press, Oxford.

Hudson, J. (2001) Responses to climate variability in the livestock sector in north west province, South Africa. MA thesis, Colorado State University, Fort Collins, Colo.

Kalkstein, L. S., Nichols, M. C., Barthel, C. D. and Greene, J. S. (1996) A new spatial synoptic classification: application to air-mass analysis. *International Journal of Climatology*, **16**: 983–1004.

Karl, T. R. and Knight, R. W. (1998) Secular trends of precipitation amount, frequency and intensity in the United States. *Bulletin of the American Meteorological Society*, **79**(2): 231–41.

Lamb, H. H. (1972) *British Isles Weather Types and a Register of the Daily Sequence of Circulation Patterns, 1861–1971*. Meteorological Office, Geophysical Memoir No. 116, HMSO, London.

Lambert, S. J. and Boer, G. J. (2001) CMIP1 evaluation and intercomparison of coupled climate models. *Climate Dynamics*, **17**: 83–106.

Lippsett, L. (2000) Beyond El Nino, *Scientific American Presents – Weather*, **11**(1): 77–83.

London Climate Change Partnership (2002)

McCarthy, J. J. *et al.* (2001) *Climate Change 2001: Impacts, adaptation and vulnerability*. Cambridge University Press, Cambridge.

Manshard, W. (1979) *Tropical Agriculture*. Longman, London.

Marsh, T. J. (2001) Climate change and hydrological stability: a look at long-term trends in south-eastern Britain. *Weather*, **56**: 319–28.

Musk, L. F. (1988) *Weather Systems*. Cambridge University Press

Neiburger, M. (1982) *Introduction to the Atmosphere*. W. H. Freeman, San Francisco.

Neiburger, M., Edinger, J. and Bonner, W. (1982) *Understanding our Atmospheric Environment*. W. H. Freeman, San Francisco.

O'Hare, G. (2001) Hurricane 07B in the Godavari delta, Andhra Pradesh, India: vulnerability, mitigation and the spatial impact. *The Geographical Journal*, **467**(1): 23–38.

O'Hare, G. (2002) Climate change and the temple of sustainable development. *Geography*, **87**(3): 234–46.

O'Hare, G. and Sweeney, J. (1986) *The Climate System*. Longman, London.

O'Hare, G. and Sweeney, J. (1993) Lamb's circulation types and British weather: an evaluation. *Geography*, **78**(1): 43–60.

O'Hare, G. and Sweeney, J. (1998) *Geography*, **78**(1) The Geographical Association. *

Oke, T. R. (1990) *Boundary Layer Climates*. Routledge, London.

Petit, J. R. *et al.* (1999) Climate and atmospheric history of the past 420,000 years from the Vostok ice core, Antartica. *Nature*, **399**: 429–36.

Pickering, K. and Owen, L. (1994) *An Introduction to Global Environmental Issues*. Routledge, London.

Pielke, Jr, R. A. and Landsea, C. W. (1998) Normalised hurricane damages in the United States, 1925–1995. *Weather and Forecasting*, **13**: 621–31.

Robinson, P. and Henderson-Sellers, A. (1999) *Contemporary Climatology*, 2nd edn. Longman, London.

Scholes, B. (1999) Will the terrestrial carbon sink saturate soon? *Global Change Newsletter*, **37** (March), International Geosphere-Biosphere Programme.

Sellers, W. D. (1965) *Physical Climatology*. University of Chicago Press, Chicago and London.

Semple, A. T. (2003) A review and unification of conceptual models of cyclogenesis. *Meteorological Applications*, **10**: 39–59.

Sheppard *et al.* (2002) The climate of the US Southwest. *Climate Research*, **21**, 219–238.

Singh, N. *et al.* (1991) Distribution and long term features of the spatial variations of the moisture regions over India. *International Journal of Climatology*, **11**: 413–27.

Stephens, G. L. *et al.* (1981) Earth radiation budgets. *Journal of Geophysical Research*, **86** (C10), 9739–60.

Strahler, A. and Strahler, A. (1989) *Elements of Physical Geography*. Wiley, New York.

Strahler, A. and Strahler, A. (1997) *Physical Geography: Science and systems of the human environment*. Wiley, New York.

Subbaramayya, I. and Naidu, C. V. (1995) Withdrawal of the south-west monsoon over India – a synoptic and climatological study. *Meteorological Applications*, **2**: 159–66.

Sweeney, J. (ed.) (1997) *Global Change and the Irish Environment*. Royal Irish Academy, Dublin.

Thomas, C. D. *et al.* (2004) Extinction risk from climate change. *Nature*, **247**: 145–8.

Thomas, D. S. G. and Middleton, N. J. (1994) *Desertification: Exploding the myth*. Wiley, Chichester.

Trenberth, K. E. and Guillemot, C. J. (1996) Physical processes involved in the 1998 drought and the 1993 floods in North America. *Journal of Climate*, **9**: 1288–98.

Van Andel, T. H. (1994) *New Views on an Old Planet: a history of global change*. Cambridge University Press, Cambridge, UK.

Webster, P. J. (1981) Monsoons. *Scientific American*, **245**(2): 70–80.

Webster, P. J. (1987) The variable and interactive monsoon, in Fein, J. S. and Stephens, P. L. (eds) *Monsoons*. Wiley, London, pp. 269–330.

Whyte, I. D. (1995) *Climate Change and Human Society*. Arnold, London.

Wick, G. (1973) Where Poseidon courts Aeolus. *New Scientist*, 18 January: 123–6.

Wigley, T. M. L. (1999) *The Science of Climate Change: Global and US perspectives*. Pew Center on Global Climate Change, Arlington, Va.

Wild, R., O'Hare, G. and Wilby, R. (1996) A historical record of blizzards/major snow events in the British Isles, 1880–1989. *Weather*, **51**(3): 88.

Wilson, M. F. *et al.* (1987) Investigation of the sensitivity of the land-surface parameterization of the NCAR community climate model in regions of tundra vegetation. *J. Climate*, **7**: 319.

Wilson, R. C. L., Drury, S. and Chapman, J. (2000) *The Great Ice Age*. Routledge, London.

Wolock, D. M. and McCabe, G. J. (1999) Estimates of runoff using water balance and atmospheric general circulation models. *Journal of the American Water Resources Association*, **35**: 1341–50.

Wolter, K. and Timlin, M. S. (1998) Measuring the strength of ENSO events: how does 1997–98 rank? *Weather*, **53**: 315–36.

GLOSSARY

Terms in *italics* are found elsewhere in this Glossary.

Absolute humidity The mass of water vapour contained per unit volume of air.

Absolute poverty A situation where poverty levels are so low in a society or group of people that lives and livelihoods are threatened.

Adaptive capacity The adaptive capacity of a system (region or group of people) refers to the potential or capability of that system to adapt to and thus cope with the effects of environmental (climate) change. Adaptive capacity depends on the severity of the climate event, the vulnerability of the system, and the technical resources available to help the system to cope with the climate event.

Adiabatic Physical changes to a system involving no heat exchanges, commonly linked to the forced or spontaneous ascent (expansion and cooling) or descent (compression and warming) of a parcel of air.

Advection The horizontal (parallel to the surface) transport of air and its properties. Most commonly refers to the horizontal transport of heat by wind action.

Aerosols Airborne solid or liquid particles, with a typical size between 0.01 and 10 μm, that reside in the atmosphere for at least several hours. Aerosols influence the *climate* directly through scattering and absorbing radiation, and indirectly through the formation and optical properties of clouds.

Airflow (index) Trigonometric measures of atmospheric circulation obtained from surface pressure or geopotential height fields. Commonly derived indices include *vorticity*, *zonal flow*, *meridional flow* and *divergence*. Certain indices have been used to replicate subjective classifications of daily *weather patterns*, or as predictor variables in *statistical downscaling* schemes.

Air mass A large mobile body of air with broadly uniform weather conditions of pressure, temperature, moisture, humidity and wind, at similar heights in the atmosphere.

Albedo The ratio of energy reflected by a surface to the energy incident or falling on that surface.

Anabatic wind Upslope movement of air in a valley, usually in daytime, when the upper valley slopes are warmed more than the valley floor. Warm air rising from the upper slopes induces an upslope wind to develop when slack regional pressure gradients exist.

Ana-front A cold *front* in which warm air is generally rising over a wedge of colder air.

Analogue A weather forecast or climate change *scenario* that closely resembles sequences of weather or *weather patterns* that have been previously experienced. Past knowledge of the accompanying meteorological conditions is used to make detailed predictions about future weather under similar patterns.

Angular momentum A measure of the amount of rotational movement of a body such as a parcel of air.

Antarctic Oscillation The AAO is a system which shows switches in the intensity of surface atmospheric pressure gradients and the associated speed of the upper westerly vortex around the south pole.

Anthropogenic Resulting from, or produced by, human beings.

Anticyclone A region of high-pressure and *divergence* at the earth's surface, in which the barometric pressure decreases progressively with distance from the core of the system.

Arctic front A largely inactive *frontal* zone marking the interface between cold air masses of the arctic and warmer air of the temperature mid latitudes, lying to the north of the *polar front*.

Arctic Oscillation The AO is a system which shows switches in the intensity of surface atmospheric pressure gradients and the associated speed of the upper westerly vortex around the north pole.

Atmosphere The gaseous envelope surrounding the earth, comprising almost entirely of nitrogen (78.1%) and oxygen (20.9%), together with several trace gases, such as argon (0.93%) and *greenhouse gases* such as carbon dioxide (0.003%).

Baroclinic Where the atmosphere exhibits variations in temperature at a particular height. A temperature gradient exists along the isobars.

Barotrophic Where the atmosphere has horizontally the same temperatures at a particular height. No temperature gradient exists along the isobars.

Bergeron-Findeisen Precipitation formation by the preferential growth of ice crystals in a cloud containing both water and ice. The transfer of water from supercooled droplets to ice crystals results from the equilibrium vapour pressure over ice being lower than that of water at the same sub zero temperature.

Bio-diversity The number or variety of biological systems and structures in an area. Usually indicated by species diversity, i.e. the number of species or species groups in a region.

Biogeochemical cycling The circulation of nutrients and minerals from the land and ocean to the atmosphere and back again.

Biological pump A mechanism whereby microscopic plant (phytoplankton) growth in the oceans can modify the temperature of the planet by altering the concentration of carbon dioxide in the atmosphere. When phytoplankton growth is high, carbon dioxide is removed from the atmosphere causing a cold phase condition, when growth is slow, carbon dioxide is returned to the atmosphere inducing a warm period.

Biosphere All the earth's living organisms and the environment they interact with.

Black body An ideal radiator which emits the maximum possible amount of radiant energy to its surroundings.

Black box Describes a system or model for which the inputs and outputs are known, but intermediate processes are either unknown or not prescribed. See *regression*.

Bond cycles A linked series of progressively colder Dansgaard-Oeschger events culminating in a Heinrich event.

Bottom up development In terms of disaster management, the use of local community awareness and participation in risk and vulnerability reduction.

Break monsoon A period of varying length (one week to several weeks) during the wet Indian summer monsoon when the rains fail and dry conditions prevail.

Centrifugal force For a body moving in a curve, the apparent force that appears to be trying to deflect the moving body outward from its centre of curvature.

Chlorofluorocarbons (CFCs) Synthetic chemical substances containing Chlorine, Fluorine and Carbon which were widely used as propellants and refrigerants until their role in stratospheric ozone depletion became evident.

Circumpolar Vortex The Circumpolar Vortex (CPV) is another name for the upper westerlies which is a fast-moving belt of upper westerly winds moving around both the polar regions in the high latitudes.

Climate The 'average weather' described in terms of the mean and variability of relevant quantities over a period of time ranging from months to thousands or millions of years. The classical period is 30 years, as defined by the World Meteorological Organisation (WMO).

Climate change Statistically significant variation in either the mean state of the *climate*, or in its variability, persisting for an extended period (typically decades or longer). Climate change may be due to natural internal processes or to *external forcings*, or to persistent *anthropogenic* changes in the composition of the atmosphere or in land use.

Climate model A numerical representation of the climate system based on the physical, chemical and biological properties of its components, their interactions and *feedback* processes, and accounting for all or some of its known properties.

Climate prediction An attempt to produce a most likely description or estimate of the actual evolution of the climate in the future, e.g. at seasonal, inter-annual or long-term timescales.

Climate projection A projection of the response of the climate system to emission or concentration scenarios of *greenhouse gases* and *aerosols*, or *radiative forcing* scenarios, often based on simulations by *climate models*. As such climate projections are based on assumptions concerning future socio-economic and technological developments.

Climate scenario A plausible and often simplified representation of the future climate, based on an internally consistent set of climatological relationships, that has been constructed for explicit use in investigating the potential consequences of anthropogenic *climate change*.

Climate variability Variations in the mean state and other statistics (such as standard deviations, the occurrence of extremes, etc.) of the climate on all temporal and spatial scales beyond that of individual weather events.

Coalescence The merging of two or more cloud droplets into a larger droplet.

Cold air drainage The flow of cold dense air (mostly at night under calm conditions and clear skies) into low-lying depressions.

Collision Precipitation formation as a result of larger cloud droplets falling more quickly than smaller ones and collecting them and absorbing them thereby enlarging further the falling droplets concerned.

Condensation nuclei Microscopic particles in the atmosphere that act as a focus or site where water vapour can condense into water droplets thus producing cloud growth.

Conditional instability Where the Environmental Lapse Rate is greater than the Saturated Adiabatic Lapse Rate but less than the Dry Adiabatic Lapse Rate. An atmospheric temperature profile exists whereby a displaced saturated air parcel is unstable while a displaced unsaturated air parcel is stable.

Conduction The transfer of sensible heat through a body by molecular collisions without any overall change in the position of the molecules.

Continentality A measure of the extent to which the climate (e.g., the annual temperature range or precipitation regime) of a location is affected by its distance from the ocean in the direction of the prevailing wind.

Convection The transfer of sensible heat by the actual movement of the heated substance, such as air or water. In meteorology, convection also means vertical transport through density imbalance, transporting mass, water vapour and aerosols as well as heat.

Coping range This is the capacity of a system (regions and groups of people) to adjust to climate variation from year to year.

Coping strategy The methods and strategies adopted by individuals and groups of people in the aftermath of a climate hazard.

Coriolis force The force that appears to act on a moving body as a result of the rotation of the earth to modify its direction of movement. In the northern hemisphere deflection is to the right of the line of motion, and in the southern hemisphere to the left of the line of motion.

Cumulonimbus clouds Tall, rain-bearing cumulus clouds reaching heights of up to 10–12 km in the atmosphere.

Cyclogenesis The process of formation of a cyclone or low pressure centre.

Cyclone (or depression) A system of winds with anticlockwise motion (in the northern hemisphere) round a low pressure centre, often associated with strong winds, thunderstorms and heavy rainfall.

Dansgaard-Oeschger events D-O events are fairly rapid and distinctive temperature changes between warm interstadial and cold stadial events.

Degree days The cumulative temperature difference above or below an arbitrary threshold temperature. Widely used for estimating space-heating and air-conditioning demand, or for growth rates/behaviour in temperature sensitive flora and fauna.

Desertification Process that involves the destruction of the quality of soils and vegetation in semi-arid areas, leading to the formation and expansion of deserts. Climate drought may or may not be a factor in this process.

Deterministic A process, physical law or model that returns the same predictable outcome from repeat experiments when presented with the same initial and boundary conditions, in contrast to *stochastic* processes.

Diatoms A class of microscopic unicellular algae that live in the surface waters of lakes, rivers and the ocean and form silica-rich flinty shells.

Disaster mitigation The various methods adopted by government and non government organisations in the relief and rehabilitation of regions and societies affected by climate disasters. They include physical approaches (e.g. hurricane shelters), early warning and evacuation procedures and government rehabilitation measures.

Divergence If a constant volume of fluid has its horizontal dimensions increased it experiences divergence and, by conservation of mass, its vertical dimension must decrease.

Domain A fixed region of the earth's surface and over-lying atmosphere represented by a *Regional Climate Model*. Also, denotes the grid box(es) used for statistical *downscaling*. In both cases, the downscaling is accomplished using pressure, wind, temperature or vapour information supplied by a host *GCM*.

Downscaling The development of climate data for a point or small area from regional climate information. The regional climate data may originate either from a *climate model* or from observations. Downscaling models may relate processes operating across different time and/or space scales.

Drought A complex multifaceted concept implying an absence of precipitation (climate drought) long enough for rivers and reservoirs to run dry (hydrological drought) and for serious moisture deficits to develop in the soil (soil/agricultural drought). As such drought is an unexpected failure of the rains and should not be confused with expected seasonal deficiencies in water supply.

Dynamical See *Regional Climate Model*.

Easterly wave A low-pressure system which forms on the poleward side of the Inter Tropical Convergence Zone. These disturbances which move westwards under the influence of the Trade Winds can develop with the right conditions into hurricanes.

Eccentricity of the earth's orbit A measure of the departure of the earth's orbit from a perfect circle. Eccentricity is measured by the distance between the two foci of an ellipse divided by the major axis. Presently the earth's orbit has an eccentricity of 0.018.

Effective rainfall This is the available rainfall for agriculture and other land uses when losses from evapotranspiration, runoff and groundwater seepage are considered.

El Niño A warm phase event in the Pacific with warm waters and high rainfall in the coastal regions of Peru and Equador and cool waters and dry conditions in eastern Australia and Indonesia. El Niño events also affect the weather and climate of distant regions around the globe.

El Niño-Southern Oscillation ENSO events are quasi regular oscillations or switches in the state of the ocean–atmosphere system in the Pacific equatorial region. One mode of this oscillation is the warm phase El Niño and the other mode is the cold phase La Niña.

Electromagnetic spectrum The full envelope or range of different wavelengths of energy emitted by a body.

Emission scenario A plausible representation of the future development of emissions of substances that are potentially radiatively active (e.g. *greenhouse gases*, *aerosols*), based on a coherent and internally consistent set of assumptions about driving forces and their key relationships.

Energy balance The net amount of energy (radiation and sensible/latent heat fluxes) present at any region in unit time. It is the difference between the net input and the net output of energy in any given region. See also *Net radiation budget*.

Ensemble (member) A set of simulations (members) in which a deterministic *climate model* is run for multiple *climate projections*, each with minor differences in the initial or boundary conditions. Conversely, *weather generator* ensemble members differ by virtue of random outcomes of successive model simulations. In either case, ensemble solutions can be grouped and then compared with the ensemble mean to provide a guide to the *uncertainty* associated with specific aspects of the simulation.

Environmental determinism A philosophy which attributed a dominant role to the physical environment in explaining the social, cultural and economic development of people.

Environmental lapse rate The general fall in temperature with altitude in the atmosphere. This is not a fixed rate but varies in time and space according to circumstances, i.e. degree of solar heating, wind strength, time of day etc.

Equinoxes The dates (21 March, 21 September) when the noon-day sun is directly overhead at the equator.

Evapotranspiration The combined loss of water from the soil and surface to the atmosphere by evaporation and plant transpiration.

External forcing A set of factors that influence the evolution of the climate system in time (and excluding natural internal dynamics of the system). Examples of external forcing include volcanic eruptions, solar variations and human-induced forcings such as changing the composition of the atmosphere and land use change.

Extreme weather event An event that is rare within its statistical reference distribution at a particular place. Definitions of 'rare' vary from place to place (and from time to time), but an extreme event would normally be as rare or rarer than the 10th or 90th percentile.

Famine Concept is usually defined by a prolonged food deficit that eventually leads to widespread and substantial increases in ill health and mortality. Most famines are human induced although climate change (e.g. drought) can be a factor.

Feedback The action by which the output of a process becomes the stimulus for the next response. Positive feedback results in a snowballing effect with system behaviour being amplified on each cycle; negative feedback has a damping effect on the system.

Föhn A warm dry wind blowing down leeward slopes of a mountain chain. The air is warmed *adiabatic*ally as it descends and can be accompanied by rapid temperature rises and snow melt.

Food insecurity A situation where there is a poor availability in the food supply for a region or society. If severe enough, chronic food insecurity can make people vulnerable to hunger and famine.

Food security A situation which indicates a regular or assured level in the availablity of food for a region or society.

Full glacial maximum A period when ice was at its most extensive and temperatures lowest during a glacial phase.

Gaia hypothesis The hypothesis that both the biological and physical systems of the earth interact with each other in a complex manner to produce self regulation which maintains the integrity of the biosphere.

General circulation model (GCM) A three-dimensional representation of the earth's atmosphere using four primary equations describing the flow of energy (first law of thermodynamics) and momentum (Newton's second law of motion), along with the conservation of mass (continuity equation) and water vapour (ideal gas law). Each equation is solved at discrete points on the earth's surface at fixed time intervals (typically 10–30 minutes), for several layers in the atmosphere defined by a regular *grid* (of about 200 km resolution). Couple ocean–atmosphere general circulation models (O/AGCMs) also include ocean, land-surface and sea-ice components. See *climate model*.

Geopotential height The work done when raising a body of unit mass against gravity (i.e., acceleration due to gravity at a given level in the atmosphere multiplied by distance) divided by the value of gravity at the earth's surface.

Geostrophic wind A wind direction and speed that reflects a balance between the Coriolis and Pressure Gradient Forces. The geostrophic wind blows parallel to the isobars in the free air above frictional influences from the surface.

Glacial cycle The period covering one successive advance and retreat of ice sheets across the polar and high latitudes.

Glacial period A cold period when ice sheets advanced from the polar regions over much of the high latitudes.

Global warming The current acceleration in the greenhouse effect.

Global warming potential The time integrated commitment to global warming of a given emission of a greenhouse gas. This takes into account both the radiative effect of the particular gas concerned and also its residence time in the atmosphere. Global Warming Potential is usually expressed relative to carbon dioxide.

Gradient wind The wind velocity that would result in an air parcel following a curved isobar. The speed would be such that the sum of the Coriolis Force and Pressure Gradient Force would produce the acceleration required to turn the trajectory of the air parcel to match the curvature of the isobar concerned.

Greenhouse effect The raising of global surface temperatures by the atmosphere which allows short wave radiation from the sun to pass through relatively unchecked while outgoing long wave heat radiation from the earth's surface is trapped. The main gases involved in this natural process of earth heating include water vapour, carbon dioxide, methane and nitrogen dioxide.

Greenhouse gas Gaseous constituents of the atmosphere, both natural and anthropogenic, that absorb and emit radiation at specific wavelengths within the spectrum of infrared radiation emitted by the earth's surface, the atmosphere and clouds. The primary greenhouse gases are water vapour (H_2O), carbon dioxide (CO_2), nitrous oxide (N_2O), methane (CH_4) and ozone (O_3).

Grid The co-ordinate system employed by *GCM* or *RCM* to compute three-dimensional fields of atmospheric mass, energy flux, momentum and water vapour. The grid spacing determines the smallest features that can be realistically resolved by the model. Typical resolutions for GCMs are 200 km and for RCMs 20–50 km.

Gross primary production The total amount of carbohydrate produced by a plant or vegetation type as a result of photosynthesis.

Gulf stream A warm salty current of water that transfers heat and moisture from the Caribbean to the central north Atlantic.

Hadley cell A direct thermally driven circulation which comprises an upward motion of air at the ITCZ and a downward movement of air in the subtropics. The atmospheric cell is completed by a polewards movement of air at high levels in the atmosphere and an equatorwards motion at the surface.

Hazard (climate) A climate change becomes a climate hazard when the severity (magnitude, frequency, duration) of the climate event falls beyond the coping range of a system. Thus the term captures those climate extremes, in addition to changes in annual averages, to which a system under study is vulnerable.

Heinrich events Very cold stadials associated with ice rafting and the movement of vast armadas of icebergs across the north Atlantic.

Hurricane A severe topical storm with an intense circular vortex and wind speeds in excess of 33 m/s or about 74 mph.

Hydrological cycle The continuous circulation of all the forms of water (water vapour, liquid water, ice) on and below the surface of the earth and from there by evaporation, transpiration and condensation to the atmosphere.

Hydrosphere The total water realm of the earth, encompassing all the land sea and atmospheric components.

Hydrostatic equilibrium The balance between the upward directed pressure vertical gradient force and the downward directed force of gravity.

Ice age A period of time between 2.5 million years ago and the present time when large continental ice sheets spread out from both polar regions but especially the north pole to cover much of the high latitudes.

Ice-albedo effect A positive feedback mechanism linking the progressive growth of ice sheets to atmospheric cooling via the ability of the growing ice sheets to reflect increasing amounts of incoming solar radiation back to space. A positive feedback ice-albedo effect is also thought to work in the opposite direction when the melting and shrinking ice sheets reflect less solar radiation back to space causing regional warming and further ice melt.

Ice storm A rare meteorological event where rain falling from higher warmer layers in the atmosphere freezes onto colder surfaces like roads, telegraph poles, houses, vegetation.

Inter tropical convergence zone The ITCZ is a low pressure region in the vicinity of the equator where the north easterly and south easterly trade winds converge.

Interannual monsoon variation Annual fluctuations in the monsoon rains closely associated with El Niño events (poor rainfall years) and La Niña events (good rainfall years).

Interglacial period A period of relatively warm and stable climate conditions lasting 10,000–15,000 years when ice sheets retreated from large areas of the high latitudes towards the polar ice caps. They occur between successive longer colder glacial periods (90,000–100,000 years) when the ice sheets advance. We are presently living in the most recent interglacial called the Holocene which began about 11,500 year ago.

Interstadial events Relatively frequent but short lived warm periods during a glacial period although they are not as warm as the warmest part (interglacial) of a glacial cycle.

Isotope One of two or more forms of an element having the same atomic number but differences in atomic weight and nuclear properties. The relative proportions of the isotopes in a sample can be used to date the material or to infer the environmental conditions under which the material formed.

Jetstream A narrow band of very strong winds aloft. In the mid latitudes this occurs in the upper westerlies.

Katabatic wind A downslope valley wind produced on clear nights by preferential cooling of air in contact with cold upper valley slopes.

Kata-front A cold *front* in which warm air is generally flowing downwards over a wedge of colder air.

Köppen system A climatic classification based on the climate needs of vegetation in six zones: tropical, subtropical, temperate, cold, polar and mountain. Sub-classes of the main groups are based on more refined precipitation and temperature characteristics.

La Niña A cold phase event in the Pacific with very cold waters and dry conditions in coastal Peru and Equador and very warm waters and heavy rainfall in eastern Australia and Indonesia. La Niña conditions can also induce weather and climate changes around the world although less dramatically than El Niño events.

Last glacial maximum The LGM is the coldest period of the last glacial advance about 18,000 years ago.

Latent heat Heat released or absorbed during a change in state of a substance. The temperature and pressure remains constant during this process.

Lee waves A wave formation in an airflow in the wake of a topographic barrier. A lenticular cloud sometimes marks the cool crest of the lee wave.

Little Ice Age A downturn in climate from approximately 1400–1850 during which cold winters and cool summers were experienced. The Little Ice Age predominantly affected the northern hemisphere.

Long-wave radiation Radiation from the atmosphere and the surface of the earth with longer wavelengths mostly in the range 3 μm to about 70 μm.

Main storm tracks The mean or most favoured pathways or trajectories of storms (hurricanes, thunderstorms, tornadoes).

Maunder minimum A state of low sunspot number between about 1645 and 1715 coinciding with one of the coldest periods of the *Little Ice Age*.

Megadrought Major *drought* episodes in North America that persisted for multiple decades, and sometimes even centuries, with no precedents in the modern instrumental record.

Meridional flow An atmospheric circulation in which the dominant flow of air is from north to south, or from south to north, across the parallels of latitude, in contrast to *zonal flow*.

Mesopause The boundary between the mesosphere and the thermosphere.

Mesosphere The atmospheric layer above the stratosphere in which the temperature decreases with height.

Milankovitch theory The hypothesis which suggests that variations in the earth's orbit around the sun hold the key to the timing of the main ice age events.

Mixing ratio The ratio of the mass of water vapour to the mass of dry air in a sample of air.

Monsoon A monsoon is a wind system which undergoes a marked seasonal change in direction. A monsoon climate is one that is determined by marked seasonal changes in wind direction and thus weather and climate.

Monsoon trough A low pressure region over India in the summer associated with the northwards movement of the ITCZ.

NCEP The acronym for the National Center for Environmental Prediction. The source of re-analysis (climate model assimilated) data widely used for dynamical and statistical *downscaling* of the current climate.

Net primary production The rate at which biomass is accumulated in the tissues of plants.

Net radiation budget The difference between the absorbed incoming solar radiation and the net loss of outgoing long-wave or thermal radiation. The concept can be applied to either the whole earth-atmosphere system or a major component of it, i.e. the atmosphere alone, earth surface alone. It can also be applied to regions, i.e. the oceans, the northern hemisphere, the Sahara.

Newton The force required to accelerate one kilogram by 1 metre per second.

Nimbostratus Extensive rain bearing relatively low lying layered clouds.

Noctilucent clouds High altitude clouds at approximately 80 km. These are so thin they can only be seen by the light scattered by air molecules on summer nights in the high middle latitudes.

Normalisation A statistical procedure involving the standardisation of a data set (by subtraction of the mean and division by the standard deviation) with respect to a predefined control period. The technique is widely used in statistical *downscaling* to reduce systematic biases in the mean and variance of climate model output.

North Atlantic Oscillation The NAO is a dual switching ocean–atmosphere mechanism in the north Atlantic. When in positive mode, high pressure differences are found between the Azores High and the Atlantic Low and westerly winds and Gulf Stream advection are strong; when in negative mode weak pressure gradients occur with weak westerlies and Gulf Stream advection.

North Atlantic Drift The NAD is a warm salty current of water that transfers heat and moisture from the southern parts of the north Atlantic to the higher latitudes of the north Atlantic.

Nuclear winter A hypothetical catastrophic deterioration in climate resulting in prolonged sub zero temperatures that might result from a major exchange of nuclear weapons.

Obliquity The tilt of the earth's axis, i.e. the angle between the plane of the axis and the plane in which the planets lie (ecliptic). This tilt varies from 22–24.5° over a period of about 40,000 years.

Occlusion A situation in which a warm *front* is overtaken by a cold front, causing a sector of warm air to be raised from the earth's surface.

Ocean currents Major flows of water in the surface layers of the oceans that transport heat (and salt) across the world.

Ocean–atmosphere oscillation A flip or oscillation from one phase or state of the ocean–atmosphere system to another phase or state.

Orbital monsoon hypothesis The theory linking long-term monsoon performance to Milankovitch type variations in solar heating.

Orographic Caused by or relating to mountains. Orographic rainfall is produced when moist air is forced to ascend a mountain slope, cools and releases some of its water vapour load.

Ozone A form of oxygen having three atoms instead of the usual two in the molecule. Formed naturally in the stratosphere at heights of 30 to 80 km above the earth's surface due to the absorption of ultraviolet radiation by oxygen. Formed in the troposphere by photochemical reactions taking place under stable summer *anticyclonic* weather.

Pacific Decadal Oscillation The PDO is a dual ocean–atmosphere system in the northern Pacific which switches from cold to warm phases and back again every 20–30 years.

Pacific North American (PNA) Patterns describing the relative position and sinuosity of the jet stream over North America.

Palmer drought severity index (PDSI) A cumulative index of soil moisture deficit that is widely used in North America to classify hydrological *drought* severity.

Parameter A numerical value representing a process or attribute in a model. Some parameters are readily measurable climate properties; others are known to vary but are not specifically related to measurable features. Parameters are also used in climate models to represent processes that poorly understood or resolved.

Pascal A pressure of 1 newton per square metre.

Phenology The science which studies the timing or seasonal activity of plant and animal behaviour.

Planetary wave See *Rossby wave.*

Polar front The *frontal* zone marking the interface between the polar maritime and tropical maritime air masses, over the North Atlantic and North Pacific oceans.

Polar stratospheric clouds PSCs are high level exceptionally cold (–70 to –80 degrees centigrade) stratospheric clouds over the polar regions which act as sites for the chemical reactions which destroy stratospheric ozone.

Potential evapotranspiration The maximum possible amount of water lost by evaporation and transpiration. This measure is based on potential losses when there is an unrestricted supply of water available.

Precession of the equinoxes The point in the earth's orbit of closest approach to the sun varies, causing the time of the equinoxes to alter over long periods. Two periodicities of 23,000 years and 18,000 years are detectable.

Predictand A variable that may be inferred through knowledge of the behaviour of one or more *predictor* variables.

Predictor A variable that is assumed to have predictive skill for another variable of interest, the *predictand.* For example, day-to-day variations in atmospheric pressure may be a useful predictor of daily rainfall occurrence.

Proxy climate A non-instrumental record of past climate conditions reconstructed from micro-fossils, isotopes, tree rings, captured by the regular accumulations of ice cores, lake and ocean sediments, peat bogs, etc.

Quaternary ice age The ebb and flow of large continental ice sheets across the high latitudes of the northern and southern hemispheres over the last 2.5 million years or so.

Radiation The transmission of energy by electromagnetic waves which may be propagated through a substance or through a vacuum at the speed of light.

Radiative forcing The change in net vertical irradiance (expressed as Watts per square metre) at the *tropopause* due to an internal change or a change in the *external forcing* of the climate system, such as, for example, a change in the concentration of carbon dioxide, or the output of the sun.

Radiosonde A balloon-borne instrument which measures temperature, pressure and humidity at various levels as it ascends through the atmosphere. These values are transmitted by radio to a ground station.

Rain shadow The area in the lee of a relief obstacle where precipitation is markedly reduced relative to the windward side.

Random See *stochastic.*

Regional climate model (RCM) A three-dimensional, mathematical model that simulates regional scale climate features (of 20–50 km resolution) given time-varying, atmospheric properties modelled by a *General Circulation Model.* The RCM *domain* is typically 'nested' within the three-dimensional *grid* used by a GCM to simulate large-scale fields (e.g. surface pressure, wind, temperature and vapour).

Regression A statistical technique for constructing empirical relationships between a dependent (*predictand*) and set of independent (*predictor*) variables. See also *black box, transfer function.*

Relative humidity A relative measure of the amount of moisture in the air to the amount needed to saturate the air at the same temperature expressed as a percentage.

Resolution The *grid* separation of a climate model determining the smallest physical feature that can be realistically simulated.

Risk This is defined as the probability or likelihood of occurrence of a climate hazard multiplied by the vulnerability (expected losses) of the target group.

Rossby wave Large-scale westerly movements in the airflow of the middle and upper troposphere forming oscillations with wavelengths of some 2000 km length stretching between sub-polar and tropical latitudes.

Sahel A semi arid region in north Africa sandwiched between the Sahara desert to the north and a wetter sub humid region to the south.

Sapropels 'Stinky' layers of black organic rich mud deposited on the Mediterranean sea floor as a result of strong inflow from the Nile which reduces the concentration of oxygen and normal processes of decay in the deeper parts of the basin.

Scenario A plausible and often simplified description of how the future may develop based on a coherent and internally consistent set of assumptions about driving forces and key relationships. Scenarios may be derived from projections, but are often based on additional information from other sources, sometimes combined with a 'narrative story-line'.

Sea breeze A local wind caused by the differential heating of land and water bodies such as a lake or sea. More rapid heating of the land by day causes the overlying warm air to rise, drawing inland cooler air that is overlying the sea. That pattern of flow reverses at night as air overlying the land cools more rapidly than that over the water body.

Sensible heat Heat experienced by our senses and which can be measured by a thermometer.

Short-wave radiation Radiation from the sun with short wavelengths mostly in the range between 0.4 μm–3.0 μm (1 μm is a millionth of a metre).

Singularity A recurrent weather feature or event that occurs around specific calendar dates in the year.

Solar constant The amount of energy received at the top of the atmosphere on a unit of area perpendicular to the sun's rays at the average distance of the earth from the sun. A value of 1,370 W/m^2 is presently the best estimate.

Solar radiation Short wave radiation or energy from the sun.

Solstices The dates (21st June, 21st December) respectively when the noon-day sun is directly overhead at the Tropic of Cancer (23.5 degrees north) and the Tropic of Capricorn (23.5 degrees south).

Species extinction The permanent loss of a species from the world's gene pool

Specific heat capacity The amount of energy required to raise the temperature of one gram of a substance by one degree Celsius.

Specific humidity The ratio of the mass of water vapour (in grams) to the mass of moist air (in kilograms) in a given volume of air.

Squalls A series of thunderstorm cells aligned at right angles to their direction of movement.

Stable equilibrium The condition of the atmosphere when the environmental lapse rate of an air mass is less than the dry *adiabatic* lapse rate. An unsaturated parcel of air at the earth's surface cools at the dry *adiabatic* lapse rate as it moves upwards and, in the process, becomes colder and denser than the surrounding air. This causes the parcel of air to sink to its original level, conditions typical in an *anticyclone*.

Stadial events Distinctive short-lived cold periods during an ice age though not as cold as the coldest part of the glacial cycle.

Stochastic A process or model that returns different outcomes from repeat experiments even when presented with the same initial and boundary conditions, in contrast to *deterministic* processes. See *weather generator*.

Storm surge The high tidal wave that forms in front of a hurricane as it makes landfall.

Stratopause The boundary between the stratosphere and the mesosphere.

Stratosphere The atmospheric layer above the troposphere, extending up to approximately 50 km, where the temperature is constant or rises with increasing height.

Subsidence inversion A temperature inversion in the mid and lower layers of the atmosphere caused by subsiding and therefore warming air.

Sustainable development Sustainable development for the poorer groups in society can be defined as the wise or optimal use of local resources by a community so that its long term standard of living can be raised and sustained.

Teleconnections Applied to the atmosphere (atmospheric teleconnections) this term signifies the statistical relationship or connectivity between the world's weather systems. It is the apparent ability of one weather system to affect other weather systems locally, regionally and even globally.

Temperature Mean kinetic energy or average speed per molecule of all the molecules in a substance.

Temperature anomaly (world) On world maps these reflect east–west, i.e. regional changes of temperature rather than north–south (latitudinal) differences.

Temperature inversion The increase of temperature with elevation in the atmosphere.

Thermal radiation All heat emissions by the earth and atmosphere.

Thermocline An upper level boundary zone in the ocean which divides warmer lighter top water from colder denser deep water.

Thermohaline conveyor A deep, slow moving, but globally extensive current of water linking all oceans and surface currents. This deep or bottom current is driven by temperature and salinity differences being much colder and heavier (saltier) than the surface currents.

Top down development In terms of disaster management, the application as solution of (often external) physical measuring and monitoring techniques and large-scale engineering works.

Tornado A rapidly rotating small-scale atmospheric vortex associated with hurricanes and thunderstorms and reaching in the latter maximum speeds of up to 300–400 mph.

Trade wind inversion A persistent subsidence inversion between the ITCZ and the sub tropics. The inversion is highest and weakest near the ITCZ and strongest and closest to the surface at the sub tropics.

Transpiration (plant) The movement or loss of water from the soil to plants and eventually to the atmosphere.

Tropical cyclones These include tropical lows (wind speed less than 8 m/s), tropical depressions (wind speed between 8–17 m/s), tropical storms (wind speed between 17–32 m/s) and severe tropical storms with hurricane wind speeds (wind speeds greater than 33 m/s).

Tropopause The boundary between the lowest part of the atmosphere, known as the troposphere, and the highly stratified region of the atmosphere, known as the stratosphere. The tropopause is typically located 10 km above the earth's surface.

Troposphere The lowermost layer of the atmosphere in which temperature falls with increasing height. The troposphere extends up to approximately 8 km at the poles to approximately 17 km at the equator.

Uncertainty An expression of the degree to which a value (e.g. the future state of the climate system) is unknown. Uncertainty can result from a lack of information or from disagreement about what is known or knowable. It can also arise from poorly resolved climate model parameters or boundary conditions.

Unsustainable development With respect to the poorer groups in society, a situation where already low living standards are threatened with further and continuous reductions. This situation results from an over-exploitation or a severe and often prolonged damage to local resources.

Upper easterlies Strong high altitude flows of westward moving air in the sub tropics which encourage among other things air ascent and rainfall over India during the summer monsoon.

Upwelling A process in the oceans where normally cold nutrient rich deep waters rise to the surface displacing warmer surface waters.

Urban heat island The dome of heat which hovers over many cities especially in the high latitudes.

Vorticity Twice the angular velocity of a fluid particle about a local axis through the particle. In other words, a measure of rotation of an air mass.

Vulnerability The vulnerability of a system depends on the inherent ability of that system to adapt to and thus cope with the effects of environmental (climate) change. In terms of individuals and groups, it is their capacity to anticipate, cope with, resist and recover from the impact of a natural hazard.

Weather generator A model whose stochastic (random) behaviour statistically resembles daily weather data at a location. Unlike *deterministic* weather forecasting models, weather generators are not expected to duplicate a particular weather sequence at a given time in either the past or the future. Most weather generators assume a link between the precipitation process and secondary weather variables such as temperature, solar radiation and humidity.

Weather pattern (or type) An objectively or subjectively classified distribution of surface (and/or upper atmosphere) meteorological variables, typically daily mean sea level pressure. Each atmospheric circulation pattern should have distinctive meteorological properties (e.g. chance of rainfall, sunshine hours, wind direction, air quality, etc). Examples of subjective circulation typing schemes include the European Grosswetterlagen, and the British Isles Lamb Weather Types.

Wind shear A condition where there is significant changes in the speed and/or direction of wind throughout the vertical atmosphere.

Younger Dryas A *Heinrich event* of 12,700–11,500 BP which produced a return to glacial conditions in many of the mountainous zones of Europe.

Zonal flow An atmospheric circulation in which the dominant flow of air follows the lines of latitude (e.g. the westerlies), in contrast to *meridional flow*.

Where appropriate, some of the above definitions were drawn from the Glossary of terms in the Summary for Policymakers, A Report of Working Group I of the Intergovernmental Panel on Climate Change (IPCC, 2001), and the Technical Summary of the Working Group I Report. Others were adapted from the *Dictionary of Geography* (Clark, 1990) and *Understanding Our Atmospheric Environment* (Neiburger *et al.*, 1982).

INDEX